The 10th International Electronic Conference on Sensors and Applications—Volume I

The 10th International Electronic Conference on Sensors and Applications—Volume I

Editors

Stefano Mariani
Francisco Falcone
Stefan Bosse
Jean-Marc Laheurte

MDPI

Basel • Beijing • Wuhan • Barcelona • Belgrade • Novi Sad • Cluj • Manchester

Editors
Stefano Mariani
Politecnico di Milano
Milan
Italy

Francisco Falcone
Public University of Navarre
Pamplona
Spain

Stefan Bosse
University of Bremen
Bremen
Germany

Jean-Marc Laheurte
Université Gustave Eiffel
Champs-sur-Marne
France

Editorial Office
MDPI AG
Grosspeteranlage 5
4052 Basel, Switzerland

This is a reprint of articles from the Proceedings published online in the open access journal *Engineering Proceedings* (ISSN 2673-4591) (available at: https://www.mdpi.com/2673-4591/58/1).

For citation purposes, cite each article independently as indicated on the article page online and as indicated below:

Lastname, A.A.; Lastname, B.B. Article Title. *Journal Name* **Year**, *Volume Number*, Page Range.

Volume I
ISBN 978-3-7258-1285-1 (Hbk)
ISBN 978-3-7258-1286-8 (PDF)
doi.org/10.3390/books978-3-7258-1286-8

Set
ISBN 978-3-7258-1283-7 (Hbk)
ISBN 978-3-7258-1284-4 (PDF)

Contents

About the Editors

Stefano Mariani

Dr. Stefano Mariani received an M.S. degree (cum laude) in civil engineering in 1995 and a Ph.D. degree in structural engineering in 1999; both of his degrees were awarded by the Polytechnic University of Milan. He is currently a Full Professor in the Department of Civil and Environmental Engineering at the Polytechnic University of Milan. He was a research scholar at the Danish Technical University in 1997, an adjunct professor at Penn State University in 2007, and a visiting professor at the Polytechnic Institute of New York University in 2009. He is a member of the Editorial Boards of several journals, including *Algorithms*, the *International Journal on Advances in Systems and Measurements*, *Inventions*, *Machines*, *Micro and Nanosystems*, *Micromachines*, and *Sensors*. He was a recipient of the Associazione Carlo Maddalena Prize for graduate students (1996), and of the Fondazione Confalonieri Prize for PhD students (2000). His main research interests are the reliability of MEMS that are subject to shocks and drops; the structural health monitoring of composite structures through MEMS sensors; numerical simulations of ductile fracture in metals and of quasi-brittle fracture in heterogeneous and functionally graded materials; extended finite element methods; the calibration of constitutive models via extended and sigma-point Kalman filters; and multi-scale solution methods for dynamic delamination in layered composites.

Francisco Falcone

Dr. Francisco Falcone received both his Telecommunication Engineering Degree (1999) and his PhD in Communication Engineering (2005) from the Public University of Navarre in Spain. From 1999 to 2000, he was Microwave Network Engineer, Siemens-Italtel, Málaga. From 2000 to 2008, he was Mobile Access Network Engineer, Telefónica Móviles, in Pamplona. In 2009 he co-founded Tafco Metawireless, spin-off of the UPNA, of which he was the first manager. In parallel, from 2003 to 2009 he was Assistant Lecturer, EEC Dept., UPNA. In June 2009 he became Associate Professor at UPNA and since September 2022, Full Professor. From January 2012 to July 2018 and from July 2019 to November 2021 he was Head of the EEC Dept., UPNA. In 2018 he was Visiting Professor at Kuwait College of Science and Technology, Kuwait, for three months. He is also affiliated with the Smart Cities Institute-UPNA, a multidisciplinary research institute with over 100 researchers, being Head of the Institute since May 2021, working on contextual and interactive environments solutions, through the integration of heterogeneous wireless communications networks. Since June 2022, he is Distinguished Visiting Professor in Telecommunications, School of Engineering and Science, Tecnologico de Monterrey, Mexico. His research area is artificial EM media, complex electromagnetic scenarios and wireless system analysis, with applications to context aware environments, Smart Cities and Smart Regions. He has over 500 contributions to journal and conference publications. He has been awarded several research awards: CST Best Paper Award 2003/2005, COIT 2005 Best Doctoral Thesis, UPNA PhD Award 2004-2006, 1st Prize Juan López de Peñalver 2010, Real Academia de Ingeniería de España, XII Talgo Foundation Award for Technological Innovation, ECSA-2 Best Paper Award (2015), Best Paper Award IISA (2015), ECSA Award -3 Best Paper Award (2016), ECSA-4 Best Paper Award (2018), Best Paper Award ISSI (2019) and IIoT 2020 Best Paper Award.

Stefan Bosse

Dr. Stefan Bosse studied physics at the University of Bremen. He received a PhD/doctoral degree in physics in the year 2002 from the University of Bremen, and a post-doctoral degree (Habilitation) and the Venia Legendi in Computer Science in the year 2016 from the University of

Bremen. Since 2016 he is teaching and researching as a Privatdozent (assoc. prof.) at the University of Bremen, Department of Computer Science. From 2018 to 2019 he was an interim professor at the University of Koblenz-Landau, Faculty Computer Science, Institute of Software Technologies. Since 2022 he is also a lecturer at the University of Siegen, Department of Mechanical Engineering. His research is dedicated to distributed artificial intelligence in general and sensor-driven data processing and sensor networks in special. This covers the design of embedded systems, virtualization, distributed and massive parallel system design, distributed and tiny machine learning, agent-based methods including simulation and material-integrated sensing systems with a high interdisciplinary and technological background. Typical applications are Structural Health Monitoring (SHM) and Mobile Crowd Sensing (MCS) at the boundaries of technical systems, humans and society.

He published more than 100 journal and conference papers and four books, and he acts as a reviewer and a Guest Editor for several international journals and is a member of a broad range of international conference program and organizing committees. He is member of several scientific committees and chairs the International Electronic Conference on Sensors and Applications since 2019. Since 2019 he is also a principal investigator of the transregional and interdisciplinary DFG founded research unit 3022 and conducts the subproject "Data-driven Damage Diagnostics", with its origin in the scientific center "Integrated Solutions in Sensorial Structure Engineering" (ISIS) as well as membership in the MAPEX Center for Materials and Processes.

Jean-Marc Laheurte

Dr. Jean-Marc Laheurte received M.Sc. and Ph.D. degrees in electrical engineering and a Habilitation to Supervise Research from the University of Nice, France, in 1989, 1992 and 1997, respectively. From 1989 to 1990, he was a research assistant at the École Polytechnique Fédérale de Lausanne, Switzerland. In 1992, he served as a postdoctoral researcher at the University of Michigan, Ann Arbor, MI, USA. From 1993 to 2002 he was an Associate Professor at the University of Nice Sophia Antipolis, France. Since 2002, he has been a Professor at the University Gustave Eiffel, France. In 2012, he spent a year as a senior RF engineer at Tagsys, La Ciotat, France. For the past eight years he has been the director of the 75-member Electronics, SYstèmes de COmmunications and Microsystems Laboratory (ESYCOM Laboratory). He is the author or co-author of two books, two book chapters, ninety technical papers in international journals and more than a hundred conference papers. He holds two patents on RFID technologies. His current research interests include the design of antennas in matter, RFID technologies, RFID localization, body array antennas (BANs).

Preface

This Issue of *Engineering Proceedings* assembles the papers presented at the 10th International Electronic Conference on Sensors and Applications (ECSA-10), held online on 15–30 November 2023, through the sciforum.net platform developed by MDPI. The annual ECSA conference was initiated in 2014 on an online basis to allow for global participation without concerns for travel and related expenditures. This year's edition saw significantly increased collaboration among the authors and the audience through live session presentations of a limited number of contributions that were selected among the over 200 submissions, which were all reviewed and accepted by the chairs and/or by the conference committee members.

After the success of the nine editions of the conference that were held from 2014 to 2022, this year's edition, ECSA-10, collected contributions concerning five thematic areas which are deeply affected by the rapid development of sensors and data processing: Chemo- and Biosensors; Physical Sensors; Sensor Networks, IoT and Structural Health Monitoring; Sensor Data Analytics; and Sensors and Artificial Intelligence. Five specific sessions were also included: Smart Agriculture Sensors; Materials for Sensing Applications; Electronic Sensors, Devices and Systems; Wearable Sensors and Healthcare Applications; and Robotics, Sensors and Industry 4.0. Three keynote presentations were delivered by Sabina Merlo, a Full Professor of Electrical and electronic measurements at the Department of Electrical, Computer and Biomedical Engineering, University of Pavia, Pavia (Italy), by Leyre Azpilicueta, from the Department of Electrical, Electronic and Communication Engineering, Institute of Smart Cities, Public University of Navarre (UPNA), and by Dirk Lehmhus, from the Department of Casting Technology and Lightweight Construction, Fraunhofer IFAM.

Stefano Mariani, Francisco Falcone, Stefan Bosse, and Jean-Marc Laheurte
Editors

engineering proceedings

MDPI

Editorial

Statement of Peer Review †

Stefano Mariani [1,*], Francisco Falcone [2,3,*], Stefan Bosse [4,*] and Jean-Marc Laheurte [5,*]

1 Department of Civil and Environmental Engineering, Politecnico di Milano, Piazza Leonardo da Vinci 32, 20133 Milano, Italy
2 Department of Electrical, Electronic and Communication Engineering & Institute for Smart Cities (ISC), Public University of Navarre, 31006 Pamplona, Spain
3 School of Engineering and Sciences, Tecnologico de Monterrey, Monterrey 64849, Mexico
4 Department of Computer Science, University of Bremen, 28359 Bremen, Germany
5 Laboratoire Electronique, SYstèmes de COmmunication et Microsystèmes (ESYCOM), Université Gustave Eiffel, 77420 Champs-sur-Marne, France
* Correspondence: stefano.mariani@polimi.it (S.M.); francisco.falcone@unavarra.es (F.F.); sbosse@uni-bremen.de (S.B.); jean-marc.laheurte@univ-eiffel.fr (J.-M.L.)

† All papers published in this volume are presented at the 10th International Electronic Conference on Sensors and Applications (ECSA-10), 15–30 November 2023; Available online: https://ecsa-10.sciforum.net/.

In submitting conference proceedings to *Engineering Proceedings*, the volume editors of the proceedings certify to the publisher that all papers published in this volume have been subjected to peer review administered by the volume editors. Reviews were conducted by expert referees to the professional and scientific standards expected of a proceedings journal.

- Type of peer review: single-blind
- Conference submission management system: 229
- Number of submissions sent for review: 229
- Number of submissions accepted: 141
- Acceptance rate (number of submissions accepted/number of submissions received): 62%
- Average number of reviews per paper: 1
- Total number of reviewers involved: 177
- Any additional information on the review process: N/A

Conflicts of Interest: The authors declare no conflict of interest.

Disclaimer/Publisher's Note: The statements, opinions and data contained in all publications are solely those of the individual author(s) and contributor(s) and not of MDPI and/or the editor(s). MDPI and/or the editor(s) disclaim responsibility for any injury to people or property resulting from any ideas, methods, instructions or products referred to in the content.

Citation: Mariani, S.; Falcone, F.; Bosse, S.; Laheurte, J.-M. Statement of Peer Review. *Eng. Proc.* **2023**, *58*, 133. https://doi.org/10.3390/ecsa058133

Published: 15 April 2024

engineering proceedings

MDPI

Editorial

Preface: Proceedings of the 10th International Electronic Conference on Sensors and Applications †

Stefano Mariani [1,*], Francisco Falcone [2,3,*], Stefan Bosse [4,*] and Jean-Marc Laheurte [5,*]

1 Department of Civil and Environmental Engineering, Politecnico di Milano, Piazza Leonardo da Vinci 32, 20133 Milano, Italy

2 Department of Electrical, Electronic and Communication Engineering & Institute for Smart Cities (ISC), Public University of Navarre, 31006 Pamplona, Spain

3 School of Engineering and Sciences, Tecnologico de Monterrey, Monterrey 64849, Mexico

4 Department of Computer Science, University of Bremen, 28359 Bremen, Germany

5 Laboratoire Electronique, SYstèmes de COmmunication et Microsystèmes (ESYCOM), Université Gustave Eiffel, 77420 Champs-sur-Marne, France

* Correspondence: stefano.mariani@polimi.it (S.M.); francisco.falcone@unavarra.es (F.F.); sbosse@uni-bremen.de (S.B.); jean-marc.laheurte@univ-eiffel.fr (J.-M.L.)

† All papers published in this volume are presented at the 10th International Electronic Conference on Sensors and Applications (ECSA-10), 15–30 November 2023; Available online: https://ecsa-10.sciforum.net/.

Citation: Mariani, S.; Falcone, F.; Bosse, S.; Laheurte, J.-M. Preface: Proceedings of the 10th International Electronic Conference on Sensors and Applications. *Eng. Proc.* **2023**, *58*, 134. https://doi.org/10.3390/ecsa058134

Published: 15 April 2024

This Issue of *Engineering Proceedings* assembles the papers presented at the 10th International Electronic Conference on Sensors and Applications (ECSA-10), held online on 15–30 November 2023 through the sciforum.net platform developed by MDPI. The annual ECSA conference was initiated in 2014 on an online basis to allow global participation without concerns over travel and related expenditures. This type of conference is particularly appropriate and useful because research related to sensors is still developing rapidly, and a platform for a direct exchange of information about the latest findings may provide further advancements in the development of novel ideas.

This year's edition saw a continuously increased collaboration among the authors and the audience through live session presentations of a limited number of contributions selected among the 229 submissions, all formaly reviewed and accepted by the chairs and/or by the conference committee members. Video recordings of the sessions and open access to papers and presentations can be found online at https://ecsa-10.sciforum.net/.

ECSA-10 collected contributions concerning five thematic areas which are deeply affected by the rapid development of sensors and data processing: Chemo- and Biosensors; Physical Sensors; Sensor Networks, IoT and Structural Health Monitoring; Sensor Data Analytics; and Sensors and Artificial Intelligence. Five specific sessions were included: Smart Agriculture Sensors; Materials for Sensing Applications; Electronic Sensors, Devices and Systems; Wearable Sensors and Healthcare Applications; as well as Robotics, Sensors and Industry 4.0. Three keynote presentations were also delivered by Sabina Merlo, Full Professor of Electrical and Electronic Measurements in the Department of Electrical, Computer and Biomedical Engineering, University of Pavia, Pavia (Italy); by Leyre Azpilicueta, from the Department of Electrical, Electronic and Communication Engineering, Institute of Smart Cities, Public University of Navarre (UPNA); and by Dirk Lehmhus, from the Department of Casting Technology and Lightweight Construction, Fraunhofer IFAM.

Beyond these events, a student session gathered the contributions of eleven PHD students. After the conference, the chairs unanimously awarded the competition prize to Nicky Andre Prabatama from Universite Gustave Eiffel for his contribution entitled "Development of A Zigbee-Based Wireless Sensor Network of MEMS Accelerometers for Pavement Monitoring". Another award was established for the Best Contribution to the Conference and was granted ex aequo to Thiago Almeida Teixeira from Amazon State University for his contribution entitled "Development of a Monitoring System against

Illegal Deforestation in the Amazon Rainforest Using Artificial Intelligence Algorithms", and to Elisabetta Bodo from University of Pavia for her contribution entitled "Artificial Nutrition Monitoring Through An Optofluidic Platform". Each winner was awarded CHF 500. Our sincere congratulations go out to the awardees for the great results achieved in their research activities and for their enthusiastic presentations during the live sessions.

A companion Special Issue of *Sensors*, hosted at https://www.mdpi.com/journal/sensors/special_issues/06XXC4L7D7, will present full-length versions of the selected papers.

Conflicts of Interest: The authors declare no conflict of interest.

engineering proceedings

Proceeding Paper

Microfabricated Gold Aptasensors for the Label-Free Electrochemical Assay of Oxytetracycline Residues in Milk †

Vassilis Machairas [1], Andreas Anagnostoupoulos [1], Dionysios Soulis [1], Anastasios Economou [1,*], Kristóf Jakab [2], Nikitas Melios [1,3], Zsófia Keresztes [2], George Tsekenis [3], Joseph Wang [4] and Thanassis Speliotis [5]

1 Department of Chemistry, National and Kapodistrian University of Athens, 157 71 Athens, Greece; vasilismax42@gmail.com (V.M.); adreasanagno@gmail.com (A.A.); dsoulis1@gmail.com (D.S.); nmelios@bioacademy.gr (N.M.)
2 Functional Interfaces Research Group, Institute of Materials and Environmental Chemistry, Research Centre for Natural Sciences, Magyar tudósok krt. 2., 1117 Budapest, Hungary; jakab.kristof@ttk.hu (K.J.); keresztes.zsofia@ttk.hu (Z.K.)
3 Biomedical Research Foundation, Academy of Athens, 4 Soranou Ephessiou Street, 115 27 Athens, Greece; gtsekenis@bioacademy.gr
4 Department of Nanoengineering, University California San Diego (UCSD), La Jolla, CA 92093-0448, USA; josephwang@ucsd.edu
5 Institute of Nanoscience and Nanotechnology, NCSR "Demokritos", Aghia Paraskevi, 15 310 Athens, Greece; t.speliotis@inn.demokritos.gr
* Correspondence: aeconomo@chem.uoa.gr; Tel.: +30-210-7274298
† Presented at the 10th International Electronic Conference on Sensors and Applications (ECSA-10), 15–30 November 2023; Available online: https://ecsa-10.sciforum.net/.

Abstract: In this work, we describe a new type of electrochemical aptasensor for the label-free detection of oxytetracycline (OTC). Thin-film gold electrodes were fabricated through sputtering gold on a Kapton film, followed by the immobilization of a thiol-modified aptamer on the electrode surface. The selective capture of OTC at the aptamer-functionalized electrodes was monitored electrochemically with the use of the $[Fe(CN)_6]^{4-}/[Fe(CN)_6]^{3-}$ redox probe. Different experimental variables were studied, through which the metrological features for OTC determination were derived. Finally, the developed sensor was implemented to achieve the detection of OTC in a spiked milk sample.

Keywords: aptasensor; electrochemical; oxytetracycline; label-free; microfabrication; gold electrodes

Citation: Machairas, V.; Anagnostoupoulos, A.; Soulis, D.; Economou, A.; Jakab, K.; Melios, N.; Keresztes, Z.; Tsekenis, G.; Wang, J.; Speliotis, T. Microfabricated Gold Aptasensors for the Label-Free Electrochemical Assay of Oxytetracycline Residues in Milk. *Eng. Proc.* **2023**, *58*, 1. https://doi.org/10.3390/ecsa-10-16018

Academic Editor: Stefano Mariani

Published: 15 November 2023

1. Introduction

Oxytetracycline (OTC) is a commonly used antibiotic in veterinary medicine; residues of OTC can be present in animal-derived food and are, therefore, consumed by humans [1]. The consumption of, and long-term exposure to, antibiotics may induce antibiotic resistance and can be harmful to human health [2]; therefore, maximum residue limits (MRLs) in foodstuff of animal origin have been set by the European Union for antibiotics [3]; in particular, an MRL of 100 μg/L for OTC residues in milk has been established.

Therefore, it is essential to develop methods for the sensitive and specific detection of OTC in the environment and food products [4]. To achieve this goal, chromatographic methods have been predominantly used, but these require expensive, laboratory-based instrumentation and trained personnel that preclude their implementation in the field for on-site analysis [5]. On the other hand, biosensors offer an alternative and highly attractive approach for antibiotic residue monitoring, due to their portability and low cost. Amongst the different types of bioreceptors used in various biosensing devices, aptamers (single-stranded oligonucleotides capable of binding to an analyte with high affinity due to the 3D structural arrangement they conform to) present numerous advantages. As a result, many aptasensors have been already developed for the detection of several antibiotics, including

OTC [6,7]. Furthermore, the implementation of electrochemical sensing in aptamer-based assays offers some important advantages, such as a high sensitivity, low cost and portable equipment, applicability to on-site analysis, and scope for miniaturization [8].

Herein, the fabrication and application of electrochemical gold-based aptasensors for the label-free assay of OTC is described. Thin-film gold electrodes were fabricated through sputtering gold on a Kapton film. Subsequently, thiol-modified aptamers were immobilized onto the electrodes via the exploitation of the interaction of sulfur with gold. The selective capture of OTC to the aptamer-functionalized electrodes was monitored electrochemically using cyclic voltammetry (CV), differential-pulse voltammetry (DPV), and electrochemical impedance spectroscopy (EIS), with $[Fe(CN)_6]^{4-}/[Fe(CN)_6]^{3-}$ as a redox probe (Figure 1). The decrease in the charge transfer of the redox probe could be related to the OTC concentration.

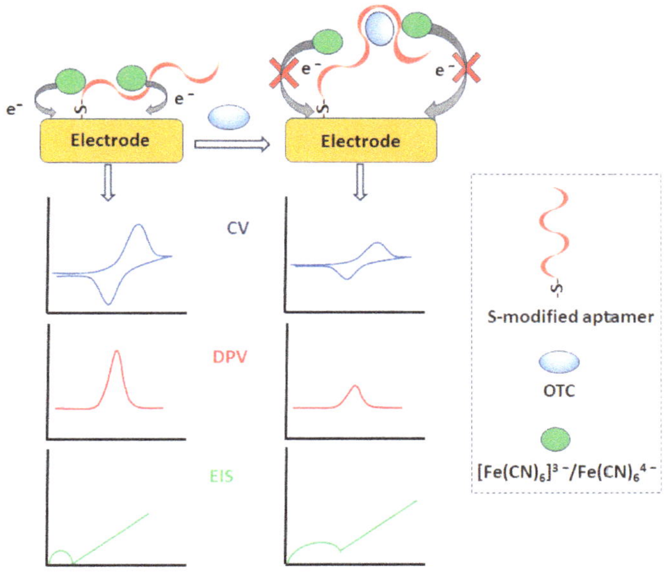

Figure 1. The principle of the aptamer-based assay for OTC at the microfabricated gold electrodes.

2. Experiment

2.1. Reagents and Materials

Reagents were of analytical grade and obtained from Sigma-Aldrich (Burlington, MA, USA). The aptamer sequence was: 5′-GGA ATT CGC TAG CAC GTT GAC GCT GGT GCC CGG TTG TGG TGC GAG TGT TGT GTG GAT CCG AGC TCC ACG TG/3ThioMC3-D/-3′ and was purchased from Integrated DNA Technologies Inc. (Coralville, IA, USA). The aptamer was diluted in phosphate buffer (PB) (10 mM, pH 7.4) containing 1 mM $MgCl_2$. OTC was purchased from Thermo Fisher Scientific (Waltham, MA, USA), and a stock solution of 100 mg/L was prepared in DMSO:PB (50:50 *v*/*v*).

OTC standard calibration solutions containing 0, 25, 50, 100, 200, 400, and 600 ng/mL of OTC were prepared in PB (pH 7.4). A milk matrix was prepared by dissolving 1.00 g of low-fat dried milk powder in 50 mL of PB (pH 7.4), centrifuging, and reconstituting the supernatant solution to a final volume of 100.0 mL with PB. OTC matrix-matched standard calibration solutions containing 0, 25, 50, 100, 200, 400, and 600 ng/mL of OTC were prepared in milk matrix.

A spiked milk sample was prepared as follows: 1.00 g of low-fat dried milk powder was spiked with 10 µg of OTC, dissolved in 50 mL of PB (pH 7.4), and centrifuged; the supernatant solution was reconstituted to a final volume of 10.0 mL with PB.

All electrochemical measurements were performed using a solution containing 5 mM $[Fe(CN)_6]^{4-}/[Fe(CN)_6]^{3-}$ in 0.1 M KCl.

2.2. Instrumentation

A Metrohm Autolab PGSTAT12 Potentiostat/Galvanostat equipped with the GPES Software v. 4.9 (Metrohm, Switzerland) was used for all the electrochemical measurements. Measurements were carried out in a standard electrochemical cell consisting of a Ag/AgCl reference electrode, a Pt counter electrode, and the microfabricated thin-film gold electrode as the working electrode. The measurement conditions were: DPV: scan from −0.1 to +0.4 V, scan rate 10 mV/s, step 5 mV, pulse amplitude 25 mV, modulation time 50 ms, interval time 0.5 s; CV: scan from −0.5 to +0.5 V, scan rate 50 mV/s; EIS: DC potential +0.25 V, AC potential 10 mV, frequency range 100,000 Hz to 0.1 Hz.

Circular dichroism (CD) spectra were recorded (200–320 nm range, 2 nm bandwidth, 50 nm/min scan rate) using a Jasco J-1500 CD spectrometer (Tokyo, Japan) thermostated at 25 °C.

Sputtering was performed with a CV401 system (Cooke Vacuum Products, W. Redding, CT, USA), and atomic force microscopy (AFM) images were taken with an SPM SMENA instrument.

2.3. Fabrication of the Thin-Film Gold Sensors

The thin-film gold electrodes were fabricated via metal sputtering. A film of flexible Kapton® HN polyimide film (50 μm thicknesses, from RS) was covered with a metal mask with the electrode pattern, and Cr and Au were sputtered on the wafer at a nominal thickness of 5 and 100 nm, respectively. A schematic diagram of the fabrication process of the gold sensors and an array of gold electrodes are illustrated in Figure 2a,b, respectively.

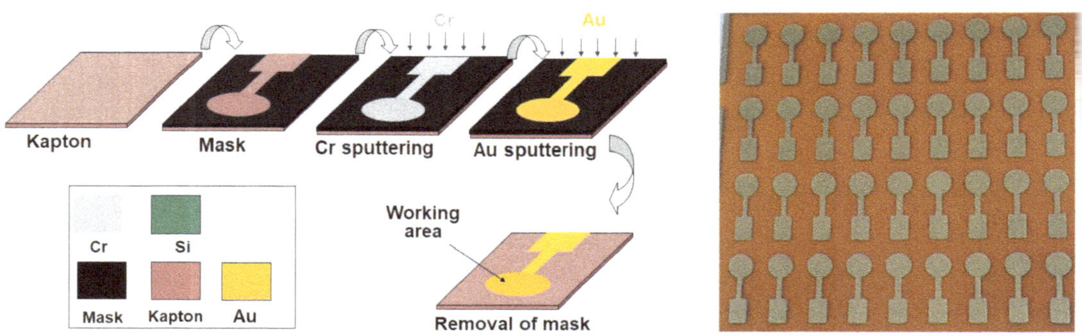

(a) (b)

Figure 2. (**a**) Schematic diagram of the fabrication process of the thin-film gold electrodes, (**b**) photograph of an array of 36 electrodes.

2.4. Experimental Protocol

The aptamer stock solution was heated for 5 min at 95 °C and allowed to cool down to room temperature. Aptamer solutions of 5, 10, 20, and 40 μM were prepared in PB containing 1 mM $MgCl_2$. An amount of 10 μL of the aptamer solution was drop-casted onto the thin-film gold working electrode surface and was left for 12 h at 4 °C in a humidity chamber. The aptamer-modified electrodes were thoroughly washed with PB to remove unbound aptamers. Subsequently, 5 μL of mercaptoethanol solution (1 mM in 1:2 (v/v) water/ethanol) was added at the aptamer-modified electrode and was left for 30 min at room temperature.

OTC detection was accomplished through incubating 60 μL of the standard solution or milk sample onto the sensor surface for 30 min at room temperature. Then, the electrode was washed thoroughly with PB to remove unbound OTC and subjected to electrochemical analysis.

3. Results and Discussion

AFM imaging of the thin-film gold electrodes indicates that the gold film is grainy in structure, with a roughness of ca. 9 nm and a grain diameter of ca. 15 nm (Figure 3a). These data demonstrate an exceptionally smooth surface morphology, which is suitable for the reproducible and uniform attachment of the aptamer, as opposed to the commonly used thick-film screen-printed gold deposits that exhibit a much rougher surface and, consequently, lower device-to-device uniformity.

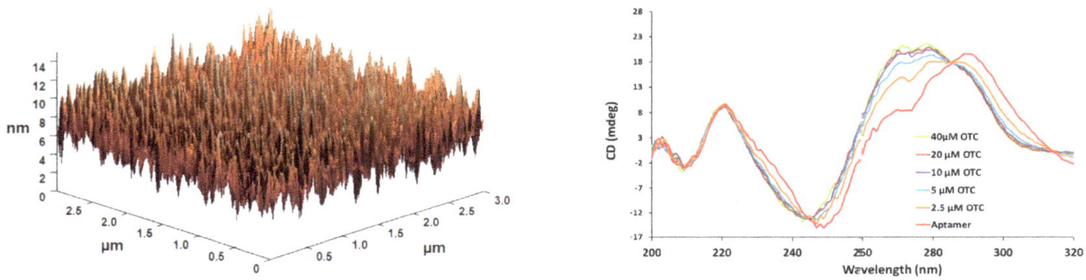

(**a**) (**b**)

Figure 3. (**a**) AFM image of the thin-film gold electrode surface, (**b**) CD spectrum of the aptamer before and after the addition of OTC.

The CD spectrum of the aptamer before and after the addition of different concentrations of OTC is illustrated in Figure 3b. The spectrum reveals that the aptamer adopts a parallel G-quadruplex conformation (maximum at 260 nm, minimum at 240 nm), while incubation with increasing OTC concentrations results in a shift of the peak minimum from ca. 240 nm to ca. 250 nm, coupled with a decrease in the peak maximum at ca. 270 nm and the appearance of a peak at ca. 290 nm, indicative of an analyte-induced folding of the aptamer into a hybrid G quadruplex or a mixture of hybrid and antiparallel quadruplexes [9].

The different steps of the aptasensor preparation and the assay were studied using EIS and CV (Figure 4). EIS demonstrated that the charge transfer resistance of the bare electrode increased upon successive immobilization of the aptamer, mercaptoethanol, and OTC (Figure 4a). This was corroborated by the respective CVs, which showed that the redox current decreased as the bare electrode was successively treated with aptamer, mercaptoethanol, and OTC (Figure 4b). These results suggest that the aptamer is successfully immobilized on the electrode surface and that the OTC is effectively bound to the immobilized aptamer. The concentration of the aptamer was studied, and a 20 µM solution was found to produce the highest sensitivity and repeatability.

Analytical measurements were performed via recording the DPV oxidation current of the $[Fe(CN)_6]^{4-}/[Fe(CN)_6]^{3-}$ redox probe at different OTC concentrations in the range 0–600 µg/L. Increasing concentrations of OTC induced a reduction in the DPV current as a result of the aptamer blocking of the electrode surface. The % signal reduction, $I\%$, was calculated as: $I\% = (i_o - i)/i_o$ (where i_o is the DPV current at the aptamer-modified electrode in the absence of OTC, and i is the DPV current at the aptamer-modified electrode in the presence of OTC).

In order to assess possible milk matrix interference, the matrix effect was calculated as $ME = I\%_{,st}/I\%_{,mm}$ (where = $I\%_{,mm}$ is the signal reduction in a matrix-matched OTC standard, and $I\%_{,st}$ is the signal reduction in the same OTC standard in PB) (Table 1). These data indicate that a more pronounced matrix effect existed at lower OTC concentrations, so a matrix-matched calibration plot is recommended for quantitative assays.

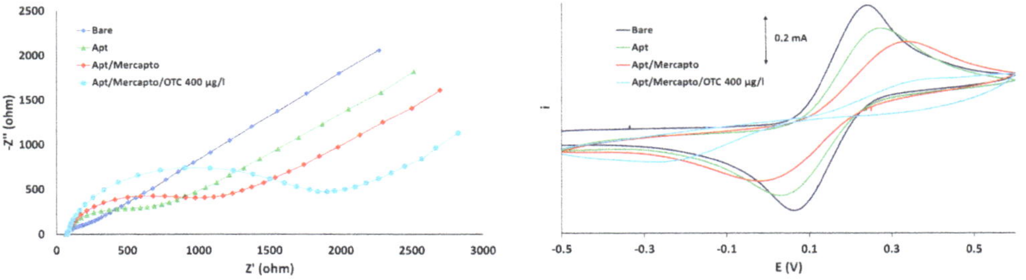

(a) (b)

Figure 4. (**a**) EIS spectra, (**b**) CVs at a bare gold electrode and gold electrodes after the successive immobilization of aptamer, conditioning with mercaptoethanol, and incubation with 200 µg/L OTC. Aptamer concentration: 20 µM.

Table 1. Study of the milk matrix effect.

[OTC] (µg/L)	ME
25	1.25
50	1.23
100	1.15
200	1.15
400	1.11
600	1.05

Typical DPV traces are illustrated in Figure 5, and the linear–linear ($I\%_{,mm}$ *vs.* [OTC]) and linear–log ($I\%_{,mm}$ *vs.* log[OTC]) calibration plots are shown as inserts. The limit of detection (LOD) of OTC was ~5 µg/L (calculated as the OTC concentration that produced a statistically different signal from the blank). The repeatability between sensors (calculated as the % relative standard deviation from five different aptasensors) was 16% at the 100 µg/L OTC level.

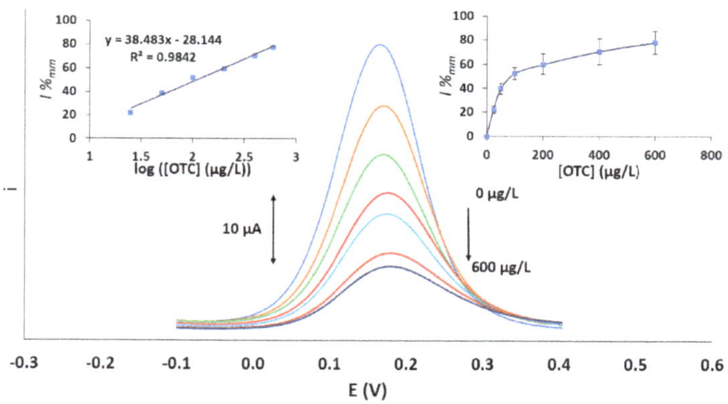

Figure 5. DPV traces obtained with the aptasensor at different matrix-matched OTC standards in the concentration range 0–600 µg/L. The linear–linear and linear–log calibration plots are shown as inserts.

Preliminary experiments were carried out for the determination of OTC in a milk sample spiked with 100 μg/L OTC. The recovery was calculated as $R\% = I\%_{,sp}/I\%_{mm,100}$ (where $I\%_{,sp}$ is the % signal reduction in the spiked milk sample, and $I\%_{mm,100}$ is the % signal reduction in a 100 μg/L matrix-matched standard of OTC). The mean recovery ($n = 3$) was calculated as 95%.

Author Contributions: Conceptualization, A.E., G.T., J.W., T.S. and Z.K.; methodology, A.E., G.T., J.W., T.S. and Z.K.; validation, V.M., A.A., K.J., N.M. and D.S.; investigation, V.M., A.A., K.J., N.M. and D.S.; data curation, K.J., N.M. and D.S.; writing—original draft preparation, A.E. and D.S.; writing—review and editing, A.E., G.T., J.W. and Z.K.; supervision, A.E., G.T., J.W. and Z.K.; project administration, A.E.; funding acquisition, A.E., G.T., J.W. and Z.K. All authors have read and agreed to the published version of the manuscript.

Funding: This work was supported by the European Union's Horizon 2020 Research and Innovation Programme under the Marie Sklodowska-Curie Grant Agreement No 101007299.

Institutional Review Board Statement: Not applicable.

Informed Consent Statement: Not applicable.

Data Availability Statement: Data sharing is not applicable to this article.

Conflicts of Interest: The authors declare no conflict of interest.

References

1. Landers, T.F.; Cohen, B.; Wittum, T.E.; Larson, E.L. A Review of Antibiotic Use in Food Animals: Perspective, Policy, and Potential. *Public Health Rep.* **2012**, *127*, 4–22. [CrossRef] [PubMed]
2. Virto, M.; Santamarina-García, G.; Amores, G.; Hernández, I. Antibiotics in Dairy Production: Where Is the Problem? *Dairy* **2022**, *3*, 541–564. [CrossRef]
3. Commission Regulation (EU), No. 37/2010. Available online: https://eur-lex.europa.eu/legal-content/EN/TXT/?uri=CELEX:32010R0037 (accessed on 13 November 2023).
4. Sversut, R.A.; da Silva, A.A.; Mazon Cardoso, T.F.; Kassab, N.M.; do Amaral, M.S.; Nunes Salgado, H.R. A Critical Review of Properties and Analytical Methods for the Determination of Oxytetracyline in Biological and Pharmaceutical Matrices. *Crit. Rev. Anal. Chem.* **2017**, *47*, 154–171. [CrossRef] [PubMed]
5. Peris-Vicente, J.; Peris-García, E.; Albiol-Chiva, J.; Durgbanshi, A.; Ochoa-Aranda, E.; Carda-Broch, S.; Bose, D.; Esteve-Romero, J. Liquid chromatography, a valuable tool in the determination of antibiotics in biological, food and environmental samples. *Microchem. J.* **2022**, *177*, 107309. [CrossRef]
6. Mehlhorn, A.; Rahimi, P.; Joseph, Y. Aptamer-Based Biosensors for Antibiotic Detection: A Review. *Biosensors* **2018**, *8*, 54. [CrossRef] [PubMed]
7. Liang, G.; Song, L.; Gao, Y.; Wu, K.; Guo, R.; Chen, R.; Zhen, J.; Pan, L. Aptamer Sensors for the Detection of Antibiotic Residues—A Mini-Review. *Toxics* **2023**, *11*, 513. [CrossRef] [PubMed]
8. Evtugyn, G.; Porfireva, A.; Tsekenis, G.; Oravczova, V.; Hianik, T. Electrochemical Aptasensors for Antibiotics Detection: Recent Achievements and Applications for Monitoring Food Safety. *Sensors* **2022**, *22*, 3684. [CrossRef] [PubMed]
9. Kejnovská, I.; Renčiuk, D.; Palacký, J.; Vorlíčková, M. CD Study of the G-Quadruplex Conformation. *Methods Mol. Biol.* **2019**, *2035*, 25–44. [CrossRef] [PubMed]

engineering proceedings

MDPI

Proceeding Paper

Gas-Sensitive Properties of β-Ga_2O_3 Thin Films Deposited and Annealed at High Temperature †

Nikita Yakovlev [1,*], Aleksei Almaev [1], Alexander Korchemagin [1], Mukesh Kumar [2] and Damanpreet Kaur [2]

1 Research and Development Center for Advanced Technologies in Microelectronics, National Research Tomsk State University, 634050 Tomsk, Russia; almaev_alex@mail.ru (A.A.); kladovshik_90@hotmail.com (A.K.)

2 Functional and Renewable Energy Materials Laboratory, Department of Physics, Indian Institute of Technology Ropar, Rupnagar 140001, India; mkumar@iitrpr.ac.in (M.K.); 2018phz0010@iitrpr.ac.in (D.K.)

* Correspondence: nik_mr_x@mail.ru

† Presented at the 10th International Electronic Conference on Sensors and Applications (ECSA-10), 15–30 November 2023; Available online: https://ecsa-10.sciforum.net/.

Abstract: The gas-sensitive properties of thin films of β-Ga_2O_3 deposited via RF magnetron sputtering while heating the substrate to 650 °C were studied. Some of the samples were subjected to additional high-temperature annealing at a temperature of 900 °C. As a result, for samples subjected to additional annealing, the response when exposed to 1% H_2 increased by five once sensitivity to hydrogen-containing gases appeared. These samples are also characterized by good long-term stability compared to samples without high-temperature annealing. The improvement in gas-sensitive characteristics is explained by a decrease in oxygen vacancies and a decrease in current density by four orders of magnitude.

Keywords: β-Ga_2O_3 thin films; gas sensors; high-temperature annealing; magnetron sputtering

Citation: Yakovlev, N.; Almaev, A.; Korchemagin, A.; Kumar, M.; Kaur, D. Gas-Sensitive Properties of β-Ga_2O_3 Thin Films Deposited and Annealed at High Temperature. *Eng. Proc.* **2023**, *58*, 2. https://doi.org/10.3390/ecsa-10-16015

Academic Editor: Stefano Mariani

Published: 15 November 2023

1. Introduction

The active development of resistive gas sensors based on β-Ga_2O_3 began in the 1980s and 1990s [1,2]. Thanks to advances in the synthesis of semiconductor materials, resistive gas sensors based on β-Ga_2O_3 have been widely developed. Compared to many metal–oxide semiconductors (SnO_2, In_2O_3, WO_3 and ZnO), the use of Ga_2O_3 as a sensitive layer makes it possible to create gas sensors that are stable at high operating temperatures and low oxygen concentrations, weakly affected by environmental humidity and characterized by high stability [3–5]. These advantages are key in the development of gas analytical systems for extreme operating conditions (elevated ambient temperatures, high concentrations of water vapor, and changes in oxygen concentration over a wide range) [4,5].

When developing resistive gas sensors based on metal–oxide semiconductors, it is important to reduce the contribution of bulk conductivity. For this reason, thin-film technology is promising for creating sensors. The magnetron sputtering method is widely used to form thin films of metal–oxide semiconductors. In particular, thin films of Ga_2O_3 are formed using RF magnetron sputtering. RF magnetron sputtering has several advantages over other methods, such as the following: low cost, ease of control, good adhesion, wide variety of materials and high sputtering speed. To improve the crystallinity of the films obtained via this method, it is necessary to anneal at T > 800 °C. When annealing T < 800 °C, amorphous and a mixture of Ga_2O_3 phases are observed [6,7], another way to obtain crystalline films is by heating the substrate during deposition. This method is poorly widespread and researched. In a number of works [8–10], crystalline layers are obtained at a substrate temperature ranging from 300 to 700 °C. The authors also note that the crystallinity of the film also depends on other factors, such as the presence of oxygen in the reactor, magnetron power, pressure in the reactor, etc. The gas-sensitive properties of Ga_2O_3 thin films obtained via high-temperature magnetron sputtering have not yet been studied. In this work, β-Ga_2O_3 samples obtained via RF magnetron sputtering at a substrate temperature of 650 °C and subsequent high-temperature annealing of 900 °C were studied.

2. Materials and Methods

β-Ga_2O_3 with a thickness of 300 nm was deposited via radio frequency magnetron sputtering of a (5N) oxide target in an argon plasma onto sapphire with a thickness of 0.43 mm. The substrate temperature during film deposition was 650 °C. This series was designated GO-650. The plate with the Ga_2O_3 film was divided into two parts, and the second part was subjected to additional annealing at a temperature of 900 °C for 30 min in an air atmosphere. These samples are designated GO-650+900. Pt contacts of various topologies were formed on the surface of the films.

To measure the electrically conductive and gas-sensitive properties, platinum plane-parallel contacts were deposited on the surface of a thin GO-650 film through a shadow mask. The interelectrode distance was 600 μm. Interdigitated contacts with an interelectrode distance of 315 μm were deposited on the surface of a thin GO-650+900 film using vacuum deposition and photolithography.

Measurements of the current-voltage (I-V) characteristics and time dependences of the sample current when exposed to various gases were carried out with a Keithley 2636A source-meter manufactured by Keithley Instruments, Inc., Cleveland, OH, USA in a sealed Nextron MPS-CHH microprobe station manufacturer NEXTRON, Busan, South Korea. This microprobe station allows for the measurement of the electrically conductive characteristics of films in the temperature range from room temperature to 750 °C with an accuracy of T ± 0.1 °C. The measurements were carried out in dark conditions, in a flow of clean dry air or in a gas mixture of pure dry air + target gas. As the target gases H_2, CO_2, CO, NO_2, and O_2 were chosen to study the effect of oxygen on the properties of thin Ga_2O_3 films, a mixture of N_2 and O_2 gases was pumped through the chamber. The flow rate of the gas mixtures through the measuring chamber (volume 100 cm^3) was maintained at 1000 cm^3/min. A special generator served as a source of pure dry air. The concentration of the target gas in the mixture was controlled by a gas mixture generator with Bronkhorst mass flow regulators. The relative error of gas flow did not exceed 1.5%. The applied voltage to the sample electrodes was 5 V.

3. Gas-Sensitive Properties of Ga_2O_3 Thin Films

The main samples, GO-650 and GO-650+900, have high resistance, even at T < 300 °C, and the current in the samples at 5 V is no more than pA. The current increases with increasing temperature from 300 to 600 °C, and from 400 to 750 °C for samples GO-650 and GO-650+900 by three orders of magnitude, respectively.

Figure 1 shows the dependence of responses to fixed H_2 concentrations on temperature for GO-650 and GO-650+900 thin films. Exposure to reducing gases H_2, CH_4, NH_3 and CO results in a reversible increase in current through the samples.

The following relationship was chosen as a reaction to reducing gases:

$$S_g = I_g / I_{air}, \tag{1}$$

where I_g is the current of a thin film in a gas mixture of pure dry air + reducing gas, and I_{air} is the thin film current in pure dry air. Exposure to O_2, NO_2 leads to a reversible decrease in the current in the GO-650 and GO-650+900 samples. The following relationship was used to examine the response to these gases:

$$S_{ox} = I_{air} / I_{ox}, \tag{2}$$

where I_{ox} is the current of a thin film in a gas mixture of pure dry air + oxidizing gas. The curves in Figure 1 are characterized by the presence of maxima S_{MAX} at a certain temperature T_{MAX}. It is recommended to select T_{MAX} as the operating temperature.

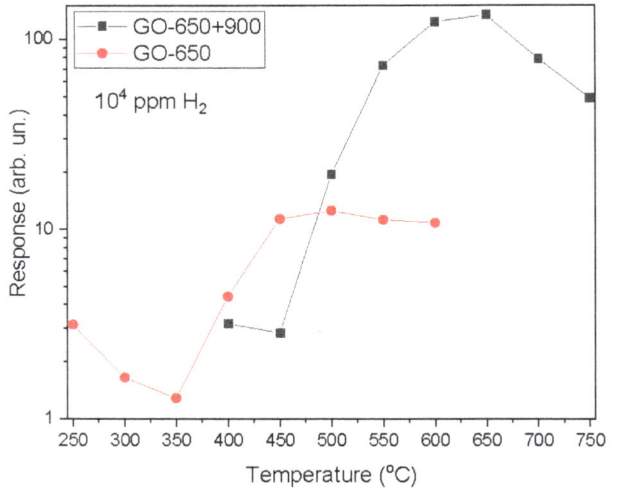

Figure 1. 25]Dependence of the response of GO-650 and GO-650+900 at 10^4 ppm H_2 on the operating temperature.

The response of GO-650 films when stored in a sealed bag after exposure to H_2 changes significantly. This results in a significant increase in response. Such a drift of gas-sensitive characteristics over time is typical for thin films of metal oxides and semiconductors. For sample GO-650+900, such a pattern is not observed, and the response deviates from the average value in small areas (Figure 2).

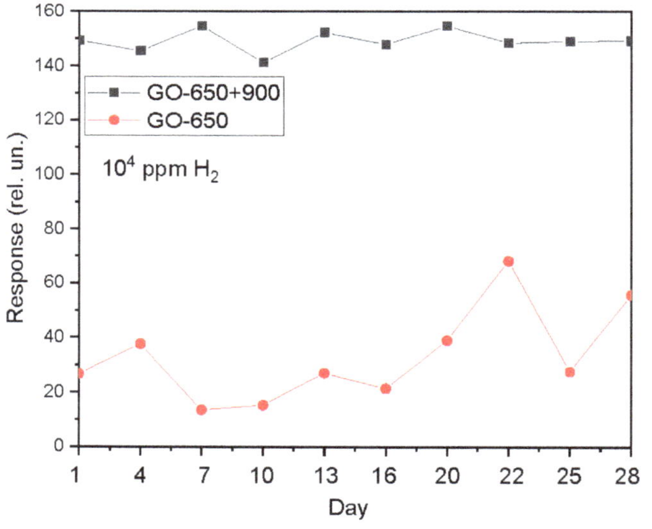

Figure 2. Long-term stability under the exposure to a fixed hydrogen concentration.

Despite the poor long-term stability of the GO-650 samples, all samples have good temporary stability over the course of a single experiment. Figure 3 shows the time dependencies of the current under five-fold exposure to 1% of H_2.

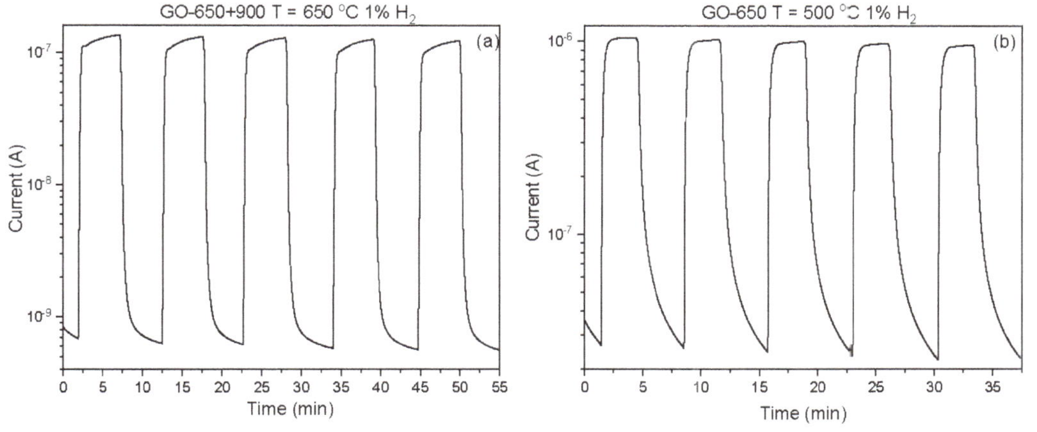

Figure 3. Time dependencies of the current through the samples under fivefold exposure to 1% of H_2 for samples with additional annealing (**a**) and without it (**b**).

The selectivity of thin films was measured. The temperature dependencies of the response for various gases for GO-650+900 thin films are presented in Figure 4.

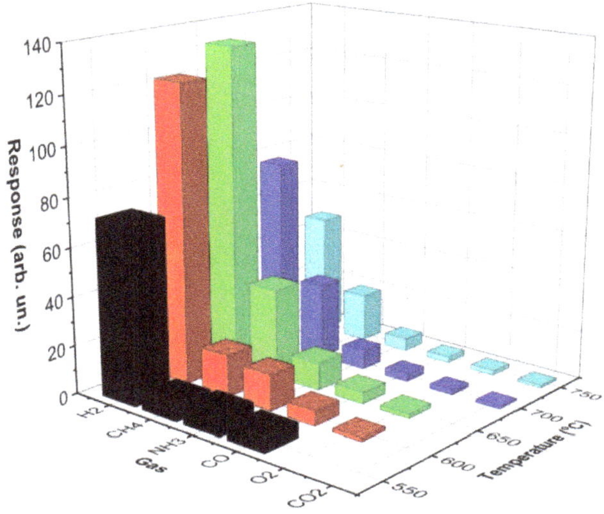

Figure 4. Temperature dependencies of the response to various gases and for thin films GO-650+900.

The concentration of CH_4, CO_2 and O_2 was 1 vol. %, the concentration of CO and NH_3 was 0.2 vol. %. GO-650 thin films do not respond well to other gases. GO-650+900 samples have a high response to H_2, NH_3 and CH_4.

The improvement in the gas-sensitive characteristics of GO-650+900 thin films is mainly due to annealing in an air atmosphere. As a result of annealing, the number of oxygen vacancies decreased, as evidenced by a decrease in the current density. The current density of the GO-650 samples is 0.10 A/cm^2, and the current density of the GO-650+900 samples is 6.27×10^{-5} A/cm^2. The current density was calculated at T = 500 °C. The stability of the GO-650+900 samples was also affected by the formation of interdigitated contacts.

4. Mechanism of Gas Sensitivity

In the air, oxygen is chemisorbed on the surface of gallium oxide, which captures electrons from the conduction band of gallium oxide, forming in the near-surface region of the semiconductor, a layer depleted of major charge carriers and electrons Oxygen chemisorbed on the surface of β-Ga_2O_3 films is presented in molecular O_2^-, atomic O^- and O^{2-} forms. The atomic form of chemisorbed oxygen O^- is the most reactive when interacting with molecules of reducing gases. In the temperature range of 200–650 °C, O^- predominates on the surface of β-Ga_2O_3.

Chemisorption of oxygen can be described by using the following expression:

$$O_2 + 2e \rightarrow 2O^-_{ch}, \quad (3)$$

where O^-_{ch} is a chemisorbed oxygen ion. Superstoichiometric gallium atoms Ga^{3+} on the surface of β-Ga_2O_3 can act as adsorption centers for oxygen. In the region of selected operating temperatures, when reducing gases appear in the air, their molecules interact with previously chemisorbed O^-_{ch}. These interactions on the surface of β-Ga_2O_3 can be described by the following reactions:

$$H_2 + O^-_{ch} \rightarrow H_2O\uparrow + e, \quad (4)$$

$$2NH_3 + 3O^-_{ch} \rightarrow N_2\uparrow + 3H_2O\uparrow + 3e, \quad (5)$$

$$CH_4 + 4O^-_{ch} \rightarrow CO_2\uparrow + 2H_2O\uparrow + 4e, \quad (6)$$

When the surface of a semiconductor is exposed to oxidizing gases NO_2 and NO, the following reactions may occur:

$$NO_2 + e^- \rightarrow NO_2^-, \quad (7)$$

$$NO_2 + O^-_{ch} \rightarrow NO_3^-, \quad (8)$$

Reactions (7) occur without the participation of O^-_{ch} ions; NO_2 and NO molecules are chemisorbed onto a free adsorption center and capture electrons from the conduction band of β-Ga_2O_3. An additional negative charge on the surface of β-Ga_2O_3 leads to a greater increase in $e\varphi_s$ and, consequently, to a decrease in conductivity and charge carrier current in β-Ga_2O_3.

5. Conclusions

For the first time, the gas-sensitive properties of β-Ga_2O_3 structures obtained via RF magnetron sputtering with heating of the substrate and subsequent high-temperature annealing have been studied. Thin films of β-Ga_2O_3 without subsequent high-temperature annealing had a 3–4 times lower response to hydrogen and had a long drift of characteristics compared to the β-Ga_2O_3 samples with subsequent high-temperature annealing. As a result of high-temperature annealing and the formation of interdigitated contacts, GO-650+900 thin films became more stable, and the responses increased several times. As a result of annealing in air at T = 900 °C in samples followed by high-temperature annealing, the number of vacancies decreases, the current density decreases by four orders of magnitude, and stability increases.

Author Contributions: Conceptualization A.A., M.K., D.K. and N.Y.; methodology A.A. and N.Y.; software A.A. and N.Y.; validation A.K., M.K. and N.Y.; formal analysis A.A. and N.Y.; investigation A.A. and N.Y.; resources A.A., M.K. and N.Y.; data curation N.Y. and D.K.; writing—original draft preparation N.Y.; writing—review and editing N.Y.; visualization N.Y.; supervision A.A. and N.Y.;

project administration A.A.; funding acquisition A.A. All authors have read and agreed to the published version of the manuscript.

Funding: This work was supported by the Russian Science Foundation (project # 20-79-10043-P).

Institutional Review Board Statement: Not applicable.

Informed Consent Statement: Not applicable.

Data Availability Statement: All data that support the findings of this study are included within the article.

Conflicts of Interest: The authors declare no conflict of interest.

References

1. Fleischer, M.; Hanrieder, W.; Meixner, H. Stability of semiconducting gallium oxide thin films. *Thin Solid Films.* **1990**, *190*, 93–102. [CrossRef]
2. Fleischer, M.; Giber, J.; Meixner, H. H_2-Induced changes in electrical conductance of β-Ga_2O_3 thin-film systems. *Appl. Phys.* **1992**, *A54*, 560–566. [CrossRef]
3. Afzal, A. β-Ga_2O_3 nanowires and thin films for metal oxide semiconductor gas sensors: Sensing mechanisms and performance enhancement strategies. *J. Mater.* **2019**, *5*, 542–557. [CrossRef]
4. Hoefer, U.; Frank, J.; Fleischer, M. High temperature Ga_2O_3-gas sensors and SnO_2-gas sensors: A comparison. *Sens. Actuators B Chem.* **2001**, *78*, 6–11. [CrossRef]
5. Lampe, U.; Fleischer, M.; Meixner, H. Lambda measurement with Ga_2O_3. *Sens. Actuators B Chem.* **1994**, *17*, 187–196. [CrossRef]
6. Pech, S.; Kim, S.; Kim, N.H. Magnetron Sputter-Deposited-Ga_2O_3 Films on c-Sapphire Substrate: Effect of Rapid Thermal Annealing Temperature on Crystalline Quality. *Coatings* **2022**, *12*, 140. [CrossRef]
7. Singh, A.K.; Gupta, M.; Sathe, V.; Katharria, Y.S. Effect of annealing temperature on β-Ga_2O_3 thin films deposited by RF sputtering method. *Superlattices Microstruct.* **2021**, *156*, 106976. [CrossRef]
8. Ramana, A.K.B.C.V. Mechanical Properties of Nanocrystalline and Amorphous Gallium Oxide Thin Films. *Adv. Eng. Mater.* **2018**, *20*, 1701033. [CrossRef]
9. Saikumar, A.K.; Nehate, S.D.; Sundaram, K.B. Review—RF Sputtered Films of Ga_2O_3. *ECS J. Solid State Sci. Technol.* **2019**, *8*, Q3064–Q3078. [CrossRef]
10. Vega, E.; Isukapati, S.B.; Oder, T.N. Microstructure and optical properties of sputterdeposited Ga_2O_3 films. *J. Vac. Sci. Technol. A* **2021**, *39*, 033412. [CrossRef]

engineering proceedings

MDPI

Proceeding Paper

Optimizable Ensemble Regression for Arousal and Valence Predictions from Visual Features †

Itaf Omar Joudeh [1,*], Ana-Maria Cretu [1] and Stéphane Bouchard [2]

[1] Department of Computer Science and Engineering, University of Quebec in Outaouais, Gatineau, QC J8Y 3G5, Canada; ana-maria.cretu@uqo.ca

[2] Department of Psychoeducation and Psychology, University of Quebec in Outaouais, Gatineau, QC J8X 3X7, Canada; stephane.bouchard@uqo.ca

* Correspondence: joui02@uqo.ca

† Presented at the 10th International Electronic Conference on Sensors and Applications (ECSA-10), 15–30 November 2023; Available online: https://ecsa-10.sciforum.net/.

Abstract: The cognitive state of a person can be categorized using the Circumplex model of emotional states, a continuous model of two dimensions: arousal and valence. We exploit the Remote Collaborative and Affective Interactions (RECOLA) database, which includes audio, video, and physiological recordings of interactions between human participants to predict arousal and valance values using machine learning techniques. To allow learners to focus on the most relevant data, features are extracted from raw data. Such features can be predesigned or learned. Learned features are automatically learned and utilized by deep learning solutions. Predesigned features are calculated before machine learning and inputted into the learner. Our previous work on video recordings focused on learned features. In this paper, we expand our work onto predesigned visual features, extracted from video recordings. We process these features by applying time delay and sequencing, arousal/valence labelling, and shuffling and splitting. We then train and test regressors to predict arousal and valence values. Our results outperform those from the literature. We achieve a root mean squared error (RMSE), Pearson's correlation coefficient (PCC), and concordance correlation coefficient (CCC) of 0.1033, 0.8498, and 0.8001 on arousal predictions; and 0.07016, 0.8473, and 0.8053 on valence predictions, using an optimizable ensemble, respectively.

Keywords: regression; machine learning; cognitive/emotional state; visual features

Citation: Joudeh, I.O.; Cretu, A.-M.; Bouchard, S. Optimizable Ensemble Regression for Arousal and Valence Predictions from Visual Features. *Eng. Proc.* **2023**, *58*, 3. https://doi.org/10.3390/ecsa-10-16009

Academic Editor: Stefano Mariani

Published: 15 November 2023

1. Introduction

The cognitive state of a person can be categorized using the Circumplex model of emotional states [1], a continuous model of two dimensions: arousal and valence, where arousal measures the energy level and valence measures the positivity level of a person's emotion. In this model, emotions are divided into four categories: happy, angry, sad, and relaxed. Each of these emotions is associated with a quadrant of the circumplex model. Happy emotions have high valence and high arousal, anger—low valence and high arousal, sad—low valence and low arousal, and relaxed—high valence and low arousal. The arousal and valence values can be estimated via machine learning regression.

We use the RECOLA database [2] which includes audio, video, and physiological recordings of online interactions between human participants to predict arousal and valance values using machine learning techniques. We previously predicted arousal and valence values using the physiological [3,4] and video [4,5] recordings of RECOLA. Features are attributes that describe the data. They can be predesigned or learned [6]. Learned features are attributes that are automatically extracted and utilized by deep machine learning solutions during the learning process. On the other hand, predesigned features are attributes that are calculated on the data before the learning process and provided as input to the machine learner. Our previous work on the video recordings of RECOLA focused on learned

features from convolutional neural networks (CNNs), such as ResNet-18 and MobileNet-v2. MobileNet-v2 achieved a root mean squared error (RMSE), Pearson's correlation coefficient (PCC), and concordance correlation coefficient (CCC) of 0.1220, 0.7838, and 0.7770 on arousal predictions; and 0.0823, 0.7789, and 0.7715 on valence predictions, respectively. In this paper, we expand our work to analyze and assess the predesigned visual features, extracted from the video recordings of RECOLA. We propose a novel combination of processing steps to prepare the visual features for regression. We leverage machine learning solutions such as regression trees, kernel regression, and ensemble regressors to predict the arousal and valence values of cognitive states. Our goal is to find the model(s) with the best prediction performance to later integrate into a virtual reality (VR) system that runs cognitive remediation exercises for users with mental health disorders (e.g., schizophrenia).

Solutions for the prediction of cognitive states ideally consist of two components: parametrization and recognition of facial expressions [6]. Parametrization is the process of specifying the visual features and coding schemes to describe the involved facial expressions. The visual features used for the prediction of cognitive states can be appearance or geometric features [7]. Geometric features represent the geometry of the face. Local Gabor Binary Patterns from Three Orthogonal Planes (LGBP-TOP) [8] is one method that is used in the extraction of appearance features, while facial landmarks [9] are usually used for geometric features. Examples of geometric features include the derivatives of the detected facial landmarks, the speed and direction of motion in facial expressions, the head pose, and the direction of the eye gaze. Appearance features represent the overall texture resulting from the deformation of the neutral facial expression. Appearance features depend on the intensity of an image, whereas geometrical features determine distances, deformations, curvatures, and other geometric properties [6]. Coding schemes can either be descriptive or judgmental [6]. Descriptive coding schemes depend on surface properties and what the face can do to describe facial expression. Judgmental coding schemes depend on the latent emotions or affects that produce them to parameterize facial expressions. The facial action coding system (FACS) [10] is one example of descriptive systems. FACS is a system that describes all visually evident facial movements [10,11]. It divides facial expressions into individual components of muscle movement, called Action Units (AUs). Coding schemes, such as facial AUs, as well as geometric and/or appearance features can then be treated as input parameters to machine learning regressors or classifiers for the prediction of cognitive states.

In the remainder of this paper, we will provide a literature review (Section 2), followed by a description of the methods used in our solution (Section 3). Then, we include a discussion of our results (Section 4). Finally, we will conclude this paper with some closing remarks (Section 5).

2. Literature Overview

RECOLA [2] is a multimodal database of natural emotions that is often used in studies on the prediction of cognitive states. It contains video, audio, and physiological recordings. It also provides predesigned features for these recordings. Arousal and valence annotations were provided by 6 raters every 40 ms of recording. The mean of the six ratings was used to label the data in our work. The database contains 5-min video recordings of 27 participants, where only data from 23 participants are publicly available. Since some of the data modalities in RECOLA contain records for 18 of the participants, we only used these 18 recordings from the RECOLA database to prove our concept.

The authors of the original RECOLA database [2] further extended their work in [11], where they performed experiments on the database for the prediction of arousal and valence values. They extracted 20 visual features on each video frame in the video recordings of RECOLA, along with their first order derivates. They then deployed a bidirectional long short-term memory recurrent neural network (BiLSTM RNN) to predict arousal and valence measures. They compared the prediction performance of the RNN between mean ratings (average of annotations from all 6 raters) and all six ratings, using both single-task and multi-task learning techniques. For arousal, they achieved a CCC of 0.427 using multi-task learning over all six ratings. For valence, they achieved a CCC of 0.431 using single-task learning over all six ratings. The authors of RECOLA [2,11] later introduced the Audio/Visual Emotion Challenge and Workshop (AVEC) in 2015 [12]. In AVEC 2018 [13], they experimented with the different types of visual features: appearance, geometric, 17 facial AUs, and bags-of-words. For arousal, they achieved a CCC of 0.312 via multi-task Lasso, while using appearance features. For valence, they achieved a CCC of 0.438 via a support vector machine (SVM), while using geometric features.

Other authors have also benefited from using the RECOLA database in their research. Han et al. [14] exploited the geometric visual features provided by AVEC to predict arousal and valence values through an RNN. They implemented an implicit fusion framework for joint audiovisual training. They achieved a CCC of 0.413 and 0.527 on arousal and valence predictions, respectively. Albadawy et al. [15] used the visual features provided by AVEC 2015, which included appearance (LGBP-TOP) and geometric (Euclidean distances between 49 facial landmarks) features. For arousal and valence predictions, they proposed a joint modelling strategy using a deep BiLSTM for ensemble and end-to-end models. Their ensemble BiLSTM model achieved a CCC of 0.699 and 0.617 for arousal and valence, respectively. In our work, we used and further processed the basic features extracted by the authors of RECOLA in [11] and experimented with a variety of regressors to predict the arousal and valence values of cognitive states.

3. Methods

We processed the visual features of RECOLA by applying time delay and sequencing, arousal and valence annotation labelling, and data shuffling and splitting. We then trained and tested regressors to predict the arousal and valence values. The following sections will discuss the details about our processing steps and regression methodology.

3.1. RECOLA's Predesigned Visual Features

The video recordings of RECOLA were sampled at a sampling rate of 25 frames/s, where visual features were extracted for each video frame [11]. As predesigned visual features, RECOLA contains 20 attributes alongside their first order derivative, resulting in 40 features in total. These attributes/features include 15 facial AUs of emotional expressions, the head-pose in three dimensions (i.e., X, Y, Z), and the mean and standard deviation of the optical flow in the region around the head. The AUs are AU1 (Inner Brow Raiser), AU2 (Outer Brow Raiser), AU4 (Brow Lowerer), AU5 (Upper Lid Raiser), AU6 (Cheek Raiser), AU7 (Lid Tightener), AU9 (Nose Wrinkler), AU11 (Nasolabial Deepener), AU12 (Lip Corner Puller), AU15 (Lip Corner Depressor), AU17 (Chin Raiser), AU20 (Lip Stretcher), AU23 (Lip Tightener), AU24 (Lip Pressor), and AU25 (Lips Part) from FACS. For more information about these features and their extraction, please refer to [11]. We used these features in our work.

3.2. Time Delay and Sequencing

RECOLA's video recordings were sampled at a rate of 25 frames/s. This means that one frame was captured every 0.04 s (40 ms). The visual features were calculated on each frame, meaning that they were provided every 40 ms as well. Since other data modalities of RECOLA only started being recorded after 2 s (2000 ms), we skipped any readings that occurred before that time. As a result, the first 50 frames (2 s $\times$ 25 frames/s) of the recordings were unused in our work.

3.3. Annotation Labelling

The data in RECOLA were labelled with respect to the arousal and valence emotional dimensions. The data samples were manually annotated using ANNEMO, an annotation tool, developed by Ringeval et al. [2]. Each recording was annotated by six raters. The mean of these six ratings was used to label the data in our work. The mean arousal and valence values were also sampled every 40 ms. As we proceeded in Section 3.2, the first 50 annotations were discarded. The remaining annotations were accordingly used to label the corresponding vectors of visual features. All labelling and fusion of data samples and features were completed according to the timing of the video frames.

3.4. Data Shuffling and Splitting

Data shuffling ensures the randomization and diversity of the data. The data were shuffled and split, where 80% went towards training and validation, and 20% went towards testing. Our training and validation dataset was 106,201 frames × 40 features in size, while the testing dataset was 26,550 frames × 40 features in size.

3.5. Regression

For the prediction of arousal and valence values, we used an optimizable ensemble regressor. We also experimented with other regression models for comparison purposes: tree regressors, regression kernels, and ensemble regression. We trained and validated four tree regressors (fine, medium, coarse, and optimizable tree), two regression kernels (SVM and least squares regression kernel), and two ensembled regressors (boosted and bagged trees). A fine regression tree is small with a leaf size of 4 [16]. A medium regression tree has 12 leaves. A coarse regression tree is large and has a leaf size of 36. An optimizable regression tree optimizes the minimum leaf size through a Bayesian optimizer. Regression kernels are Gaussian regression models for nonlinear regression over large datasets. An SVM kernel maps the features into a high-dimensional space and fits a linear SVM model to the transformed features. A least squares regression kernel maps the features into a high-dimensional space and fits a least squares linear regression model to the transformed features. The boosted trees model ensembles regression trees using the LSBoost algorithm. The bagged trees model ensembles regression trees by bootstrap-aggregation. An optimizable regression ensemble optimizes training hyperparameters (ensemble method, number of learners, learning rate, minimum leaf size, and number of predictors to sample) via Bayesian optimization. We implemented 5-fold cross-validation during training to avoid overfitting.

4. Discussion of Results

After training the aforementioned models, we tested them by predicting the arousal and valence values on the testing set to evaluate the performance when presented with new data. Table 1 summarizes the validation and testing performances in terms of the RMSE, PCC, and CCC performance measures. A smaller RMSE value signifies better performance, whereas greater PCC and CCC values signify better performance. We have achieved a testing RMSE, PCC, and CCC of 0.1033, 0.8498, and 0.8001 on arousal predictions, respectively. We have achieved a testing RMSE, PCC, and CCC of 0.07016, 0.8473, and 0.8053 on valence predictions, respectively. These performances were obtained using an optimizable ensemble regressor. Our performances are better than those from the literature [11–15] (see Section 2), who performed more complex processing and feature extraction. In Table 1, the validation performances were evaluated by performing 5-fold cross validation across the training data. The testing performances were computed by using the trained model for predicting the arousal and valence values of the testing set. The rows corresponding to the best prediction performances are displayed in bold font in Table 1. Figure 1 displays a plot of the predicted arousal and valence values against the actual values, as per the best model (i.e., optimizable ensemble). In the plot of a perfect regression model, the predicted values would be the same as the actual values, resulting in a diagonal line of points [16]. Models where the points are scattered near the diagonal line represent good models, with less errors. Table 1 and Figure 1 also compare the performances of our models for learned [4] and predesigned features. Using predesigned features showed an improvement in our prediction performance.

Table 1. Summary of Prediction Performances.

Prediction	Regression Type	Validation RMSE	Testing RMSE, PCC, CCC
Arousal	Fine Tree	0.15389	0.1477, 0.6812, 0.6805
	Medium Tree	0.14601	0.1410, 0.6902, 0.6838
	Coarse Tree	0.14477	0.1410, 0.6731, 0.6516
	Optimizable Tree	0.14351	0.1396, 0.6861, 0.6719
	SVM Kernel	0.13665	0.1354, 0.7018, 0.6807
	Least Squares Kernel	0.13444	0.1331, 0.7097, 0.6633
	Boosted Trees	0.161	0.1607, 0.5463, 0.3743
	Bagged Trees	0.11285	0.1082, 0.8304, 0.7796
	Optimizable Ensemble	**0.10791**	**0.1033, 0.8498, 0.8001**
	MobileNet-v2 [4]	0.12178	0.1220, 0.7838, 0.7770
Valence	Fine Tree	0.10191	0.0981, 0.6975, 0.6967
	Medium Tree	0.097111	0.0944, 0.7011, 0.6947
	Coarse Tree	0.097623	0.0948, 0.6826, 0.6610
	Optimizable Tree	0.096525	0.0945, 0.6922, 0.6801
	SVM Kernel	0.094882	0.0943, 0.6855, 0.6495
	Least Squares Kernel	0.092417	0.0916, 0.7030, 0.6574
	Boosted Trees	0.11142	0.1104, 0.5525, 0.3467
	Bagged Trees	0.074689	0.0714, 0.8421, 0.7962
	Optimizable Ensemble	**0.073335**	**0.0702, 0.8473, 0.8053**
	MobileNet-v2 [4]	0.08309	0.0823, 0.7789, 0.7715

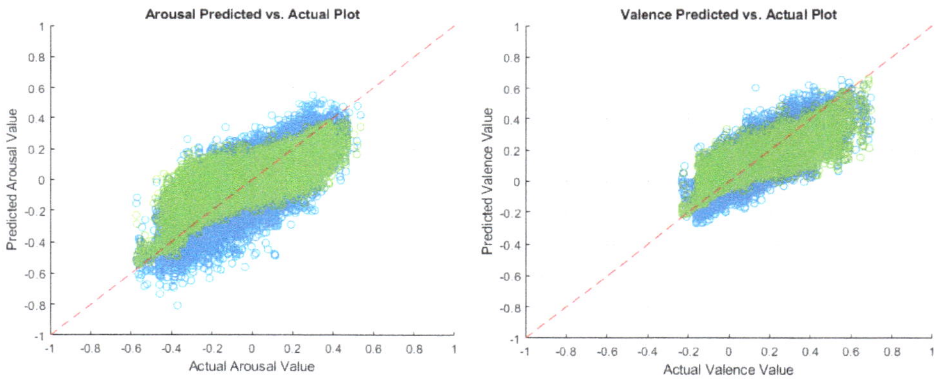

(a) (b)

Figure 1. Predicted versus actual plots of (**a**) arousal, and (**b**) valence predictions by an optimizable ensemble trained on visual features (green), and MobileNet-v2 trained on video frames (blue). The diagonal, red-dashed line represents a perfect regression model.

5. Conclusions

In conclusion, we performed arousal and valence predictions by exploiting the predesigned visual features of the RECOLA database. The feature vectors were processed and accordingly labelled with their corresponding arousal or valence annotations. We trained, validated, and tested an optimizable ensemble as well as other regressors to predict arousal and valence values. The optimizable ensemble achieved a RMSE, PCC, and CCC of 0.1033, 0.8498, and 0.8001 on arousal predictions, and 0.07016, 0.8473, and 0.8053 on valence predictions, respectively. To the best of our knowledge, our prediction performances on arousal and valence predictions are the best in comparison to the literature. Going forward, we will carry out our project with the optimizable ensemble as the prediction mechanism for predesigned visual features. Since we achieved good prediction performance using physiological [3,4] and visual data, we can work on acoustic data and start combining our solutions for the different data modalities. In the future, we will apply our findings to real data, obtained from a VR system.

Author Contributions: Conceptualization, I.O.J., A.-M.C. and S.B.; methodology, I.O.J.; software, I.O.J.; validation, I.O.J.; formal analysis, I.O.J.; investigation, I.O.J.; resources, A.-M.C. and S.B.; data curation, I.O.J.; writing—original draft preparation, I.O.J.; writing—review and editing, I.O.J., A.-M.C. and S.B.; visualization, I.O.J.; supervision, A.-M.C. and S.B.; project administration, A.-M.C. and S.B.; funding acquisition, A.-M.C. and S.B. All authors have read and agreed to the published version of the manuscript.

Funding: This study was funded by the Natural Sciences and Engineering Research Council of Canada (NSERC)'s Discovery grant, number RGPIN-2023-03415.

Institutional Review Board Statement: Not applicable.

Informed Consent Statement: Not applicable.

Data Availability Statement: No new data were created. Data was obtained from the RECOLA team and are available, upon request at https://diuf.unifr.ch/main/diva/recola/download.html.

Acknowledgments: A special thank you to the providers of the RECOLA database, F. Ringeval, A. Sonderegger, J. Sauer, and D. Lalanne.

Conflicts of Interest: Stéphane Bouchard is the President of, and owns equity in, In Virtuo Clinics and Development, a spin-off company from the university that distributes virtual environments designed for the treatment of mental disorders. The terms of these arrangements have been reviewed and approved by the University of Québec in Outaouais in accordance with its conflict of interest policies. Stéphane Bouchard has received honoraria for presenting research and providing workshops. He also receives royalties from books.

References

1. Russell, J. *Affective Space Is Bipolar*; American Psychological Association: Washington, DC, USA, 1979.
2. Ringeval, F.; Sonderegger, A.; Sauer, J.; Lalanne, D. Introducing the RECOLA multimodal corpus of remote collaborative and affective interactions. In Proceedings of the 10th IEEE International Conference and Workshops on Automatic Face and Gesture Recognition, Shanghai, China, 22–26 April 2013; pp. 1–8.
3. Joudeh, I.O.; Cretu, A.; Guimond, S.; Bouchard, S. Prediction of Emotional Measures via Electrodermal Activity (EDA) and Electrocardiogram (ECG). *Eng. Proc.* **2022**, *27*, 47.
4. Joudeh, I.O.; Cretu, A.-M.; Bouchard, S.; Guimond, S. Prediction of Continuous Emotional Measures through Physiological and Visual Data. *Sensors* **2023**, *23*, 5613. [CrossRef] [PubMed]
5. Joudeh, I.O.; Cretu, A.-M.; Bouchard, S.; Guimond, S. Prediction of Emotional States from Partial Facial Features for Virtual Reality Applications. In Proceedings of the 26th Annual CyberPsychology, CyberTherapy and Social Networking Conference (CYPSY26), Paris, France, 11–13 July 2023.
6. Corneanu, C.A.; Simon, M.O.; Cohn, J.F.; Guerrero, S.E. Survey on RGB, 3D, Thermal, and Multimodal Approaches for Facial Expression Recognition: History, Trends, and Affect-Related Applications. *IEEE Trans. Pattern Anal. Mach. Intell.* **2016**, *38*, 1548–1568. [CrossRef] [PubMed]
7. Al Osman, H.; Falk, T.H. Multimodal affect recognition: Current approaches and challenges. *Emot. Atten. Recognit. Based Biol. Signals Images* **2017**, *8*, 59–86.
8. Almaev, T.R.; Valstar, M.F. Local Gabor Binary Patterns from Three Orthogonal Planes for Automatic Facial Expressions Recognition. In Proceedings of the 2013 Humaine Association Conference on Affective Computing and Intelligent Interaction, Geneva, Switzerland, 2–5 September 2013; IEEE Computer Society: Washington, DC, USA; pp. 356–361.
9. Xiong, X.; De la Torre, F. Supervised descent method and its applications to face alignment. In Proceedings of the 2013 IEEE Conference on Computer Vision and Pattern Recognition, Portland, OR, USA, 23–28 June 2013; pp. 532–539.
10. Ekman, P.; Friesen, W.V. *Facial Action Coding System: A Technique for the Measurement of Facial Movement*; Consulting Psychologists Press: Palo Alto, CA, USA, 1978.
11. Ringeval, F.; Eyben, F.; Kroupi, E.; Yuce, A.; Thiran, J.P.; Ebrahimi, T.; Lalanne, D.; Schuller, B. Prediction of Asynchronous Dimensional Emotion Ratings from Audiovisual and Physiological Data. *Pattern Recognit. Lett.* **2015**, *66*, 22–30. [CrossRef]
12. Ringeval, F.; Schuller, B.; Valstar, M.; Jaiswal, S.; Marchi, E.; Lalanne, D.; Cowie, R.; Pantic, M. AV+EC 2015—The First Affect Recognition Challenge Bridging Across Audio, Video, and Physiological Data. In Proceedings of the AVEC'15, Brisbane, Australia, 26–27 October 2015; ACM: New York, NY, USA; pp. 3–8.

13. Ringeval, F.; Schuller, B.; Valstar, M.; Cowie, R.; Kaya, H.; Schmitt, M.; Amiriparian, S.; Cummins, N.; Lalanne, D.; Michaud, A.; et al. AVEC 2018 Workshop and Challenge: Bipolar Disorder and Cross-Cultural Affect Recognition. In Proceedings of the AVEC'18, Seoul, Republic of Korea, 22 October 2018; ACM: New York, NY, USA.
14. Han, J.; Zhang, Z.; Ren, Z.; Schuller, B. Implicit Fusion by Joint Audiovisual Training for Emotion Recognition in Mono Modality. In Proceedings of the 2019 IEEE International Conference on Acoustics, Speech and Signal Processing, Brighton, UK, 12–17 May 2019; pp. 5861–5865.
15. Albadawy, E.; Kim, Y. Joint Discrete and Continuous Emotion Prediction Using Ensemble and End-to-End Approaches. In Proceedings of the 20th ACM International Conference on Multimodal Interaction (ICMI '18), Boulder, CO, USA, 16–20 October 2018; ACM: New York, NY, USA; pp. 366–375.
16. Help Center. Help Center for MATLAB, Simulink, and Other MathWorks Products. Available online: https://www.mathworks.com/help/ (accessed on 2 September 2023).

engineering proceedings

MDPI

Proceeding Paper

Developing a Cross-Platform Application for Integrating Real-Time Time-Series Data from Multiple Wearable Sensors †

Pataranit Sirithummarak * and Zilu Liang *

Faculty of Engineering, Kyoto University of Advanced Science, Kyoto 615-8577, Japan
* Correspondence: nothing2say.develop@gmail.com (P.S.); liang.zilu@kuas.ac.jp (Z.L.)
† Presented at the 10th International Electronic Conference on Sensors and Applications (ECSA-10), 15–30 November 2023; Available online: https://ecsa-10.sciforum.net/.

Abstract: This research presents a cross-platform application, developed with Flutter, for the efficient integration and management of real-time data from wearable sensors including Apple Watch, Android Wear, and Empatica E4. Compatible with iOS and Android, the app collects various physiological signals for easy analysis by health professionals. Utilizing InfluxDB, a time-series database, our development ensures efficient data handling, even from multiple sources, and enables real-time analytics. This robust, scalable tool signifies a notable advancement in mHealth, offering seamless data integration and management for those utilizing wearable sensor technology in healthcare research and practice.

Keywords: cross-platform application; real-time data; time-series database; wearable sensors

Citation: Sirithummarak, P.; Liang, Z. Developing a Cross-Platform Application for Integrating Real-Time Time-Series Data from Multiple Wearable Sensors. *Eng. Proc.* **2023**, *58*, 4. https://doi.org/10.3390/ecsa-10-16185

Academic Editor: Stefano Mariani

Published: 15 November 2023

1. Introduction

In health research, the importance of collecting quality data is indisputable. The recent advance in consumer-grade wearable technologies further demands technical support for data integration [1–3]. However, domain experts often lack software development skills for integrating health data collected with different wearable sensors, posing a risk of data loss, compromising the quality of data, and impeding data-centered research.

Our research is inspired by the need to support researchers dealing with wearable sensor data [4]. The complexity of developing mobile applications compatible with these sensors often extends beyond the expertise of data scientists, demanding excessive time and resources that could be better spent on core research objectives. Walch et al.'s difficulty [5] in collecting and processing health data from the Apple Watch, characterized by manual submission, inconsistent sampling rates, and data loss, inspired our research to address these complexities.

To address this challenge, we developed a solution that enables domain experts to overcome technical barriers and optimize data collection and integration. We present a cross-platform application, specially designed for seamless integration of data from wearable sensors for health data research. This app streamlines the collection, management, and analysis of time-series data, effectively addressing key design considerations such as the optimal method for developing mHealth apps, the feasibility of Flutter as a cross-platform framework for these apps, and the most suitable time-series database for efficient data storage and real-time tracking in mHealth.

The proposed application transcends these challenges, offering a streamlined approach to data collection, equipped with real-time monitoring, and designed with a focus on ensuring the acquisition of quality data which is integral to insightful health studies. By circumventing the technical intricacies often associated with app development and data integration, we provide a tool that stands as a nexus between efficient data collection and impactful health research, marking a significant stride in the convergence of technology and health research.

2. Methods

We introduce "ZensorsFlow", a cross-platform application designed to seamlessly integrate and manage real-time data from multiple wearable sensors, addressing the evolving needs of mHealth and real-time health monitoring.

2.1. Development Environment and Integration Process

2.1.1. Cross-Platform Mobile Application Development

- Flutter Framework:

ZensorsFlow is grounded in Flutter for both iOS and Android, a decision inspired by the SDK's proficiency in ensuring a consistent user experience across varied platforms and its flexibility [6]. The app leverages Flutter's streamlined process of generating unified code for diverse operating systems, markedly boosting developmental efficiency. Dart, Flutter's intrinsic language, is pivotal due to its asynchronous features, ensuring adept real-time sensor data handling. Our choice of Flutter is not just a technical decision but a strategic initiative to balance performance, development speed, and user engagement seamlessly.

2.1.2. Integration with InfluxDB Cloud Service

- Real-time Data Streaming and Management:

The Flutter-based mobile application was integrated with InfluxDB cloud service to facilitate the real-time streaming and efficient management of data from wearable devices, ensuring seamless and scalable data handling. InfluxDB's adeptness at handling time-series data is underscored by its top rating in performance studies in the mHealth and IoT spheres [7–9]. This empirical backing solidified our selection of InfluxDB for optimized real-time data management in ZensorsFlow.

2.2. Wearable Devices Data Acquisition

2.2.1. iOS—Apple Watch

- Native Development Using Swift:

The Apple Watch's sensor data is accessed through a native development approach using Swift, harnessing the watch's intricate sensor capabilities to collect real-time data.

- Watch Connectivity and Platform Channels:

The Watch Connectivity framework is utilized for the seamless transmission of sensor data between iOS devices and Apple Watches. The data collected is then relayed to the Flutter application through platform channels, ensuring real-time integration and a unified data presentation.

2.3. Android—Android Wear

- Hybrid Development Using Flutter and Native Methods:

The Android Wear application is primarily developed using Flutter with the integrated Flutter Wear plugin, and Kotlin, which is employed for native development to access sensor capabilities.

- Data Layer API and Platform Channels:

The Data Layer API is instrumental in bridging the communication between the Android mobile app and wear app, comparable to the Watch Connectivity framework in iOS. This native environment tool ensures seamless data transmission between Android devices. The data from native environment are transferred to the Flutter application using platform channels.

2.4. Empatica E4 Integration

Integration of the Empatica E4 across iOS and Android is achieved through a native development approach, leveraging the E4link SDK officially provided by Empatica

(Cambridge, MA, USA). This official SDK ensures the precise and real-time acquisition of sensor data, aligning with the data collection from the Apple Watch on iOS and Android Wear on Android.

In both iOS and Android environments, the real-time data from the Empatica E4 is communicated to the main Flutter application through platform channels, ensuring seamless data integration across the platforms.

3. Data Collection and Real-Time Streaming

3.1. Data Types and Collection Methodology

- Physiological Signals:

We are equipped to collect a variety of physiological signals from Apple Watch, Android Wear, and Empatica E4, each offering unique datasets essential for comprehensive health monitoring. Apple Watch and Android Wear provide raw acceleration data and processed heart rate data. The real-time collection of these data types is optimized through native development. The E4, however, provides a richer dataset. It captures Inter-Beat Interval (IBI), Blood Volume Pulse (BVP), 3-axis acceleration, Electrodermal Activity (EDA), and skin temperature. Our application is engineered to synchronize these diverse data types and sampling rates seamlessly, enhancing the precision and reliability of collected health data.

3.2. Real-Time Data Streaming to InfluxDB

- Data Integration and Management:

An exposition of the mechanism through which real-time data, collected from the wearable devices, is streamed, integrated, and managed effectively using the InfluxDB cloud service.

In summary, our development is rooted in Flutter and unifies iOS and Android user experiences, while real-time data management is optimized with InfluxDB. Native development in Swift and Kotlin ensures precise wearable data capture, enriched by the E4link SDK for Empatica E4 integration. The application adeptly handles varied physiological signals and sampling rates. Figure 1 illustrates the systematic architecture, and Figure 2 delineates the app interaction and data flow, showcasing the seamless orchestration of data acquisition, integration, and management.

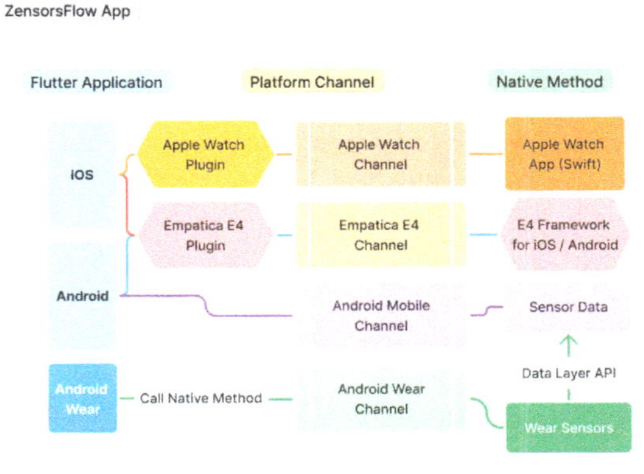

Figure 1. ZensorsFlow's cross-platform system architecture diagram.

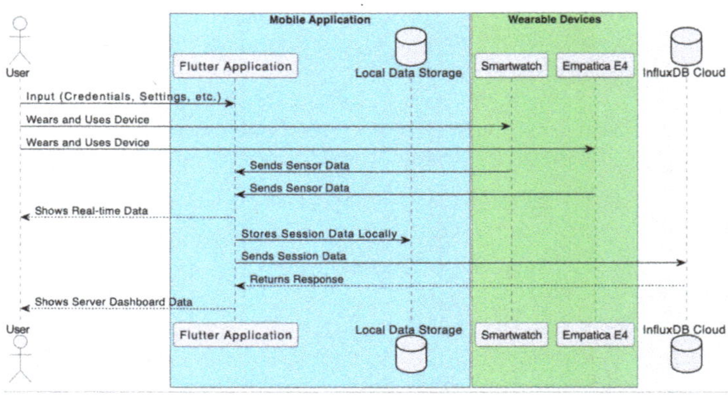

Figure 2. ZensorsFlow's interaction and data sequence diagram.

4. Results and Discussion

4.1. Results

The utility of our application, ZensorsFlow, can be delivered to end-users immediately upon installation. Users are guided through a simplified process of setting up InfluxDB and generating an API key. Following this initialization, the app invites users to a streamlined interface where the database configuration is a one-time, user-friendly process.

The application's operation is intuitive. Users easily connect their wearable devices, initiating real-time data recording. Our app distinguishes itself by supporting simultaneous data collection from multiple devices, a feature tested rigorously.

Data, once recorded, are instantly sent to the pre-configured InfluxDB, and users can monitor real-time data via the dashboard. A notable feature is the option to export data to CSV files for in-depth analysis, enhancing the app's utility for detailed health insights.

4.2. Discussions

The use of Flutter's applicability has both strengths and constraints. While we benefited from its cross-platform compatibility and the efficiency of Hot Reload, challenges like limited connectivity with wearables and extended build times surfaced.

Future research is needed to further refine and optimize the efficiency and usability of our application.

4.3. Future Works

The future development would include integrating low-code platforms like FlutterFlow to simplify and expedite the development process. We aim to expand device support, focusing on enhancing Android Wear integration while bolstering the app's security, performance, and utility.

5. Conclusions

We have presented the development and evaluation of "ZensorsFlow", a cross-platform application that we developed to streamline the process of real-time data integration from a variety of wearable sensors, including, but not limited to, Empatica E4, Android Wear, and Apple Watch. This development addresses the pressing need for efficient and user-friendly tools in the collection and management of health data.

Through rigorous testing and evaluation, we confirmed Flutter's robust capability to develop mobile health applications that seamlessly synchronize with wearable technologies. This affirms Flutter as a viable framework, marrying flexibility and performance, indicative of its potential prominence in future mHealth app developments.

In addressing the development constraints, particularly notable with the Apple Watch which necessitates native development to fully exploit its capabilities, we established an innovative and harmonious integration of platform-specific developments for both Android and iOS. We engineered platform channels, bridging the Flutter application to native functionalities, ensuring a seamless synchronization of real-time sensor data. With Android Wear, the comprehensive environment of Flutter enabled streamlined application development and data integration. Conversely, for the Apple Watch, we utilized Swift for native development, effectively tapping into its intricate sensor capabilities. These diversely sourced data are cohesively channeled to the Flutter application via platform channels, affirming a consistent and unified data integration and user experience across the diverse platforms.

Furthermore, our exploration into the realm of time-series databases led to us identifying InfluxDB as the optimal choice. Characterized by its efficiency in real-time data management, InfluxDB stands out as a pivotal tool in handling the vast and dynamic datasets generated by sensor technology, confirming our hypothesis concerning database efficacy.

In light of these findings, we hope that ZensorsFlow will serve as a cornerstone in the advancement of mHealth, paving the way for the collection of high-quality multimodal health datasets. Our contributions are poised to not only address the existing challenges but also to catalyze future innovations in the seamless and efficient integration of wearable sensor data for enhanced health research and clinical applications.

Author Contributions: P.S. conceptualized the project, Z.L. provided feedback and discussions; P.S. and Z.L. wrote and proofread the manuscript. All authors have read and agreed to the published version of the manuscript.

Funding: This research received no external funding.

Institutional Review Board Statement: Not applicable.

Informed Consent Statement: Not applicable.

Data Availability Statement: The source code of the ZensorFlow can be accessed at https://github.com/KUAS-ubicomp-lab/ZensorFlow.

Conflicts of Interest: The authors declare no conflict of interest.

References

1. Liang, Z.; Ploderer, B.; Liu, W.; Nagata, Y.; Bailey, J.; Kulik, L.; Li, Y. SleepExplorer: A visualization tool to make sense of correlations between personal sleep data and contextual factors. *Pers. Ubiquitous Comput.* **2016**, *20*, 985–1000. [CrossRef]
2. Bertrand, L.; Cleyet-Marrel, N.; Liang, Z. Recognizing Eating Activities in Free-Living Environment Using Consumer Wearable Devices. *Eng. Proc.* **2021**, *6*, 58.
3. Ploderer, B.; Rodgers, S.; Liang, Z. What's keeping teens up at night? Reflecting on sleep and technology habits with teens. *Pers. Ubiquitous Comput.* **2023**, *27*, 249–270. [CrossRef]
4. Sirithummarak, P.; Liang, Z. Investigating the Effect of Feature Distribution Shift on the Performance of Sleep Stage Classification with Consumer Sleep Trackers. In Proceedings of the 2021 IEEE 10th Global Conference on Consumer Electronics (GCCE), Kyoto, Japan, 12–15 October 2021; pp. 242–243.
5. Walch, O.; Huang, Y.; Forger, D.; Goldstein, C. Sleep Stage Prediction with Raw Acceleration and Photoplethysmography Heart Rate Data Derived from a Consumer Wearable Device. *Sleep* **2019**, *42*, zsz180. [CrossRef] [PubMed]
6. Ameen, S.Y.; Mohammed, D.Y. Developing Cross-Platform Library Using Flutter. *Eur. J. Eng. Technol. Res.* **2022**, *7*, 18–21. [CrossRef]
7. Hao, Y.; Qin, X.; Chen, Y.; Li, Y.; Sun, X.; Tao, Y.; Zhang, X.; Du, X. TS-benchmark: A Benchmark for Time Series Databases. In Proceedings of the 2021 IEEE 37th International Conference on Data Engineering, Chania, Greece, 19–22 April 2021; pp. 588–599. [CrossRef]
8. Mostafa, J.; Wehbi, S.; Chilingaryan, S.; Kopmann, A. SciTS: A Benchmark for Time-Series Databases in Scientific Experiments and Industrial Internet of Things. In Proceedings of the 34th International Conference on Scientific and Statistical Database Management, Los Angeles, CA, USA, 19–22 April 2021; pp. 1–11.
9. Gangadhar, S. The Real Time Environmental Time Series Data Analysis Using Influx DB. *Int. J. Adv. Sci. Innov.* **2020**, *1*. [CrossRef]

engineering proceedings

MDPI

Proceeding Paper

Texture Classification Based on Sound and Vibro-Tactile Data [†]

Mustapha Najib * and Ana-Maria Cretu *

Department of Computer Science and Engineering, University of Quebec in Outaouais, Gatineau, QC J8Y 3G5, Canada

* Correspondence: najm03@uqo.ca (M.N.); ana-maria.cretu@uqo.ca (A.-M.C.)

† Presented at the 10th International Electronic Conference on Sensors and Applications (ECSA-10), 15–30 November 2023; Available online: https://ecsa-10.sciforum.net/.

Abstract: This paper focuses on the development and validation of an automatic learning system for the classification of tactile data in form of vibro-tactile (accelerometer) and audio (microphone) data for texture recognition. A novel combination of features including the standard deviation, the mean, the absolute median of the deviation, the energy that characterizes the power of the signal, a measure that reflects the perceptual properties of the human system associated with each sensory modality, and the Fourier characteristics extracted from signals, along with principal component analysis is shown to obtain the best results. Several machine learning models are compared in an attempt to identify the best compromise between the number of features, the classification performance and the computation time. Longer sampling periods (2 s. vs. 1 s) provide more information for classification, leading to higher performance (average of 3.59%) but also augment the evaluation time by an average of 29.48% over all features and models. For the selected dataset, the XGBRF model was identified to represent overall the best compromise between performance and computation time for the proposed novel combination of features over all material types with an F-score of 0.91 and a computation time of 4.69 ms, while kNN represents the next best option (1% improvement in performance at the cost of 2.13 ms increase in time with respect to XGBRF).

Keywords: texture classification; accelerometer; microphone; machine learning; feature selection

Citation: Najib, M.; Cretu, A.-M. Texture Classification Based on Sound and Vibro-Tactile Data. *Eng. Proc.* **2023**, *58*, 5. https://doi.org/10.3390/ecsa-10-16082

Academic Editor: Stefano Mariani

Published: 15 November 2023

1. Introduction

The tactile perception of material properties is a difficult task, but also of great importance for the skillful manipulation of objects in fields such as robotics, virtual reality and augmented reality. Given the diversity of material properties, integrated tactile perception systems require efficient extraction and classification of features from data collected by tactile sensors. There are several publications on the topic of texture recognition in the literature. Characteristics of material textures can be retrieved using vision-based, tactile-based or sound-based data. Most publications rely uniquely on images to identify textures in various domains [1], while others are using sensor data collected by various tactile technologies (i.e., Microelectromechanical Magnetic, Angular Rate and Gravity–MARG systems, pressure sensors, accelerometers, microphone, etc.) while the surface of the sensor enters in contact with a probed textured surface. Other publications capitalize on combinations of various tactile sensory sources [2–4]. Most publications employ feature extraction techniques [5] to identify the most relevant data to focus on prior to applying machine learning solutions to classify or recognize textures or textured materials. In tactile sensing, known feature extraction techniques include principal component analysis, PCA [4,6], frequency signatures [7] and the Fast Fourier Transform (FFT) [3,8], both for sound and vibro-tactile data. Some researchers focus mostly on real-time processing [3,4], and tend to choose less complex machine learning solutions (such as the k-nearest neighbors, KNN [3,7,8], two-layer multilayer perceptron MLP [3,4], or SVM [3,6]). Others make use of convolutional neural network architectures [8] that do not require the extraction of features, as this process is embedded in their architecture. However, they come at additional computational cost due to their increased complexity.

This paper focuses on the development and validation of an automatic learning system for the classification of tactile data in form of vibro-tactile (accelerometer) and audio (microphone) data for texture recognition. We aim to identify the right balance between classification accuracy and compact, fast solutions, with potential for real-time performance. We propose a novel combination of features for this purpose. In order to reduce the dimensionality of the tactile dataset and identify the most compact models, we apply PCA as well as a process of selection of features based on their importance. Several machine learning models are compared in an attempt to identify the best compromise between the number of features, the classification performance and the computation time. We also demonstrate that the choice of the sampling length from the tactile signals is an important aspect that has a significant impact on classification performance.

2. Materials and Methods

2.1. Dataset for Texture Classification

The VibTac-12 dataset used in this paper is created by Kursun and Patooghy [9]. It is based on a vibro-tactile stimulator system to generate controlled vibrations on textured materials (i.e., sandpapers of various grits, Velcro strips with various thicknesses, aluminum foil and rubber bands of various stickiness) and an embedded system to record tactile data. Two sensors, a microphone and an accelerometer attached to a probe, capture the audio and vibro-tactile signals as the probe rubs against the surface of textured materials. The interested reader is invited to consult reference [3] for details on the experimental setup and the data collection process. In this paper, we employ the two available tactile data sources in the dataset, namely sound recordings and data collected by the accelerometer that measures the changes in acceleration and orientation of the probe in contact with the texture surface along three axes. It is important to state that we are not making use of the data processing sequence that the authors of [3] used, we only use their raw data. Our focus is to identify a set of powerful features that allow us to accurately clasify these data in the shortest time possible.

2.2. Proposed Solution for Texture Classification from Sound and Vibro-Tactile Data

Figure 1 illustrates the proposed solution for texture classification from sound and vibro-tactile data. Input data consists of sound data contained in the Sdf.csv file from the VibTac-12 dataset [9] and of accelerometer data, and contained in the Xdf, Ydf and Zdf files, respectively. This data goes through a pre-processing stage in order to normalize it, eliminate outliers and extract and select features in order to transform it into a usable format for classification. Texture class names are encoded with numerical identifiers.

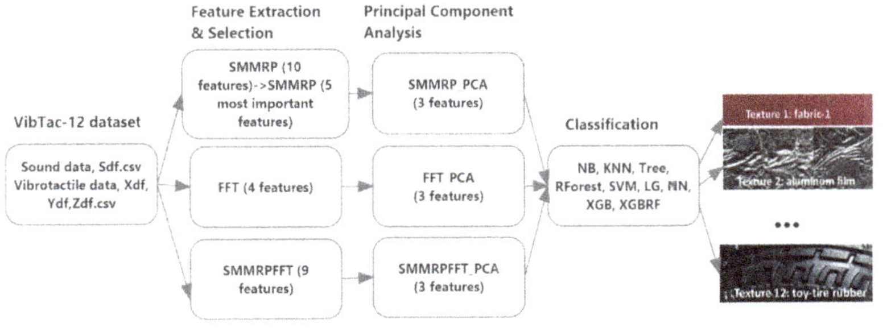

Figure 1. Proposed approach for texture classification.

To study the impact of each source (type) of data, of the chosen features and of the length of the sampling period on the performance of texture recognition, we created several datasets. We name them in Figure 1 to clearly identify their data source, features and lengths.

It is important to mention that creating these different datasets is only for making their interpretation easier. They correspond in fact to the feature extraction process in machine learning that does not require the creation of separate datasets. Our solution is implemented in Python. We make use of several open-access libraries, including Librosa [10], a library for audio signal processing, Scikit-learn for the implementation of the machine learning algorithms and the Eli5 library [11] to identify feature importance.

2.3. Data Processing

2.3.1. Data Transformation

For each texture, 20 s of multimodal recordings are available. Therefore, each line of the Xdf, Ydf and Zdf files respectively contains 4000 samples (20 s $\times$ 200 Hz sampling rate) and each line of the Sdf file contains 160,000 samples (20 s $\times$ 8 kHz sampling rate). To reduce the computational load and in order to create our datasets, we sampled, without replacement, 100 random samples from the raw data sequences, and thus each dataset contains a total of 1200 records. A unique multimodal tactile dataset (SXYZ) containing sound data (S) and accelerometer (XYZ) data is thus created for testing.

2.3.2. Feature Extraction

We extracted from the created tactile dataset commonly used features in time series analysis, including the following 10 features: (1) MEAN, representing the mean; (2) standard deviation (STD), representing the dispersion of values around the mean; (3) median absolute deviation (MAD); (4) RMSE, the energy characterizing the power of the signal; (5) CHROMA: a representation of the musical characteristics related to the tonality and harmony of an audio signal; (6) SPECTRAL_CENTROID: a measure of the position of the center of gravity of the spectral energy distribution of a signal, calculated as the weighted average of the frequencies in the signal power spectrum, where the weights are given by the spectral magnitude at each frequency; (7) SPECTRAL_BANDWIDTH: a measure of the spread of spectral energy distribution in a signal, i.e., the frequency range of a signal; (8) SPECTRAL_ROLLOFF: a feature representing the frequency below which a given percentage of the signal's total spectral energy is concentrated; (9) PERCEPTUAL: a feature reflecting the perceptual properties of the human system associated with each sensory modality useful to characterize the quality or perceptual properties of a signal; and (10) the zero crossing rate (ZCR) that measures the frequency at which a signal changes polarity. We also extracted four features obtained by applying the Fast Fourier Transform (FFT) to the S, X, Y, and Z signals. The two resulting datasets are named SMMRP (10 features) and FFT (4 features) in Figure 1. These features are extracted from the sound and vibrotactile signals for 1 s and for 2 s sampling periods.

2.3.3. Feature Selection

Feature selection is a key data preparation step aiming to reduce the number of features to be included in modeling, by selecting the most relevant features for classification. It can help determine if there are features that are less useful, and thus could be potentially removed to reduce the model complexity. Using a random forest algorithm, we identified in the 10 feature-SMMRP dataset (Figure 2a) that the features STD, MEAN, MAD, RMSE and PERCEPTUAL (Figure 2b) are those that contribute the most to predictions. As such, we chose to continue our work with these 5 features (denoted SMMRP in the rest of the paper and SMMRP (5 features) in Figure 1) along with the four FFT features. However, we noticed that some of these features are correlated. To address this issue, as well as to further reduce the complexity of the dataset, we used PCA on these features (_PCA datasets in Figure 1). As shown in Figure 1, in all the cases we have chosen the first 3 principal components that capture roughly 95% of the total variance when we only use the SMMRP features, roughly 97% for the FFT features and 100% when all features are used together (SMMRPFFT) for a sampling period of 2 s.

	MEAN	STD	MAD	CHROMA	RMSE	SPECTRAL_CENTROID	SPECTRAL_BW	ROLLOFF	ZCR	PERCEPTUAL	label_id	label
0	6.547687	1.596416	0.600997	0.653539	2.302323	526.431296	311.812666	909.375000	0.126060	-31.447525	0	fabric-1
1	6.596422	1.306557	0.742516	0.637499	2.286934	526.009975	312.480009	905.712891	0.125969	-31.465581	0	fabric-1
2	6.653673	1.252235	0.654761	0.620316	2.276366	524.287882	311.003382	903.015137	0.129395	-31.580163	0	fabric-1
3	6.614880	1.500490	0.732085	0.669408	2.316631	526.826044	312.465917	908.691406	0.129280	-31.153812	0	fabric-1
4	6.714372	1.724328	0.690445	0.560110	2.400012	528.294377	311.531106	907.373047	0.128464	-30.504923	0	fabric-1
5	6.656070	1.228024	0.631351	0.601297	2.300615	527.227991	312.234673	907.348633	0.126998	-30.722076	0	fabric-1

(a)

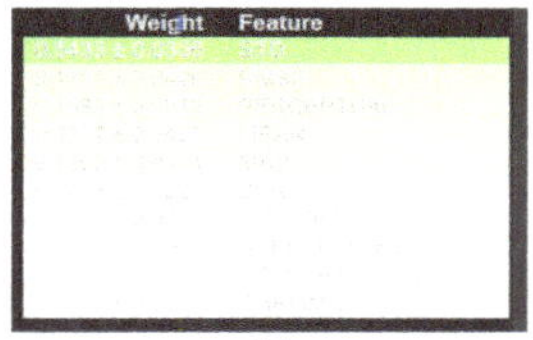

(b)

Figure 2. (**a**) Sample of data; and (**b**) feature importance in the dataset using SMMRP features.

Figure 3 shows the data dispersion for the various texture classes using PCA. One can notice that, for all datasets—SMMRP (Figure 3a), FFT (Figure 3b) and their combination SMMRPFFT (Figure 3c)—certain classes are easily separable (distinguished), for example "aluminum_film", in light red, or "fabric-3", in purple, while others, like "fabric-1" in dark blue and "toy_tire_rubber" in dark red overlap, as do "fabric-2" in dark green and "moquette-1" in dark orange. These latter classes will therefore be more difficult to classify correctly, regardless of the classifier used and of the features selected. One can also notice that the separability is improved when the combination of features (SMRRPFFT) is used, in Figure 3c.

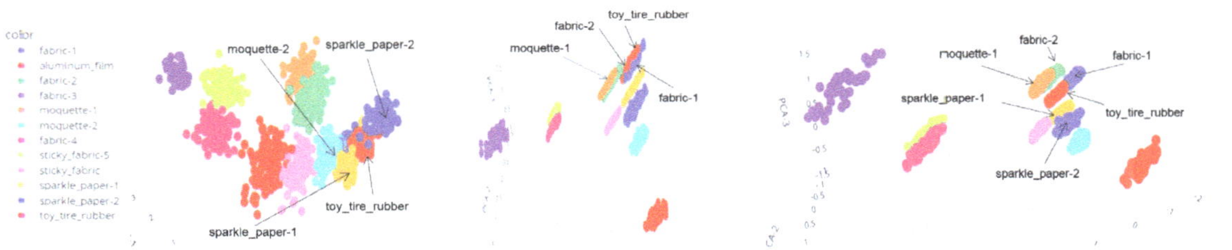

(a) (b) (c)

Figure 3. Principal component analysis for (**a**) SMMRP; (**b**) FFT; and (**c**) SMMRPFFT features (Note: the three axes of each figure represent the first 3 PCA components).

2.4. Data Classification

Available data are split in training (80%) and testing (20%) sets and the F-Score is used as a performance measure along with the computation time (in ms). For classification, we chose a series of classifiers based on the nature of the data, their use in the literature and their proven performance across various domains. These include Gaussian Naive Bayes (NB) classification, decision trees (Tree), random forest (RForest)—consisting of 1000 decision trees—support vector machines (SVM), the K-nearest-neighbors (KNN) algorithm, logistic regression (LG), neural networks (NN), a 2-layer MLP [12], XGBOOST (XGB) [13] and the Extreme Gradient Boosting with Random Forest (XGBRF).

3. Results

To evaluate the performance of our solution, we performed various tests with the chosen algorithms, mostly with default parameters and for the various combinations of features. In the first place, we studied the impact of the sampling period on the results. Table 1 shows that for all algorithms and combinations of features the performance in terms of F-Score is higher for a 2 s sampling (by 3.7%), but this comes at the price of an increased evaluation time by an average of 29.48% (6.2 ms) for all features and algorithms tested. We continued the remainder of tests with a sampling period of 2 s. In an attempt to identify the best combination of features, we compared the F-score and the evaluation time for all the chosen algorithms when using three features for all datasets (after PCA). While the fact that we use a only three features leads to slight decrease in performance (average of 1.66% over all algorithms) with respect to using the five most important features, i.e., SMMRP (five features), it also saves on average 41.5% of the time (or 15 ms), and thus represents a good compromise between the complexity of the task reflected by the number of features, the classification performance and the computation time.

Table 1. Comparison for sampling rate of 1 s (1s-) and 2 s (2s-) in terms of F-Score.

	NB	KNN	Tree	RForest	SVM	LG	NN	XGB	XGBRF	Average
1s—SMMRP-PCA (5 feat.)	0.95	0.94	0.94	0.95	0.94	0.91	0.96	0.97	0.95	0.95
1s—FFT-PCA (4 feat.)	0.95	0.95	0.94	0.95	0.93	0.94	0.95	0.95	0.94	0.94
2s—SMMRP-PCA (5 feat.)	0.98	0.99	0.97	1	0.98	0.97	1	0.98	0.99	0.984
2s—FFT-PCA (4 feat.)	0.99	0.96	0.98	0.99	0.96	0.96	1	0.99	0.99	0.980

Figure 4a shows that the use of FFT features (in orange) lead to a lower performance and the highest average performance is obtained by the combination all features (SMMRPFFT, in gray). Although normally it is not advisable to use graphs that are not to the actual scale, i.e., normally the Y axis should start at 0, we have chosen scaling in the figure to better highlight the slight differences in performance between the results obtained. According to the results in Figure 4c, the use of data from the accelerometer only (XYZ_, yellow) performed better with all models when compared to sound data only (S_, brown). Sound data only resulted in poorer performance, in particular with KNN, SVM and LG classifiers.

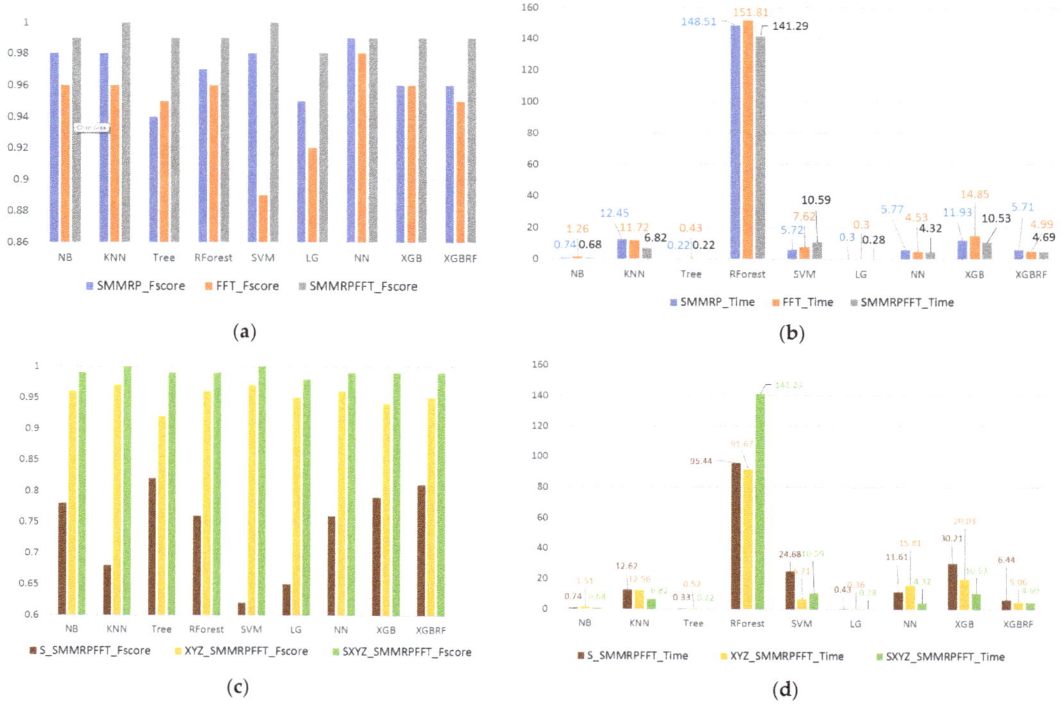

Figure 4. Comparison of: (**a**) F-score for SMMRP, FFT and SMMRPFFT features; (**b**) computation time (in ms) for SMMRP, FFT and SMMRPFFT features; (**c**) F-score for sound S_, accelerometer XYZ_ and sound and accelerometer data (SXYZ_) for SMMRPFFT features for 2 s sampling period; and (**d**) computation time for sound S_, accelerometer XYZ_ and sound and accelerometer data (SXYZ_) for SMMRPFFT features; for 2 s sampling period.

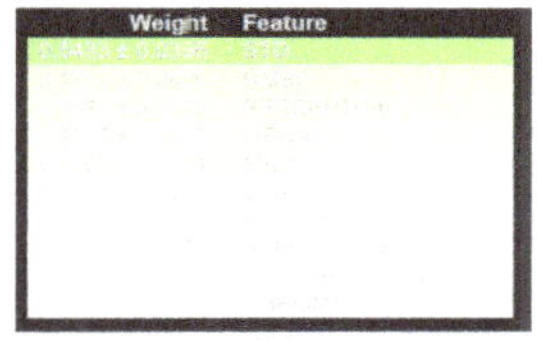

The best performance is achieved using the combination of the two sources of tactile data (SXYZ_, green; the gray and green bars represent the same information, but Figure 4a,c have different scales). The higher performance comes to the price of a slightly higher average evaluation time (Figure 4d) of the order of 2.8 ms with respect to the FFT features (the fastest on average). On average, LG and Tree offer the best compromise in terms of performance and evaluation time. The best overall performance, computed as an average over all feature combinations, is associated with XGBRF (F-score = 0.91) and the lowest time with LG and Tree (0.35 ms), while the highest time is associated with RForest. These findings suggest that the fusion of SMMRP and FFT features yields more powerful composite features capable of achieving good predictions while maintaining a low evaluation time and thus representing good candidates for real-time implementations for this specific dataset. However, it is important to verify that this performance generalizes to other similar tactile datasets to confirm the robustness of this novel feature combination.

Another series of tests was aimed at studying the performance by type of texture. We analyzed the confusion matrices obtained for all algorithms and all feature combinations. Table 2 shows the aggregate correct and wrong predictions, as a percentage over the total number of samples from each texture class and as an average over all the algorithms tested, for the SMMRPFFT features and for the 12 classes. One can notice that among the 12 classes, "toy-rubber-tire" (worst performance), "moquette-1", "fabric-2", and the two samples of "sparkle-paper" are more difficult to classify. These results are coherent with Figure 3, in which these classes are overlapping. Among the tested algorithms, Tree and NN make the most wrong classifications and XGBRF the least.

Table 2. Correct and wrong predictions per texture type (material) using SMMRPFFT features.

Material	Correct Predictions (%)	Wrong Predictions—Material Type, Algorithm (%)
fabric-1	100	
aluminium_film	100	
fabric-2	90	moquette-1, XGB (5%), toy-tire-rubber, NN (5%)
fabric-3	100	
moquette-1	77	fabric-2, NN, LG, SVM (16%), fabric-4, NN (7%)
moquette-2	100	
fabric-4	100	
sticky fabric-5	100	
sticky-fabric	100	
sparkle-paper-1	95	sparkle_paper-2, LG (5%)
sparkle-paper-2	92	sparkle_paper-1, Tree (8%)
toy-tire-rubber	55	fabric-1, LG, SVM, RForest, Tree, kNN, NB (44%)

4. Discussions and Conclusions

We have successfully implemented and validated a learning method that achieves high performance (F-score) in classifying textures measured by tactile sensors. We have demonstrated the importance of selecting and extracting features to enhance classification performance. Furthermore, we demonstrated that the choice of sample period is a significant aspect of time series classification, with an important impact on classification accuracy. Longer sampling periods (2 s. vs. 1 s) provide more information for classification, leading to higher performance (average of 3.59%) but also augment the evaluation time by an average of 29.48% over all features and models. Finally, we demonstrated that the balance between performance and evaluation time is crucial for informed decisions when selecting a classification model. For the selected dataset, we identified the XGBRF to offer the best compromise between performance over all material types and computation time, while kNN represents the next best option (1% improvement in performance at the cost of 2.13 ms increase in time with respect to XGBRF for SMMRFFT).

Author Contributions: Conceptualization, M.N. and A.-M.C.; methodology, M.N. and A.-M.C.; software, M.N.; validation, M.N. and A.-M.C.; formal analysis, M.N.; investigation, M.N.; resources, A.-M.C.; data curation, M.N.; writing—original draft preparation, M.N.; writing—review and editing, M.N. and A.-M.C.; visualization, M.N.; supervision, A.-M.C.; project administration, A.-M.C.; funding acquisition, A.-M.C. All authors have read and agreed to the published version of the manuscript.

Funding: This research was funded by the NSERC Discovery grant number DDG-2020-00045.

Institutional Review Board Statement: Not applicable.

Informed Consent Statement: Not applicable.

Data Availability Statement: Data are available in this manuscript.

Acknowledgments: Thanks to the providers of the VibTac-12 dataset, O. Kursun and A. Pathoogy.

Conflicts of Interest: The authors declare no conflict of interest. The funders had no role in the design of the study; in the collection, analyses, or interpretation of data; in the writing of the manuscript; or in the decision to publish the results.

References

1. Liu, L.; Chen, J.; Fieguth, P.; Zhao, G.; Chellappa, R.; Pietikainen, M. From BoW to CNN: Two Decades of Texture Representation for Texture Classification. *Int. J. Comput. Vis.* **2019**, *127*, 74–109. [CrossRef]
2. Luo, S.; Yuan, W.; Adelson, E.; Cohn, A.G.; Fuentes, R. Vitac: Feature sharing between vision and tactile sensing for cloth texture recognition. In Proceedings of the IEEE International Conference on Robotics and Automation (ICRA), Brisbane, QLD, Australia, 21–25 May 2018.
3. Kursun, O.; Patooghy, A. An Embedded System for Collection and Real-Time Classification of a Tactile Dataset. *IEEE Access* **2020**, *8*, 97462–97473. [CrossRef]
4. Oliveira, T.E.A.; Cretu, A.-M.; Petriu, E.M. A Multi-Modal Bio-Inspired Tactile Sensing Module for Surface Characterization. *Sensors* **2017**, *17*, 1187. [CrossRef] [PubMed]
5. Humeau-Heurtier, A. Texture feature extraction methods: A survey. *IEEE Access* **2019**, *7*, 8975–9000. [CrossRef]
6. Kerr, E.; McGinnity, T.; Coleman, S. Material recognition using tactile sensing. *Exp. Syst. Appl.* **2018**, *94*, 94–111. [CrossRef]
7. Wang, S.; Albini, A.; Maiolino, P.; Mastrogiovanni, F.; Cannata, G. Fabric Classification Using a Finger-Shaped Tactile Sensor via Robotic Sliding. *Front. Neurorobot.* **2022**, *16*, 808222. [CrossRef]
8. Huang, S.; Wu, H. Texture Recognition Based on Perception Data from a Bionic Tactile Sensor. *Sensors* **2021**, *21*, 5224. [CrossRef]
9. Kursun, O.; Patooghy, A. *VibTac-12: Texture Dataset Collected by Tactile Sensors*; IEEE Dataport: Piscataway, NJ, USA, 2020. [CrossRef]
10. McFee, B.; Raffel, C.; Liang, D.; Ellis, D.P.; McVicar, M.; Battenberg, E.; Nieto, O. Librosa: Audio and music signal analysis in python. In Proceedings of the 14th Python in Science Conference, Austin, TX, USA, 6–12 July 2015; Volume 8, pp. 18–24.
11. Korobov, M.; Lopuhin, K. ELI5 Documentation Release 0.11.0, 23 January 2021. Available online: https://readthedocs.org/projects/eli5/downloads/pdf/latest/ (accessed on 19 September 2023).
12. Géron, A. *Hands-On Machine Learning with Scikit-Learn, Keras, and TensorFlow*, 2nd ed.; O'Reilly Media: Sebastopol, CA, USA, 2019.
13. Chen, T.; Guestrin, C. XGBoost: A Scalable Tree Boosting System. In Proceedings of the 22nd ACM SIGKDD International Conference on Knowledge Discovery and Data Mining, San Francisco, CA, USA, 13–17 August 2016. [CrossRef]

engineering proceedings

MDPI

Proceeding Paper

Evaluating Compact Convolutional Neural Networks for Object Recognition Using Sensor Data on Resource-Constrained Devices †

Icaro Camelo * and Ana-Maria Cretu *

Department of Computer Science and Engineering, University of Quebec in Outaouais, Gatineau, QC J8Y 3G5, Canada

* Correspondence: veri02@uqo.ca (I.C.); ana-maria.cretu@uqo.ca (A.-M.C.)

† Presented at the 10th International Electronic Conference on Sensors and Applications (ECSA-10), 15–30 November 2023; Available online: https://ecsa-10.sciforum.net/.

Abstract: The goal of this paper is to evaluate various compact CNN architectures for object recognition trained on a small resource-constrained platform, the NVIDIA Jetson Xavier. Rigorous experimentation identifies the best compact CNN models that balance accuracy and speed on embedded IoT devices. The key objectives are to analyze resource usage such as CPU/GPU and RAM used to train models, the performance of the CNNs, identify trade-offs, and find optimized deep learning solutions tailored for training and real-time inference on edge devices with tight resource constraints.

Keywords: machine learning; compact convolutional networks; object recognition resource-constraint devices; IoT; sensor data processing

Citation: Camelo, I.; Cretu, A.-M. Evaluating Compact Convolutional Neural Networks for Object Recognition Using Sensor Data on Resource-Constrained Devices. *Eng. Proc.* **2023**, *58*, 6. https://doi.org/10.3390/ecsa-10-16202

Academic Editor: Stefan Bosse

Published: 15 November 2023

1. Introduction

Nowadays, artificial intelligence (AI) has become very prominent and impactful owing to its proficiency in accomplishing a wide variety of tasks with high levels of effectiveness and efficiency. Some of the areas where AI has demonstrated its capabilities include, but are not restricted to, visual recognition tasks like image classification, object detection, sensor data, and natural language processing. Deep learning is an advanced sub-discipline of machine learning that emphasizes refining artificial neural networks with multiple layers to apprehend intricate representations of data. It can learn useful features from raw data without manual feature engineering. In contrast, the advent of Internet-of-Things devices having inbuilt sensors opens novel prospects for implementing convolutional neural networks (CNNs) directly on resource-limited devices. However, these devices have limited memory, storage, and computing power, making extensive and complex CNNs infeasible. Implementing compact CNNs with smaller models and computational needs on IoT devices enables localized capabilities like object recognition without relying on the cloud. This reduces latency while improving privacy and reliability. To facilitate model training and inference, several types of specialized hardware have emerged such as CPUs, graphics processing units (GPUs)/tensor processing units (TPUs), and field-programmable gate arrays (FPGAs).

Researchers have been investigating the training and inference performance of models in resource-constrained devices. Ajit et al. [1] provide a broader review of CNNs without directly addressing the impact of training using different hardware. Nevertheless, the paper offers valuable context regarding the algorithmic steps and applications of CNNs across various fields. Recent studies [2] indicate that both GPUs and TPUs significantly improved the performance and accuracy of CNN models, with TPUs outperforming GPUs in certain cases. This suggests that the choice of hardware can have an important impact on model accuracy and overall performance. Other work focuses on GPU and TPU deployment for image classification tasks [3]. Only a few papers have explored the use of the NVIDIA Jetson Xavier NX platform for deploying imaging applications. Among these, Jabłoński et al. [4] evaluate the performance of Jetson Xavier NX for real-time image processing for plasma diagnostics. The authors implement several image processing algorithms on the platform and evaluate their performance in terms of speed and accuracy. They found that the platform is able to achieve good performance on the image processing tasks, and that it is well-suited for real-time applications due to its fast-processing speeds. Kortli et al. [5] propose a hybrid model that combines a CNN with a long short-term memory (LSTM) network for lane detection and implement and demonstrate the ability to achieve good performance for this task on the Jetson Xavier NX.

The goal of this paper is to evaluate various compact CNN architectures for object recognition in images trained on the NVIDIA Jetson Xavier NX. The key objectives are to analyze resource usage such as CPU/GPU and RAM used to train models, the performance of the CNNs, identify trade-offs, and find optimized deep learning solutions tailored for training and real-time inferencing on devices with tight resource constraints.

2. Materials and Methods

2.1. NVIDIA Jetson Xavier Platform

The NVIDIA Jetson Xavier NX [6] is a system-on-a-chip (SoC) developed by NVIDIA. It is designed for use in a wide range of applications, including autonomous machines, robotics, and edge computing. The Xavier NX is based on the NVIDIA Volta architecture and features a 6-core Arm Cortex-A57 processor, a 512-core NVIDIA Volta GPU, and a deep learning accelerator (DLA). It is designed to be highly energy efficient and has a small form factor, making it suitable for use in devices with limited space and power resources. The Xavier NX can deliver high performance for a range of tasks, including machine learning, image and video processing, and computer vision. It is targeted at developers and OEMs who are looking to build advanced, high-performance systems for a variety of applications.

2.1.1. Setting Up the NVIDIA Xavier NX Board

The process of installing NVIDIA SDK Manager and Jetpack [7], flashing an SD card, and installing an SSD drive while changing the root file system ('rootfs') to the SSD involves a series of specific procedures.

The first step is to download the NVIDIA SDK Manager from the NVIDIA SDK Manager download page. This step requires one to have an NVIDIA Developer account to access the download. Once the file has been downloaded, the terminal must be opened, and one must navigate to the directory where the file is saved. The permission of the file is then changed to make it executable with a specific command. The SDK Manager is then installed by running a particular command in the terminal.

After the SDK Manager has been installed, it can be executed by typing 'sdkmanager' into the terminal and then logging in with the NVIDIA Developer account credentials. Within the SDK Manager, one must select the appropriate hardware configuration in the 'Target Hardware' section. Then, the desired Jetpack version is selected in the 'SDKs' section. One must then follow the prompts to complete the installation process.

Flashing the SD card is a task handled by the SDK Manager during the Jetpack installation process, requiring the SD card to be connected to the host machine. If a manual flash of the SD card is needed, a tool like Etcher [8] can be utilized. One can download and install Etcher, select the image file they want to flash, select the SD card, and start the flashing process.

The installation of the SSD drive and the changing of the 'rootfs' to point to the SSD first necessitates physically connecting the SSD to the device. After this, the SSD must be formatted, which, in Linux, can be performed using a specific command, replacing 'sdX' with the appropriate device id. One is then guided through a series of prompts to create a new partition and format it. After the SSD has been formatted, it is mounted by running a specific command. The contents from the SD card are then copied to the SSD using the 'rsync' command. One then must edit the '/boot/extlinux/extlinux.conf' file on the SD card to point to the SSD, changing 'root=/dev/mmcblk0p1' to 'root=/dev/sdX1'. The device is then rebooted, after which the system should boot from the SSD. It is crucial to remember to replace 'sdX' with the user's SSD drive id and '/dev/mmcblk0p1' with the actual root partition. Additionally, it is of paramount importance that one backs up any vital data before proceeding with these steps and proceeds with caution when modifying system files or disk partitions.

2.1.2. Deploying CNNs on NVIDIA Jetson Xavier NX

Once the above steps are completed, one can begin creating and training machine learning models. Code development can be made more effective by installing the proper Integrated Development Environment (IDE) on the Jetson Xavier NX or by establishing a remote connection over SSH. Pytorch and Tensorflow are both available in the NVIDIA Jetson SDK. The PyTorch library was our choice for implementation. Moreover, we also use Torchvision, which is a PyTorch add-on library that provides datasets, model architectures, and image transformations for computer vision, NumPy for numerical operations, and Scikit Learn that provides utilities for machine learning, including model evaluation metrics. The Pytorch profiler, torch.profiler, is also used for profiling model inference to analyze GPU/CPU usage and memory consumption. For maximizing performance and obtaining the best out of the NVIDIA Xavier NX, we activated all 8 CPUs, enabling its 6 cores, which causes the board to consume 20 Watts of power. NVIDIA provides a script called 'jetson_clocks'. The script is provided by NVIDIA to optimize the board performance through the implementation of static maximum frequency settings for CPU, GPU, and EMC clocks. It is also recommended to activate fans, but we found that the temperatures are not high when the board is managed with the default values.

2.2. *Datasets for Experimentation*

In this paper, we focus on 2D object recognition in images. We chose two datasets for experimentation. The first one is CIFAR-10 [9], a well-known dataset in computer vision for object recognition. It contains 60,000 32 × 32 color images, all of which contain one of the 10 distinct object classes. Each class comprises 6000 images, rendering a grand total of 10 unique object classes. The test batch contains 1000 randomly selected from each class. The second one, STL-10 [10], contains 96 × 96 color images across 10 classes with 500 training and 800 test images per class, totaling 5000 labeled training images and 8000 labeled test images. It is commonly employed to benchmark machine learning models and provides 800 test images per class compared to 1000 for CIFAR-10, a moderate difference. As the STL-10 dataset has fewer labeled training images with higher resolution (32 × 32 for CIFAR-10 vs. 96 × 96 for STL-10), we wanted to observe how well models can generalize to different quantity of data and deal with different image sizes.

2.3. *Methodology*

A series of compact, lightweight CNN architectures, namely AlexNet, ShuffleNet v2, SqueezeNet, ResNet50, and MobileNet v2, are implemented and evaluated on the Jetson Xavier NX platform. The performance is compared either when training the algorithms directly from scratch on the platform or when using a transfer learning process.

2.3.1. Tested Architectures

We have chosen 5 compact architectures for testing: AlexNet [11] comprises 8 layers with trainable parameters, including 5 convolutional layers paired with max pooling layers, followed by 3 fully connected layers. Each layer uses a ReLU activation function, except for the output layer. It also uses dropout layers, which prevent the model from overfitting. ShuffleNetV2 [12] is an efficient, lightweight CNN architecture designed for mobile and embedded vision applications with limited computational resources. Its architecture is composed of 50 layers and incorporates two operations: pointwise group convolution and channel shuffle, which significantly reduce computational costs while still preserving accuracy. The architecture of SqueezeNet [13] is based on a shuffle operation that enables channel interleaving, reducing the number of computations required by the network. ResNet [14] is a pioneering CNN architecture that utilizes residual connections to enable training of very deep networks. Skip connections allow gradients to flow directly to earlier layers. ResNet's key components include residual blocks, stacked together to form the network, and a bottleneck design for deeper versions. This architecture enabled the training of extremely deep networks, from 48 up to 152 layers. In this paper, we use Resnet50, which is a ResNet variant that has 50 layers. Finally, MobileNetV2 [15] uses depthwise and pointwise separable convolutions to reduce parameters and computations needed while incurring a slight decrease in performance. The architecture introduces inverted residual blocks, a modification of the standard residual block found in the ResNet architectures, which allows for efficient training on limited computation power. When training these models, we resized and normalized the input image size for each architecture and we shuffled the training datasets. We also applied data augmentation techniques, i.e., cropping and horizontal flipping. We froze the hidden layers to both avoid relearning generic features and improve training performance.

2.3.2. Test Design and Performance Evaluation

In order to train models and assess their performance, we used the two datasets in Section 2.2. Each model is trained initially with 10 epochs and the number of epochs is increased to 30, 50, 60, 100, 150, and 200 epochs, for a fixed batch size of 64. The loss function is set as cross-entropy loss, and the optimization algorithm is Stochastic Gradient Descent (SGD) with a learning rate of 0.001 and momentum of 0.9. For monitoring the training process, a script runs in background collecting CPU/GPU/RAM utilization from the board. Also, the loss is printed every 200 batches. The best model is identified by the highest F1-score with less computation cost. However, the time required to train, and accuracy of a model should also be considered depending on the use case. To test our model, we train from scratch directly on the board and also use transfer learning. The latter takes advantage of knowledge previously learned from models trained on large datasets, which in our case, is the ImageNet dataset [16]. Transfer learning reduces the time required to train a model as it freezes the hidden layers that contain general knowledge (i.e., uses pre-trained weights obtained during learning on ImageNet dataset) and retrains only a limited number of layers, particularly those toward the output layer that contains the specific knowledge of the target task. It achieves a good performance quicker with reduced computation costs. After training, the model's performance is evaluated on the test sets mentioned in Section 2.2. The precision, recall, and F1-score for each model are calculated using the Scikit Learn library's functions.

3. Results

Tables 1 and 2 summarize the results we obtained on the two datasets using the five tested CNN architectures when trained from scratch and when using the pre-trained weights computed via transfer learning, with the best performance highlighted in bold. On the CIFAR-10 dataset, the AlexNet model trained from scratch improved its F-score from a modest 0.640 after only 10 epochs to an impressive F-score of 0.824 after 50 epochs and finally to an optimal 0.845 after 200 full epochs of training. Using transfer learning, the performance of the model increases significantly with fewer epochs, achieving an F-score of 0.911 for 100 training epochs. The pre-trained ShuffleNet model achieved an F-score of 0.916 with only 10 epochs. Unlike the model trained from scratch, this model did not see significant improvement as the number of epochs increased and peaked at an F-score of 0.924 for 30 epochs. The SqueezeNet with transfer learning scored very well (an F-score of 0.902) with only 3 h of training. The pre-trained ResNet50 achieved 0.841 F-score with 30 epochs. On the other hand, MobileNetV2 showed a modest improvement of 0.073 in

F-score when using pre-trained weights vs. training from scratch, but in less than half the training time.

Table 1. Summary of the best model performance on CIFAR-10, with the best model performance highlighted in bold.

Model		Epochs	Precision	Recall	F-Score	Time to Train
AlexNet	Scratch	200	0.846	0.846	0.845	21 h 46 min
	Pre-trained	100	0.912	0.911	0.911	10 h 47 min
ShuffleNet	Scratch	100	0.740	0.741	0.741	13 h 02 min
	Pre-trained	**30**	**0.924**	**0.924**	**0.924**	**4 h 07 min**
SqueezeNet	Scratch	50	0.767	0.763	0.761	11 h 40 min
	Pre-trained	30	0.902	0.902	0.902	3 h
Resnet50	Scratch	100	0.655	0.647	0.649	38 h 15 min
	Pre-trained	30	0.842	0.842	0.841	1 h 24 min
MobileNetV2	Scratch	150	0.750	0.751	0.750	5 h
	Pre-trained	100	0.828	0.830	0.828	2 h 12 min

Table 2. Summary of the best model performance on STL-10, with the best model performance highlighted in bold.

Model		Epochs	Precision	Recall	F-Score	Time to Train
AlexNet	Scratch	100	0.985	0.984	0.984	1 h 33 min
	Pre-trained	**30**	**0.995**	**0.995**	**0.995**	**30 min**
ShuffleNet	Scratch	100	0.475	0.475	0.474	1 h 20 min
	Pre-trained	100	0.914	0.913	0.913	1 h 15 min
SqueezeNet	Scratch	100	0.613	0.567	0.571	2 h 30 min
	Pre-trained	10	0.862	0.862	0.861	8 min
Resnet50	Scratch	200	0.449	0.452	0.464	3 h 42 min
	Pre-trained	10	0.914	0.915	0.914	10 min
MobileNetV2	Scratch	150	0.328	0.276	0.257	33 min
	Pre-trained	150	0.786	0.781	0.781	32 min

Overall, the pre-trained models achieved an increased average F-score of 13.2% and an average decrease in computational time of 75.8%. For this dataset, the best performance is associated with ShuffleNet and the fastest model to train is Resnet50, but for a decrease in performance of 10.38%. The best compromise between performance and training time seems to be achieved by the pre-trained SqueezeNet, with a decrease of only 2.2% in performance with respect to the best model, but for double the time with respect to the fastest model.

As shown in Table 2, on the STL-10 dataset, consistent with the previous dataset, AlexNet is the one achieving the best performance when trained from scratch on the board. The pre-trained AlexNet and ResNet50 models only require very few epochs to achieve excellent results on this dataset. The pre-trained ShuffleNet scored almost twice better than its from scratch version with the same number of epochs. MobileNetV2 from scratch struggled to learn even after 150 epochs. For this dataset, the pre-trained models achieved an increased average F-score of 34.2% and an average decrease in computational time of 73.4%. With an F-score of 0.995, the AlexNet with transfer learning achieves the best performance on this dataset, while the fastest model (one third of the time required by AlexNet) is SqueezeNet for a decrease of 13.5% in performance.

4. Discussion and Conclusions

As expected, the pre-trained models performed very well compared to their counterparts trained from scratch (average of 23.7% over the two datasets), as the base model trained on ImageNet is suitable for the task. Also, the pretrained model took less training time (average of 74.6% shorter over the two datasets). The pretrained AlexNet performed better (8.4% improvement) on STL-10 compared to CIFAR-10, with only 30 epochs instead of 100 epochs on CIFAR-10. The same remains true for the counterpart trained from scratch, i.e., a 13.9% improvement on CIFAR for half the training epochs. This suggests that AlexNet can learn well with fewer images but with higher resolution samples. The pretrained ShuffleNet performed well on CIFAR-10 with only 30 epochs and on STL-10 took more time (100 epochs) to achieve an F-score greater than 0.90. The ShuffleNet model trained from scratch obtained subpar performance (less than 0.5) on STL-10, suggesting that the model struggles with less data. SqueezeNet achieved an F-score of 0.861 with only 10 epochs on the STL-10 dataset and the performance did not improve with more epochs. However, it achieves an F-score of 0.902 with 30 epochs on CIFAR-10. Similar to ShuffleNet, the SqueezeNet model trained from scratch obtained a low performance (0.57). This implies that SqueezeNet requires more data to increase performance. ResNet50 achieves an F-score of 0.914 with only 10 epochs on the STL-10 dataset but only scores 0.842 with 30 epochs on CIFAR-10. Like AlexNet, ResNet50 does well with few data but with higher resolution images. MobileNetV2 did not show a big discrepancy in terms of F-score across the two datasets, performing slightly better (4% improvement) on the CIFAR-10 dataset with 50 fewer epochs.

As previously mentioned, the best performing model on the CIFAR-10 dataset was the pre-trained ShuffleNet trained on 30 epochs with an F-score of 0.924, whereas the pre-trained AlexNet model achieved an F-score of 0.995 on the STL-10 dataset when trained with 30 epochs. On average, ShuffleNet achieved an F-score of 0.918 over the two datasets, whereas AlexNet achieved 0.953. Even though AlexNet scored slightly better than ShuffleNet, the training time is another determining factor. On average, ShuffleNet only needed 2 h 41 min to achieve good performance on both datasets, while AlexNet took 5 h 30 min, making ShuffleNet the preferred model. Regarding the speed training, the pre-trained ResNet50 model was the fastest to train on CIFAR-10, taking 1 h 24 min to obtain an F-score of 0.841, which is approximately half the time. SqueezeNet, the second fastest model, took 3 h to achieve an F-score of 0.902 on CIFAR-10. On STL-10, the pre-trained SqueezeNet was the fastest, showing an F-score of 0.861 in 8 min. On the other hand, the pretrained ResNet50 model took only two extra minutes (10 min total) to reach an impressive F-score of 0.914.

It is worth mentioning that during the training time, Jetson Xavier NX was overall extremely efficient in terms of resource utilization, using almost 100% of CPU, GPU, and RAM available.

The primary limitations encountered while training CNNs on the NVIDIA Xavier NX board stemmed from the constrained memory resources available. We encountered some challenges while training larger models such as VGG-11 and VGG-19 on CIFAR-10 with the container repeatedly exiting due to out-of-memory errors. After several attempts, we managed to find an appropriate batch size that works for these models. While the Xavier NX platform enables training in many moderate CNN architectures, memory constraints impose clear limits on model and dataset scale versus high-end GPUs or cloud-based accelerators with abundant RAM. In summary, RAM availability represents the primary bottleneck for more advanced deep learning tasks on this embedded hardware. In our future work, we will be exploring alternative optimizers and loss functions that could potentially improve convergence speed, model performance, and robustness. Additionally, leveraging hardware-specific libraries such as Nvidia's TensorRT could also improve inference performance on the Xavier NX via strategies tailored to the GPU architecture

Author Contributions: Conceptualization, I.C. and A.-M.C.; methodology, I.C. and A.-M.C.; software, I.C.; validation, I.C. and A.-M.C.; formal analysis, I.C.; investigation, I.C.; resources, A.-M.C.; data curation, I.C.; writing—original draft preparation, I.C.; writing—review and editing, I.C. and A.-M.C.; visualization, I.C.; supervision, A.-M.C.; project administration, A.-M.C.; funding acquisition, A.-M.C. All authors have read and agreed to the published version of the manuscript.

Funding: This research was funded in part by the NSERC Discovery grant number DDG-2020-00045 and by the NSERC UTILI grant number CREAT-2019-528123.

Institutional Review Board Statement: Not applicable.

Informed Consent Statement: Not applicable.

Data Availability Statement: Data are available in this manuscript.

Conflicts of Interest: The authors declare no conflicts of interest. The funders had no role in the design of the study; in the collection, analyses, or interpretation of data; in the writing of the manuscript; or in the decision to publish the results.

References

1. Ajit, A.; Acharya, K.; Samanta, A. A Review of Convolutional Neural Networks. In Proceedings of the 2020 International Conference on Emerging Trends in Information Technology and Engineering (ic-ETITE), Vellore, India, 24–25 February 2020; pp. 1–5.
2. Ravikumar, A.; Sriraman, H.; Saketh, P.M.S.; Lokesh, S.; Karanam, A. Effect of neural network structure in accelerating performance and accuracy of a convolutional neural network with GPU/TPU for image analytics. *PeerJ Comput. Sci.* **2022**, *8*, e909. [CrossRef] [PubMed]
3. Bochkovskiy, A.; Wang, C.; Liao, H. YOLOv4: Optimal Speed and Accuracy of Object Detection. *arXiv* **2020**, arXiv:2004.10934. [CrossRef]
4. Jabłoński, B.; Makowski, D.; Perek, P.; Nowakowski, P.N.V.; Sitjes, A.P.; Jakubowski, M.; Gao, Y.; Winter, A. The W-X Team Evaluation of NVIDIA Xavier NX Platform for Real-Time Image Processing for Plasma Diagnostics. *Energies* **2022**, *15*, 2088. [CrossRef]
5. Kortli, Y.; Gabsi, S.; Jridi, M.; Voon LF, L.Y.; Atri, M. Hls-based hardware acceleration on the Zynq SoC: A real-time face detection and recognition system. In Proceedings of the 2022 IEEE 9th International Conference on Sciences of Electronics, Technologies of Information and Telecommunications, SETIT 2022, Hammamet, Tunisia, 28–30 May 2022; pp. 61–64.
6. NVIDIA Jetson Xavier NX. Available online: https://www.nvidia.com/en-sg/autonomous-machines/embedded-systems/jetson-xavier-nx/ (accessed on 27 September 2023).
7. NVIDIA SDK Manager and Jetpack. Available online: https://developer.nvidia.com/embedded/jetpack (accessed on 27 September 2023).
8. Etcher Software. Available online: https://etcher.download/about/ (accessed on 27 September 2023).
9. The CIFAR-10 Dataset. Available online: https://www.cs.toronto.edu/~kriz/cifar.html (accessed on 19 September 2023).
10. STL-10 Dataset. Available online: https://cs.stanford.edu/~acoates/stl10/ (accessed on 19 September 2023).
11. Krizhevsky, A.; Sutskever, I.; Hinton, G.E. Imagenet classification with deep convolutional neural networks. *Commun. ACM* **2017**, *60*, 84–90. [CrossRef]
12. Zhang, X.; Zhou, X.; Lin, M.; Sun, J. ShuffleNet: An Extremely Efficient Convolutional Neural Network for Mobile Devices. In Proceedings of the 2018 IEEE/CVF Conference on Computer Vision and Pattern Recognition (CVPR), Salt Lake City, UT, USA, 18–23 June 2018; pp. 6848–6856.
13. Iandola, F.N.; Han, S.; Moskewicz, M.W.; Ashraf, K.; Dally, W.J.; Keutzer, K. SqueezeNet: AlexNet-level accuracy with 50x fewer parameters and <0.5 MB model size. *arXiv* **2016**, arXiv:1602.07360.
14. He, K.; Zhang, X.; Ren, S.; Sun, J. Deep Residual Learning for Image Recognition. In Proceedings of the 2016 IEEE Conference on Computer Vision and Pattern Recognition (CVPR), Las Vegas, NV, USA, 27–30 June 2016; pp. 770–778. [CrossRef]
15. Sandler, M.; Howard, A.; Zhu, M.; Zhmoginov, A.; Chen, L.-C. MobileNetV2: Inverted Residuals and Linear Bottlenecks. In Proceedings of the 2018 IEEE/CVF Conference on Computer Vision and Pattern Recognition, Salt Lake City, UT, USA, 18–23 June 2018; pp. 4510–4520. [CrossRef]
16. Russakovsky, O.; Deng, J.; Su, H.; Krause, J.; Satheesh, S.; Ma, S.; Huang, Z.; Karpathy, A.; Khosla, A.; Bernstein, M.; et al. ImageNet Large Scale Visual Recognition Challenge. *Int. J. Comput. Vis.* **2015**, *115*, 211–252. [CrossRef]

engineering proceedings

Proceeding Paper

Sheathless Dielectrophoresis-Based Microfluidic Chip for Label-Free Bio-Particle Focusing and Separation †

Reza Vamegh, Zeynab Alipour * and Mehdi Fardmanesh

Department of Electrical Engineering, Sharif University of Technology, Tehran 1458889694, Iran; reza.vamegh@ee.sharif.edu (R.V.); fardmanesh@sharif.edu (M.F.)
* Correspondence: alipour_zeynab@ee.sharif.edu
† Presented at the 10th International Electronic Conference on Sensors and Applications (ECSA-10), 15–30 November 2023; Available online: https://ecsa-10.sciforum.net/.

Abstract: This paper presents a novel microfluidic dielectrophoresis (DEP) system to focus and separate cells of similar size based on their structural differences, which is more challenging than separation by size. Because, in this case, the DEP force is only proportional to the polarizabilities of cells, we used live and dead yeast cells as bio-particles to investigate the chip efficiency. Our designed chip consists of three sections. First, we focused on cells at the center of the microchannel by employing a negative DEP phenomenon. After that, cells were separated due to the different deflection from high-electric-field areas. Finally, a novel outlet design was utilized to facilitate separation by increasing the gap between the two groups of cells. The proposed sheath-free design has one inlet for target cell injection requiring only one pump to control the flow rate, which reduces costs and complexity. Successful discrimination of the particles was achieved by using DEP force as a label-free and highly efficient technique. As an accessible and cost-effective method, soft lithography with a 3D-printed resin mold was used to fabricate the microfluidic parts. The microchannel was made of polydimethylsiloxane (PDMS) material that is biocompatible. The electrodes were made of gold due to its biocompatibility and non-oxidation, and a titanium layer was sputtered as the buffer layer for the adhesion of the sputtered gold layer to the glass. A standard microfabrication process was employed to create the electrode pattern. O_2 plasma treatment yielded leakage-free bonding between the patterned glass and PDMS structure containing the microfluidic channel. The maximum voltage applied to the electrodes (26 V) was lower than the threshold value for cell electroporation. The simulations and experimental results both confirm the effectiveness of the proposed microfluidic chip.

Keywords: dielectrophoresis; microfluidics; planar electrodes; cell separation; active focusing

Citation: Vamegh, R.; Alipour, Z.; Fardmanesh, M. Sheathless Dielectrophoresis-Based Microfluidic Chip for Label-Free Bio-Particle Focusing and Separation. *Eng. Proc.* **2023**, *58*, 7. https://doi.org/10.3390/ecsa-10-16255

Academic Editor: Stefano Mariani

Published: 15 November 2023

1. Introduction

Microfluidics is a rapidly growing field that concerns the manipulation and study of minute amounts of fluids flowing through microchannels [1]. Due to its precise control over fluids, microfluidics presents unique features, including reduced sample size requirements, enhanced speed and sensitivity, and the ability to integrate multiple analytical and laboratory functions onto a single chip [2]. This interdisciplinary technology has led to astonishing advances in various domains, including drug discovery [3], medical diagnostics [4], biosensing [5], environmental monitoring [6], and food safety [7].

The ability to analyze cells is the cornerstone of numerous scientific investigations, including diagnostic and therapeutic applications and biomedical research. The study of cells provides essential information on their structure, function, and behavior, and aids in understanding the physiology of organisms. The information obtained through cell analysis has significant applications in fields like oncology, neurology, and immunology [8]. In this regard, isolating specific cell populations from a heterogeneous sample with high purity and yield is a vital step in several experimental protocols [9]. Cell separation processes are necessary for gene expression analysis, proteomics, and cell-based therapies.

The combination of microfluidics and cell separation has unlocked new prospects in studying intricate cellular processes. Based on their operational principles, microfluidic cell separation techniques can be classified into passive and active methods [10]. Passive methods harness the power of meticulous channel structures, hydrodynamic forces, and steric interactions to manipulate particles via various mechanisms, such as deterministic lateral displacement (DLD) [11], pinch flow fractionation (PFF), hydrodynamic filtration, and inertial and secondary flow [12]. Conversely, active microfluidic separation methods rely on external fields to propel particles toward specific locations for separation. Fluorescence-activated cell sorting (FACS) [13], acoustophoresis [14], magnetophoresis [15], and dielectrophoresis (DEP) [16] are some of these active methods.

One of the most prominent microfluidic cell separation techniques is the DEP separation method, a non-invasive and label-free approach. Its high flexibility ensures the selective manipulation of cells based on their diverse bio-physical characteristics, including their size, shape, and dielectric properties. Moreover, it is highly proficient in processing large volumes of cells, resulting in high-purity samples with exceptional throughput. Lastly, DEP is a highly versatile tool that can integrate with other microfluidic techniques, including microfluidic imaging, single-cell analysis, and microfluidic sensors [17,18].

In this work, we present the design and fabrication of an integrated microfluidic chip capable of separating cells of similar size based on their structural differences via the DEP mechanism. The novel feature of our microchip lies in utilizing a single applied frequency to achieve cell focusing, without using sheath flows and separation, while positioning electrodes on both sides of the microchannel. We used the negative DEP phenomenon to narrow the streamline of cells in the middle of the microchannel. Hence, all cells experienced the same electric field, with the sole differentiating factor being the structure-based variation in the DEP forces they encountered. The subtle arrangement and size of the focusing and separating microelectrodes attained optimal non-uniformity of the electric field while minimizing the voltage requirements to promote cell viability. Our easy, low-cost, and rapid fabrication method allows cost-effective mass production of our device. We simulated flow field distribution, electric field distribution, and particle trajectories to optimize the device operation. Finally, we validated our microchip's performance by conducting experiments using live and dead yeast cells as bio-particles with similar sizes.

2. Materials and Methods

2.1. Theoretical Background

The DEP phenomenon occurs when non-uniform electric fields interact with neutral particles possessing a dipole moment. When such particles are placed in an inhomogeneous electric field, charges will start accumulating at the interface between the medium and the particle, creating dipoles that interact with the electric field. Consequently, the particles undergo a net force and begin to move. The fundamental principle of DEP relies on the difference in polarizabilities between the particles and their suspending medium. If the particle has a higher polarizability than the medium, it will experience a net force towards areas of high electric fields. Conversely, if the particle's polarizability is lower than that of the medium, the DEP force will be directed in the opposite direction.

Utilizing the dipole moment method on a homogenous spherical particle of radius r suspended in a dielectric medium with permittivity ε_m, the time-average DEP force acting on the particle can be determined [19]:

$$\left\langle \vec{F}_{DEP}(t) \right\rangle = 2\pi\varepsilon_m r^3 Re\{f_{CM}(\omega)\} \nabla E_{rms}^2, \tag{1}$$

where $Re\{\}$ is the real part of a complex number, E_{rms} is the root mean square of the applied electric field, and $f_{CM}(\omega)$ is the Clausius–Mossotti factor [19]:

$$f_{CM}(\omega) = \frac{\varepsilon_p^*(\omega) - \varepsilon_m^*(\omega)}{\varepsilon_p^*(\omega) + 2\varepsilon_m^*(\omega)}, \tag{2}$$

where $\varepsilon_p{}^*(\omega)$ and $\varepsilon_m{}^*(\omega)$ are the effective complex permittivity of the particle and medium, respectively, which is given by $\varepsilon - j\sigma/\omega$, where ε is the permittivity of the material, σ is the material conductivity, and ω is the angular frequency.

From the DEP force equation, it is evident that the DEP force direction depends on the value of $Re\{f_{CM}(\omega)\}$. When $Re\{f_{CM}(\omega)\} > 0$, the DEP force acts in the direction of the high electric field, causing the particle to be attracted towards the field's region, which is called positive DEP (PDEP). Conversely, for $Re\{f_{CM}(\omega)\} < 0$, particles are deflected from the region of higher field intensity to a region with lower field intensity, known as negative DEP (NDEP).

Using the two-shell model (Figure 1a) for live and dead yeasts [20], we plotted the $Re\{f_{CM}(\omega)\}$ (Figure 1b) and determined 10 kHz to be our optimal operating frequency.

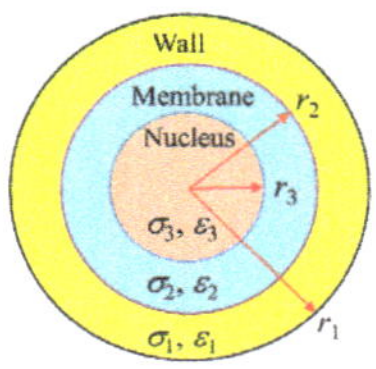

Symbol	Values	
	Live cells	Dead cells
r_1 (μm)	3	2.5
σ_1 (μS/cm)	140	15
ε_1	60	60
r_2 (μm)	2.78	2.25
σ_2 (μS/cm)	2.5×10^{-3}	1.6
ε_2	6	6
r_3 (μm)	2.772	2.242
σ_3 (μS/cm)	2000	70
ε_3	50	50

(**a**)

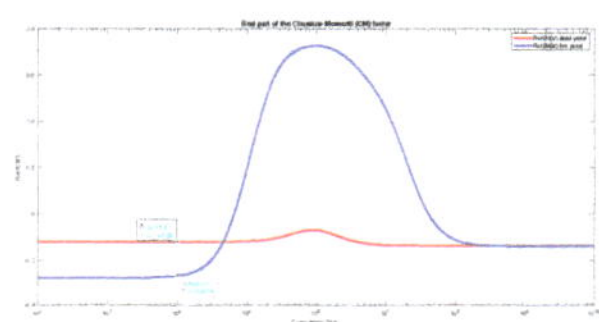

(**b**)

Figure 1. Yeast cells: (**a**) two–shell model [20] and (**b**) calculated $Re\{f_{CM}(\omega)\}$ for live and dead yeast cells.

2.2. Design and Simulation

Figure 2 presented herein illustrates our novel and inventive design strategy:

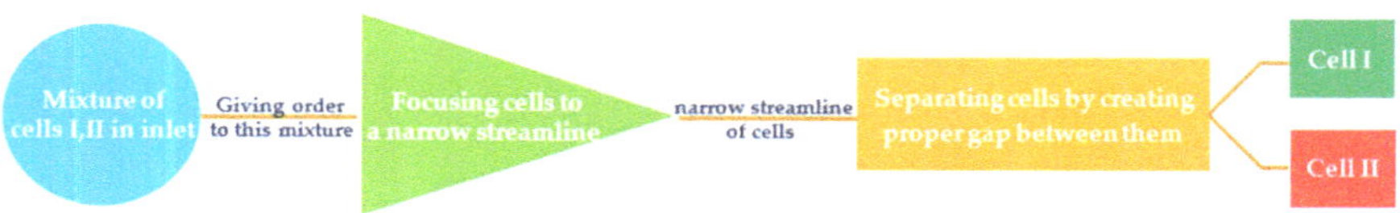

Figure 2. Strategy to design microfluidic DEP-based chip for cell separation.

Our initial objective aimed to ensure that all cells experienced the same electric field. We leveraged the NDEP phenomenon to cause the movement of particles in the same direction toward low-field regions by applying a frequency of 10 kHz to the focusing electrodes. To facilitate the concentration process and reduce the required applied voltage, we began with a smaller section with a width of 50 microns that gradually widened to 100 microns. At the 100-micron segment, we employed two sets of four electrodes to concentrate the cells towards the center of the microchannel. The width of the upper electrodes was 120 microns (Figure 3a), while the lower electrodes had an 80-micron width, creating the desired non-uniform field. Figure 3b shows the overall performance of the focusing section consisting of three subsections, with a total length of 3400 microns.

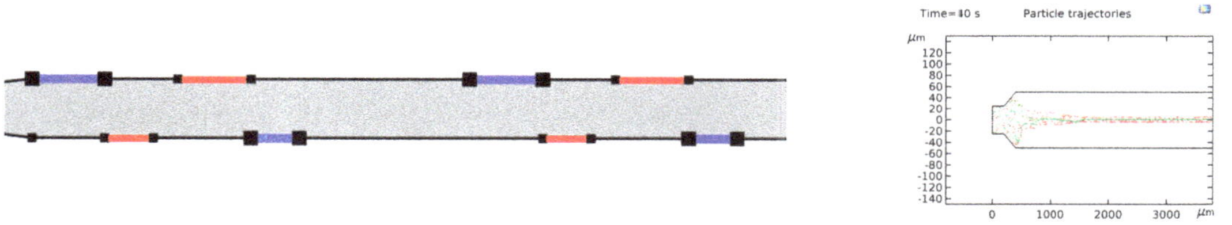

(**a**) (**b**)

Figure 3. Focusing section: (**a**) electrode arrangement; (**b**) overall performance (live and dead yeast cells are shown in green and red, respectively).

After cell focusing, we went on to separate the cells by gradually increasing the gap between viable (green) and nonviable (red) cells using NDEP. The separation section was carefully optimized to ensure the maximum distance between the two clusters of cells, utilizing a 10:1 size ratio for the electrodes in an arrangement that caused the particles' trajectory to oscillate about the channel's symmetry axis. In the final part of the separation section, we incorporated a step increase in the channel width to ease separation by widening the gap between the two groups of cells. Through further simulations and optimizations, we determined that 26 volts is the minimum operating voltage for both focusing and separation electrode pairs, with the red electrodes grounded. Figure 4a displays the electric field intensity over the entire structure. Moreover, Figure 4b shows the particles' trajectory near the outlets, with a 37-micron gap between the two cell groups.

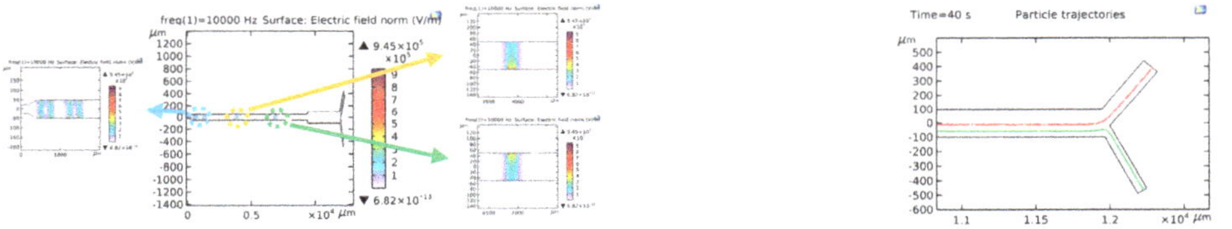

(**a**) (**b**)

Figure 4. Optimized structure: (**a**) electric field distribution; (**b**) particles' trajectory near the outlets.

2.3. Fabrication

Figure 5 represents the fabrication steps, comprising three phases: microelectrode formation on glass, microfluidic channel creation, and bonding. Gold was selected for the microelectrodes due to its suitable electrical conductivity, non-oxidizing properties, and biocompatibility. A sputtering process was employed to deposit a titanium layer to improve the adhesion of gold to glass, followed by Au layer deposition. To form the pattern of our microelectrodes, we performed standard photolithography by utilizing S1813 positive photoresist. Afterward, the excessive Au and Ti were etched, leaving only the microelectrode details. For the microfluidic part, we employed soft lithography, which involved using an LCD 3D printer to produce a 3D-printed mold to cast the PDMS. Subsequently, the resulting microchannel part was cured and perforated to create inlet and outlet ports. Next, we plasma-bonded the PDMS-based microchannel part to the patterned substrate containing electrodes to form an integrated microfluidic chip. Lastly, the inlet and outlet hoses were placed.

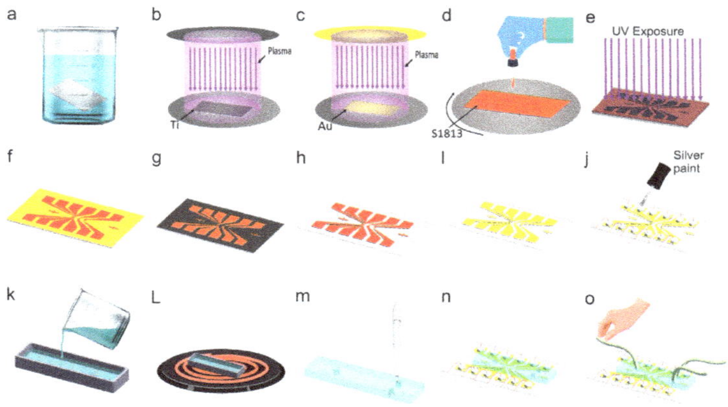

Figure 5. Fabrication Process: (**a**) Cleaning glass substrate in acetone and isopropyl alcohol (IPA). (**b**) DC plasma sputtering of titanium layer. (**c**) DC plasma sputtering of Au layer. (**d**) Spin coating S1813 positive photoresist. (**e**) Photolithography process for electrode patterning. (**f**) Developing excessive photoresist in the NaOH. (**g**) Etching the gold layer. (**h**) Etching the Ti layer. (**i**) Removing the photoresist using acetone. (**j**) Placing the electrical contact wires using the silver paint. (**k**) Pouring the PDMS in the 3D-printed mold for the microchannel part. (**L**) Curing the PDMS on the hotplate for 60 min at 80 °C. (**m**) Punching the inlet and outlet ports. (**n**) Plasma bonding the PDMS-based microchannel part to the substrate containing electrodes. (**o**) Placing inlet and outlet hoses.

Figure 6 shows the integrated microfluidic chip.

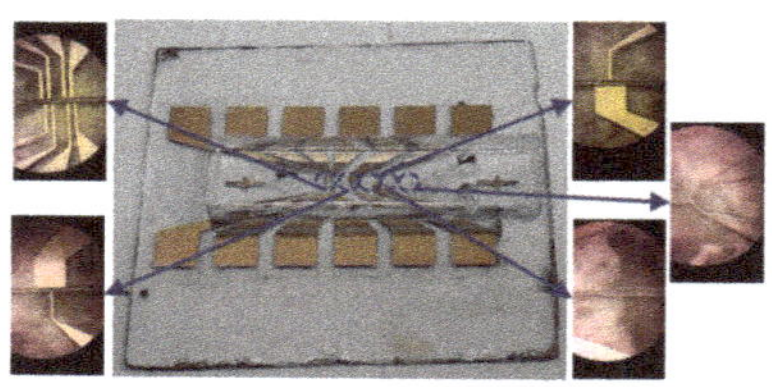

Figure 6. Fabricated chip structure.

2.4. Sample Preparation

Since we wanted to develop a yeast solution with minimal electrical conductivity, we chose deionized water as the base solvent. Initially, we heated 50 milliliters of deionized water to 40 °C for 30 min, followed by the addition of 3.5 g of sugar, which we stirred to dissolve. We then added 1.5 g of dry yeast powder and stirred the solution. The resulting solution containing cells was diluted in a 1:4 ratio with the base sugar water solution. We subjected half of the original yeast solution to a temperature of 100 °C for 30 min to obtain a solution with dead yeast cells.

Using the electrochemical impedance spectroscopy (EIS) system, we determined the relative permittivity and conductivity of the base solution to be 243.6 and 78.64 uS/cm.

3. Results and Discussion

The results presented in Figure 7a–c exhibit the successful implementation of particle focusing using the NDEP phenomenon. After entering the channel and passing through the two pairs of concentrating electrodes, each consisting of four electrodes, cells are positioned in the center of the microchannel. By exploiting the NDEP phenomenon, particles tend to migrate toward regions with lower electric field intensity, experiencing oscillatory and sinusoidal motion along the channel's symmetry axis before ultimately settling at the microchannel centerline.

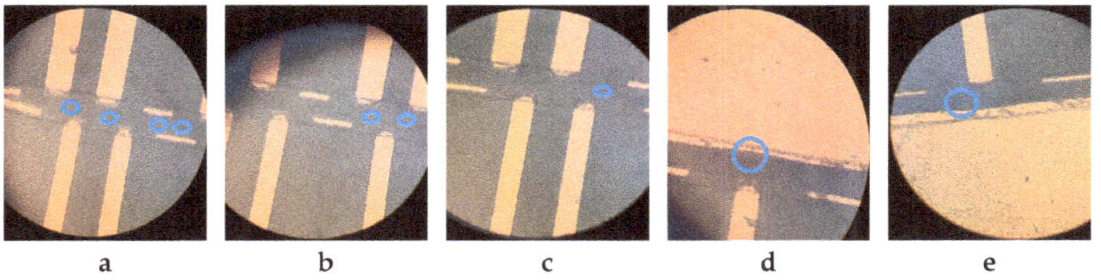

Figure 7. Particles' trajectories (cells are marked in blue circles) (**a**) near 1st focusing pair, (**b**) between 1st and 2nd focusing pair, (**c**) near 2nd focusing pair, (**d**) near 1st separating pair, and (**e**) near 2nd separating pair.

Figure 7d,e indicate the cells' trajectory near the first and second separating electrodes, respectively. As observed, particles move away from small electrodes where electric field lines accumulate and move towards the regions of lower electric field intensity.

4. Conclusions

We introduced an integrated microfluidic chip for separating bio-particles of similar size. Our innovative microelectrode design and microchannel architecture enable the precise separation of cells based on their distinct dielectric properties. Leveraging a single frequency in our microelectrode configuration, we achieved exceptional efficiency and accuracy in sorting bio-particles. Our microfluidic chip comprises two parts—a focuser and a separator. Our NDEP force-driven focuser actively aligns particles in the center of the channel without any additional pump or sheath flow. We employed an electric field of 26 volts at a 10 kHz frequency to accomplish this task. In the separator part, we utilized the NDEP phenomenon to separate particles through varying deviations in the electric field, using the electric field conditions as the focuser part. Through a step increase in channel width, we maximized the gap between the two target particles to improve the effectiveness of the separation process. Our microdevice is easy to fabricate, requires low voltage, and is compatible with many types of cells, making it an attractive option for research and clinical applications. Our novel chip design has the potential to drastically change existing cell separation methods, opening the door for further advancements in various fields of life sciences and biomedical research.

Author Contributions: Conceptualization, Z.A.; methodology, Z.A. and R.V.; software, R.V.; validation, R.V., Z.A., and M.F.; investigation, Z.A. and R.V.; resources, M.F.; data curation, R.V. and Z.A.; writing—original draft preparation, R.V. and Z.A.; writing—review and editing, R.V., Z.A., and M.F.; visualization, R.V., Z.A., and M.F.; supervision, M.F.; project administration, Z.A. All authors have read and agreed to the published version of the manuscript.

Funding: This research received no external funding.

Institutional Review Board Statement: Not applicable.

Informed Consent Statement: Not applicable.

Data Availability Statement: Data are available in this manuscript

Conflicts of Interest: The authors declare no conflict of interest.

References

1. Whitesides, G.M. The origins and the future of microfluidics. *Nature* **2006**, *442*, 368–373. [CrossRef] [PubMed]
2. Bahnemann, J.; Grünberger, A. Microfluidics in Biotechnology: Overview and Status Quo. *Adv. Biochem. Eng. Biotechnol.* **2022**, *179*, 1–16. [CrossRef]
3. Liu, Y.; Sun, L.; Zhang, H.; Shang, L.; Zhao, Y. Microfluidics for Drug Development: From Synthesis to Evaluation. *Chem. Rev.* **2021**, *121*, 7468–7529. [CrossRef]
4. Pandey, C.M.; Augustine, S.; Kumar, S.; Kumar, S.; Nara, S.; Srivastava, S.; Malhotra, B.D. Microfluidics Based Point-of-Care Diagnostics. *Biotechnol. J.* **2018**, *13*, 1700047. [CrossRef] [PubMed]

5. Kulkarni, M.B.; Ayachit, N.H.; Aminabhavi, T.M. Biosensors and Microfluidic Biosensors: From Fabrication to Application. *Biosensors* **2022**, *12*, 543. [CrossRef] [PubMed]
6. Pol, R.; Céspedes, F.; Gabriel, D.; Baeza, M. Microfluidic lab-on-a-chip platforms for environmental monitoring. *TrAC Trends Anal. Chem.* **2017**, *95*, 62–68. [CrossRef]
7. Mu, R.; Bu, N.; Pang, J.; Wang, L.; Zhang, Y. Recent Trends of Microfluidics in Food Science and Technology: Fabrications and Applications. *Foods* **2022**, *11*, 3727. [CrossRef] [PubMed]
8. Christodoulou, M.I.; Zaravinos, A. Single-Cell Analysis in Immuno-Oncology. *Int. J. Mol. Sci.* **2023**, *24*, 8422. [CrossRef] [PubMed]
9. Hu, P.; Zhang, W.; Xin, H.; Deng, G. Single cell isolation and analysis. *Front. Cell Dev. Biol.* **2016**, *4*, 116. [CrossRef] [PubMed]
10. Shiri, F.; Feng, H.; Gale, B.K. *Passive and Active Microfluidic Separation Methods*; Elsevier Inc.: Amsterdam, The Netherlands, 2022.
11. Sherbaz, A.; Konak, B.M.K.; Pezeshkpour, P.; Di Ventura, B.; Rapp, B.E. Deterministic Lateral Displacement Microfluidic Chip for Minicell Purification. *Micromachines* **2022**, *13*, 365. [CrossRef] [PubMed]
12. Zhao, Q.; Yuan, D.; Zhang, J.; Li, W. A review of secondary flow in inertial microfluidics. *Micromachines* **2020**, *11*, 461. [CrossRef] [PubMed]
13. Drescher, H.; Weiskirchen, S.; Weiskirchen, R. Flow cytometry: A blessing and a curse. *Biomedicines* **2021**, *9*, 1613. [CrossRef] [PubMed]
14. Yang, A.H.J.; Soh, H.T. Acoustophoretic sorting of viable mammalian cells in a microfluidic device. *Anal. Chem.* **2012**, *84*, 10756–10762. [CrossRef] [PubMed]
15. Poudineh, M.; Sargent, E.H.; Pantel, K.; Kelley, S.O. Profiling circulating tumour cells and other biomarkers of invasive cancers. *Nat. Biomed. Eng.* **2018**, *2*, 72–84. [CrossRef] [PubMed]
16. Valijam, S.; Salehi, A.; Andersson, M. Design of a low-voltage dielectrophoresis lab-on-the chip to separate tumor and blood cells. *Microfluid. Nanofluidics* **2023**, *27*, 22. [CrossRef]
17. Zhang, H.; Chang, H.; Neuzil, P. DEP-on-a-chip: Dielectrophoresis applied to microfluidic platforms. *Micromachines* **2019**, *10*, 423. [CrossRef] [PubMed]
18. Pesch, G.R.; Du, F. A review of dielectrophoretic separation and classification of non-biological particles. *Electrophoresis* **2021**, *42*, 134–152. [CrossRef] [PubMed]
19. Gascoyne, P.R.C.; Vykoukal, J. Particle separation by dielectrophoresis. *Electrophoresis* **2002**, *23*, 1973–1983. [CrossRef] [PubMed]
20. Patel, S.; Showers, D.; Vedantam, P.; Tzeng, T.R.; Qian, S.; Xuan, X. Microfluidic separation of live and dead yeast cells using reservoir-based dielectrophoresis. *Biomicrofluidics* **2012**, *6*, 1–12. [CrossRef] [PubMed]

engineering proceedings

MDPI

Proceeding Paper

Investigation of Rectifier Responses Affecting Operational Bandwidth in an Electromagnetic Vibration Energy Harvester †

Rui Zhong [1,2], Xueying Jin [3], Beichen Duan [3] and Chung Ket Thein [1,2,*]

1 Department of Mechanical, Materials and Manufacturing Engineering, University of Nottingham Ningbo China, Ningbo 315100, China; rui.zhong@nottingham.edu.cn
2 Nottingham Ningbo China Beacons of Excellence Research and Innovation Institute, Ningbo 315100, China
3 Department of Electrical and Electronic Engineering, University of Nottingham Ningbo China, Ningbo 315100, China; ssyxj3@nottingham.edu.cn (X.J.); ssybd1@nottingham.edu.cn (B.D.)
* Correspondence: chungket.thein@nottingham.edu.cn

† Presented at the 10th International Electronic Conference on Sensors and Applications (ECSA-10), 15–30 November 2023; Available online: https://ecsa-10.sciforum.net/.

Abstract: Energy harvesters provide excellent solutions for the power supply problem of wireless sensor nodes (WSNs), and energy harvesters with a wider bandwidth will clearly better serve WSNs and assist in the construction of Industry 4.0. However, the bearing of rectifiers on the load bandwidth of energy harvesters has rarely been investigated. This paper focuses on the impact of diverse rectifiers on the load electrical response of an electromagnetic energy harvester in the sweep mode of experiments, especially on the load bandwidth. The rectifiers were set as a half-wave rectifier and a full-bridge rectifier, respectively, and two different rectifier diodes were adopted in the experiment. The experimental results suggest that the half-wave rectifier exhibited certain advantages especially in the bandwidth field. If a full-bridge rectifier using high-speed switching diodes is replaced with a half-wave rectifier using Schottky diodes under the load resistance of 100 Ω, the load bandwidth will increase by almost 1.9 times. A preliminary analysis of the experimental results is provided at length.

Keywords: rectifiers; diode; vibration energy harvesting; wireless sensor nodes

Citation: Zhong, R.; Jin, X.; Duan, B.; Thein, C.K. Investigation of Rectifier Responses Affecting Operational Bandwidth in an Electromagnetic Vibration Energy Harvester. *Eng. Proc.* **2023**, *58*, 8. https://doi.org/10.3390/ecsa-10-16016

Academic Editor: Stefano Mariani

Published: 15 November 2023

1. Introduction

Wireless sensor nodes (WSNs) have received widespread attention as emerging miniaturized intelligent devices due to their ability to monitor the health of industrial equipment structures [1]. One of the important factors limiting the large-scale application of WSNs is power supply, and vibration energy harvesters (VEHs) provide suitable solutions to avoid the inconvenience caused by battery applications. Among various VEHs, electromagnetic vibration energy harvesters (EMVEHs) that can collect energy generated by the human body or the vibration of destination equipment are eye-catching, and have the advantage of a simple structure that facilitates large-scale manufacturing [2–4]. Meanwhile the rectifier, a crucial module, plays a pivotal role in the energy harvesting system. It can convert the mechanical energy generated by vibration into AC power, and further convert AC power into DC power that can be used to power the load through a rectifier [5–8]. It is self-evident that as a major module of the energy harvester system, the rectifier will inevitably have a significant impact on the electrical parameters of the load, which is worthy of in-depth investigation and analysis.

The half-wave rectifier composed of a single diode and the full-bridge rectifier composed of four diodes (H-bridge) are the two most frequently adopted rectifiers in the field of power electronics. Based on these two rectifiers, a wide variety of advanced rectifiers have been derived for energy harvesters. However, for low-power harvesters, the complex structure of advanced rectifiers means that the proportion of energy loss in the circuit will be significant, and the sophisticated control strategy will inevitably lead to an increase in cost. Therefore, advanced rectifiers will not be discussed here. Half-wave rectifiers are usually at a disadvantage in competition with full-bridge rectifiers due to their inability to harvest half of the AC energy. The existing energy harvester circuits involving rectifiers often take the power of the load and the electromechanical conversion efficiency of the system as evaluation points. Research results have shown that half-wave rectifiers can also occupy a place in low-power energy harvesting systems, but the evaluation of bandwidth as a parameter has not been mentioned [9]. A wider bandwidth signifies that the harvester will be able to deliver more electrical energy to the load within a larger applied frequency range. Therefore, this paper compares the responses of half-wave rectifiers and full-bridge rectifiers applied to EMVEH from multiple perspectives, especially bandwidth, through experiments.

2. Methodology

2.1. Governing Equations

Here, WSNs can be simplified as a pure load resistor without parasitic parameters, and the electrical parameters of the load resistor are of concern. The primary link is to establish a circuit model of the harvester containing a half-wave rectifier and a circuit model of the harvester with a full-bridge rectifier, as shown in Figure 1.

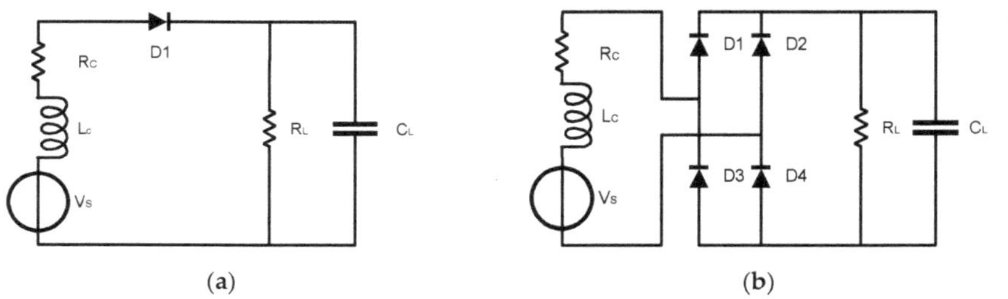

Figure 1. (**a**) Circuit model of the EMVEH containing a half-wave rectifier; (**b**) Circuit model of the EMVEH containing a full-bridge rectifier.

In the circuit model, R_c and L_c represent coil resistance and coil inductance, respectively, while D_1 to D_4 represent rectifier diodes. R_L is the load resistor, and in parallel with it is the filter capacitor C_L. As a sinusoidal voltage source, V_s is used to characterize the electrical output parameters of EMVEH, attributed to the conversion relationship between mechanical and electrical domains. Here, the performance of the load in the frequency domain is investigated, so the expression for V_s can be given as:

$$V_s = 4Nbl_c^2 c_f x\omega \tag{1}$$

where N is the turn number of the coil, b is the average magnetic flux density, l_c is the standard diameter of the coil, c_f is the coil filling coefficient, x represents the coil displacement, and ω represents the applied frequency [4,10]. By listing a set of node voltage equations, the voltage value of the load resistor in the frequency domain can be obtained. It should be noted that diode voltage drops, typically considered as a constant, needs to be considered here as a series combination of an equivalent resistor R_d and a voltage source V_d. Therefore, the voltage of the load resistor in half-wave circuit can be expressed as:

$$V_{half}(\omega) = \frac{R_L(V_s - V_d)}{(1 + j\omega R_c C_L)(R_c + j\omega L_c + R_d) + R_L} \tag{2}$$

where $j^2 = 1$. For a full-bridge rectifier circuit, the difference between it and a half-wave rectifier circuit in the frequency domain is that there are two diodes running in each half cycle. Therefore, the expression for the voltage of the load resistor in a full-bridge rectifier circuit is:

$$V_{full}(\omega) = \frac{R_L(V_s - 2V_d)}{(1 + j\omega R_c C_L)(R_c + j\omega L_c + 2R_d) + R_L} \quad (3)$$

2.2. Experimental Setup

A single-degree-of-freedom cantilever beam EMVEH was used for experiments, as shown in Figure 2a. The free end of the cantilever beam has a coil mass, while the other end is fixed on a bracket. When the harvester is forced to undergo harmonic vibration, the coil will cut the magnetic induction line regularly, generating sinusoidal AC power under the domination of Faraday's electromagnetic induction law. Figure 2b reveals the overall experimental system setup, including the MS-50 shaker, DR08 resistor box, RX7/N37 precision capacitor box, and NR HA-08 data acquisition (DAQ) card.

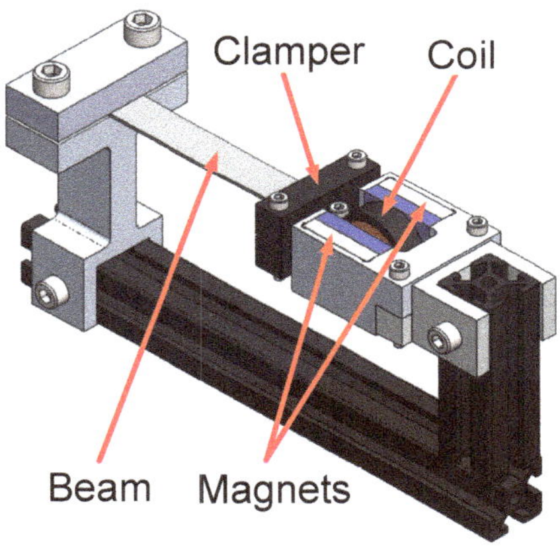

(**a**)

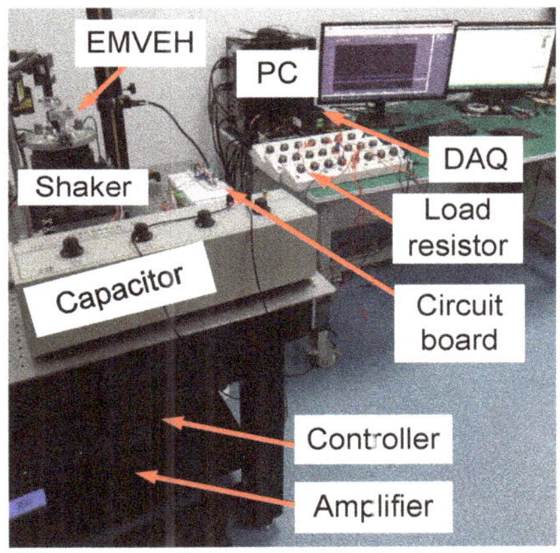

(**b**)

Figure 2. (**a**) SDOF electromagnetic energy harvester; (**b**) Overall experimental system setup.

By inputting the correct control instructions on the PC, the shaker can be given a preset acceleration with the assistance of the controller and amplifier, while the probe from the DAQ card is connected to both ends of the load resistor to obtain voltage data. Some key parameters related to the experiments can be seen in Table 1.

Table 1. Key parameter settings related to the experiment.

Descriptions	Parameters
Acceleration	0.1 g
Beam length	77 mm
Beam width	12.7 mm
Beam thickness	1 mm
Coil turns	570
Coil resistance	7.6 Ω
Coil inductance	4.7 mH
Filter capacitor	470 μF

The examination of the bandwidth and maximum power of the load resistor led to the adoption of the sweep mode (18–21 Hz) in this experiment, with the resistor values ranging from 100 to 800 Ω in steps of 100 Ω. The generally used high speed switching diode 1N4148 and Schottky diode MBR745 are both considered in experiments to identify the impact of different rectifier diode models on the responses.

3. Results and Analysis

For basic half-wave circuits and full-bridge circuits, it is difficult to avoid the occurrence of ripples in the load–voltage curves at relatively low frequencies, which to some extent may cause interference in the investigation of the parameters of concern. Therefore, it is imperative to filter the voltage–frequency curve, which means it is necessary to convert the generated sawtooth waves into relatively smooth waves for comparison and analysis. Figure 3 demonstrates the voltage–frequency curves under multiple load resistors, rectifiers, and types of rectifier diodes.

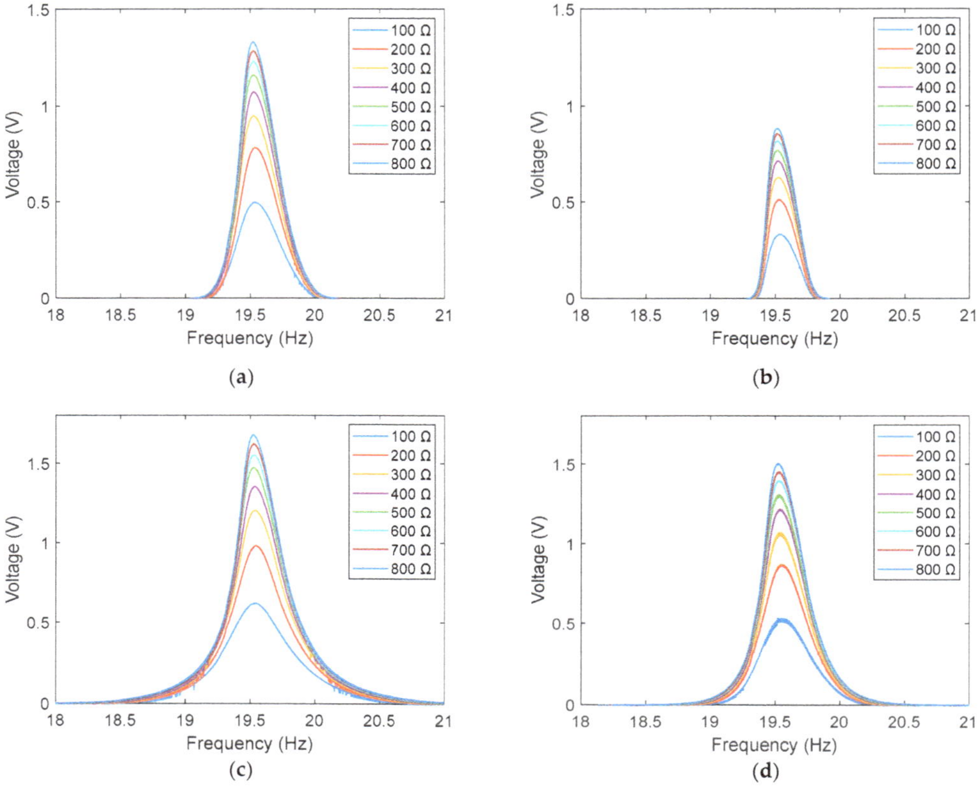

Figure 3. Performances of load voltage versus frequency under different load resistors: (**a**) Half-wave circuit using 1N4148 diodes; (**b**) Full-bridge circuit using 1N4148 diodes; (**c**) Half-wave circuit using MBR745 diodes; (**d**) Full-bridge circuit using MBR745 diodes.

From Figure 3, it can be preliminarily observed that diverse load resistor values, circuit structures, and rectifier diode models have no effect on the resonant frequency of these load resistors, which undoubtedly paves the way for obtaining the maximum voltage point in these load voltage–frequency curves. Once the load voltage and load resistance values are known, the power of the load can be obtained according to methods in circuit theory. Moreover, the bandwidth of each curve can be derived from the maximum voltage point. Table 2 declares the maximum load power value and bandwidth of each curve obtained from the experiment.

Table 2. Peak power and bandwidth performance of load resistors under distinct circuits in the experiment.

Load Resistance (Ω)	1N4148				MBR745			
	Have-Wave		Full-Bridge		Have-Wave		Full-Bridge	
	Peak Power (mW)	Bandwidth (Hz)	Peak Power (mW)	Bandwidth (Hz)	Peak Power (mW)	Bandwidth (Hz)	Peak Power (mW)	Bandwidth (Hz)
100	2.48	**0.27**	1.11	0.19	3.88	**0.36**	2.86	0.30
200	3.08	**0.25**	1.31	0.18	4.83	**0.31**	3.78	0.27
300	3.00	**0.24**	1.32	0.19	4.85	**0.29**	3.81	0.27
400	2.88	**0.24**	1.28	0.18	4.62	**0.28**	3.72	0.26
500	2.70	**0.24**	1.18	0.19	4.35	**0.28**	3.44	0.25
600	2.53	**0.23**	1.12	0.18	4.02	**0.27**	3.26	0.25
700	2.36	**0.23**	1.05	0.18	3.76	**0.27**	3.01	0.25
800	2.21	**0.23**	0.98	0.18	3.52	**0.27**	2.83	0.25

The experimental results indicate that the curves plotted from the experimental data exhibit some similarities. In each case, as the value of the load resistor increases, the load voltage reveals an upward trend, but the growth rate gradually decreases, resulting in a trend of first increasing and then decreasing in the variation of load power. It is worth noting that these changing trends are not only applicable to circuits containing half-wave rectifiers, but also to circuits with full-bridge rectifiers, and are independent of the type of rectifier diode adopted. Simultaneously, the increment in the value of the load resistor results in a slight decrease in the bandwidth of the load voltage–frequency curve in each case, although this shift is slightly relative to the change in the load voltage value.

Certainly, diverse voltage–frequency curves also exhibit noticeable differences. Under each set of load resistor values, if the ripple coefficient is not evaluated, the support of the half-wave rectifier will enable the load resistor to exhibit greater advantages, that is, compared to the load in a full-bridge circuit under the same conditions, the load served by the half-wave rectifier can harvest more energy, occupy larger peak power and wider bandwidth. Here, in order to better investigate the impact of rectifiers on load bandwidth, the bandwidth ratio can be defined as the ratio of load bandwidth in a half-wave rectifier circuit to that in a full-bridge rectifier circuit. Take the case where the load resistance value is 100 Ω. If the rectifier diodes in the harvester circuit are Schottky diodes, the bandwidth ratio is 1.2. If the model of the rectifier diode is set to 1N4148, the half-wave rectifier brings visible advantages to the load bandwidth, with a bandwidth ratio of 1.42. Obviously, regardless of the rectifier diode type, the half-wave rectifier provides better service for the load in the bandwidth field. Reviewing the experimental data, it can be observed that as the load resistance increases, the bandwidth ratio shows a falling tendency. Compared to high-speed switching diodes, Schottky diodes should be selected and placed in energy harvester circuits. The presence of Schottky diodes will broaden the bandwidth of the voltage frequency curve under other conditions that are consistent. Still considering the case where the load resistance value is 100 Ω, if a half-wave rectifier is used in the circuit, the operation of the Schottky diodes will result in a curve bandwidth of 1.33 times that of 1N4148 diodes when they are adopted, and this multiple in the full-bridge rectifier circuit reaches 1.58. If a half wave rectifier using MBR745 diodes is replaced by a full bridge rectifier using 1N4148 diodes, the load bandwidth will be dramatically increased by 1.9 times.

For EMVEHs with a tiny harvesting power, the diode loss is significant, so the fewer diodes in half-wave rectifiers has become a key factor in this load bandwidth evaluation competition that outperforms full-bridge rectifiers. Meanwhile, the superiorities brought by Schottky diodes for load bandwidth transmit new evidence for the necessity of their promotion. Nevertheless, in addition to peak power and bandwidth, the energy harvesting interface circuit for WSNs should also consider other factors such as ripple coefficient and power factor. Therefore, balancing the pros and cons among multiple electrical parameters is a challenge that harvester circuit designers must face.

4. Conclusions

The investigation of rectifiers responses based on electromagnetic energy harvesters reveals the advantages that half-wave rectifiers can bring to the load in term of bandwidth compared to full-bridge rectifiers, which are perspectives that few researchers have previously focused on. The merit of certain harvesters containing half-wave rectifiers in providing power to WSNs needs to be made a profound study. Schottky diodes are prominent electronic components as rectifier diodes for low-power harvesters, otherwise peak power and bandwidth will be unnecessarily reduced.

Author Contributions: Conceptualization, C.K.T.; methodology, R.Z. and C.K.T.; software, R.Z.; validation, R.Z. and X.J.; formal analysis, R.Z.; investigation, X.J., B.D. and R.Z.; resources, C.K.T.; writing— original draft preparation, R.Z.; writing—review and editing, C.K.T.; visualization, C.K.T.; supervision, C.K.T.; project administration, C.K.T. All authors have read and agreed to the published version of the manuscript.

Funding: The authors wish to thank the Li Dak Sum Innovation Fellowship 2023 from Educational Foundation of the University of Nottingham Ningbo China for their funding the project under project No. E06221200002.

Institutional Review Board Statement: Not applicable.

Informed Consent Statement: Not applicable.

Data Availability Statement: The data presented in this study are available upon request from the corresponding author.

Conflicts of Interest: The authors declare no conflict of interest.

References

1. Dziadak, B.; Kucharek, M.; Starzyński, J. Powering the WSN Node for Monitoring Rail Car Parameters, Using a Piezoelectric Energy Harvester. *Energies* **2022**, *15*, 1641. [CrossRef]
2. Wang, S.W.; Ke, Y.W.; Huang, P.C.; Hsieh, P.H. Electromagnetic Energy Harvester Interface Design for Wearable Applications. *IEEE Trans. Circuits Syst. II Express Briefs* **2018**, *65*, 667–671. [CrossRef]
3. Glynne-Jones, P.; Tudor, M.J.; Beeby, S.P.; White, N.M. An Electromagnetic, Vibration-Powered Generator for Intelligent Sensor Systems. *Sensors Actuators A Phys.* **2004**, *110*, 344–349. [CrossRef]
4. Toluwaloju, T.I.; Thein, C.K.; Halim, D.; Yang, J. Dynamic Responses of the 2DOF Electromagnetic Vibration Energy Harvester through Different Electrical Coil Connections. *Mech. Syst. Signal Process* **2023**, *184*, 109709. [CrossRef]
5. Moon, J.; Leeb, S.B. Power Electronic Circuits for Magnetic Energy Harvesters. *IEEE Trans. Power Electron.* **2015**, *31*, 270–279. [CrossRef]
6. Wang, X.; Liang, X.; Wei, H. A Study of Electromagnetic Vibration Energy Harvesters with Different Interface Circuits. *Mech. Syst. Signal Process.* **2015**, *58*, 376–398. [CrossRef]
7. Mitcheson, P.D.; Toh, T.T.; Wong, K.H.; Burrow, S.G.; Holmes, A.S. Tuning the Resonant Frequency and Damping of an Electromagnetic Energy Harvester Using Power Electronics. *IEEE Trans. Circuits Syst. II Express Briefs* **2011**, *58*, 792–796. [CrossRef]
8. Wang, X.; Liang, X.; Hao, Z.; Du, H.; Zhang, N.; Qian, M. Comparison of Electromagnetic and Piezoelectric Vibration Energy Harvesters with Different Interface Circuits. *Mech. Syst. Signal Process.* **2016**, *72–73*, 906–924. [CrossRef]
9. Clare, L.R.; Burrow, S.G. Half-Wave Rectifiers Offer Advantages for Vibration Energy Harvesters. *Electron. Lett.* **2010**, *46*, 1623–1624. [CrossRef]
10. Zhong, R.; Thein, C.K.; Halim, D.; Xu, J. Effect of Multiple Connection Modes of Capacitors on the Response of Vibration Energy Harvester. In Proceedings of the 2023 3rd International Conference on Electrical, Computer, Communications and Mechatronics Engineering (ICECCME), Canary Islands, Spain, 19–21 July 2023; IEEE: Piscataway, NJ, USA, 2023; pp. 1–6.

engineering proceedings

Proceeding Paper

The IPANEMA Project: Underwater Acoustic Structure for Volcanic Activity and Natural CO_2 Emissions Monitoring †

Letizia S. Di Mauro [1,*], Dídac Diego-Tortosa [1], Giorgio Riccobene [1], Carmelo D'Amato [1], Emanuele Leonora [2], Fabio Longhitano [2], Angelo Orlando [1] and Salvatore Viola [1]

1 Istituto Nazionale di Fisica Nucleare (INFN), Laboratori Nazionali del Sud (LNS), Via S. Sofia 62, 95123 Catania, Italy; didac.diegotortosa@lns.infn.it (D.D.-T.); giorgio.riccobene@lns.infn.it (G.R.); damatoc@lns.infn.it (C.D.); aorlando@lns.infn.it (A.O.); sviola@lns.infn.it (S.V.)

2 Istituto Nazionale di Fisica Nucleare (INFN), Sezione di Catania, Via Santa Sofia 64, 95123 Catania, Italy; emanuele.leonora@ct.infn.it (E.L.); fabio.longhitano@ct.infn.it (F.L.)

* Correspondence: letizia.dimauro@lns.infn.it

† Presented at the 10th International Electronic Conference on Sensors and Applications (ECSA-10), 15–30 November 2023; Available online: https://ecsa-10.sciforum.net/.

Abstract: Carbon dioxide produced by human activities (i.e., use of fossil fuels, deforestation, and livestock farming) is the main greenhouse gas causing global warming. In 2020, the concentration in the atmosphere exceeded the pre-industrial level by 48% (before 1750). The study of natural CO_2 (carbon dioxide) emissions due to volcanic activity through innovative measurement techniques is the main goal of the IPANEMA project. These studies are both essential for the evaluation of natural CO_2 emissions and for the development of future carbon capture and storage in underground geological formations to ensure that there are no leaks from the storage sites. Through the installation of two underwater acoustic stations, i.e., one in Panarea and one in the Gulf of Catania, we want to investigate techniques for estimating the flux of CO_2 emitted by natural sources, locating emission sources, and, in general, monitoring volcanic activity.

Keywords: underwater acoustics; monitoring; carbon dioxide; volcanic activity; Panarea; Catania

Citation: Di Mauro, L.S.; Diego-Tortosa, D.; Riccobene, G.; D'Amato, C.; Leonora, E.; Longhitano, F.; Orlando, A.; Viola, S. The IPANEMA Project: Underwater Acoustic Structure for Volcanic Activity and Natural CO_2 Emissions Monitoring. *Eng. Proc.* **2023**, *58*, 9. https://doi.org/10.3390/ecsa-10-16169

Academic Editor: Jean-marc Laheurte

Published: 15 November 2023

1. Introduction

The European Carbon Dioxide Capture and Storage Laboratory (ECCSEL-ERIC) is a distributed pan-European research infrastructure that connects the main existing laboratories in Europe working on CO_2 capture and storage (CCS). The IPANEMA project provides for the implementation of the ECCSEL NatLab-Italy laboratory and involves OGS (National Institute of Oceanography and Experimental Geophysics), INGV (National Institute of Geophysics and Volcanology), and INFN (National Institute for Nuclear Physics). The goal is the creation of a network of tools for monitoring and studying natural CO_2 emissions, mainly due to the volcanic activity present in the Mediterranean area, with particular attention to the Aeolian Islands and the Ionian Sea.

The IPANEMA project is conducted in two different geographical areas of Sicily (see Figure 1). On the one hand, an autonomous station will be installed on Panarea Island, and on the other hand, another seafloor observatory will be connected to the onshore in the Gulf of Catania.

(a) (b) (c)

Figure 1. (**a**) The two study sites of the IPANEMA project: the location for the Panarea observatory in the Aeolian Islands (North-East Sicily) and the location for the observatory in the Gulf of Catania, at the end of the Test Site cable (East Sicily). (**b**) Active tectonic faults that form an extensive CO_2-emitting fumarolic field in Panarea. (**c**) Deep underwater multidisciplinary laboratory since 2005 in the Gulf of Catania.

1.1. Panarea Island Observatory

The first study area, in the shallow waters of Panarea (24 m depth), is part of a large quiescent submarine strato-volcano, characterized by active tectonic faults that form an extensive CO_2-emitting fumarolic field. Fumaroles are underwater openings on the seabed that allow the escape of high-temperature gases. They are typical of volcanic and geothermal areas and represent an important source of information on the volcanic activity of the area to which they are closely related. This site has therefore already been chosen by OGS to investigate the effects of an increase in carbon dioxide on the biogeochemical and ecological functions of the marine ecosystem. The study of natural CO_2 emissions in Panarea represents an optimal "test bench" for analyzing CO_2 storage monitoring and control methods. Among several proposed techniques to reduce CO_2 concentration in the atmosphere, there is carbon capture and storage (CCS), which allows for the removal of CO_2 from the atmosphere and its permanent storage under the Earth's surface. Monitoring of CO_2 storage is a fundamental phase that guarantees the safety and sustainability of this methodology and consists of checking that there are no leaks from the storage site. In this context, underwater acoustics represent a non-invasive and efficient method to monitor CO_2 emissions in the marine environment from underground sources [1–6]. In the framework of the IPANEMA project, an autonomous observatory hosting a four-hydrophone acoustic array will be installed off Panarea coasts, aiming at the development of acoustic techniques for bubble localization and the measurement of the CO_2 flux.

1.2. Gulf of Catania Observatory

The second study area is located in the Gulf of Catania (25 km east off the coast at a depth of 2050 m), where a deep underwater multidisciplinary laboratory has been operative since 2005. The site serves the OnDE, SMO, SN 1, and FOCUS-ERC experiments [7–10]. The area is suitable for the study of numerous natural hazards due to its high seismicity and the presence of the Etna volcano, whose roots sink down to the seafloor. In this area, the possible presence of acoustic signals (tremors and cracks) related to the volcanic activity of Etna will be investigated.

2. The Underwater Stations

The Panarea shallow water site will be acoustically monitored by an array of four large-bandwidth hydrophones (model Ocean Sonics IcListen Smart Hydrophone) fed by a battery pack. In order to locate CO_2 emissions acoustically, the four hydrophones are time-synchronized. Time synchronization has been obtained by using the internal clock of one of the acoustic sensors as a master clock for the other sensors. Synchronization was tested and verified during tests on the dry bench. From considerations on the internal memory capacity of the hydrophones (497 GB), on the electrical consumption of the sensor, and on the energy capacity of the battery pack (408 Ah), a data-taking mode that provides 6-min recordings every hour was set. For the first measurement campaign, acoustic data are acquired with a sampling frequency of 128 kHz. This choice guarantees the acoustic monitoring of the study area for about 6 months. Given the shallow depth at which the Panarea station is located, it was deemed necessary to create a protective structure to safeguard the instrumentation from wave and atmospheric phenomena and from fishing activities. This structure, designed at the INFN-LNS, consists of a fixed part to be anchored to the seabed and a removable part that houses the sensors and battery pack and will be cyclically replaced at the end of each data collection period. The battery pack that powers the hydrophones is placed inside a bentosphere that has been subjected to a vacuum leak test. The hydrophones are fixed on supports whose respective distances are 1 m (see Figure 2).

(a) (b) (c) (d)

Figure 2. Panarea acoustic station during the mechanical assembly phase: (**a**) the commercial hydrophone forming the array, (**b**) the battery pack, (**c**) the IPANEMA station with the open deck, and (**d**) the IPANEMA station with the closed deck where the four hydrophones are shown.

On 9th May 2023, the installation of the Panarea acoustic station was completed in an area rich in submerged hydrothermal springs where the INGV has a monitoring site (see Figure 3). After six months, the removable part will be replaced by an identical one, while continuing to collect data without big lost chunks and being able to download the data taken. This will allow the start of the analysis of the months already recorded, checking the data, and testing the analysis algorithms to automate future deployments.

Figure 3. Panarea acoustic station during its first deployment: (**a**) a crane boat submerging the station in water; and (**b**) a submerged diver releasing the station from the crane.

The study area in the Gulf of Catania will be monitored by a cabled underwater acoustic array consisting of four hydrophones and oceanographic instrumentation for the measurement of the main seawater parameters, such as pressure, conductivity, temperature, depth, dissolved oxygen, salinity, and pH (see Figure 4).

Figure 4. The IPANEMA Catania station: (**a**) the mechanical frame (front view), (**b**) the mechanical frame (side view), and (**c**) the electronics container.

Hydrophones having a sensitivity variation as a function of the depth better than 1 dB (model SMID TR-401) were selected for the Catania deep sea study area. They are installed at the vertices of a tetrahedron with an edge of 1 m. All acoustic sensors and oceanographic instrumentation are Global Positioning System (GPS) time-synchronized. Data transmission takes place through optical fibers, and all data is streamed in real-time to shore, which will allow for their analysis at the same time. The deployment of the Catania acoustic station is foreseen in spring/summer 2024.

3. The Data Analysis

The effectiveness of passive acoustic monitoring techniques in the context of a controlled CO_2 gas release experiment was demonstrated in the QICS (Quantifying and Monitoring Potential Ecosystem Impacts of Geological Carbon Storage) project conducted off the west coast of Scotland [1]. Further studies conducted in the North Sea have demonstrated the possibility of determining the diameter of CO_2 bubbles acoustically and estimating their position using beamforming techniques [2,3]. Passive acoustic has also been applied to the monitoring of natural gas seeps [4–6].

Bibliographic research has revealed the presence of various scientific works dedicated to the study of the fumaroles of Panarea using acoustic methodologies. In [5], the variations in the flow of CO_2 and the presence of any anomalies were studied, and in [6], an approach to quantifying bubble size and gas flux was developed.

The primary goal of the research in Panarea is the size estimation of the bubbles starting from their acoustic spectrum, thus making it possible to determine the flow of gas emitted. Size estimation of the CO_2 bubbles can be carried out using the Minnaert formula [11], which correlates its fundamental peak frequency with its radius r and the environmental pressure P (see Equation (1)).

$$f = \frac{1}{2\pi r\sqrt{\frac{3P\gamma}{\rho}}} \tag{1}$$

where γ is the ratio of gas-specific heat at constant pressure and volume, and ρ is the water density.

This formula is only suitable for cases where the rate of bubble generation is low and it is possible to detect bubble signals automatically from ambient background noise. Instead, in the case of high gas flux conditions, sounds will appear as a continuous, stationary random process, and it will be necessary to apply the inversion method to the measured acoustic spectrum [2,6].

The gas flux estimate is derived from the Probability Density Function (PDF) of the bubble equilibrium radius, considering acoustic propagation to determine the acoustic range. Isolated CO_2 emissions will be tracked acoustically by using multi-lateration techniques based on the difference in the times of arrival of the sound to the sensors. Continuous emissions will be located through beamforming and cross-spectrum methods [3].

4. Conclusions and Next Steps

The IPANEMA project aims to study the natural emissions of carbon dioxide in the Mediterranean region due to volcanic activity. The project is part of ECCSEL-ERIC, which has a broader purpose of monitoring artificial CO_2 emissions to combat climate change and testing the feasibility of carbon capture and storage techniques. During this project, two underwater stations have been designed for underwater data recording: Panarea, an autonomous station replaceable every 6 months, and Catania, a station connected and fully controlled from onshore.

As a start to the installation phase, the Panarea station has already been deployed and is operational. It is planned to periodically replace the hydrophones and the battery pack (a removable pack from the base). After that, efforts will be made to retrieve the already collected data and analyze it with the aim of quantifying the emitted carbon dioxide flux and the location of the fumaroles.

Author Contributions: Conceptualization, S.V.; methodology, L.S.D.M., D.D.-T., G.R. and S.V.; software, L.S.D.M., D.D.-T. and S.V.; investigation, L.S.D.M., D.D.-T. and S.V.; resources, C.D., E.L. and F.L.; data curation, L.S.D.M., D.D.-T. and S.V.; writing—original draft preparation, L.S.D.M. and D.D.-T.; writing—review and editing, L.S.D.M., D.D.-T., G.R. and S.V.; visualization, L.S.D.M., D.D.-T. and S.V.; supervision, G.R. and S.V.; project administration, A.O. and S.V.; funding acquisition, A.O. and S.V. All authors have read and agreed to the published version of the manuscript.

Funding: This work has been carried out in the framework of the IPANEMA PROJECT PO-FESR Asse—Azione: Asse II—Azione Progetti Tematici II.1—Infrastrutture di ricercar PIR01_00018 and ECCSEL NatLab Italy.

Institutional Review Board Statement: Not applicable.

Informed Consent Statement: Informed consent was obtained from all subjects involved in the study.

Data Availability Statement: Data sharing is not applicable to this article.

Conflicts of Interest: The authors declare no conflict of interest.

References

1. Bergès, B.J.; Leighton, T.G.; White, P.R. Passive acoustic quantification of gas fluxes during controlled gas release experiments. *Int. J. Greenh. Gas Control.* **2015**, *38*, 64–79. [CrossRef]
2. Li, J.; White, P.R.; Roche, B.; Bull, J.M.; Leighton, T.G.; Davis, J.W.; Fone, J.W. Acoustic and optical determination of bubble size distributions—Quantification of seabed gas emissions. *Int. J. Greenh. Gas Control.* **2021**, *108*, 103313. [CrossRef]
3. Li, J.; White, P.R.; Bull, J.M.; Leighton, T.G.; Roche, B.; Davis, J.W. Passive acoustic localisation of undersea gas seeps using beamforming. *Int. J. Greenh. Gas Control.* **2021**, *108*, 103316. [CrossRef]
4. Leifer, I.; Tang, D. The acoustic signature of marine seep bubbles. *J. Acoust. Soc. Am.* **2006**, *121*, EL35–EL40. [CrossRef]
5. Longo, M.; Lazzaro, G.; Caruso, C.G.; Corbo, A.; Sciré Scappuzzo, S.; Italiano, F.; Gattuso, A.; Romano, D. Hydro-acoustic signals from the Panarea shallow hydrothermal field: New inferences of a direct link with Stromboli. *Geol. Soc.* **2021**, *519*, SP519-2020. [CrossRef]
6. Li, J.; Roche, B.; Bull, J.M.; White, P.R.; Davis, J.W.; Deponte, M.; Gordini, E.; Cotterle, D. Passive acoustic monitoring of a natural CO_2 seep site—Implications for carbon capture and storage. *Int. J. Greenh. Gas Control.* **2019**, *93*, 102899. [CrossRef]
7. Riccobene, G. Long-term measurements of acoustic background noise in very deep sea. *Nucl. Instruments Methods Phys. Res. Sect. A Accel. Spectrometers Detect. Assoc. Equip.* **2009**, *604*, S149–S157. [CrossRef]
8. Simeone, F.; Viola, S. The SMO Project: A Submarine Multidisciplinary Observatory in Deep-Sea. In Proceedings of the 2011 IEEE 8th International Conference on Mobile Ad-Hoc and Sensor Systems (MASS), València, Spain, 17–22 October 2011; pp. 898–903. [CrossRef]
9. Favali, P.; Chierici, F.; Marinaro, G.; Giovanetti, G.; Azzarone, A.; Beranzoli, L.; De Santis, A.; Embriaco, D.; Monna, S.; Lo Bue, N.; et al. NEMO-SN1 Abyssal Cabled Observatory in the Western Ionian Sea. *IEEE J. Ocean. Eng.* **2013**, *38*, 358–374. [CrossRef]
10. Gutscher, M.-A.; Royer, J.-Y.; Graindorge, D.; Murphy, S.; Klingelhoefer, F.; Aiken, C.; Cattaneo, A.; Barreca, G.; Quetel, L.; Riccobene, G.; et al. Fiber optic monitoring of active faults at the seafloor: I the FOCUS project. *Photoniques* **2019**, 32–37. [CrossRef]
11. Minnaert, M. XVI. On musical air-bubbles and the sounds of running water. *Lond. Edinb. Dublin Philos. Mag. J. Sci.* **1933**, *16*, 235–248. [CrossRef]

engineering proceedings

MDPI

Proceeding Paper

Golomb–Rice Coder-Based Hybrid Electrocardiogram Compression System †

Sachin Himalyan [1,*] and Vrinda Gupta [2]

1 School of VLSI Design and Embedded System, National Institute of Technology Kurukshetra, Kurukshetra 136119, Haryana, India

2 Department of ECE, National Institute of Technology Kurukshetra, Kurukshetra 136119, Haryana, India; vrindag16@gmail.com

* Correspondence: himalyan22@gmail.com

† Presented at the 10th International Electronic Conference on Sensors and Applications (ECSA-10), 15–30 November 2023; Available online: https://ecsa-10.sciforum.net/.

Abstract: Heart-related ailments have become a significant cause of death around the globe in recent years. Due to lifestyle changes, people of almost all age brackets face these issues. Preventing and treating heart-related issues require the electrocardiogram (ECG) monitoring of patients. The study of patients' ECG signals helps doctors identify abnormal heart rhythm patterns by which screening problems like arrhythmia (irregular heart rhythm), myocardial infarction (heart attacks), and myocarditis (heart inflammation) is possible. The need for 24 h heart rate monitoring has led to the development of wearable devices, and the constant monitoring of ECG data leads to the generation of a large amount of data since wearable systems are resource-constrained regarding energy, memory, size, and computing capabilities. The optimization of biomedical monitoring systems is required to increase their efficiency. This paper presents an ECG compression system to reduce the amount of data generated, which reduces the energy consumption in the transceiver, which is a significant part of the overall energy consumed. The proposed system uses hybrid Golomb–Rice coding for data compression, a lossless data compression technique. The data compression is performed on the MIT BIH arrhythmia database; the achieved compression ratio of the compression system is 2.75 and 3.14 for average and maximum values, which, compared to the raw ECG samples, requires less transmission cost in terms of power consumed.

Keywords: ECG compression; power management; data compression; MIT BIH arrhythmia; Golomb–Rice encoder

Citation: Himalyan, S.; Gupta, V. Golomb–Rice Coder-Based Hybrid Electrocardiogram Compression System. *Eng. Proc.* **2023**, *58*, 10. https://doi.org/10.3390/ecsa-10-16209

Academic Editor: Stefano Mariani

Published: 15 November 2023

1. Introduction

Advancements in sensory systems, VLSI technology, and wireless sensor networks (WSN) have opened up new avenues of technological applications. Wearable technology has emerged as a promising market, a combination of various technologies catering to multiple applications. With the healthcare landscape increasingly embracing personalized medicine, the global wearable sensor market is projected to experience a robust compound annual growth rate (CAGR) of around 38% between 2017 and 2025. Notably, the development of smartwatches is anticipated to exhibit an exceedingly rapid rate of expansion during this period [1]. Any wearable technology has a standard building block, e.g., sensors, processors, and communication units. These technologies rely on a basic unit, i.e., "data". Every wearable technology aims to collect, process, and communicate acquired data from the sensors [2]. Some primary design constraints every wearable technology aspires to overcome are size, memory management, power management, latency, and computational efficiency. Out of these metrics, power management is the most sought-after area in which optimizations are performed, and this is because wearable technologies have a limited size, resulting in fewer batteries [3]. The communication system consumes most of the energy from the various subsystems discussed. The prime reason is the limited computational capacity of these systems; hence, the acquired data need to be transmitted to a central system, resulting in energy consumption. The extent of transmission is directly proportional to the amount of data being sent, which, for a physiological monitoring system, is very large when constant information about body vitals is needed.

2. Literature Survey

Data compression techniques aim to reduce the extent of generated data to minimize time and power consumption due to transmission and memory. Various data compression schemes have been devised. Major bifurcation among these techniques is based on data retrieval after compression, which constitutes lossless data compression and lossy data compression methods. All compressed data can be retrieved with lossless methods, but these methods result in smaller compression ratios (CRs), which is defined as the ratio of original ECG data to compressed data. On the other hand, complete data retrieval is not possible with lossy methods, but when compared with lossless methods, these can produce a greater CR. This category's most commonly used schemes are transforming coding, vector quantization, and fractal compression. The selection of data compression methods is application-dependent. Lossy methods are generally used in applications where a specific amount of loss in data does not affect the performance of the systems, e.g., audio compression, video compression, gaming, and multimedia streaming. In comparison, lossless methods are used in data-critical applications like databases, scientific data compression, biomedical data, and communication systems. Hybrid methods use predictive and run-length encoding to balance the CR and quality [4]. Various hardware implementations for ECG data compression have been developed, aiming at low-power applications. Y. Zou et al. proposed a hardware model for ECG acquisition based on the wavelet transform; this implementation uses a high frequency of operation and is a lossy method [5]. As a result, this method is not viable for wearable sensor systems. C.J. Deepu et al. used a prediction-based hybrid algorithm for data compression [6]. F. Nasimi et al., Lin Y et al., and Chen Y. et al. implemented ECG hardware using lossless methods producing high CR values, but these methods require a large number of complex computations for data retrieval, which degrades the energy efficiency of the system [7,8]. Tsung-Han Tsai et al. developed a low-power data compression system for multichannel data using predictive and entropy coding [9]. Another similar study by Sarma J. et al. devised a hardware implementation of lossless data compression for wearable nodes; this method uses linear filtering, run-length encoding, and Golomb–Rice coding for data compression [10]. Tsai and Kuo implemented a lossless compression scheme that uses linear prediction for prediction accuracy and GRC for entropy coding. This method uses basic digital circuits to implement the subsystems, resulting in power-efficient operation and few logic gates, in turn achieving less chip area [10].

3. Methodology

This section discusses the methodology used in the system. Various subsystems are discussed below, and Figure 1 depicts the steps involved in the ECG compression system:

A derivative block is required to obtain the first difference in the ECG values taken from the MIT BIH arrhythmia dataset, and it finds the difference between the present value and the previous sample value. This step fulfils two purposes. Firstly, it reduces the amplitude values of the data samples since the ECG signals have values close to each other. This also minimizes the number of iterations of the compression algorithm since a zero value of difference results in no additional computational cost for the calculation of the quotient and remainder. After the derivative block, the magnitude of the first difference values is taken, and a packet of 8 values is chosen to find the respective means. This operation is performed to maintain the amplitude values and to provide immunity towards noise levels. To achieve better compression ratios, it is necessary to have the minimum value of amplitudes possible as it requires the minimum number of bits for representation.

The acquired mean values are compared with a threshold value1, which is selected based on the amplitude regions in the ECG values, which are mainly divided into three central regions: low, medium, and high amplitude. The obtained mean values are subjected to threshold comparison with the chosen threshold values T1, T2, and T3. The output of this comparison determines the factor by which the 8-bit packet will be divided. This packet is the same one that was chosen earlier to find the mean. The division operation further reduces the amplitude values. Further steps involve the encoding of values based on the division.

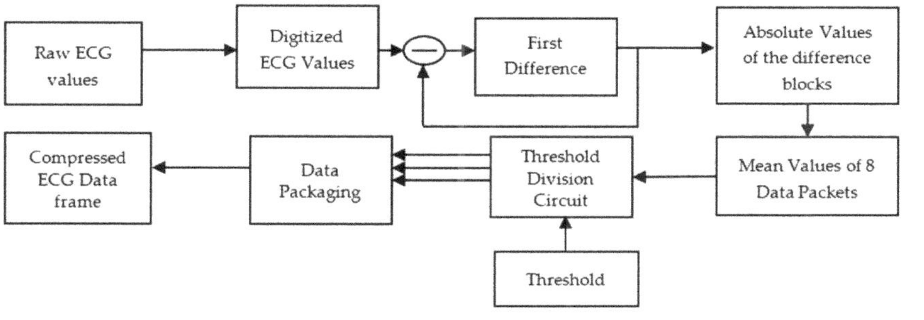

Figure 1. Epileptic seizure detection methodology.

The Golomb–Rice encoder performs the encoding of the ECG values, and this block is the most essential part of the data compression system. The Golomb–Rice coding (GRC) method is usually employed where the amplitudes are very low in value. First, the samples are divided into groups of symbols, and then these symbols are assigned a code word, which is usually equal to the number of parameters subtracted by the decided coder parameter. This parameter can be decided based on various signal metrics like variance or geometric mean to incorporate the maximum number of reoccurring sample values.

In GRC, the sample values are segregated into two parts upon parameter division, i.e., quotient and remainder. For these obtained quotients and remainders, separated coding schemes are used. Quotients are coded in a unary scheme. GRC is popular due to numerous reasons, such as its low complexity for hardware and software implementations, its viability for various sample data types, and it being a lossless scheme; therefore, it is suitable for applications where sample drops can degrade the efficiency. Here, the value of k is chosen based on the threshold comparison with the obtained mean values. It is observed that the quotient values of samples have a very high frequency over zero values, which can result in a further reduction in the number of bits used for encoding. If the number of consecutive zeros is obtained for a run of encoding, then a binary number can signify the run length rather than sending that many zeros. This method further reduces the CR.

$$Quotient = \left\lfloor \frac{D(n)}{2^k} \right\rfloor, \text{ where, } \mathrm{k} = 3,4,5 \tag{1}$$

$$Remainder = D(n) \bmod 2^k, \text{ where, } \mathrm{k} = 3,4,5 \tag{2}$$

The obtained quotient values are then coded further to enhance the CR run-length encoding (RLE), which is used to encode the consecutive zero quotient values. RLE aims to reduce the redundancy due to the repeated number of characters. RLE is useful in applications where there are repeating sample values in succession to each other. A particular marker bit needs to be used at the decoder end to understand that the RLE code has arrived; in this case, "000" is used as a marker to identify that the runs of zeros have arrived. The data obtained after encoding contains various values, i.e., raw ECG, mean, absolute values, quotients, and remainders. The last step remains to combine the essential information for transmission. The packaging is performed in two different ways, one for zero values and the other for non-zero values of the quotient. The packaging starts with the initial value of the 11-bit ECG signal, which is followed by the initial value of the difference. The next block is the marker indicating the factor used for the division of values specified. This block also indicates the zero quotient values by indicating a distinct marker. After this, quotient values are sent, which are variable in length due to a unary coding scheme. In the end, the remainder is sent over in binary form; hence, it is shown as a variable in the data frame. For a run of zeros, frame 3,4 is replaced by marker 000, and the variable run length is coded in binary format. The explained packaging format used here is for 8-bit packets, and this process is repeated for all 450 packets of 8 bits, resulting in a total of 3600 values. The overall hardware of the compression system is shown in Figure 2.

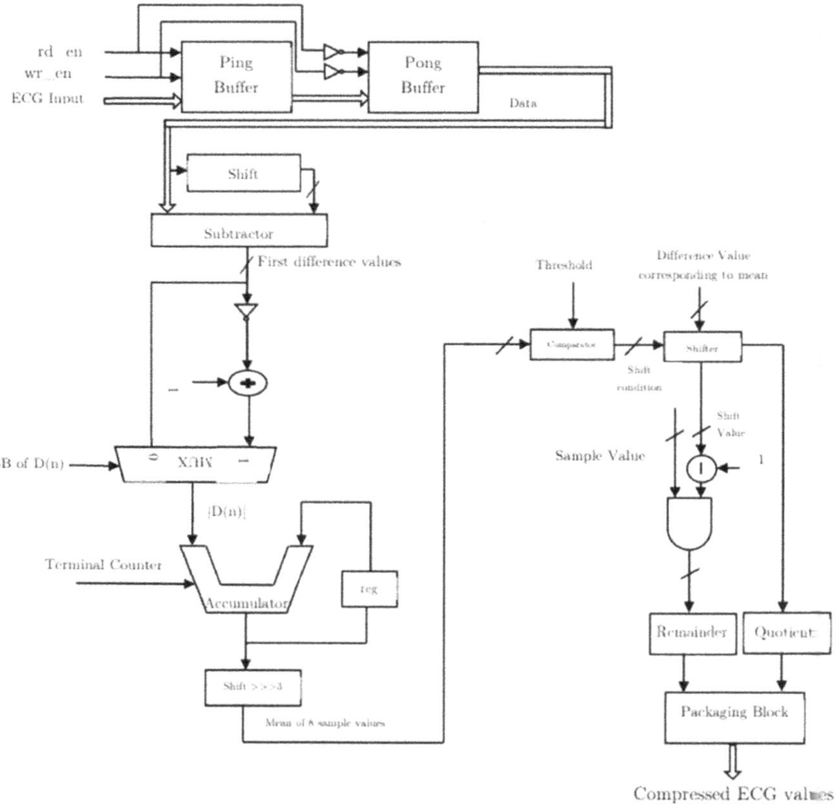

Figure 2. The overall architecture of the compression system.

4. Results

The results obtained from the waveforms and RTL schematic from the compression system are shown in this section. At the positive edge of the clock cycle and low reset value, the addresses of the memory locations from where the values were to be taken were loaded. The data outputted the corresponding value of the ECG signal from the buffer. The clock period was 2, and the complete data took 7.2 microseconds to reach the system. Figure 3 shows the output of the shift operation and the subtraction to find the value of the first difference. It can be observed that some values were exceptionally large. This is due to 2's complement representation of negative numbers in binary format. In the next step, absolute values were taken for the first difference. After that, eight sample values were taken to find the sum and shifted right by 3 to obtain the mean of the values. The obtained values were then given to the threshold comparator, which decided the factor from which the sample

values corresponding to the mean should be divided, and quotients and remainders were obtained. Quotient values for respective data samples were taken, and the packaging for the 8-bit samples was performed. It was observed that for the first eight data samples containing 88 bits of data, a total of 24 bits were generated. Subsequently, the range lay between 24 and 32 bits for other samples. The average CR for the compression system was found to be 2.75, and the maximum CR was found to be 3.14. With a power consumption of 2.9 W, around 92% of the power was utilized in I/O operation from memory to fetch the data, which can be reduced in the actual design because real-time data are acquired in the latter case, as seen in Figure 3. The logic power utilization was 0.2W for the logical operations performed in the compression system. The total number of lookup tables (LUTs) used was 408, and the flip flops used numbered 51. Table 1 shows a summary of the results.

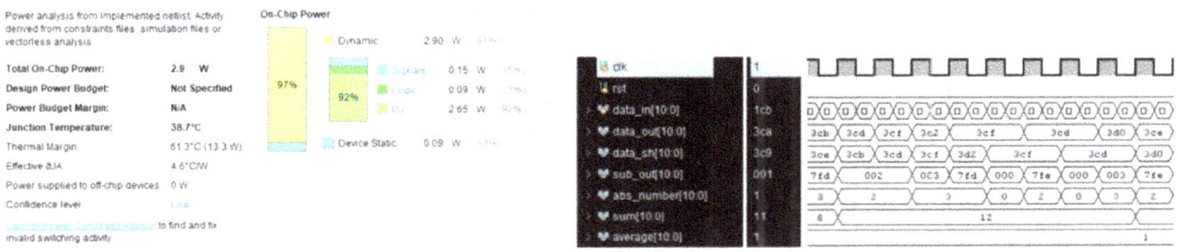

Figure 3. Power report and outputs of various subsystems.

Table 1. Summary of results.

Metric	Value
Board used	• Family: Artix-7 low voltage; • Package: csg324; • Part number: xc7a100tlcsg324-2L.
Mean CR	2.89
Highest CR	3.6
Lookup tables used	408
Flip flops used	51
Power (logical, I/0)	0.2 W, 2.7 W
Delay	7.2 microseconds

5. Conclusions and Future Scope

In this paper, a lossless ECG data compression system was presented. The compression system used the Golomb–Rice coding method to encode the ECG signals. The MIT BIH arrhythmia dataset was used, which contains 11-bit raw ECG sample values. The CR attained by the compression system was 2.89 and 3.6 for average and maximum values. The design implementation was tested on Nexys DDR4 FPGA, which is of the Artix-7 low-voltage family. The design consumed 408 LUTs and 51 FFs at a clock frequency of 0.5 KHz. The system's logical power consumption was 0.2 W, and the I/O consumption was 2.7 W. It was observed that there is a trade-off between the transmitted power and the processing power in the sensor node if we aim to decrease the power consumption due to the transmission of extensive data.

Table 2 shows a comparison between different ECG compression techniques. The achieved compression ratio resulted in lower energy requirements to send data and a lower storage space requirement, which help achieve two critical wearable metrics, i.e., power and memory management, which help in the development of a better and optimized wearable system. Additional computations must be performed via the processor used in the sensor node. However, due to advances in VLSI technology, the processor design is highly optimized and can provide better savings. The implemented compression system can be further extended for physical design implementation. The changes at this step, like clock gating, power gating, and algorithmic level changes, can further reduce the system's power consumption. The proposed system can also be modelled for sensor node simulations to map power savings due to the compression methods applied. The system can also be used for various biomedical signals for data compression.

Table 2. Comparison of results.

Parameter	This Work	2017 [11]	2017 [12]	2016 [13]	2017 [14]	2020 [15]
Method	Golomb–Rice Encoder	Context-Aware Compression	Entropy Coding	JointCoding Package	Wavelet Shrinkage	Lossless ECG Compression
Compression ratio	2.75	2.15	2.15	2.1	2.70	2.77

Author Contributions: S.H. was responsible for the conceptualization and methodology of the paper. The design and simulation of the experiments and the drafting of the manuscript were performed by S.H. as well. V.G. performed the manuscript review, proofreading, and reference collection for the research. All authors have read and agreed to the published version of the manuscript.

Funding: This research received no external funding.

Institutional Review Board Statement: Not Applicable.

Informed Consent Statement: Not applicable.

Data Availability Statement: Data sharing is not applicable to this article.

Conflicts of Interest: The authors declare no conflicts of interest.

References

1. WebCite Query Result. webcitation.org. Available online: https://webcitation.org/73HUXmOKl (accessed on 25 October 2023).
2. Dargie, W.; Poellabauer, C. *Fundamentals of Wireless Sensor Networks: Theory and Practice*; Wiley: Berlin, Germany, 2010; p. 31[illegible].
3. King, P.H. Wearable Sensors: Fundamentals, Implementation and Applications. *IEEE Pulse* **2021**, *12*, 30–31. [CrossRef]
4. Li, R.; Wang, L.; Yin, L. Materials and Devices for Biodegradable and Soft Biomedical Electronics. *Materials* **2018**, *11*, 2108. [CrossRef] [PubMed]
5. World Health Statistics. Monitoring Health for the SDGs, Sustainable Development Goals. 2022. Available online: https://www.who.int/publications/i/item/9789240051157 (accessed on 16 May 2023).
6. Alam, M.S.; Rahim, N.M.S. Compression of ECG signal based on its deviation from a reference signal using discrete cosine transform. In Proceedings of the ICECE (2008)–5th International Conference on Electrical and Computer Engineering, Dhaka, Bangladesh, 20–22 December 2008; pp. 53–58. [CrossRef]
7. Nasimi, F.; Khayyambashi, M.R.; Movahhedinia, N.; Law, Y.W. Exploiting similar prior knowledge for compressing ECG signals. *Biomed. Signal Process. Control* **2020**, *60*, 101960. [CrossRef]
8. Zou, Y.; Han, J.; Xuan, S.; Huang, S.; Weng, X.; Fang, D.; Zeng, X. An energy-efficient design for ECG recording and R-peak detection based on wavelet transform. *IEEE Trans. Circuits Syst. II Express Briefs* **2015**, *62*, 119–123. [CrossRef]
9. Deepu, C.J.; Zhang, X.; Liew, W.S.; Wong, D.L.T.; Lian, Y. An ECG-SoC with 535nW/channel lossless data compression for wearable sensors. In Proceedings of the 2013 IEEE Asian Solid-State Circuits Conference, A-SSCC 2013, Singapore, 11–13 November 2013; pp. 145–148. [CrossRef]
10. Tsai, T.H.; Kuo, W.T. An Efficient ECG Lossless Compression System for Embedded Platforms with Telemedicine Applications. *IEEE Access* **2018**, *6*, 42207–42215. [CrossRef]
11. Chen, S.-L.; Villaverde, J.F.; Lee, H.-Y.; Chung, D.W.-Y.; Lin, T.-L.; Tseng, C.-H.; Lo, K.-A. A Power-Efficient Mixed-Signal Smart ADC Design With Adaptive Resolution and Variable Sampling Rate for Low-Power Applications. *IEEE Sens. J.* **2017**, *17*, 3461–3469. [CrossRef]
12. Deepu, C.J.; Heng, C.-H.; Lian, Y. A Hybrid Data Compression Scheme for Power Reduction in Wireless Sensors for IoT. *IEEE Trans. Biomed. Circuits Syst.* **2017**, *11*, 245–254. [CrossRef] [PubMed]
13. Deepu, C.J.; Zhang, X.; Heng, C.H.; Lian, Y. A 3-Lead ECG-on-Chip with QRS Detection and Lossless Compression for Wireless Sensors. *IEEE Trans. Circuits Syst. II Express Briefs* **2016**, *63*, 1151–1155. [CrossRef]

14. Ieong, C.-I.; Li, M.; Law, M.-K.; Mak, P.-I.; Vai, M.I.; Martins, R.P. A 0.45 V 147–375 nW ECG Compression Processor With Wavelet Shrinkage and Adaptive Temporal Decimation Architectures. *IEEE Trans. Very Large Scale Integr. (VLSI) Syst.* **2017**, *25*, 1307–1319. [CrossRef]
15. Tsai, T.-H.; Hussain, M.A. VLSI Implementation of Lossless ECG Compression Algorithm for Low Power Devices. *IEEE Trans. Circuits Syst. II Express Briefs* **2020**, *67*, 3317–3321. [CrossRef]

engineering proceedings

MDPI

Proceeding Paper

Low-Cost Environmental Monitoring Station to Acquire Health Quality Factors †

Ioannis Christakis *, Vasilios A. Orfanos, Pavlos Chalkiadakis and Dimitrios Rimpas

Department of Electrical and Electronics Engineering, University of West Attica, P. Ralli & Thivon 250, 12244 Egaleo, Greece; vorfanos@uniwa.gr (V.A.O.); pchalk@uniwa.gr (P.C.); drimpas@uniwa.gr (D.R.)

* Correspondence: jchr@uniwa.gr

† Presented at the 10th International Electronic Conference on Sensors and Applications (ECSA-10), 15–30 November 2023; Available online: https://ecsa-10.sciforum.net/.

Abstract: With the exponential development of MEMS (Micro-Electromechanical Systems) in the last decade, emphasis has been placed on the construction of IoT devices in conjunction with an appropriate information system to assist citizens in various fields (transportation, trade, etc.). More specifically, in the health sector,, there are specific IoT devices which can monitor a patient's health condition or provide environmental data for the area, information which affects health quality conditions. In densely populated areas and especially in large cities, in terms of environmental pollution, as well as the known issue of air pollution, citizens are also exposed to solar radiation (ultraviolet UVA UVB radiation), as well as to noise pollution in areas where people live and work. Ultraviolet radiation, especially during the summer months, is responsible for skin cancer and various eye diseases, while noise pollution can create mental disorders in humans, especially in children. In this article, a low-cost solar radiation and noise pollution monitoring station is presented. The parts that compose the station and its implementation are a microcontroller (TTGO-OLED32) with an integrated LoRa device, an ultraviolet radiation sensor and sound sensors. In addition, a mini ups device is used in case of power failure and a GPS device is utilized for the location point. The measurements are obtained by the sensors every ten minutes and are transmitted via the LoRa network to an application server in which the user has direct access to the environmental data of a specific area. In conclusion, the data obtained from such IoT devices help in the study of cities to optimize factors in people's lives.

Keywords: IoT health system; ultraviolet radiation; noise pollution; sensors

Citation: Christakis, I.; Orfanos, V.A.; Chalkiadakis, P.; Rimpas, D. Low-Cost Environmental Monitoring Station to Acquire Health Quality Factors. *Eng. Proc.* **2023**, *58*, 11. https://doi.org/10.3390/ecsa-10-16096

Academic Editor: Stefano Mariani

Published: 15 November 2023

1. Introduction

Issues regarding environmental pollution and health quality, specifically in developed large cities, have been the subject of various studies in recent years [1–4]. In general, the term health is linked to terms like the wellness of the body and the effects on human life [5–8]; however, mental health also proves to be equally significant [9–11]. In the everyday environment where someone lives and works, there is a set of pollutants, such as atmospheric pollutants including increased greenhouse gases, but also other forms of pollution such as noise and light. Regarding atmospheric pollution, the air quality in densely populated areas has been shown to have an impact on human health, both from gaseous pollutants and from particulate matter [12–14], which are the result of human activities such as the development of industry, residential heating systems, traffic, etc. In addition, there are other forms of environmental pollution, like the acoustic noise of an area originating from industrial or residential activities, leading to the degradation of life quality, as well as the incidence of mental disorders. It greatly contributes to human mental disturbance [15], and although it has not been proven to be connected with mental diseases, a correlation between residences near airports and the prevalence of strong symptoms of depression is evident [16]. Subjective health symptoms, such as fatigue and headaches, are consistently reported more often by children who live near an airport facility or go to school in noisy areas [17]. There are issues reported where the light pollution of an area with ultraviolet radiation can be harmful to humans. Research has shown the harmful effects of human exposure to ultraviolet radiation from the sun, demonstrating that it causes skin cancer in a very large percentage [18]. UV radiation can also affect vision as it carries higher energy than visible light and high dose exposure to UV radiation causes direct cell damage, which plays an important role in cancer development [19]. In the last decade, the rapid development and construction of electronic circuits has resulted in the creation of reliable microcontrollers and low-cost sensors. This aspect offers feasibility to build affordable environmental monitoring systems, both in terms of air quality monitoring [20] and environmental conditions monitoring such as acoustic noise and light pollution of an area, giving citizens invaluable information for their residential area. As a communication carrier, a LoRaWAN [21] network can be used to ensure the precise linkage of monitoring stations without fees in a spatial coverage of 3 km. This paper presents the study and implementation of a low-cost environmental monitoring station which includes both noise and light pollution sensors. The retrieved values of the noise level are measured at dB for noise pollution, while the level of ultraviolet radiation is measured as a UV index of light pollution. Due to rapid technological developments, there are many reliable low-cost systems that meet low power consumption and excellent processing standards such as the TTGO@ESP32 microprocessor [22]. These systems offer sufficient local data analysis with low power consumption and LoRa wireless connectivity. This article is organized into the following sections: Section 2 describes the materials and methods, Section 3 comprises the results and discussion, and Section 4 offers the conclusions.

2. Materials and Methods

Three sections compose the IoT architecture [23], perception, network, and application, as seen in Figure 1.

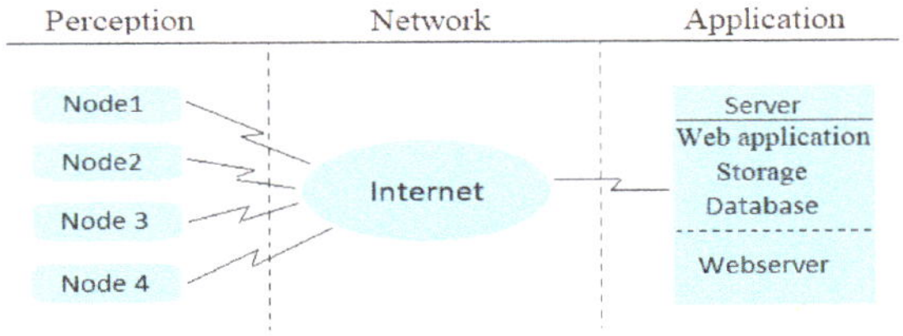

Figure 1. IoT architecture topology.

In the case of this work, the low-cost environmental monitoring station constitutes the perception part, including the microcontroller and sensors. The microprocessor unit (CPU) is the TTGO@ESP32, and as sensors, a GYML8511 UV [24] sensor for light pollution and a sound sensor module [25] for sound pollution were used. The network section, also referred to as the transparent segment, aims to connect the perception and application sections. This module is responsible for transmitting data from the low-cost stations to a central station (server) that contains the application section. In this work, we utilize the built-in LoRa network function contained in TTGO@ESP32. Data transfer is conducted using "The Things Network (TTN)" LoRaWAN network. The application section is the final stage of the IoT system, which provides the services to the end user. This section supports many applications for the development of IoT (Internet of Things) systems. In this work, the application section is supported by the Cayenne application server web portal [26]. The components of the low-cost environmental conditions monitoring station are the microprocessor and the sensors. In addition, an expansion board has been constructed for the interconnection between the CPU and sensors.

- TTGO@ESP32

The main CPU is the TTGO@ESP32 [22] (Figure 2a), which is an open-source Arduino-based firmware for IoT implementations. It is an ideal module for IoT devices as it remains highly affordable with a plethora of features, such as processing capabilities with a fast time response. It can be programmed using the open-source Arduino Software (IDE), providing convenience in coding and uploading to the board while being a familiar and user-friendly software. This processor satisfies the requirements of communication for the sensors array (25 I/O ports, UART, I2C, SPI interfaces). The CPU data processing speed further satisfies this implementation with low power consumption. The microprocessor integrates the data transfer communication which can be implemented over a long-range (LoRa) network or a wireless network (Wi-Fi). In addition, a battery (type 18650) is connected to the integrated charger of the main board for uninterrupted operation in the event of a power failure and voltage stability.

(a) (b) (c) (d)

Figure 2. The CPU and sensors of low-cost station: (**a**) the TTGO@ESP32 CPU board; (**b**) the UV sensor GYML8511; (**c**) the sound sensor; (**d**) the GPS module.

- Sensors

For the environmental conditions, light and noise pollution sensors were utilized; the GYML8511 [24] ultraviolet sensor (Figure 2b) outputs an analog signal in relation to the amount of the UV light, as the sensor is capable of detecting wavelengths of 280–390 nm light with high precision. For noise pollution validation, a sound sensor module [25] (Figure 2c) of a dynamic microphone (electromagnetic microphone) and amplifier circuit with analog output was exploited. In addition, a GPS module (Figure 2d) was used for the station location.

- Station implementation

The diagram of implementation and the final construction of the environmental conditions monitoring station are shown in Figures 3a and 3b, respectively. The whole device is integrated in a waterproof box type IP66 (with dimensions of 115 mm × 150 mm), which is rather small. The total cost of the station is approximately EUR 80. Every ten minutes, a measurement is conducted and transferred over the LoRa network via the Internet to the application server.

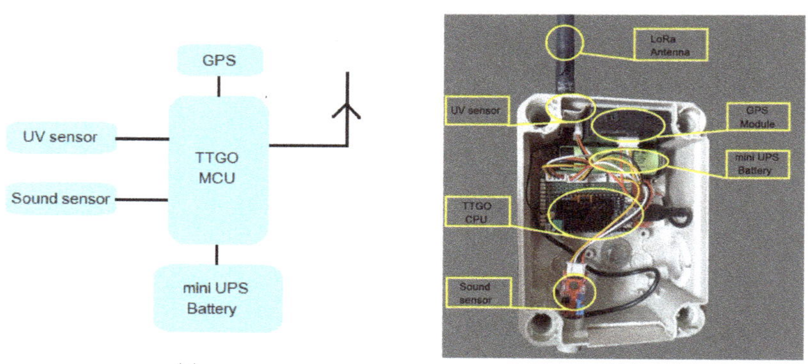

(a) (b)

Figure 3. The implementation and construction of the low-cost station: (**a**) the station components diagram; (**b**) the final construction of the low-cost monitoring station.

3. Results and Discussion

The evaluation and accuracy of the measurements from the low-cost environmental conditions monitoring station are presented in this section. This custom-made monitoring layout has been installed on the roof of a building, directly in the sun, at a height of 1.5 m, located in eastern Attica in Greece, specifically, in the municipality of Agia Paraskevi. In the evaluation of the UV low-cost sensor, the common known measurement of the UV pollution is the UV index, the calculation of the output of the low-cost UV sensor is given in miliWatt/square centimeter, and the investigation of the transformation of the output sensor to the UV index has taken place. According to the datasheet of the sensor [24], specifically, the graph of the output voltage, UV intensity characteristics can extract the slope of the sensor response. The proposed UV index (Equation (1)) is the result, as the output of the sensor is multiplied by the slope and the result is divided by a correction factor (A).

$$UVI = \frac{\left(calculate\ output\left(\frac{\mathrm{mW}}{\mathrm{cm}^2}\right) \cdot slope\left(\frac{\mathrm{cm}^2}{\mathrm{mW}}\right) \right)}{A} \tag{1}$$

The reference data are received by the official monitoring station in cooperation with the National Observatory of Athens [27]. The distance between the low-cost station and the reference station is 3 km. The following Figure 4 shows the time-series measurements and the correlation of the UV index of the low-cost monitoring station and the reference station on the last day of June, July and August of 2023.

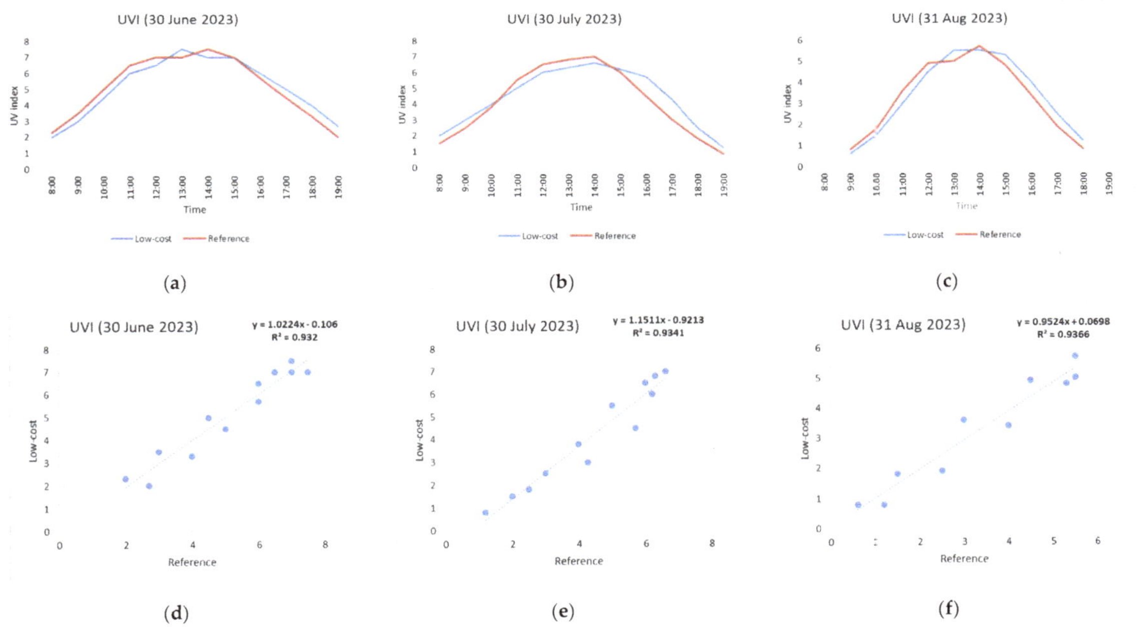

Figure 4. Time series and correlations of measurements of low-cost sensor and reference of UV index: (**a**) time series of 30 June 2023; (**b**) time series of 30 July 2023; (**c**) time series of 31 August 2023; (**d**) correlations of 30 June 2023; (**e**) correlations of 30 July 2023; (**f**) correlations of 31 August 2023.

Regarding noise pollution, the results of the measurements are shown indicatively as there are no reference measurements. The calibration of the microphone module was performed in a laboratory where the noise level decibel (dB) was detected by official instruments. Figure 5 shows the time-series measurements of noise pollution for the low-cost monitoring station on the last day of June, July and August of 2023.

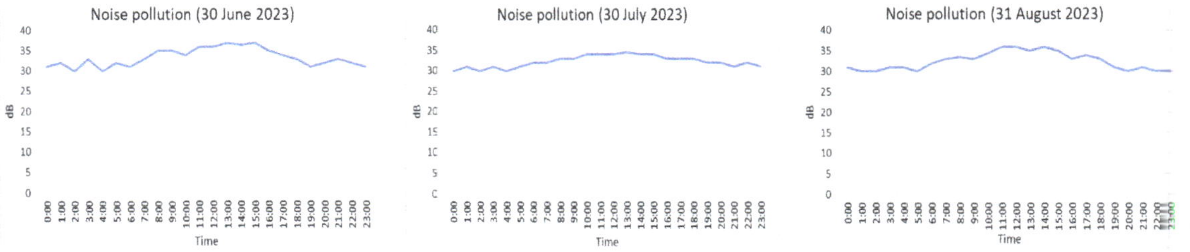

(a) (b) (c)

Figure 5. Time-series measurements of noise pollution (dB): (**a**) time period of 30 June 2023; (**b**) time period of 30 July 2023; (**c**) time period of 31 August 2023.

4. Conclusions

In the environment, alongside atmospheric pollution, there are other forms of pollution, such as light and noise, which affect human health and wellness. In this article, a low-cost UV radiation and noise pollution monitoring station has been presented. Data transfer has been conducted through the LoRa network, while visualization has been supported by Cayenne, which is an IoT application server on the Internet. The measurements have been retrieved in vivo and the measured values have been corrected using a proposed equation to extract the UV index from the UV sensor and the sound output in decibels (dB) from the sound sensor. The results are encouraging as the low-cost UV sensor shows a correlation coefficient R^2 greater than 90% with respect to the reference data, while the noise value accuracy shows satisfactory results, as the station was installed in a quiet neighborhood of northeast Attica. The use of low-cost sensors and their utilization with microcontrollers of new technology can satisfy a wide range of detection for environmental conditions. The aim is to attract more and more people to actively participate in actions such as public health monitoring which is a common good for all. However, it is a good and affordable solution to inform the general public about the environmental conditions where they live and work.

Author Contributions: Conceptualization, I.C. and D.R.; methodology, I.C., V.A.O., P.C. and D.R.; software, I.C.; validation, V.A.O. and P.C.; formal analysis, I.C., V.A.O., P.C. and D.R.; investigation, I.C. and D.R.; resources, V.A.O. and P.C.; data curation, V.A.O. and P.C.; writing—original draft preparation, I.C.; writing—review and editing, D.R. and V.A.O.; visualization, V.A.O.; supervision, I.C. and D.R.; project administration, I.C. All authors have read and agreed to the published version of the manuscript.

Funding: This research received no external funding.

Institutional Review Board Statement: Not applicable.

Informed Consent Statement: Not applicable.

Data Availability Statement: All of the data created in this study are presented in the context of this article.

Conflicts of Interest: The authors declare no conflicts of interest.

References

1. Manisalidis, I.; Stavropoulou, E.; Stavropoulos, A.; Bezirtzoglou, E. Environmental and health impacts of air pollution: A review. *Front. Public Health* **2020**, *8*, 14. [CrossRef] [PubMed]
2. Zhang, X.; Chen, X.; Zhang, X. The impact of exposure to air pollution on cognitive performance. *Proc. Natl. Acad. Sci. USA* **2018**, *115*, 9193–9197. [CrossRef] [PubMed]
3. Lin, B.; Zhu, J. Changes in urban air quality during urbanization in China. *J. Clean. Prod.* **2018**, *188*, 312–321. [CrossRef]
4. Liang, L.; Wang, Z.; Li, J. The effect of urbanization on environmental pollution in rapidly developing urban agglomerations. *J. Clean. Prod.* **2019**, *237*, 117649. [CrossRef]

5. Jiang, Y.; Niu, Y.; Xia, Y.; Liu, C.; Lin, Z.; Wang, W.; Ge, Y.; Lei, X.; Wang, C.; Cai, J.; et al. Effects of personal nitrogen dioxide exposure on airway inflammation and lung function. *Environ. Res.* **2019**, *177*, 108620. [CrossRef]
6. Nuvolone, D.; Petri, D.; Voller, F. The effects of ozone on human health. *Environ. Sci. Pollut. Res.* **2018**, *25*, 8074–8088. [CrossRef]
7. Momtazan, M.; Geravandi, S.; Rastegarimehr, B.; Valipour, A.; Ranjbarzadeh, A.; Yari, A.R.; Dobaradaran, S.; Bostan, H.; Farhadi, M.; Darabi, F.; et al. An investigation of particulate matter and relevant cardiovascular risks in Abadan and Khorramshahr in 2014–2016. *Toxin Rev.* **2019**, *38*, 290–297. [CrossRef]
8. Goudarzi, G.; Alavi, N.; Geravandi, S.; Idani, E.; Behrooz, H.R.A.; Babaei, A.A.; Alamdari, F.A.; Dobaradaran, S.; Farhadi, M.; Mohammadi, M.J. Health risk assessment on human exposed to heavy metals in the ambient air PM10 in Ahvaz, southwest Iran. *Int. J. Biometeorol.* **2018**, *62*, 1075–1083. [CrossRef] [PubMed]
9. Ventriglio, A.; Bellomo, A.; di Gioia, I.; Di Sabatino, D.; Favale, D.; De Berardis, D.; Cianconi, P. Environmental pollution and mental health: A narrative review of literature. *CNS Spectr.* **2021**, *26*, 51–61. [CrossRef]
10. Mueller, M.A.; Flouri, E.; Kokosi, T. The role of the physical environment in adolescent mental health. *Health Place* **2019**, *58*, 102153. [CrossRef] [PubMed]
11. Ventriglio, A.; Torales, J.; Castaldelli-Maia, J.M.; De Berardis, D.; Bhugra, D. Urbanization and emerging mental health issues. *CNS Spectr.* **2021**, *26*, 43–50. [CrossRef] [PubMed]
12. World Health Organization. *WHO Global Air Quality Guidelines: Particulate Matter (PM2.5 and PM10), Ozone, Nitrogen Dioxide, Sulfur Dioxide and Carbon Monoxide*; World Health Organization: Geneva, Switzerland, 2021.
13. Ritz, B.; Hoffmann, B.; Peters, A. The effects of fine dust, ozone, and nitrogen dioxide on health. *Dtsch. Ärzteblatt Int.* **2019**, *116*, 881. [CrossRef] [PubMed]
14. Orellano, P.; Reynoso, J.; Quaranta, N.; Bardach, A.; Ciapponi, A. Short-term exposure to particulate matter (PM10 and PM2.5), nitrogen dioxide (NO_2), and ozone (O_3) and all-cause and cause-specific mortality: Systematic review and meta-analysis. *Environ. Int.* **2020**, *142*, 105876. [CrossRef] [PubMed]
15. Kou, L.; Kwan, M.P.; Chai, Y. Living with urban sounds: Understanding the effects of human mobilities on individual sound exposure and psychological health. *Geoforum* **2021**, *126*, 13–25. [CrossRef]
16. Beutel, M.E.; Brähler, E.; Ernst, M.; Klein, E.; Reiner, I.; Wiltink, J.; Michal, M.; Wild, P.S.; Schulz, A.; Munzel, T.; et al. Noise annoyance predicts symptoms of depression, anxiety and sleep disturbance 5 years later. Findings from the Gutenberg Health Study. *Eur. J. Public Health* **2020**, *30*, 487–492. [CrossRef] [PubMed]
17. Clark, C.; Head, J.; Haines, M.; van Kamp, I.; van Kempen, E.; Stansfeld, S.A. A meta-analysis of the association of aircraft noise at school on children's reading comprehension and psychological health for use in health impact assessment. *J. Environ. Psychol.* **2021**, *76*, 101646. [CrossRef]
18. Leiter, U.; Keim, U.; Garbe, C. Epidemiology of skin cancer: Update 2019. *Sunlight Vitam. D Ski. Cancer* **2020**, *1268*, 123–139.
19. Yam, J.C.; Kwok, A.K. Ultraviolet light and ocular diseases. *Int. Ophthalmol.* **2014**, *34*, 383–400. [CrossRef] [PubMed]
20. Christakis, I.; Hloupis, G.; Stavrakas, I.; Tsakiridis, O. Low cost sensor implementation and evaluation for measuring NO_2 and O3 pollutants. In Proceeding of the 9th International Conference on Modern Circuits and Systems Technologies (MOCAST), Bremen, Germany, 7–9 September 2020.
21. The Things Network. Available online: https://www.thethingsnetwork.org/ (accessed on 31 August 2023).
22. ESP32 Series—Espressif Systems. Available online: https://www.espressif.com/sites/default/files/documentation/esp32_datasheet_en.pdf (accessed on 31 August 2023).
23. Said, O.; Masud, M. Towards internet of things: Survey and future vision. *Int. J. Comput. Netw.* **2013**, *5*, 1–17.
24. ML8511—SparkFun Electronics. Available online: https://cdn.sparkfun.com/datasheets/Sensors/LightImaging/ML8511_3-8-13.pdf (accessed on 31 August 2023).
25. KY-038 Datasheet. Available online: https://datasheetspdf.com/pdf-file/1402048/Joy-IT/KY-038/1 (accessed on 31 August 2023).
26. Cayenne Community Forum, Cayenne. Available online: https://cayenne.mydevices.com/cayenne/forum_login (accessed on 31 August 2023).
27. Latest Conditions in Nomismatokopeio. Available online: https://penteli.meteo.gr/stations/nomismatokopeio/ (accessed on 31 August 2023).

engineering proceedings

Proceeding Paper

Evaluating Urban Topography and Land Use Changes for Urban River Management Using Geospatial Techniques †

Ashish Mani [1,2,*], Maya Kumari [1] and Ruchi Badola [2]

1 Amity School of Natural Resources and Sustainable Development, Amity University, Sector-125, Noida 201303, India; mkumar10@amity.edu
2 Wildlife Institute of India, Chandrabani, Dehradun 248001, India; ruchi@wii.gov.in
* Correspondence: ashish.mani@s.amity.edu; Tel.: +91-9717623547

† Presented at the 10th International Electronic Conference on Sensors and Applications (ECSA-10), 15–30 November 2023; Available online: https://ecsa-10.sciforum.net/.

Abstract: This study focused on the urban river management using geospatial techniques of the Dehradun Municipal Corporation (DMC) and its associated watersheds of the Bindal River and Rispana River. Shutter Radar Topography Mission (SRTM) Digital Elevation Model (DEM) data with a spatial resolution of 30 m was used for the delineation of watershed boundaries and drainage networks and for identifying topographic features. Additionally, Sentinel-2 data with a spatial resolution of 10 m were utilized to analyze change in land use in 2017 and 2022. The drainage patterns in the Bindal and Rispana watersheds were dendritic in shape with moderate relief. The study found a significant decline in agricultural land from 17.94% in 2017 to 14.66% in 2022. This decline was accompanied by an increase in built-up area from 32.53% to 35.44%. The increased biotic pressure poses a critical threat to river health and biodiversity. This study highlights the urgent need for comprehensive river management strategies to efficiently monitor biotic pressure due to transformations in land use. This research will be beneficial to diverse stakeholders, including decision-makers and urban planners engaged in the sustainable management of water resources and urban development.

Keywords: remote sensing; geographic information system (GIS); digital elevation model (DEM); land use changes; watershed; urban river management

Citation: Mani, A.; Kumari, M.; Badola, R. Evaluating Urban Topography and Land Use Changes for Urban River Management Using Geospatial Techniques. *Eng. Proc.* **2023**, *58*, 12. https://doi.org/10.3390/ecsa-10-16004

Academic Editor: Stefano Mariani

Published: 15 November 2023

1. Introduction

The world is moving toward a freshwater crisis which affects river systems and their surrounding environments [1]. Urbanization, climate change, the over-exploitation of natural resources, and an increasing population pose a threat to the ecosystem services and biodiversity of freshwater resources [2,3]. With increasing land development and human activities in nearby and distant watersheds, the supplies of drinking water for cities are stressed due to potential contamination. India's 35% urban population generates almost 65% of the country's wastewater [4]. Urban river management in India is crucial for controlling the effects of growing anthropogenic pressure on the river system [5]. Urban river management refers to the comprehensive planning, development, and sustainable utilization of river resources within urban areas. The maintenance and restoration of these resources are essential elements of effective urban river management. This may include maintaining water quality, managing river flow, taking steps to prevent flooding, reviving riverfront areas for recreation and biodiversity, and making sure adequate sanitation and wastewater treatment. The roles of digital elevation models (DEMs) and multispectral sensor-based geospatial techniques are significant for urban river management studies. Earlier geospatial techniques are tedious, expensive, and more error-prone than the latest geospatial techniques. Spatial and temporal analyses of an area will provide us with a comprehensive view of the topography and changes in the land use pattern of an area [6]. In this study, we evaluate the land use changes and topographical factors of the Dehradun Municipal Corporation (DMC) and its associated watersheds. This study's main concern is the impact of humans on nearby river systems. Urban planners and water resource managers who are involved in urban river management and sustainable development will benefit from the findings of this study.

2. Methods

2.1. Study Area

The Dehradun Municipal Corporation (DMC) is the lifeline of the Doon Valley in Uttarakhand, India. It is the Central Business District (CBD) of Dehradun City. After originating from the Lesser Himalayas, the Bindal River, with a total length of ~23 km, and the Rispana River, with a total length of ~27 km, flow through the DMC before merging to form the Suswa River near Mothrowala, which then flows to the Song River and eventually joins the Ganga River [7]. The study area coordinates fall between a latitude of 30°13′45″ N to 30°27′30″ N and a longitude of 77°55′00″ E to 78°08′45″ E. For the total study area, the watershed areas of the Bindal and Rispana Rivers. Viz., 44.40 km^2 and 58.09 km^2, are combined with the DMC area of 183.70 km^2 [8], excluding the watershed area which falls inside the DMC area. Finally, the study covers a total area of 204.23 km^2 and has elevations that range from 549 m to 2278 m. The region receives ~2000 mm of average annual rainfall each year. The winters are extremely cold, with a temperature of 2 °C, and the summers are mostly hot, with a temperature of 43 °C. A map of the study area is providedd in Figure 1 below.

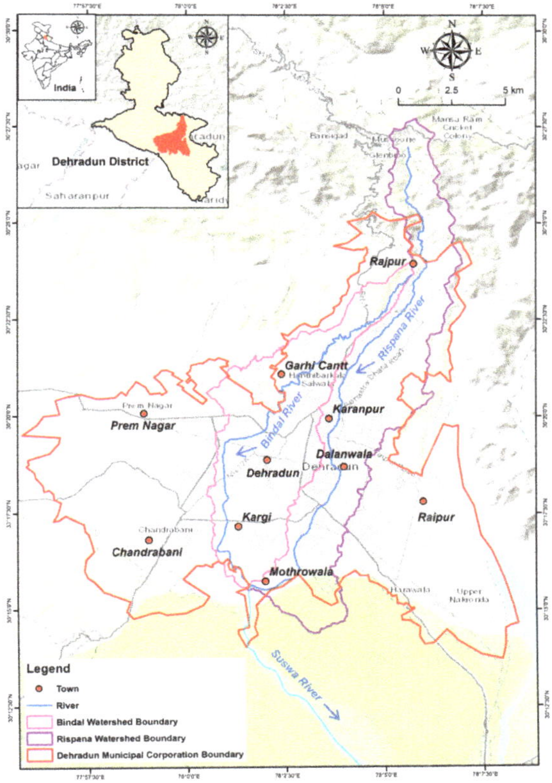

Figure 1. Study area map.

2.2. Materials and Methods

Digital elevation model (DEM) data and satellite-based multispectral sensor data were integrated to understand the study area. Shutter Radar Topography Mission (SRTM) DEM data at a 30 m resolution for the year 2014 were used for a topographical evaluation [9]. These data were further processed using ArcGIS desktop software version 10.6 and spatial analysis tools for the delineation of watershed boundaries and drainage networks. The Sentinel 2 multispectral sensor data at a 10 m resolution for the years 2017 and 2022 were used for land use changes [10]. Land use changes were performed through the supervised classification method in ERDAS Imagine software version 16.6. The supervised classification method involves selecting specific representative samples for each land use class, referred to as training samples, and the accuracy of the classification largely relies on the quality of these training samples [11]. These samples are employed by image classification algorithms to recognize land use classes across the entire image. In this study, we applied the widely used Maximum Likelihood Classifier (MXL) as our supervised classification algorithm. Further, the statistical process was completed using ArcGIS desktop software. Additionally, elevation, slope, drainage network, and land use change maps of the study area were prepared.

3. Results and Discussion

Topographical factors such as elevation, slope, and drainage network, along with land use changes, are essential for urban river management studies. These factors are explained below.

3.1. Elevation

The elevation of the DMC and its associated watershed ranges from 549 m to 2278 m (Figure 2a). The higher the elevation, the greater the slope [12]. The DMC area has a low-to-moderate elevation range.

3.2. Slope

The area's steepness is indicated by its slope. According to Figure 2b, the slope was classified into five classes: very gentle (0° to <=5°), gentle (>5° to =10°), moderate (>10° to =20°), steep (>20° to =40°), and very steep (>40° to 65°). The majority of the area has a relatively gentle to medium slope. A gentle and moderate slope has high groundwater infiltration and less runoff, whereas a steep slope has low groundwater infiltration and less runoff [13]. Additionally, steep slope regions are vulnerable to soil erosion.

3.3. Drainage Network

A drainage network comprises all the paths of streams that flow toward a reference point. It is constrained by a topographically defined drainage divide. Rainfall falling on the far side of the drainage divide runs downslope into an adjoining drainage network. The drainage pattern is dendritic in shape. All the drainage inside the Bindal River and Rispana River watersheds drains through the DMC area in a south direction toward an outlet in Mothrowala (Figure 2c). Both rivers bring municipal households' sewage waste and industrial effluent into the Suswa River, which joins the Song River near Jhabrawala and ultimately merges with the Ganga River near Tehari Farm, Raiwala, Uttarakhand, India. The drainage network's role is crucial for understanding the anthropogenic threats near urban rivers.

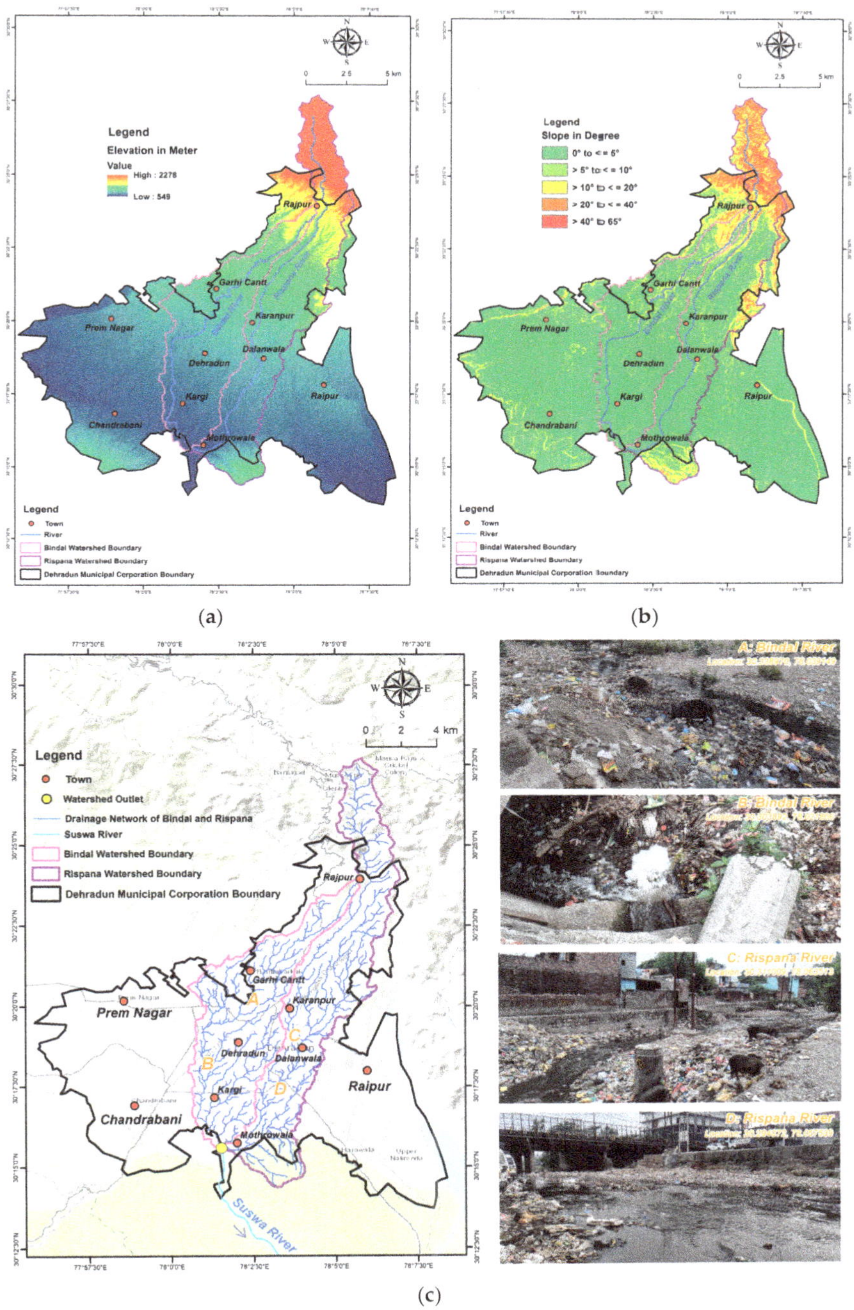

Figure 2. (**a**) Elevation map; (**b**) slope map; (**c**) drainage network map.

3.4. Land Use Changes

Multidisciplinary research on land use changes includes economics, geography, urban planning, and environmental studies [14,15]. It plays a crucial role in informing policies and practices that can help balance human development with the preservation of natural resources and ecosystems [16]. The supervised classification approach in ERDAS Imagine software was used in this study to create land use changes for the years 2017 and 2022. A statistical analysis was also performed using ArcGIS desktop software. The dominating classes in the area are dense vegetation and built-up land (Figure 3a,b). According to the analysis of land use changes, from 2017 to 2022, there was a 2.91% increase in built-up area and a 3.28% decrease in agricultural land area (Table 1). The changes in agricultural land area and built-up area depend on increasing population growth, the expansion of smart cities, and rapid socio-economic growth [17]. Also, increasing tourist pressure and the Dehradun climate are other significant factors for the rise in built-up area [18]. River health and biodiversity are under serious threat due to the increasing built-up area and other anthropogenic factors [19].

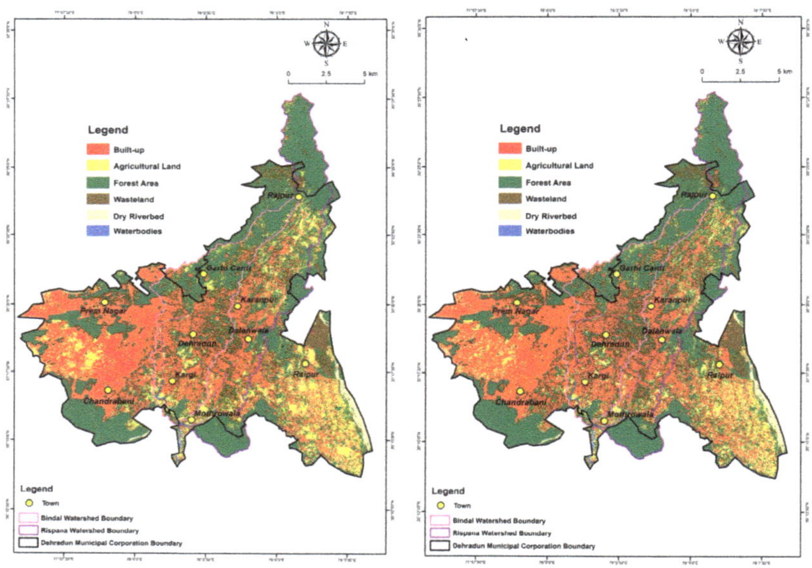

(a) (b)

Figure 3. (**a**) Land use map, 2017; (**b**) land use map, 2022.

Table 1. Land Use changes analysis.

S. No	Land Use Classes	2017 Area in km^2	2022 Area in km^2	2017 Area in %	2022 Area in %	Area Change in km^2	Area Change in %
1	Built-up	66.43	72.37	32.53	35.44	5.94	2.91
2	Agricultural Land	36.64	29.95	17.94	14.66	−6.69	−3.28
3	Barren Land	10.58	10.98	5.18	5.38	0.4	0.2
4	Dense Vegetation	86.59	86.88	42.4	42.54	0.29	0.14
5	Dry Riverbed	3.23	3.15	1.58	1.54	−0.08	−0.04
6	Waterbodies	0.76	0.9	0.37	0.44	0.14	0.07
	Total Area	204.23	204.23	100	100		

4. Conclusions

This study investigates the possible impacts of topographical factors and land use changes on urban river management using DEM and multispectral sensor based geospatial techniques. A comprehensive evaluation and interpretation implied that from 2017 to 2022, there was a change in the land use: the agricultural land area is decreasing, and the built-up area is increasing. The dendritic pattern of the watershed's drainage system makes it easier to understand various topographical factors, including runoff and infiltration rate. The findings of this study demonstrate the increasing biotic pressure on urban rivers driven by municipal households' sewage waste and industrial effluent which flow from small river systems to large river systems. In order to effectively monitor the biotic pressure caused by land use changes, this study emphasizes the urgent necessity for proper urban river management planning. This study will be useful to urban planners and policymakers who are involved in urban development and water resource management. Also, in the near future, topography and land use change-based studies will be significant for valuable and systematic urban river management.

Author Contributions: Conceptualization, A.M.; methodology, A.M., M.K. and R.B.; resources, A.M.; data curation, A.M.; writing—original draft preparation, A.M.; writing—review and editing, A.M., M.K. and R.B. All authors have read and agreed to the published version of the manuscript.

Funding: This research received no external funding.

Institutional Review Board Statement: Not applicable.

Informed Consent Statement: Not applicable.

Data Availability Statement: The data presented in this study are available upon request from the corresponding author.

Acknowledgments: The authors would like to thank their respective institutes for their constant encouragement and support.

Conflicts of Interest: The authors declare no conflicts of interest.

References

1. Kummu, M.; Guillaume, J.H.; de Moel, H.; Eisner, S.; Flörke, M.; Porkka, M.; Siebert, S.; Veldkamp, T.I.; Ward, P.J. The world's road to water scarcity: Shortage and stress in the 20th century and pathways towards sustainability. *Sci. Rep.* **2019**, *6*, 38495. [CrossRef] [PubMed]
2. Grill, G.; Lehner, B.; Thieme, M.; Geenen, B.; Tickner, D.; Antonelli, F.; Babu, S.; Borrelli, F.; Cheng, L.; Crochetiere, H.; et al. Mapping the world's free-flowing rivers. *Nature* **2019**, *569*, 215–221. [CrossRef] [PubMed]
3. Huggins, X.; Gleeson, T.; Kummu, M.; Zipper, S.C.; Wada, Y.; Troy, T.J.; Famiglietti, J.S. Hotspots for social and ecological impacts from freshwater stress and storage loss. *Nat. Commun.* **2022**, *13*, 439. [CrossRef] [PubMed]
4. NITI Aayog Report. Urban Wastewater Scenario in India. Available online: https://www.niti.gov.in/sites/default/files/2022-09/Waste-Water-A4_20092022.pdf (accessed on 28 August 2023).
5. National Mission for Clean Ganga (NMCG) Report. Urban River Management Plan (URMP). A Strategic Framework for Managing Urban River Stretches in the Ganga River Basin. Available online: https://nmcg.nic.in/writereaddata/fileupload/48_Urban%20River%20Management%20Plan%20framework.pdf (accessed on 28 August 2023).
6. Chin, A.; Gregory, K.J. Managing urban river channel adjustments. *Geomorphology* **2005**, *69*, 28–45. [CrossRef]
7. Mani, A.; Kumari, M.; Badola, R. Morphometric analysis of Suswa River Basin using geospatial techniques. *Eng. Proc.* **2022**, *27*, 65.
8. Mussoorie Dehradun Development Authority (MDDA) Report. Dehradun Master Plan-2041 (Draft). Available online: http://mddaonline.in/wp-content/uploads/2023/04/Dehradun-Master-Plan-2.0A.pdf (accessed on 29 August 2023).
9. U.S. Geological Survey. Earth Explorer. Available online: https://earthexplorer.usgs.gov/ (accessed on 31 March 2023).
10. European Space Agency (ESA) Copernicus Open Access Hub. Available online: https://scihub.copernicus.eu/ (accessed on 31 March 2023).
11. Salah, M. A survey of modern classification techniques in remote sensing for improved image classification. *J. Geomat.* **2017**, *11*, 21.
12. Mani, A.; Bansal, D.; Kumari, M.; Kumar, D. Land Use Land Cover Changes and Climate Change Impact on the Water Resources: A Study of Uttarakhand State. In *River Conservation and Water Resource Management*; Rai, P.K., Ed.; Springer Nature: Singapore, 2023; pp. 1–16.

13. Mani, A.; Kumar, D. Morphometric Analysis of Manali Watershed of Beas River Basin for Watershed Management. *VayuMandal* **2020**, *46*, 21–29.
14. Sterling, S.M.; Ducharne, A.; Polcher, J. The impact of global land-cover changes on the terrestrial water cycle. *Nat. Clim. Chang.* **2013**, *3*, 385–390. [CrossRef]
15. Zhao, A.; Zhu, X.; Liu, X.; Pan, Y.; Zuo, D. Impacts of land use change and climate variability on green and blue water resources in the Weihe River Basin of northwest China. *Catena* **2016**, *137*, 318–327. [CrossRef]
16. Jacobson, C.R. Identification and quantification of the hydrological impacts of imperviousness in urban catchments: A review. *J. Environ. Manag.* **2011**, *92*, 1438–1448. [CrossRef] [PubMed]
17. Deng, X.; Huang, J.; Rozelle, S.; Zhang, J.; Li, Z. Impact of urbanization on cultivated land changes in China. *Land Use Policy* **2015**, *45*, 1–7. [CrossRef]
18. Dey, J.; Sakhre, S.; Gupta, V.; Vijay, R.; Pathak, S.; Biniwale, R.; Kumar, R. Geospatial assessment of tourism impact on land environment of Dehradun, Uttarakhand, India. *Environ. Monit. Assess.* **2018**, *190*, 181. [CrossRef] [PubMed]
19. Rana, D.; Bisht, A.; Mushtaq; Bhadula, S. Ichthyo-Faunal Diversity of Suswa River, Doon Valley, Uttarakhand, India. *N. Y. Sci. J.* **2017**, *10*, 106–112.

engineering proceedings

MDPI

Proceeding Paper

Design and Development of Internet of Things-Based Smart Sensors for Monitoring Agricultural Lands †

Dhiya Sabu [1], Paramasivam Alagumariappan [1,*], Vijayalakshmi Sankaran [2] and Pavan Sai Kiran Reddy Pittu [1]

1 Department of Biomedical Engineering, Vel Tech Rangarajan Dr. Sagunthala R&D Institute of Science and Technology, Chennai 600062, India; vtu16805@veltech.edu.in (D.S.); vtu18106@veltech.edu.in (P.S.K.R.P.)
2 Department of Electronics and Communication Engineering, Vel Tech Rangarajan Dr. Sagunthala R&D Institute of Science and Technology, Chennai 600062, India; drvijayalakshmis@veltech.edu.in
* Correspondence: parama.ice@gmail.com or draparamasivam@veltech.edu.in; Tel.: +91-984-378-0801

† Presented at the 10th International Electronic Conference on Sensors and Applications (ECSA-10), 15–30 November 2023; Available online: https://ecsa-10.sciforum.net/.

Abstract: In recent years, the demand for efficient and sustainable agricultural practices has been leveraged, leading to smart farming practices. These practices aim to enhance agricultural processes and productivity and minimize resource waste. One of the crucial challenges faced by farmers is the uneven distribution of soil humidity and pH across their agricultural lands. Further, the irregularity in soil moisture content and pH can lead to poor crop performance, water wastage, and increased resource utilization. In this work, an Internet of Things-based smart sensor node was developed, which consists of humidity and pH sensors to ensure the efficient management of water and soil conditions across an entire farm. Also, an array of humidity and pH sensors were placed across the farm, and these units worked independently as they have their own controller and battery unit. The developed device was integrated with a solar cell, which charged the battery. Further, the data acquired from these sensors were wirelessly transmitted to the base station, and it gathered the information of each unit, including their humidity levels, pH values, signal strength, and energy supply. This information was processed at the base station, and a graphical overview of the farm with the acquired information was represented, which provides farmers with real-view insight to identify areas with poor humidity and pH conditions. These data were transmitted to an IoT cloud, offering the farmer the ability to monitor their farm from a remote location. In cases where humidity levels dropped drastically and remained unchecked for more than two hours, the system triggered an alert. This mechanism makes sure that farmers are notified of potential issues, allowing them to prevent crop damage and optimize their resource usage.

Keywords: agriculture; cloud platform; Internet of Things; smart farming; humidity; pH

Citation: Sabu, D.; Alagumariappan, P.; Sankaran, V.; Pittu, P.S.K.R. Design and Development of Internet of Things-Based Smart Sensors for Monitoring Agricultural Lands. *Eng. Proc.* **2023**, *58*, 13. https://doi.org/10.3390/ecsa-10-16207

Academic Editor: Francisco Falcone

Published: 15 November 2023

1. Introduction

Agriculture is one of the prominent components of human civilization, supplying food, prosperity, and an integral means of survival for inhabitants worldwide. As the world continued to grow, the farming sector and the overall agricultural industry were faced with significant challenges, as they had to produce more food and supplements for the growing population with limited resources [1]. At the same time, they had to make sure that it did not have a negative impact on the environment. Agriculture in many developing countries is confronted with a lot of complications, such as nutrient deficiencies, imbalance of pH and humidity, and multiple subterranean pests, along with water scarcity [2]. Often, the hidden adverse effects of crops go unrecognized or undetected until it is too late, and these pose a hidden threat to the farmers and the foundation of the whole population that relies on agriculture for survival. All the efforts and hard work made by the farmers go in complete vain, which, in turn, affects their livelihood badly as it becomes a more expensive fix. Abnormalities of the pH in the soil can also permanently impair crop productivity and yield and, at the same time, affect the overall quality [3]. The implementation of cutting-edge and groundbreaking technologies in agriculture is serving as a feasible solution for a lot of issues and has helped advance the fields of precision farming and pest disease management. The IoT continues to play a major role in this journey [4].

State-of-the-art technologies, specifically the Internet of Things (IoT), have helped to be an effective collaborator in the attempt to track, regulate, and evaluate in order to enhance the overall health of the crop and yield [5]. One major advantage of the IoT in agriculture is that it provides access to remote monitoring, which can be beneficial to farmers handling large areas of land [6]. They can easily receive updates on their smartphones and other similar devices.

The accessibility and availability of nutrients, vitamins, minerals, activity of microbial species, etc., are vital for plant growth. The pH of the soil has a direct impact on factors such as the efficiency and yield of the crops [3]. Furthermore, the deviations from normal levels of humidity can have an undesirable impact on the intake of water, transpiration, and general crop health. These imbalances can lead to various problems, such as stunted growth, increased susceptibility to various diseases and pests, and even pollination, and photosynthesis. These humidity fluctuations are often overlooked until and unless the crop or harvest shows signs of being in distress. In the pre-technology era, farmers made use of the conventional and manual techniques to keep an eye on their crops. This consisted of making decisions based on experience, manual irrigation, and visual examination [7]

In recent years, sensor technology has come into existence, and a single sensor can very well provide the required data, but relying on just one sensor has its own limitations due to a lack of context and inefficiency. Investigations and experiments have been conducted concerning topics such as the health of the soil, tracking and surveillance of diseases, insect management, and climate-related challenges [8]. The aforementioned initiatives provided valuable insights and led to the creation of numerous monitoring systems and methodologies. To avoid the mentioned drawbacks, scientists have developed highly sophisticated sensors and data acquisition systems that can gauge moisture and humidity levels, soil wetness, and systems that can foresee potential hazards to crops. Predictive models and disease detection techniques have also been established [9].

Amidst other challenges, the major difficulty lies in making farmers from rural areas understand these datasets. The data have to be unambiguous so that the farmers can easily perceive the data and make wise judgments, as it can be a difficult task for them due to their low literacy rate and lack of access to technologies [10,11]. So, there is a need for a user-friendly interface and assistance tools. These systems should help farmers convert raw data into understandable and practical insights so that they can improve their farming approaches [11].

The objective of this paper was to develop an advanced IoT-based smart sensor system that is customized for agricultural practices. This article aims to enhance decision making in order to improve crop yields and resource utilization and, at the same time, intends to eradicate the shortcomings of wired technology and single-sensor solutions.

2. Materials and Methods

Figure 1 shows a graphical diagram of the proposed approach. The proposed IoT-based smart sensor system consists of more than two smart sensor modules, which are represented as device 1, device 2, and device n. Furthermore, these devices were displaced randomly on the agricultural land, as shown in Figure 1.

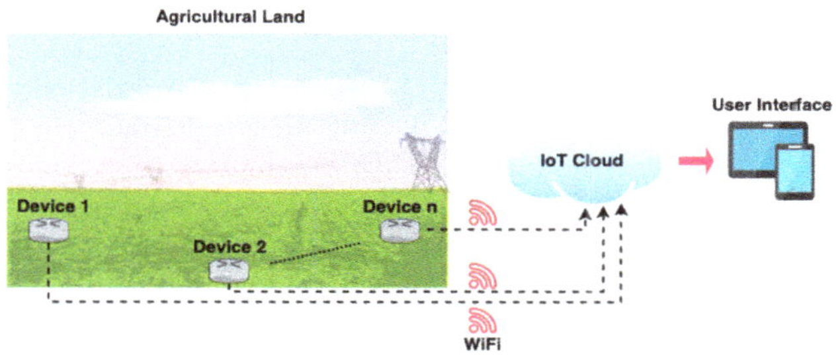

Figure 1. Proposed IoT-based smart sensor system.

2.1. The Proposed System

Figure 2 shows the overall hardware block diagram of the proposed device. Also, the proposed IoT-based smart sensor device is a standalone device, which is self-powered using a solar cell and a battery unit.

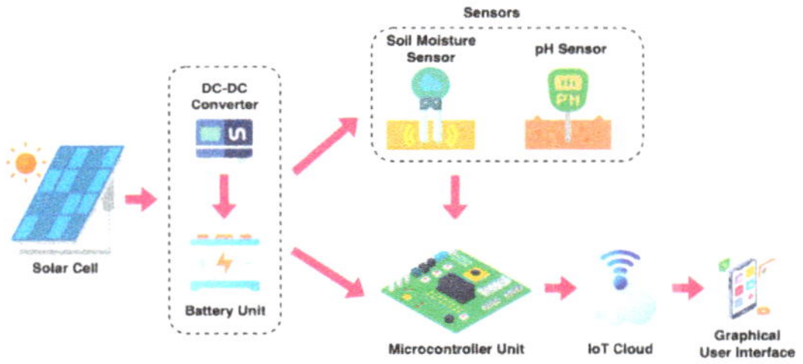

Figure 2. Overall hardware block diagram of the proposed device.

The entire device components are arranged in a three-stack structure, in which the solar cell is placed at the top stack; the battery, microcontroller unit, DC-to-DC converter, and battery charger unit are mounted on the middle stack; and the sensors, namely the pH sensor and the soil moisture sensor, are arranged on the bottom stack. Also, the foam floats were attached to the bottom stack to make the device float if the land was filled with water.

2.1.1. Solar Cell

The proposed device was a made as a standalone device, in which the electrical energy is generated with the help of a solar cell. In general, a solar cell is a device that converts light energy, due to solar irradiance, into electricity. In this work, a compact solar cell with a dimension of no more than 14 cm was used, which generated an output voltage of 12 V and a current of 200 mA. Also, the adopted solar cell was capable of generating the 2.4 watts of power.

2.1.2. DC-to-DC Converter

The output of the solar cell was 12 volts, and all the device components utilized in this work required a maximum of 5 volts. So, it was essential to step down the voltage level of the solar cell, and to achieve this operation, a DC-to-DC converter was used [12]. The XY-3606-based DC-to-DC converter was used as a step-down converter, which reduced the voltage from 12 volts to 5 volts and supplied a maximum current of 5 A.

2.1.3. Battery

In this work, a solar cell was utilized to generate electrical energy. Also, the solar irradiance may not be constant all the time, and, in turn, the electrical energy generated varies with time. To supply a constant source to all the device components, the battery was utilized. Further, a lithium polymer (LiPo) battery with 3.7 volts (1000 mAh) was used. Also, the LiPo battery was charged using the LiPo battery charger circuit.

2.1.4. Sensors

In this work, two different sensors were used, namely the pH sensor and the soil moisture sensor. Further, the acidity and basicity of the soil were measured with the help of the pH sensor. In general, its value ranges from 1 to 14, where 1 indicates that the soil is more acidic and 14 indicates that the soil is more basic in nature. Also, a pH value of 7 indicates that the soil is neither acidic nor basic. Furthermore, the proposed pH sensor was capable of operating from 3.3 to 5.5 volts. Also, the same pH sensor module was utilized to measure the temperature of the soil. The soil moisture sensor module has a pair of electrodes and a comparator board that operates from 3.3 to 5 volts.

2.1.5. Microcontroller Unit

An ESP8266 microcontroller unit, otherwise known as the Node MCU, was utilized in this work to collect sensor data. Further, the Node MCU operates on a 5-volt power supply, which can be powered using a LiPo battery through the Pololu U3V70F5 board. The Node MCU has an in-built WiFi module that helps the user to feed their sensor data to the IoT cloud. In general, the ESP8266 has one analog pin to which one sensor can be connected. So, the external analog-to-digital converter, namely the ADS1115, was connected to the utilized Node MCU. Further, the sensor data were converted into digital data, and these digital data were fed to the Node MCU via the inter-integrated circuit (I2C) protocol.

2.1.6. IoT Cloud Platform

Two different IoT clouds, namely the ThingSpeak platform and the custom-designed IoT platform, were used in this work to log the parameters of the agricultural land. The ThingSpeak IoT cloud platform stores the data with respect to time. Also, the custom-designed IoT platform has various features to monitor the wetness, temperature, and pH of the soil.

3. Results and Discussion

Figure 3 shows the ThingSpeak IoT cloud platform. A ThingSpeak account can be created by any person for free, and the sensor data can be logged to the appropriate account.

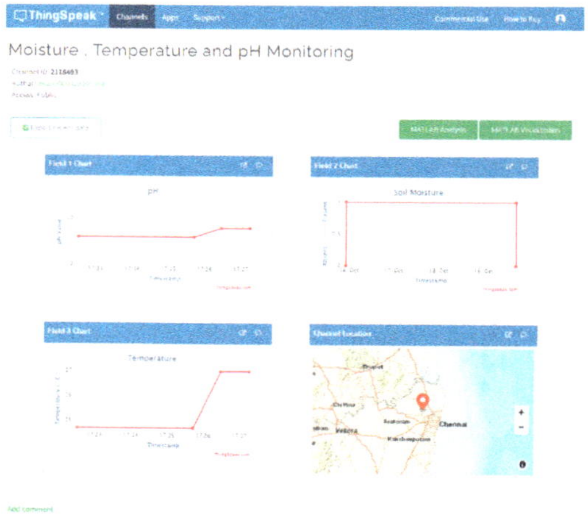

Figure 3. ThingSpeak IoT cloud platform.

Every account was created with an individual read-and-write application programmable interface (API) key. It was observed that the three different sensor values, such as the pH, temperature, and moisture of the soil, were logged as a graph in the ThingSpeak account. Also, it was seen that the pH of the soil was logged in the field 1 chart, while soil moisture was logged in the field 2 chart. In general, the soil moisture sensor provided values in the form of analog signals especially from 0 to 5 volts, depending on the level of soil moisture. In this work, a threshold of 60% was set, and once the moisture sensor sensed beyond 60%, it produced 5 volts, which was considered one. Also, if the soil moisture sensor sensed below 60%, it produced 0 volts, which was considered 0. Both of these cases were logged in the field 2 chart of Figure 3. The field 3 chart was utilized to log the temperature of the soil. Additionally, the location of the farm is shown as a channel location. Figure 4a shows the custom-designed IoT cloud platform that was designed for smart farming. It was demonstrated that the smart sensor modules were located at distinct places, as shown in the farm layout of Figure 4a. Also, each smart sensor module was capable of measuring three different parameters, namely the pH, temperature, and moisture. It was observed that the sensor data of all the smart sensor modules were logged. Furthermore, the logged data can be visualized by pressing the logs icon, the water pump can be activated by pressing the pump button, and the pump will be switched off once all the three smart sensor modules sense soil moisture in three different places of farmland.

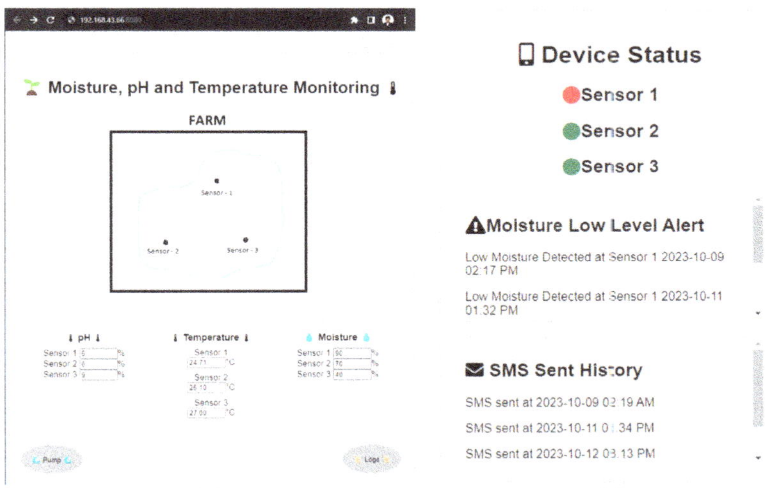

(**a**) (**b**)

Figure 4. (**a**) Custom-designed IoT cloud platform for smart farming. (**b**) Custom-designed IoT cloud platform to monitor the status of the smart device.

Typically, the pH of the soil changes due to factors such as rainfall, weather conditions, etc. In this regard, it is essential to monitor the pH of the soil so that the wellness of the crops can be ensured. Additionally, the biological activities of the soil depend on the soil temperature. Figure 4b shows the custom-designed IoT cloud platform to monitor the status of smart sensor modules. Further, it was observed that if the smart sensor module sensed soil moisture, it will alert the farmer with a notification, as shown in Figure 4b. Also, the short messaging service (SMS) was sent with the help of the Twilio service since the smart sensor modules were connected to the internet. It was clearly seen that the device provides the low-level moisture alert via SMS to the farmer. Also, the proposed IoT-based smart sensor module was found to be extremely efficient, which helps the farmer to take care of their agricultural parameters without any deviation.

4. Conclusions

Generally, multi-modal farming systems consist of wired data transmission; these wires can be damaged by rodents or weather conditions, which would make it hard to assess the fault. Also, the data visualization of the agricultural IoT devices fails when people have low literacy. In this work, a smart sensor-based agricultural land monitoring system was designed and developed in which farmers can visualize agricultural parameters, such as the pH, temperature, and moisture of the soil, through the ThingSpeak or custom-designed IoT cloud platforms. The results demonstrate that the proposed smart sensor modules are capable of monitoring agricultural parameters, such as the pH, temperature, and moisture of the soil, which helps the farmer to take the necessary actions towards the growth of their crops. This work appears to be of high societal relevance since it can be easily implemented and maintained. Also, this system is less expensive and self powered using solar-based renewable energy resources, which can be utilized to monitor all the agricultural parameters under an uneven landscape.

Author Contributions: D.S. and P.A. conceptualized the idea for this work. V.S. provided the required resources. P.S.K.R.P. designed and developed the hardware. P.S.K.R.P. and D.S. carried out the investigation and data curation. V.S. designed the visualization. P.A. validated the acquired results and prepared the original draft. D.S. and V.S. reviewed and edited the original draft. V.S. supervised, and P.A. administered the work. All authors have read and agreed to the published version of the manuscript.

Funding: This research received no external funding.

Institutional Review Board Statement: Not applicable.

Informed Consent Statement: Not applicable.

Data Availability Statement: Data are contained within the article.

Conflicts of Interest: The authors declare no conflicts of interest.

References

1. Kour, V.P.; Arora, S. Recent developments of the internet of things in agriculture: A survey. *IEEE Access* **2020**, *8*, 129924–129957. [CrossRef]
2. Dhanaraju, M.; Chenniappan, P.; Ramalingam, K.; Pazhanivelan, S.; Kaliaperumal, R. Smart farming: Internet of Things (IoT)-based sustainable agriculture. *Agriculture* **2022**, *12*, 1745. [CrossRef]
3. Husnain, A.; Hussain, I.; Ijaz, A.B.; Zafar, A.; Ch, B.I.; Zafar, H.; Sohail, M.D.; Niazi, H.; Touseef, M.; Khan, A.A.; et al. Influence of soil pH and microbes on mineral solubility and plant nutrition: A review. *Int. J. Agric. Biol. Sci.* **2021**, *5*, 71–81.
4. Tzounis, A.; Katsoulas, N.; Bartzanas, T.; Kittas, C. Internet of Things in agriculture, recent advances and future challenges. *Biosyst. Eng.* **2017**, *164*, 31–48. [CrossRef]
5. Idoje, G.; Dagiuklas, T.; Iqbal, M. Survey for smart farming technologies: Challenges and issues. *Comput. Electr. Eng.* **2021**, *92*, 107104. [CrossRef]
6. Marković, D.; Koprivica, R.; Pešović, U.; Randić, S. Application of IoT in monitoring and controlling agricultural production. *Acta Agric. Serbica* **2015**, *20*, 145–153. [CrossRef]
7. Ayaz, M.; Ammad-Uddin, M.; Sharif, Z.; Mansour, A.; Aggoune, E.H. Internet-of-Things (IoT)-based smart agriculture: Toward making the fields talk. *IEEE Access* **2019**, *7*, 129551–129583. [CrossRef]
8. Devi, G.; Sowmiya, N.; Yasoda, K.; Muthulakshmi, K.; Balasubramanian, K. Review on application of drones for crop health monitoring and spraying pesticides and fertilizer. *J. Crit. Rev.* **2020**, *7*, 667–672.
9. dos Santos, U.J.; Pessin, G.; da Costa, C.A.; da Rosa Righi, R. AgriPrediction: A proactive internet of things model to anticipate problems and improve production in agricultural crops. *Comput. Electron. Agric.* **2019**, *161*, 202–213. [CrossRef]
10. Sreekantha, D.K.; Kavya, A.M. Agricultural crop monitoring using IOT–A study. In Proceedings of the 2017 11th International Conference on Intelligent Systems and Control (ISCO), Coimbatore, India, 5–6 January 2017; pp. 134–139.
11. Lova Raju, K.; Vijayaraghavan, V. IoT technologies in agricultural environment: A survey. *Wirel. Pers. Commun.* **2020**, *113*, 2415–2446. [CrossRef]
12. Paramasivam, A.; Bhaskar, K.B.; Madhanakkumar, N.; Vanchinathan, C. Analysis of an Enhanced Positive Output Super-Lift Luo Converter for Renewable Energy Applications. In *Advances in Smart Grid Technology: Select Proceedings of PECCON 2019—Volume I*; Springer: Singapore, 2020; pp. 127–136.

engineering proceedings

MDPI

Proceeding Paper

Celestial Body Surface Mapping for Resource Discovery Using Satellites †

Yuxian Li *, Cesar Vargas-Rosales * and Rafaela Villalpando-Hernandez

Tecnologico de Monterrey, School of Engineering and Sciences, Monterrey 64849, Mexico; rafaela.villalpando@tec.mx

* Correspondence: a00829851@tec.mx (Y.L.); cvargas@tec.mx (C.V.-R.)

† Presented at the 10th International Electronic Conference on Sensors and Applications (ECSA-10), 15–30 November 2023; Available online: https://ecsa-10.sciforum.net/.

Abstract: Exploring the solar system provides new possibilities for the survival and development of humankind. Although we do not yet have the technology to migrate on a large scale, we must first find proper celestial bodies, understand their environment, and utilize them. In the future, with highly developed space technology, planets and natural satellites will become "islands" in the solar system for human settlements. However, the lack of information on this matter is one of the greatest challenges in this field. And, for the purpose of geographic data collection, improvements in surface precision and resolution with highly accurate 3D modeling and 2D maps are needed, and the surface of celestial bodies should be mapped by making full use of geometry and different types of map projection methods. Different techniques should be used for the accurate localization of rovers using satellites, thus combining both to create a map of the resource distribution. In this article, a projection method based on conventional techniques for better accuracy and efficiency of mapping and projections of a celestial body is proposed. For further research, hardware enhancements, improvements in multimodal techniques, and the development of communication protocols for rovers and satellites could be pivotal.

Keywords: projection mapping; localization algorithm; surface construction

Citation: Li, Y.; Vargas-Rosales, C.; Villalpando-Hernandez, R. Celestial Body Surface Mapping for Resource Discovery Using Satellites. *Eng. Proc.* **2023**, *58*, 14. https://doi.org/10.3390/ecsa-10-15998

Academic Editor: Francisco Falcone

Published: 15 November 2023

1. Introduction

The observation of celestial bodies has evolved with the use of advanced instruments, and their surface projection for resource localization serves as inspiration for humankind. To discover features, projection methods used for the surface of the Earth are used, e.g., the Mercator, Behrmann, and Lambert methods [1]. Yet, they have inefficiencies; e.g., the Mercator method, which is spherically based, is not suitable for elliptical shapes (significant differences far from the Equator) [1].

In this paper, a projection method integrating the characteristics of other methods is proposed to improve accuracy and data collection and to help in the exploration of celestial bodies through a satellite system. Using already existing methods, in cooperation with a rover, a map of the distribution of elements and key features of the celestial body surface is built.

The rest of the paper is organized as follows: in Section 2, the general idea about the proposal is described, and the 2D/3D projection methods are explained. Section 3 gives details about the algorithms for the satellites. Then, the instrument requirements are explained in Section 4. Section 5 is designated for the discussion, and finally, Section 6 for the conclusions.

2. Revolutionizing Celestial Body Projection

Methods to project the surface of the Earth have been used to better understand it. However, all methods have errors due to factors such as gravitational perturbations, atmospheric conditions, and instrument errors, and by combining methods, a new approach with features from all can be obtained. The proposed method is a mixture of other methods, where the projection method uses geometry and satellites jointly to form images and to create 2D/3D maps. Algorithms, jointly with on-surface vehicles, create resource distribution maps. We use the case of Mars; however, this method is valid for similar celestial bodies.

Proposed Method

The method integrates seven low orbit satellites in the exosphere (above 200 km [2]) of Mars, where atmospheric density can be ignored, gravity is almost imperceptible, and low energy consumption is achieved [2] (page 39). The satellite orbit inclination needs to be the same as that of Mars, 25.2°. With equations for motion and dynamic force models acting on the satellite, like atmospheric drag, among others, one can calculate the orbit. Our purpose is not to obtain the orbits, since algorithms already exist [3]. Thus, without a loss of generality, we let the inclination of Mars be 0° (Figure 1A), and by dividing Mars into six parts with similar areas using seven satellites, one oversees the areas of Mars as shown in Figure 1B.

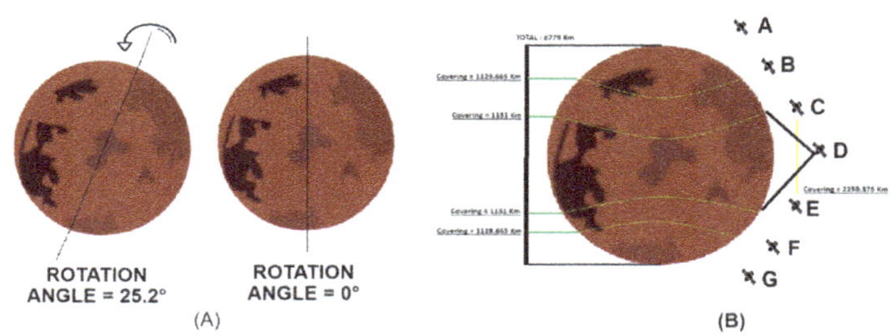

Figure 1. (**A**) Mars inclination angle. (**B**) General view of the satellite system where surface coverage is performed by seven satellites identified as A–G.

For satellite A (see Figure 2A), the distance to the surface (altitude $x = 250$ km) is $d = \sqrt{(1129.665 + x)^2 + (1129.665 + x)^2} = 1951.14$ km. Then, it can be located at approximately 1951.14 km at an angle of 45° from the origin of the surface to cover the 1129.665 km radius. Since Mars is spherical, satellite G has the same characteristics as A for the lower part of Mars. For the second area, satellite B has to be placed just at the edge of what satellite A reaches (1129.665 km). Satellites B and C have the same altitude, which defines a straight plane. Satellites are equipped with monitoring cameras, and their angles and the minimum distances they need to reach (the hypotenuse) are obtained using trigonometry, where the opposite and adjacent sides are the longitude and latitude of Mars. For B and F, the camera has a wide angle of 127.874° and a minimum distance of 1432.8194 km for a coverage of 1131 km, i.e., $\sphericalangle = tan^{-1}(1129.665 - altitude/1131) = 37.874° + 90° = 127.874°$. Also, for C and E (see Figure 2B), the camera angle is 24.2308° with a minimum distance of 1156.997 km for the coverage, i.e., $\sphericalangle = tan^{-1}(1131/(2260 + altitude)) = 24.2308°$.

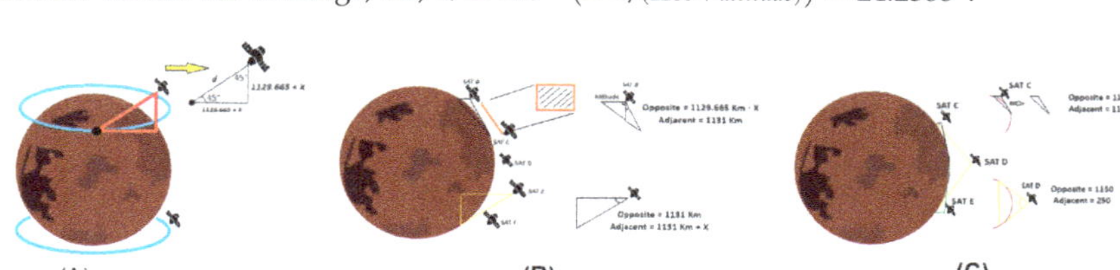

Figure 2. (**A**) Representation of satellite A and G and orbit trajectory. (**B**) Projection and coverage for the satellites B, C, E and F. (**C**) Coverage and the angle of satellite D.

Note that the higher the altitude is, the smaller the angle and the easier it is to adjust the camera position. The Mercator method projects the surface between satellites C and D with the same altitude. The orbit of satellite D forms a cylinder, inside which Mars is located. Satellites C and E will have cameras rotating for the multimodal technique, for better resolution and accuracy. Regarding Figure 2C, we obtain the condition of the camera's angle as follows: $\sphericalangle = tan^{-1}(1130/_{altitude}) = 77.524^\circ \times 2 = 155.049^\circ$, and the minimum distance satellite D covers (2260 km) is given by $Hypotenuse(SAT\ D) = 1157.3245$ km.

3. Algorithms for Enhancing Surface Condition Capture

With the coverage areas defined, each satellite, except A and G, rotates the camera for the multimodal technique. For satellites A and G, which are within a hexagon (see Figure 3A), there are six points of contact, indicating positions (separated 722.39 km) where images of the surface are taken, i.e., $C = \pi \times 1129.665 + altitude \rightarrow \frac{C}{6} = 722.39\text{km}$. For the remaining satellites, the orbits are divided into 12 points, and a dodecagon is formed (see Figure 3B). Between each point, a rotation of the camera is performed with angles already calculated and pointing directions of camera indicated by the arrows (see Table 1). For satellites B and F, the points are separated $\frac{C}{12} = 361.195\text{km}$. By rotating the camera, satellites A and C take shots with different angles for zones 1 and 2, respectively. For satellites C and E, we have $C = \pi \times 2260 + altitude \rightarrow \frac{C}{12} = 591.666\text{km}$. For satellite D, rotation is not necessary, due to its coverage angle (155.049°).

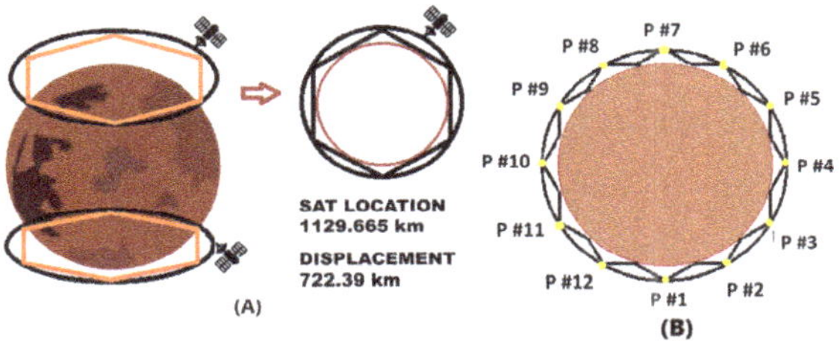

Figure 3. (**A**) Positions for SAT A and F shots. (**B**) Dodecagon and points for the remaining satellites.

Table 1. Summary of relationship between taking shots and positions for a single iteration. Arrows indicate up or down direction the camera is pointing at.

Sat/# Position													Distance between Points (km)	Up Angle	Down Angle
SAT A	1↓		2↓		3↓		4↓		5↓		6↓		722.39	None	45°
SAT B	1↑	2↓	3↑	4↓	5↑	6↓	7↑	8↓	9↑	10↓	11↑	12↓	361.195	39.310°	127.874°
SAT C	1↓	2↑	3↓	4↑	5↓	6↑	7↓	8↑	9↓	10↑	11↓	12↑	591.667	39.3165°	127.874°
SAT D	1←	2←	3←	4←	5←	6←	7←	8←	9←	10←	11←	12←	952.687	155.049°	None
SAT E	1↑	2↓	3↑	4↓	5↑	6↓	7↑	8↓	9↑	10↓	11↑	12↓	591.667	127.874°	39.3165°
SAT F	1↓	2↑	3↓	4↑	5↓	6↑	7↓	8↑	9↓	10↑	11↓	12↑	361.195	127.874°	39.3165°
SAT G	1↑		2↑		3↑		4↑		5↑		6↑		722.39	45°	None

The rotation algorithm is summarized in Table 2A. Once the positions are defined, when the system achieves a cycle of its trajectory, a map is generated due to the relatively large distance between points. The benefit of orbits with 6 and 12 points is the assurance that for each orbit, the reference of the system is always the central axis. The seven satellites will always form a straight line, which ensures that the relationship is maintained (see Figure 4) for the various shots of each satellite (see Table 2B).

Table 2. (**A**) Summary for rotation orders. (**B**) Distance between points considering 360 points.

	(A)		(B)	
SAT/# Point	**Even**	**Odd**	**Satellites**	***N***
SAT A	DOWN	DOWN	SAT A	12.0398 km
SAT B	DOWN	UP	SAT B	12.0398 km
SAT C	UP	DOWN	SAT C	21.9038 km
SAT D	CENTER	CENTER	SAT D	31.756 km
SAT E	DOWN	UP	SAT E	12.0398 km
SAT F	UP	DOWN	SAT F	12.0398 km
SAT G	UP	UP	SAT G	12.0398 km

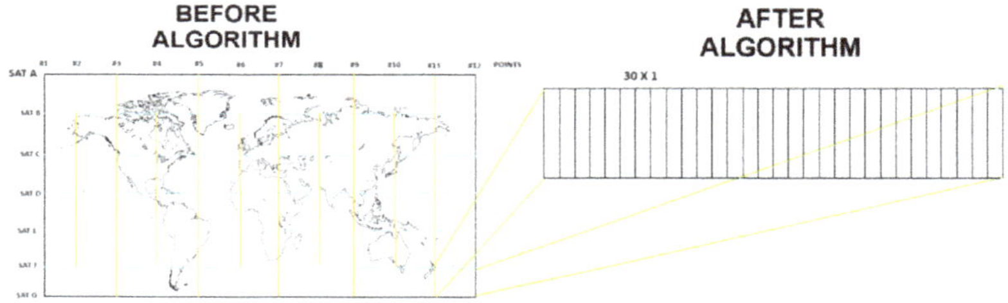

Figure 4. Before and after applying the displacement algorithm.

Displacement Algorithm

To improve the accuracy and observation of Mars' surface, satellites rotate for the hexagon and dodecagon, with orbits divided into 360 points where satellites are always located. The separation distances of the adjacent points, *N*, are noted in Table 2B. For example, regarding satellite B, after the first iteration (point 1), point 1 + *N* takes place. So, with a displacement of 30 points, point 1 is now point 2. However, the algorithm works in cooperation with the camera rotation, and it is necessary to reach 60 iterations to have at least two shots of a zone. After some iterations, Mars is classified with 360 lines (Figure 4). For A and G, only 30 iterations are needed for 360 lines, since the camera position is always the same.

4. Operation Requirements

The multimodal technique involves having several images of the same area from different angles. This results in a unified image with all the features, improving accuracy in location and aiding in data interpretation; thus, a fusion multimodal data model is proposed [4]. Initially, two images are obtained, but over time, a precise map is created through the combination of a large number of images. The key aspect is that with each iteration of the satellite, the map of the surface becomes clearer and more accurate.

4.1. Location for Rover

High-precision rover positioning is a principal factor for the safety of the rover. In planetary exploration, navigation and positioning, the absence of the Global Positioning System (GPS) and the lack of references are the most complicated parts. Thus, the use of localization algorithms is proposed, such as the SQP algorithm, where two satellites efficiently determine the location of an object [5], or the utilization of Mars rover images of ground rocks and satellite images for rock distribution pattern matching to determine the location [6]. Other techniques can be used, e.g., radio measurement positioning. After positioning and communication with satellites, the process of acquiring data begins. Information such as images of the surface is obtained. The creation of stereoscopic images, 3D point cloud data, orthophotos, contour maps and slope maps can be obtained based on the collected data, which is useful for subsequent explorations.

4.2. Hardware and Software

The rover requirements, such as the robot arms, antennas, solar panel, Navigation Camera (Navcam), Panoramic Camera (Pancam), and Obstacle Avoidance Camera (Hazcam) [7], are based on the Spirit model. Also, the CheMin Inlet is key, as it is capable of identifying the minerals and chemical composition of rocks and soil [8]. To obtain projections and mapping, this hardware plays an important role with a system with CubeSats. Apart from the hardware that all CubeSats commonly have, e.g., a camera, a power system, and a transmission system [9], some aspects needed to improve efficiency are the following:

- A small, adaptable and stable high-resolution thermal infrared imaging system [10], with a broadband detector array and a compact mechanical cryocooler. The camera includes a sensor chip with a 25 μm pitch and a resolution of 640 × 512 pixels.
- High spectral resolution cameras installed in the satellites, helping the resource discovery when the rover does not recognize resources or when it is not able to get to some location.
- The rotation system for cameras is necessary to adjust the camera to specific angles for taking photographs. It is based on an odometer and the inertial measurement unit (IMU).
- The inertial measurement unit (IMU) to measure and record acceleration, orientation, angular rates, and gravitational forces. With this, the location or number of points where the satellite is situated can be identified.
- A UHF communication link. This is a necessary transceiver for communication. It can be similar to that used in China's Mars exploration mission, for 'Tianwen One' where the Ultra-High Frequency system was tasked with delivering communication services between the Lander and Encircler during the entry, descent and landing (EDL) phases [11].

With localization and mineral inspection, it is possible to achieve both 2D and 3D mapping of surfaces and mineral distributions. This can be conducted using data collected from the rover and the satellites that are then analyzed with existing software programs such as the following:

- Celestial mapping system (CMS) used for the visualization and analysis of celestial bodies [12]. This application utilizes outdated data from NASA missions. However, we would begin by inputting the gathered information, typically presented in coordinates. As the second step, customization enables the display of specific information. The final step is generating a 2D or 3D map. It is important to note that this process involves updating existing data with new information to generate new maps.
- Nasa's solar system treks, which is a set of tools that provides access to planetary data [13]. First, the desired planet is selected, and as the second step, the different tools are selected to modify the coordinates, visualization, or other features based on the data collected using the proposed method. Ultimately, this software enables us to generate 2D/3D maps for further analysis and allows us to create virtual journeys across the surfaces of planets.

5. Discussion

All projection methods are capable of mapping a celestial body, regardless of its shape, whether spherical or elliptical. However, when compared to other traditional methods, the Berghaus star projection displays distortion mainly in the corners and edges. Also, the polar regions present significant challenges when utilizing the equidistant projection with two points, due to the great distance between the poles. Nevertheless, the proposed projection method is free of distortion, and it has the ability to become increasingly detailed and accurate, depending on the multimodal method and the number of iterations. Furthermore, it is better developed for small celestial bodies, due to limited hardware capabilities (see Table 3) based on [1]. To enhance accuracy, in addition to requiring more sensors or satellites, it is imperative to establish more shot points for a better resolution. Communication between the satellite and rover is essential, and guiding the rover to the best route for a destination could potentially extend the rover's lifespan. Also, hardware for the spectrometer analysis on the satellite could help identify minerals in areas unreachable for the rover, thereby prolonging the rover's lifespan.

Table 3. Comparison of different projection features.

Projections	Poles Zone Precision	Equator Zone Precision	Limitations	Growth Capacity
Bonne's Equator	Low	Low	Suitable for small-scale maps, and not appropriate for illustrating vast regions	No
Berghaus star	Affected by central point	Affected by central point	Complex mathematical calculations, and not for small zones	No
Two points equidistant	High	Low	Beyond two specific points, shape may lose accuracy in other regions	No
Mercator	Low	High	Unsuitable for representing global maps, due to significant distortion	No
Proposed method	High	High	Limited by hardware design, resolution depends on multimodal technique	Yes

6. Conclusions

The proposed method has extensive potential, as it allows for a time evolution beyond the consideration of several established projection methods. The distinct advantage that it has over other approaches is that it yields a higher number of iterations, which results in more precise surface imaging. However, such precision may ultimately be constrained by hardware limitations, but the problem can be optimized by periodic data transmission processes to Earth for storage and processing. Also, it could be enhanced with hardware upgrades, the implementation of the multimodal technique, and improvements in the communication between the rover and satellites. Similarly, through collaboration with the rover, a successful mineral discovery on the surface can be achieved, followed by marking specific locations on a map for future use. Future research will be focused on enhancing hardware, implementing the multimodal technique, and improving communication methods between the rover and satellites.

Author Contributions: Conceptualization, Y.L. and C.V.-R.; methodology, Y.L.; validation, Y.L. and R.V.-H.; investigation, Y.L.; resources, C.V.-R. and R.V.-H.; writing—original draft preparation, Y.L.; writing—review and editing, C.V.-R. and R.V.-H.; supervision, C.V.-R. and R.V.-H.; project administration, C.V.-R. All authors have read and agreed to the published version of the manuscript.

Funding: This research received no external funding.

Institutional Review Board Statement: Not applicable.

Informed Consent Statement: Not applicable.

Data Availability Statement: Data are contained within the article.

Acknowledgments: We would like to acknowledge the support from the Smart Digital Technologies and Infrastructure Research Group, the project "Digital Technologies to Create Adaptive Smart Cities" of the Challenge-Based Research Program, and the School of Engineering and Science at Tecnologico de Monterrey for providing the means to develop this collaborative work.

Conflicts of Interest: The authors declare no conflict of interest.

References

1. Esri. (s.f.). Mercator Projection. ArcGIS. Available online: https://desktop.arcgis.com/zh-cn/arcmap/latest/map/projections/mercator.htm (accessed on 1 September 2023).
2. Haider, S.A. *Aeronomy of Mars*; Springer Nature: Berlin/Heidelberg, Germany, 2023.
3. David, H.; Bohn, P. Precise orbit determination for low earth orbit satellites. *Ann. Marie Curie Fellowsh.* **2006**, *4*, 128–135.

4. Castillo, A.C.; Marroquin-Escobedo, J.A.; Gonzalez-Irigoyen, S.; Martinez-Santoyo, M.; Villalpando-Hernandez, R.; Vargas-Rosales, C. Recreating Lunar Environments by Fusion of Multimodal Data Using Machine Learning Models. *Eng. Proc.* **2022**, *27*, 54. [CrossRef]
5. Ma, X.; Ning, X.; Chen, X.; Liu, J. Geometric Coplanar Constraints-Aided Autonomous Celestial Navigation for Spacecraft in Deep Space Exploration. *IEEE Access* **2019**, *7*, 112424–112434. [CrossRef]
6. Kaichang, D.; Zongyu, Y.; Zhaoqin, L. A new method for Mars rover positioning based on the integration of ground images and satellite images. *Spacecr. Eng.* **2010**, *19*, 8–16.
7. NASA FACTS. Mars Exploration Rovers. Available online: https://d2pn8kiwq2w21t.cloudfront.net/documents/mars03rovers_t6lfTK7.pdf (accessed on 10 September 2023).
8. NASA. ChemCam Instrument (Chemistry and Camera). Mars Science Laboratory—Curiosity Rover. Available online: https://mars.nasa.gov/msl/spacecraft/instruments/chemin (accessed on 3 September 2023).
9. Sophia, S.; Bott, L.; Fong, A.; Morley, J. Ag2030: AWES-The Cubesat solution. 2022. Available online: https://www.nysf.edu.au/wp-content/uploads/2022/02/1_Innovation-Proposal_Jacinta-Morely-Lucy-Bott-and-Sophia-Seumahu-.pdf (accessed on 10 September 2023).
10. NASA. NASA Technology Transfer Portal. CubeSat Compatible High Resolution Thermal Infrared Imager (GSC-TOPS-138). Available online: https://technology.nasa.gov/patent/GSC-TOPS-138 (accessed on 5 September 2023).
11. Zezhou, S.; Huiping, Q.; Yu, H.; Fan, B.; Ting, Z.; Ji, X. Design of UHF band relay communication system for "Tianwen-1" Mars probe. *J. Nanjing Univ. Aeronaut. Astronaut./Nanjing Hangkong Hangtian Daxue Xuebao* **2022**, *54*, 5.
12. Celestial Mechanics and Dynamical Astronomy. Available online: https://celestial.arc.nasa.gov/ (accessed on 10 September 2023).
13. Solar System Treks. Available online: https://trek.nasa.gov/ (accessed on 10 September 2023).

engineering proceedings

Proceeding Paper

Intelligent Interplanetary Satellite Communication Network for the Exploration of Celestial Bodies †

Abraham Urieta-Ortega, Cesar Vargas-Rosales * and Rafaela Villalpando-Hernandez

Tecnologico de Monterrey, School of Engineering and Science, Monterrey 64849, Mexico; a01274805@tec.mx (A.U.-O.); rafaela.villalpando@tec.mx (R.V.-H.)
* Correspondence: cvargas@tec.mx

† Presented at the 10th International Electronic Conference on Sensors and Applications (ECSA-10), 15–30 November 2023; Available online: https://ecsa-10.sciforum.net/.

Abstract: Recently, a significant interest in space exploration has emerged, driven by the lack of resources and the quest for answers to issues like climate change. New technologies give us the possibility of exploring our solar system and its surroundings in greater detail. But current space communications operate with a lack of efficiency due to the vast distances between celestial bodies within our solar system. Also, factors such as bandwidth asymmetry contribute to disruptions in the satellite communication network. This paper proposes the definition of the infrastructure of an interplanetary communication network, built upon a communications protocol featuring dynamic routing. This infrastructure aims to optimize information transmission by adapting communications to surrounding conditions. The envisioned infrastructure involves strategically placing network nodes at key Lagrange points around each planet within the asteroid belt. The nodes will be aware of their position, integrating sensing capabilities and intelligent algorithms. Next to each planet, a node with more capabilities will collect information from nanosatellites orbiting a planet and relay the information back to Earth. This structure will allow decision-making processes based on exploration data of the most significant celestial bodies within the asteroid belt, providing valuable insights such as constant monitoring of the dark side of the moon and difficult-to-reach zones in the solar system.

Keywords: interplanetary internet; communication network; celestial bodies exploration

Citation: Urieta-Ortega, A.; Vargas-Rosales, C.; Villalpando-Hernandez, R. Intelligent Interplanetary Satellite Communication Network for the Exploration of Celestial Bodies. *Eng. Proc.* **2023**, *58*, 15. https://doi.org/10.3390/ecsa-10-15999

Academic Editor: Stefan Bosse

Published: 15 November 2023

1. Introduction

Currently, we face challenges such as global warming and overconsumption of natural resources, and it is essential to continue seeking solutions on Earth. However, it is equally important to explore alternatives for resources beyond our planet to ensure the human survival. Currently, we face another issue: space explorations today are limited by the data communication capabilities of devices in space, primarily due to the lack of a space communications infrastructure.

This paper presents an innovative solution by proposing a network infrastructure consisting of strategically positioned nodes orbiting celestial bodies to address the lack of space communication infrastructure. These nodes work cooperatively to maintain efficient and stable communications for coverage of celestial bodies within the asteroid belt, addressing the proposed features of a unified communication architecture [1–3].

The paper is organized as follows. Section 2 describes the proposed communication network infrastructure, with hardware descriptions and their placement on celestial bodies. Section 3 outlines the network architecture, proposing routing protocols for different hierarchy levels within the network. Finally, the conclusions are presented in Section 4.

2. Communication Network Infrastructure

The infrastructure of the communication network aims to establish a physical link between network devices by using wireless optical communications technology, which is challenging since it relies on the line of sight (LOS).

The proposed infrastructure at the level of a planet, as shown in Figure 1, will be replicated for all other significant bodies within the asteroid belt. It includes:

- Area Border Router (ABR)
- Five Lagrange Subnet Routers (LSR)
- Swarm of Planet Monitoring Nanosatellites (PMN)

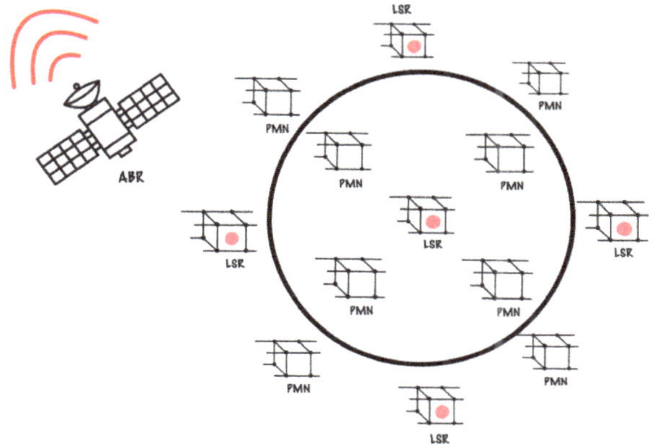

Figure 1. Arrangement of Network Devices for a Celestial Body.

2.1. Area Border Router (ABR)

This network node has the capability to establish and manage high-speed connections with the LSR satellites in the network, allowing effective interconnectivity of different subnetworks. Its high processing capacity with multiple cores is based on the RISC architecture, offering a better balance between performance and energy consumption. This enables data processing for navigation, orbit control, and protocol management with low latency.

The ABR is equipped with solar power systems for its payload systems and energy storage via supercapacitors for actions that require a higher amount of power, such as antenna orientation corrections. In addition to these energy systems, for locations and potential uncontrollable events where sunlight is limited, these satellites are equipped with radioisotope thermoelectric generators (RTGs).

Regarding navigability, to maintain a correct path, the ABR adjusts its course based on position feedback from the LSR satellites in orbit to ensure uninterrupted communication.

2.2. Lagrange Subnet Routers (LSR)

These satellites operate similarly to Low Earth Orbit Relay Satellites (LEO Relay Satellites) and are crucial as they will be responsible for relaying the information collected from the PMNs to the ABR satellite. Therefore, in terms of communication payload, they are equipped with transponders and signal amplifiers capable of fulfilling the data relay objective. The LSRs have solar power systems and batteries to ensure proper power supply for their payload and navigation system.

2.2.1. Lagrange Point Placing

To address the challenges associated with wireless optical communication technology, it is proposed that the five LSRs be positioned at Lagrange points. Lagrange points offer advantages in terms of orbital stability for devices such as satellites. Taking the Earth-Moon system as an example, these points are named L_1, L_2, L_3, L_4, and L_5, see Figure 2. L_4 and L_5 are considered stable, making it relatively straightforward to maintain a position relative to the planet without significant orbital adjustments. However, a problem arises with L_1 L_2, and L_3 points, as they are considered unstable. Despite their potential utility in maintaining a certain position for devices, these points are susceptible to disturbances in their orbital paths due to gravitational and non-gravitational forces. When a nanosatellite moves away from its equilibrium point in these regions, it continues to drift away.

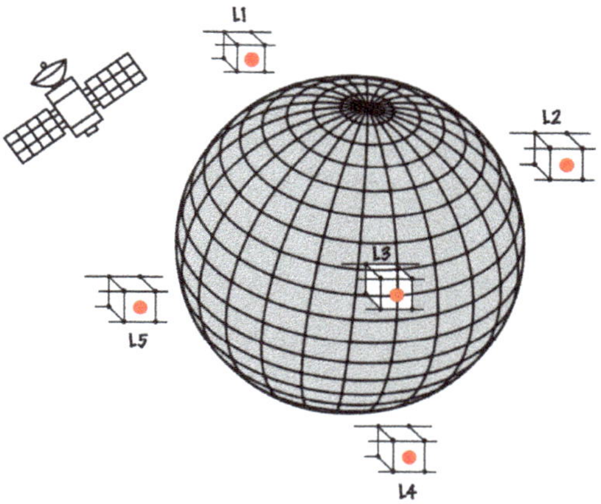

Figure 2. LSRs Arranged on a Planet.

The physical design of the LSR comes into play here. To utilize all five Lagrange points effectively, the LSR must have a significant amount of mass, enough to make them less susceptible to gravitational and non-gravitational forces that may disrupt their planned orbits.

2.2.2. Auto-Localization

Please note that, in addition to the LSR design, the rest of the swarm involves continuous orbit control and correction. For this purpose, the methodology of navigability through the analysis of local gravity maps, as proposed in [4], is suggested.

This methodology is based on the selection and analysis of gravity maps to support autonomous navigation. By converting these maps into grayscale images and performing statistical analyses based on frequency distribution (e.g., histograms, gradient values), characteristics related to gravitational anomalies in a specific area can be obtained. The authors in [4] note that larger gravity anomalies or differences within a zone enhance navigability. Consequently, this inertial navigation method is ideal for LSR satellites, given the significant difference in gravity between a Lagrange point and its surroundings. Thus, the navigation system can process the optimal route and adjust the course of the nanosatellites to maintain their positions at the Lagrange points [2].

Concerning the navigation payload, the LSR is equipped with sensors like those used in submarines [5]. These sensors collect data about gravity anomalies in their surroundings and correct their inertial navigation system (INS) without depending on external feedback.

2.3. *Planet Monitoring Nanosatellites*

The potential of a swarm of PMNs is significant. With a payload focused on science, these nanosatellites aim to explore the celestial body to which they are deployed. They can be equipped with specialized sensors such as imaging sensors, spectrometers, magnetometers, radiometers, and high-resolution cameras capable of operating in space, as illustrated in [6,7]. The instruments and sensors on each PMN will depend on the mission's objectives and the area of the celestial body where they are operating.

To capture data from different angles or locations within the deployment area, PMNs are equipped with control and maneuverability systems [7]. These systems allow them to adjust their orientation and change their orbit, which is determined based on reference points provided by the LSRs. On the other hand, due to their limited physical size, nanosatellites have constraints when it comes to energy. PMNs utilize solar panels with maximum power point tracking (MPPT) algorithms. These algorithms automatically track the maximum power point on the solar panels, maximizing the efficiency of solar energy conversion into electricity. Additionally, they work together with voltage and current sensors to always monitor energy consumption efficiency on the PMN.

3. Network Protocol

Each selected celestial body within the infrastructure design will have its own independent communication network, which will be interconnected with other networks of the same size (subnets), creating a unified network that reaches all selected celestial bodies. The entire network, including the communication subnets of the infrastructure, is modeled according to the Open Systems Interconnection (OSI) Model.

3.1. Routing Protocol

In this layer, network segmentation is defined, playing a crucial role in device interconnection. For the network proposed, the use of the Open Shortest Path First (OSPF) routing protocol is considered, which bases its costs on link bandwidth.

All routers in the network are programmed with the same protocol. This enables them to share information among themselves and learn routes from the entire network topology, not just from their neighbors. Figure 3 illustrates the structure of an OSPF area in the network, i.e., the subnet covering a celestial body.

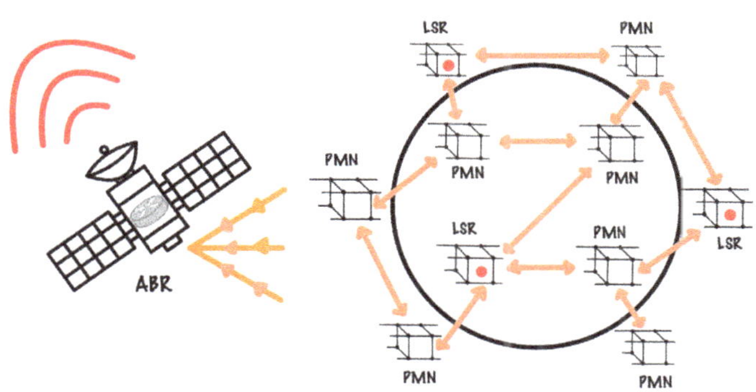

Figure 3. Representation of interconnections among ABR, LSR and PMN swarm.

Within this subnet, nanosatellites orbiting the celestial body act as devices. Satellites marked with a red dot represent those located at Lagrange points (LSR), which serve as routers within the subnet. The larger, higher-ranking satellite in the OSPF protocol is an Area Border Router (ABR), facilitating data intercommunication with other areas and providing information for routing decisions.

As the network topology is dynamic, both LSR and the ABR routers must react and select the best available route based on metrics associated with each end-to-end route interface's cost.

3.2. External Routing Protocol

Currently, there is no common standard for space-based communications. To establish communication links with other space-based devices, a public IP is required, which is where the EGP protocol comes into play.

Building on the Link State (LS) routing algorithm and scaling up within the network that reaches all celestial bodies within the asteroid belt, interconnection links are proposed to be established through router satellites ABR. These router satellites will be responsible for constructing the algorithm's database, generating their respective routing tables using the Dijkstra algorithm [8] to find the shortest path to all possible destinations within the network, and determining which interface to use to reach them. Lastly, they will update the network topology as it changes, regenerating routing tables when the state changes.

4. Conclusions

To enable more active exploration of celestial bodies, this paper proposes a satellite infrastructure that covers the most significant celestial bodies within the asteroid belt. It establishes specifications for the distribution of various satellites within the network, utilizing Lagrange points as strategic placement locations for router nanosatellites around celestial bodies, along with the technical requirements that the infrastructure must meet to achieve proper communication. Furthermore, it presents a network architecture with dynamic routing that allows for adaptation to potential changes or variations within the network, establishing both IGP and EGP routing protocols.

Author Contributions: Conceptualization, A.U.-O. and C.V.-R.; methodology, A.U.-O.; validation, A.U.-O. and R.V.-H.; investigation, A.U.-O.; resources, C.V.-R. and R.V.-H.; writing—original draft preparation, A.U.-O.; writing—review and editing, C.V.-R. and R.V.-H.; supervision, C.V.-R and R.V.-H.; project administration, C.V.-R. All authors have read and agreed to the published version of the manuscript.

Funding: This research received no external funding.

Institutional Review Board Statement: Not applicable.

Informed Consent Statement: Not applicable.

Data Availability Statement: Data sharing is not applicable.

Acknowledgments: We would like to acknowledge the support from the Smart Digital Technologies and Infrastructure Research Group, the project "Digital Technologies to Create Adaptive Smart Cities" of the Challenge-Based Research Program, and the School of Engineering and Science at Tecnológico de Monterrey for providing the means to develop this collaborative work.

Conflicts of Interest: The authors declare no conflicts of interest.

References

1. Alhilal, A.Y.; Braud, T.; Hui, P. A Roadmap Toward a Unified Space Communication Architecture. *IEEE Access* **2021**, *9*, 99633–99650. [CrossRef]
2. Carmona-Gallegos, S.Z.; Duran-Bonilla, C.E.; Abboud, K.; Gongora-Torres, J.M.; Vargas-Rosales, C. The Interplanetary Internet for Observation and Monitoring of the Solar System. *Eng. Proc.* **2022**, *27*, 63. [CrossRef]
3. Alhilal, A.; Braud, T.; Hui, P. The Sky is NOT the Limit Anymore: Future Architecture of the Interplanetary Internet. *IEEE Aerosp. Electron. Syst. Mag.* **2019**, *34*, 22–32. [CrossRef]
4. Liu, F.; Li, F.; Jing, X. Navigability Analysis of Local Gravity Map with Projection Pursuit-Based Selection Method by Using Gravitation Field Algorithm. *IEEE Access* **2017**, *7*, 75873–75889. [CrossRef]
5. New to Subsea Navigation? Available online: https://www.nortekgroup.com/knowledge-center/wiki/new-to-subsea-navigation (accessed on 11 October 2023).
6. Pajusalu, M.; Slavinskis, A. Characterization of Asteroids Using Nanospacecraft Flybys and Simultaneous Localization and Mapping. In Proceedings of the 2019 IEEE Aerospace Conference, Big Sky, MT, USA, 2–9 March 2019; pp. 1–9. [CrossRef]
7. Abboud, K.; Carmona-Gallegos, S.Z.; Duran-Bonilla, C.E.; Villalpando-Hernandez, R.; Vargas-Rosales, C. Identification of Magnetic and Gravitational Field Patterns for Localization in Space. *Eng. Proc.* **2022**, *27*, 45. [CrossRef]
8. Tanenbaum, A.S.; Feamster, N.; Wetherall, D. *Computer Networks*, 6th ed.; Pearson Education Limited: London, UK, 2021.

engineering proceedings

MDPI

Proceeding Paper

Devising an Internet of Things-Based Healthcare Medical Container for the Transportation of Organs and Healthcare Products Using Unmanned Aerial Vehicles †

Vijayalakshmi Sankaran [1], Paramasivam Alagumariappan [2,*], Balasubramanian Esakki [3], Jaesung Choi [4], Mohamed Thoufeek Kanrar Shahul Hameed [2] and Pavan Sai Kiran Reddy Pittu [2]

1 Department of Electronics and Communication Engineering, Vel Tech Rangarajan Dr. Sagunthala R&D Institute of Science and Technology, Chennai 600062, India; drvijayalakshmis@veltech.edu.in
2 Department of Biomedical Engineering, Vel Tech Rangarajan Dr. Sagunthala R&D Institute of Science and Technology, Chennai 600062, India; vtu16823@veltech.edu.in (M.T.K.S.H.); vtu18106@veltech.edu.in (P.S.K.R.P.)
3 Department of Mechanical Engineering, National Institute of Technical Teachers Training and Research (NITTTR), Ministry of Education, Government of India, Chennai 600113, India; esak.bala@gmail.com
4 Department of Computer Science and Engineering, Sunmoon University, Chungnam 31460, India; jschoi@sunmoon.ac.kr
* Correspondence: draparamasivam@veltech.edu.in or parama.ice@gmail.com; Tel.: +91-984-378-0801

† Presented at the 10th International Electronic Conference on Sensors and Applications (ECSA-10), 15–30 November 2023; Available online: https://ecsa-10.sciforum.net/.

Abstract: Every second counts when a patient who requires an organ transplant is finally matched with a donor. The organ's post-transplant performance declines with the increasing time between the organ's removal and transplantation into the recipient. Organs must be transported from point A to B as quickly and safely as possible to improve the chances of success. In addition to delivering medical goods or vaccines to hard-to-reach places, drones can help us to save lives across the world, but there, are some issues to address, one of which is maintaining container temperature and humidity and monitoring it. Further, drones carrying medical containers flying at different altitudes causes temperature changes, which may affect the organs. To tackle such difficulties, in this work a smart container embedded with a Peltier module (thermoelectric cooler) and a temperature sensor has been developed to maintain the temperature thereby providing safety for healthcare products or organs. Further, the relay module is utilized to control the Peltier module and ESP8266 WIFI Microcontroller (MCU) which also enables the user to send live data to the cloud and also allows the user to monitor and control the temperature remotely. The Blynk Internet of Things (IoT) platform is used to monitor the temperature. Results show that the proposed system is highly efficient at monitoring and controlling temperature changes accurately according to user-defined values. For demonstration purposes, the temperature of the container is maintained at 12 degrees Celsius and the performance of the system is presented. The medical cargo drone carrying healthcare products is tested in real time and at different altitude levels to examine the performance of the developed system.

Keywords: drone; healthcare; humidity; internet of things; Peltier; temperature; transportation

Citation: Sankaran, V.; Alagumariappan, P.; Esakki, B.; Choi, J.; Hameed, M.T.K.S.; Pittu, P.S.K.R. Devising an Internet of Things-Based Healthcare Medical Container for the Transportation of Organs and Healthcare Products Using Unmanned Aerial Vehicles. *Eng. Proc.* **2023**, *58*, 16. https://doi.org/10.3390/ecsa-10-16003

Academic Editor: Stefano Mariani

Published: 15 November 2023

1. Introduction

In recent years, the healthcare industry has undergone tremendous technological advancements. The scope of these advancements has no limitations, ranging from an Internet of Medical things to advancements in the point of robotic surgeries. Even after undergoing such remarkable development, traditional methods are still adopted to transport healthcare products or organs in the majority of nations [1]. Developed nations have adapted the usage of unmanned aerial vehicles for the transportation of organs as to speed up the delivery process and enhance the chances of successful medical outcomes, maintaining the organ's function until transplanting it into the patient is ensured through effective organ conservation and protection, which is a crucial part in transplantation and also these UAV's or Medical Cargo Drones are able to deliver medical goods at hard to reach places which could help us to save lives during calamities [2–4]. But these UAV's, or Medical Cargo Drones, are subjected to various external factors such as temperature, pressure, humidity, etc., due to the varying altitudes at which these UAV'S or Medical Cargo Drones operate, these changes would have a great impact on the healthcare products or organs which are being carried [3,5].

The various storage temperatures of healthcare products and organs are as follows: RBCs (Red Blood Cells) or whole blood at 2 to 10 °C, FFP (Fresh Frozen Plasma) at 2 to 6 °C, cryoprecipitate 2 to 4 °C, platelets at 20 to 24 °C, frozen vial transport at −25 °C to −15 °C, refrigerated vials transport at 2 °C to 8 °C, and organs at 4 °C to 8 °C. In this context, temperatures below 2 °C significantly increase the risk of cold injury with some proteins denaturing below 0 °C, at temperatures above 12 °C the higher metabolic demand for oxygen leads to irreversible hypoxic injury, and thus significantly impairs organ function [6–8]. Temperature ranges of 2 to 8 °C or 4 to 8 °C are recommended for use with organ preservation solutions [6,9,10]. Medicines are transported in 3 different temperature ranges based on them <−20 °C, 2–8 °C, and 15–25 °C [8].

To maintain these temperature ranges and to ensure the safety of healthcare products and organs this problem needed to be addressed. The aim and objective of this work is to design and develop an IoT-Based Healthcare Medical Container for Transportation of Organs and Healthcare Products using Unmanned Aerial Vehicle.

2. Literature Survey

Over the recent years, organ and vaccine preservation has been a challenging work during transportation. Liao et al. (2021) [5] presented a study on blood preservation during storage and transportation. Further, the authors discussed the various factors that affects the quality of stored blood, such as temperature fluctuations, oxygenation, and the presence of certain additives. Also, the authors described various methods for monitoring the temperature and oxygen levels of stored blood, including temperature-sensitive labels and oxygen sensors. Jing et al. (2018) [7] discussed an overview of organ preservation techniques, highlighting the importance of temperature control during organ preservation. Nyemba et al. (2019) [11] discussed the evaluation and feasibility assessment of refrigeration systems devoid of harmful refrigerants for the storage of vaccines.

The authors discussed the demerits of traditional cooling methods and suggested better preservation techniques of organs and vaccines below, Minasian et al. (2015) [3] discussed the state of the art in heart preservation technology which highlights the importance of temperature control during organ preservation. Also, the authors discussed the limitations of traditional cooling methods and suggested that Peltier cooling modules could provide a better solution for organ preservation. Nyemba et al. (2019) [11], the authors discussed the performance characteristics of various refrigeration systems, including the cooling capacity, the power consumption, and the temperature stability. Also, the authors concluded that the Peltier cooling is a promising technology for vaccine storage due to its low power consumption, compact size, and lack of harmful refrigerants. Sathyan et al. (2021) [1] discussed the importance of efficient organ transportation, highlighting the challenges faced in the preservation of organs during transportation. Further, the authors designed a 3D CAD model of a novel approach that utilizes a combination of Peltier cooling technology and a vacuum insulated container for organ transportation.

Chang et al. (2019) [12] reported on the development and testing of a thermoelectric air-cooling module for electronic devices. The authors tested the performance of the system in terms of the cooling capacity, the power consumption, and the temperature drop across the module. The results of the study showed that the thermoelectric air-cooling module had a cooling capacity of up to 17.7 W, a power consumption of up to 31.5 W, and a temperature drop of up to 14.5 °C. The study also showed that the cooling performance of the system was affected by various factors, such as the ambient temperature, the air flow rate, and the geometry of the heat sink. Afshari et al. (2020) [13] presented a review of Peltier cooling devices and their applications. The authors discussed the basic principles of Peltier cooling and the factors that affect their performance, such as the thermoelectric materials used, the temperature difference across the device, and the electrical input power. Zeng et al. (2020) [14] presented a theoretical analysis of temperature control for a thermal insulation box. The authors demonstrated the various factors that can affect the temperature stability of the box, such as the thermal conductivity of the insulation material, the thermal conductivity of the surrounding environment, and the heat transfer coefficient and also described the various temperature monitoring devices that can be used to ensure proper temperature control during transportation. Remeli et al. (2020) [15] conducted an experimental study of a mini cooler using a Peltier thermoelectric cell. The authors utilized a Peltier thermoelectric cell as the cooling mechanism and tested its performance in terms of the cooling capacity and the coefficient of performance. The authors also investigated the effects of various parameters, such as the input power, the heat load, and the ambient temperature, on the cooling performance of the system. The results of the study showed that the mini cooler had a cooling capacity of up to 7.38 W and a coefficient of performance of up to 0.07. Also, the study showed that the input power, the heat load, and the ambient temperature all had significant effects on the cooling performance of the system. Here, the authors discussed the performance and other factors affecting the Peltier module, providing an analysis how to achieve a better performance.

Schubert et al. (2019) [16] discussed an overview of sensor technologies for monitoring the cold chain during transportation. The authors discussed the various types of sensors that can be used to monitor temperature, humidity, and other factors that can affect the quality and safety of perishable goods during transportation.

3. Materials and Methods

3.1. The Proposed System

In this work, a IoT-based smart healthcare medical box with a temperature control mechanism is designed and developed. Figure 1 shows the overall block diagram of IoT-based smart healthcare medical box. It is observed that the proposed system consists of various components such as a temperature sensor, Peltier module, battery, microcontroller, and relay units.

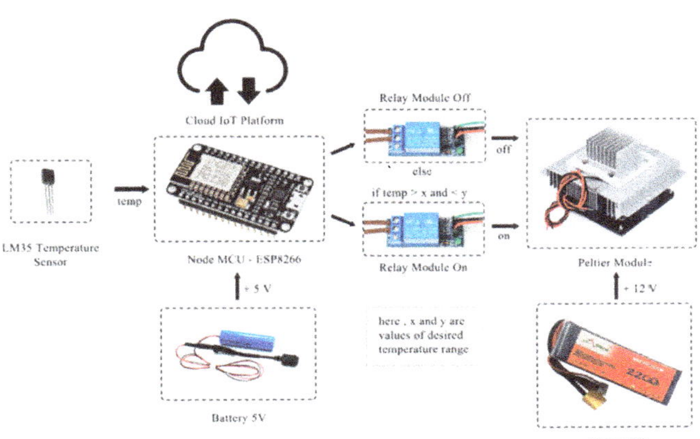

Figure 1. Overall block diagram of IoT-based smart healthcare medical box.

3.1.1. Temperature Sensor

In this work, an LM35 temperature sensor was utilized to measure the temperature of the medical box. LM35 is a commonly used temperature sensor which can be used to measure temperature in degree Celsius. Also, it measures the temperature more accurately than the thermistors. The LM35 is very cheap, compact, and it is a three terminal device which operates on a voltage range of 4 to 30 V. Further, it produces 10-mV per degree Celsius, which is linear and directly calibrated in Celsius. Also, the LM35 temperature sensor can measure the temperature range of −55 degree Celsius to 150 degrees Celsius which is more suitable for the proposed work.

3.1.2. Node MCU-ESP8266

In this work, an ESP8266 node MCU microcontroller was utilized. Further, the ESP8266 has inbuilt WiFi (WiFi protocol—IEEE 802.11 b/g/n) and operates on 5 V supply which can be supplied using batteries. Also, the microcontroller has an analog to digital converter pin in which the LM35 is connected and the analog signals from the temperature sensor is converted into digital signal for further processing. Once the input signal is converted into digital, the microcontroller calibrates the signal directly to Celsius by properly coding it. Also, the ESP8266 microcontroller is capable of feeding/reading the data directly to/from the IoT cloud with the help of read/write API keys.

3.1.3. Freezer Arrangement

The freezer arrangement is constructed with the help of the Peltier, which is otherwise known as a thermoelectric cooler. The Peltier operates on a supply voltage of 12 V to 15 V and it draws the maximum current of 10 A. Once the Peltier is powered, it creates the temperature difference, i.e., a hot and cool side on each side of it. The heat produced on the hot side of the Peltier is continuously removed with the help of fan arrangement, and on the cool side, continuous cooling is utilized by the medical box of the proposed system.

3.1.4. Relay Unit

In general, the relay module acts as an automatic switch which can be switched ON/OFF without any manual intervention by control electrical control signals. In this work, a 3.3 Volt relay module was utilized to drive the Peltier module and the coolant fan arrangement. The relay module is capable of switching the Peltier module and coolant fan ON/OFF by applying a 3.3 V/5 V control signal which shall be supplied by the microcontroller.

3.1.5. Power Supply

The power supply of this work is simply battery technology. Two different batteries were used for the proposed system, whereas one battery was utilized to power the Peltier module and another battery was utilized to power the microcontroller and other electronic components. Also, the battery utilized to power the microcontroller and other components provided a 5 V supply, which can be commonly seen in smartphone battery banks. A Lithium Polymer battery utilized to power Peltier module has a charge capacity of 2200 mAh, and voltage range of 11.1 V which is more sufficient to source the Peltier module.

Firstly, the temperature of the medical box was read by the LM35 sensor and the temperature values were given to the ESP8266 microcontroller. Since the ESP8266 microcontroller is a WiFi controller, the temperature values were updated to the Blynk page. Blynk is an IoT (Internet of Things) platform that allows users to remotely control and monitor their devices over the internet. Blynk provides an API that allows developers to integrate their devices and applications with the Blynk platform. It also provides a range of functionalities, including device authentication and data streaming, and it consists of a mobile app and a cloud-based server that communicate with each other to enable remote device control.

Figure 2 shows the flowchart for the workflow of the smart healthcare medical box. Also, the microcontroller checks the medical box temperature with the set temperature, whereas the set temperature can be set with the help of Blynk IoT platform. Further, if there is any temperature decrease/increase in the medical box comparable to the set temperature, the microcontroller switches the Peltier module ON/OFF with the help of relay unit and the medical box temperature will be maintained in line with the set temperature. The medical box was fixed to the readymade drone. This drone's thermo-regulating medical container provided a reliable and efficient system for transporting medical products, ensuring the safety and efficacy of the medical products, which will ultimately save lives across the world.

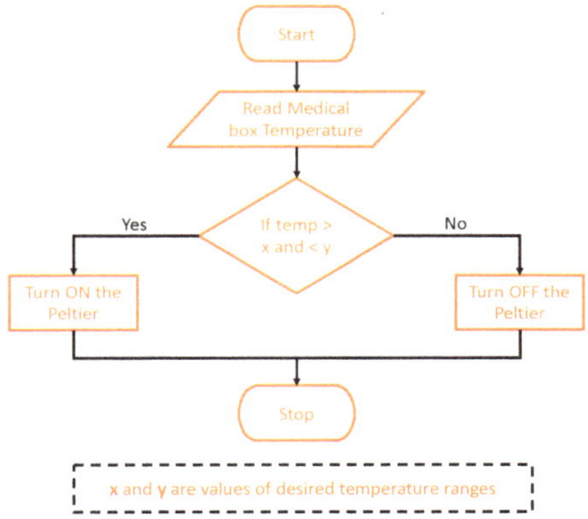

Figure 2. Flowchart for working of smart healthcare medical box.

3.2. Internet of Things Framework

In this proposed work, a custom-designed Blynk page was developed and which acted as a user interface when using the internet. Further, the Blynk page was used to monitor and control the temperature of the medical compartment. The medical compartment was capable of carrying any healthcare products/organs, etc. and the appropriate menu had to be set through a user interface. The various storage temperatures are: RBC (Red Blood Cells) or whole blood at 2 to 10 °C, FFP (Fresh Frozen Plasma) at 2 to 6 °C, cryoprecipitate 2–4 °C, platelets 20–24 °C, frozen vials are transported at −25 °C to −15 °C, refrigerated vials are transported at 2 °C to 8 °C, organs at 4 °C to 8 °C. As temperatures decrease to below 2 °C, there is a significant increase in the risk of cold injury with some proteins denaturing below 0 °C. At temperatures above 12 °C, the higher metabolic demand for oxygen leads to irreversible hypoxic injury, and thus significantly impairs organ function. To balance the two tendencies, 4 °C to 8 °C is said to be ideal. The page is designed in such a way and for the aforementioned temperature ranges, the appropriate menu has to be selected.

Generally, the Raspberry PI and other advanced microcontrollers would have the capability to transmit/update the data to the Blynk page. So, the designed page can be accessed with the help of Application Programming Interface (API) keys and can be accessed from anywhere. The page has the option to select the type of healthcare product/organ accordingly as the compartment temperature is maintained. Also, the present compartment temperature, humidity, vibration, altitude and GPS can be monitored in the developed webpage. Since, the microcontroller was connected to the internet, with the help of open access communication platforms such as Twilio, etc., mail/Short Messaging Service (SMS) can be sent to the sender of the medical healthcare products once it has been delivered to the receiver.

4. Results and Discussion

Figure 3a shows the prototype of a smart medical box integrated with an unmanned aerial vehicle.

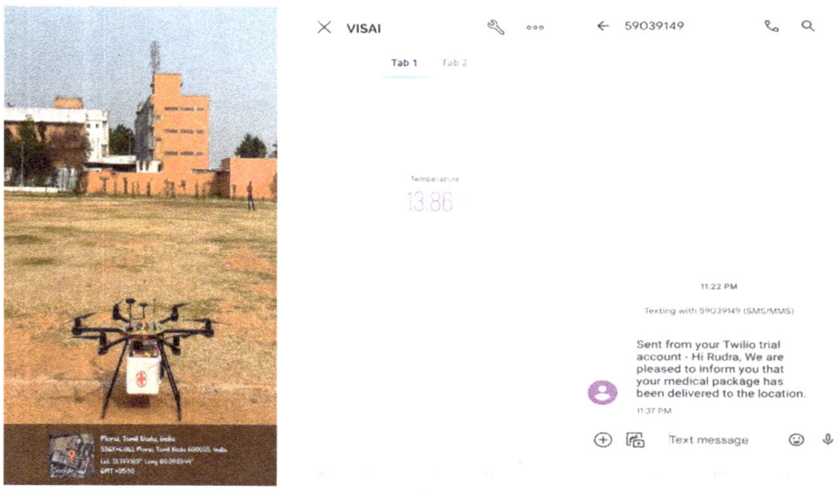

(**a**) (**b**) (**c**)

Figure 3. (**a**) Smart medical box integrated with an unmanned aerial vehicle; (**b**) the user page of Blynk IoT platform; and (**c**) Twilio messenger service.

The container box was attached to a drone, and trial tests were conducted at high altitudes to simulate actual delivery conditions, as shown in Figure 3a. Figure 3b shows the custom designed user page of Blynk IoT platform. The proof of concept was evaluated and it is evident that, by using the Peltier module, the temperature of the medical compartment can be maintained at the set values. The Peltier module was used to maintain a constant temperature within the container during delivery. Also, trial tests were conducted to assess the efficiency of our container. The temperature inside the container was recorded continuously with the help of the Blynk IOT platform. The trial tests showed that the container with the Peltier module successfully maintained the temperature range for 6 to 8 h. This indicates that the container is suitable for transporting medical goods that require a constant temperature during delivery.

The proposed system is integrated with Twilio API. The Twilio API sends a message to notify the user that the package has been successfully delivered to the destination. The container with the Peltier module proved to be efficient in maintaining a constant temperature during delivery, which is crucial for medical goods that require specific temperature conditions. The integration of the Twilio API allows us to notify the user of successful delivery, which adds an extra layer of convenience and security to the system which is shown in Figure 3c.

Also, Figure 4 shows the QGround Control user interface which was used to monitor the drone parameters and control the drone. Further, it demonstrated that the altitude, target distance and spacing of the drone can be fixed using QGround Control user interface. Also, it was clearly observed that the path of the drone can be configured using the same user interface, whereas the drone will travel through the specified path, as shown in Figure 4. Hence, the QGround Control provides an efficient solution for path planning, mission planning, and monitoring of drone.

Figure 4. QGround control user interface.

Figure 5 shows the temperature data of the developed container box. It was observed that the temperature of the container is almost maintained constant. Further, the temperature values were recorded during flight which was demonstrated that the temperature can be maintained constant with the help of Peltier. For demonstration purposes, the set temperature of 12 degrees was set and it is clearly shown that the Peltier module maintains the container temperature in line with the set temperature. The future work includes conducting trials in different environmental conditions to assess the container's efficiency and further improvements to the system, such as adding a humidity sensor and non-contact type temperature sensor. Also, the non-contact type temperature sensor would be highly beneficial to measure temperature of organs and liquids present inside the container.

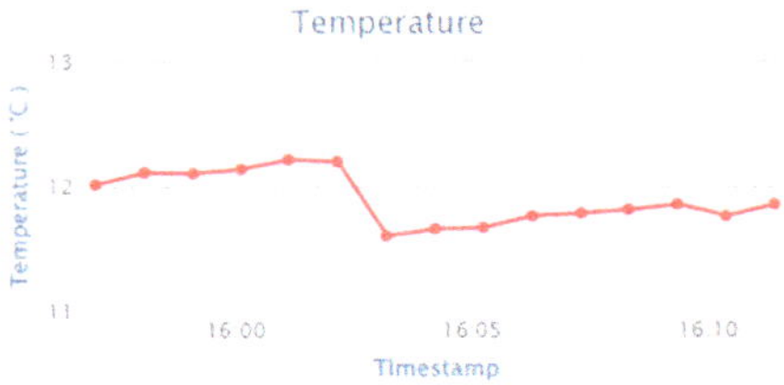

Figure 5. Temperature data.

5. Conclusions

A novel medical container box embedded with a Peltier module, temperature sensor, ESP8266 WIFI Microcontroller, relay board, and battery power banks to maintain and monitor the temperature has been proposed. The integrated IoT framework of the medical container allows the user to monitor and control the temperature remotely. The entire module was included in an UAV and field test, which was performed at various altitude levels and at a range of 2 km. The developed device was able to maintain the desired temperature while being transmitted via an IoT platform. The existing road transport has been replaced by advanced drone technology, which results in reduced transportation time. The developed device ensures high-level safety for transportation of healthcare products, including organs, medical goods, and vaccines and can also quickly and safely deliver medical products to hard-to-reach places, saving lives across the world. Additionally, the proposed solution does not require cryogenic systems, which can be expensive and challenging to maintain. This will help to reduce the risk of product damage, spoilage, or loss during transportation. Overall, the developed device could have a significant impact on the healthcare sector, ultimately improving quality of life.

Author Contributions: V.S., P.A. and B.E. conceptualized the idea of this manuscript. M.T.K.S.H and P.S.K.R.P. designed and developed the medical container. P.A., J.C. and B.E. conducted the test flights and performed the formal analysis. All the authors contributed towards preparing the manuscript. All authors have read and agreed to the published version of the manuscript.

Funding: This research received no external funding.

Institutional Review Board Statement: Not applicable.

Informed Consent Statement: Not applicable.

Data Availability Statement: Data are contained within the article.

Conflicts of Interest: The authors declare no conflict of interest.

References

1. Sathyan, A.; Bhadresha, K.S.; Rajan, D.; Destin, R.M.; Scaria, R.; Renjith, R. An advanced approach for an efficient mode of organ transportation. *Mater. Today Proc.* **2021**, *47*, 5358–5363. [CrossRef]
2. Stehlik, J.; Edwards, L.B.; Kucheryavaya, A.Y.; Benden, C.; Christie, J.D.; Dipchand, A.I.; Dobbels, F.; Kirk, R.; Rahmel, A.O.; Hertz, M.I. The Registry of the International Society for Heart and Lung Transplantation: 29th official adult heart transplant report—2012. *J. Heart Lung Transpl. Transpl.* **2012**, *31*, 1052–1064. [CrossRef] [PubMed]
3. Minasian, S.M.; Galagudza, M.M.; Dmitriev, Y.V.; Karpov, A.A.; Vlasov, T.D. Preservation of the donor heart: From basic science to clinical studies. *Interact. Cardiovasc. Thorac. Surg.* **2015**, *20*, 510–519. [CrossRef] [PubMed]
4. Peltier, J.; Iriart, X.; Mouton, E.; Benoit, G. Best practices in organ preservation for transplantation. *J. Visc. Surg.* **2020**, *157*, S37–S43.
5. Liao, J.; Ma, S.; Lv, Z.; Li, D.; Li, Y. Advances in blood preservation during storage and transportation. *Blood Transfus. Transfus. Sangue* **2021**, *19*, 257–265.
6. Michel, S.G.; LaMuraglia, G.M., II; Madariaga, M.L.L.; Anderson, L.M. Innovative cold storage of donor organs using the Paragonix Sherpa PakTM devices. *Heart Lung Vessel.* 2015, *in press*.
7. Jing, L.; Yao, L.; Zhao, M.; Peng, L.P.; Liu, M. Organ preservation: From the past to the future. *Acta Pharmacol. Sin.* **2018**, *39*, 845–857. [CrossRef] [PubMed]
8. Beutel, G.; Goyal, R.K. Temperature control in vaccine storage and distribution. *J. Pharm. Policy Pract.* **2021**, *14*, 1–14.
9. Fahy, G.M.; Hirsch, A. Prospects for organ preservation by vitrification. In *Organ Preservation: Basic and Applied Aspects A Symposium of the Transplantation Society*; Springer: Dordrecht, The Netherlands, 1982.
10. de Vries, R.J.; Yarmush, M.; Uygun, K. Systems engineering the organ preservation process for transplantation. *Curr. Opin. Biotechnol.* **2019**, *58*, 192–201. [CrossRef]
11. Nyemba, W.R.; Chinguwa, S.; Marango, B.L.; Mbohwa, C. Evaluation and feasibility assessment of the sustainability of refrigeration systems devoid of harmful refrigerants for storage of vaccines. *Procedia Manuf.* **2019**, *35*, 291–297. [CrossRef]
12. Chang, Y.W.; Chang, C.C.; Ke, M.T.; Chen, S.L. Thermoelectric air-cooling module for electronic devices. *Appl. Therm. Eng.* **2009**, *29*, 2731–2737. [CrossRef]
13. Afshari, F.; Afshari, F.; Ceylan, M.; Ceviz, M.A. A review study on peltier cooling devices; applications and performance. In Proceedings of the on 3nd International Conference on Technology and Science, Antalya, Turkey, 18–20 December 2020; pp. 31–34.
14. Zeng, Y.; Wang, M.; Zhang, H.; Qian, L. Theoretical analysis of temperature control for a thermal insulation box. *J. Therm. Anal. Calorim.* **2020**, *139*, 1503–1509.
15. Remeli, M.F.; Bakaruddin, N.E.; Shawal, S.; Husin, H.; Othman, M.F.; Singh, B. Experimental study of a mini cooler by using Peltier thermoelectric cell. *IOP Conf. Ser. Mater. Sci. Eng.* **2020**, *788*, 012076. [CrossRef]
16. Schubert, C.; Hagedorn, P.; Langer, G. Sensor technologies for monitoring the cold chain. *Curr. Opin. Food Sci.* **2019**, *30*, 1–8.

engineering proceedings

MDPI

Proceeding Paper

Design and Development of a Low-Cost and Compact Real-Time Monitoring Tool for Battery Life Calculation †

Dimitrios Rimpas *, Vasilios A. Orfanos, Pavlos Chalkiadakis and Ioannis Christakis

Department of Electrical and Electronic Engineering, University of West Attica, P. Ralli & Thivon 250, 12244 Egaleo, Greece; vorfanos@uniwa.gr (V.A.O.); pchalk@uniwa.gr (P.C.); jchr@uniwa.gr (I.C.)

* Correspondence: drimpas@uniwa.gr

† Presented at the 10th International Electronic Conference on Sensors and Applications (ECSA-10), 15–30 November 2023; Available online: https://ecsa-10.sciforum.net/.

Abstract: Lithium-ion batteries are utilized everywhere from electronic equipment, smartphones and laptops to electric vehicles; however, certain disadvantages are inherited including high-cost and low-temperature range, caused by high currents. In this paper, a compact and low-cost battery management system is presented which can measure the voltage and current for both the battery and power supply. Two NTC thermistors, 10k and 100K Ohm each, are exploited for collecting battery temperature in two different spots of the socket for direct comparison and validation while an additional sensor measures external temperature and humidity. A charging socket is provided for charging the cell through an external source with dynamic voltage output to test the battery response. Finally, an Arduino-compatible device is implemented in order to protect the battery from overcharging. This system collects parameters at a 10 s time rate and calculates precious parameters of the battery-like State of Charge (SoC), State of Health (SoH) and State of Power (SoP). Keeping the operation within a safe zone of 20–80% SoC maximizes longevity and ensures that it can provide even the maximum power to cover the load required; hence, these three parameters are considered collaborative. Afterwards, the collected data are sent over Wi-Fi on the internet application server for real-time monitoring, in an efficient, portable and low-cost setup.

Keywords: lithium; battery; monitoring; validation; arduino; aging; sensors

Citation: Rimpas, D.; Orfanos, V.A.; Chalkiadakis, P.; Christakis, I. Design and Development of a Low-Cost and Compact Real-Time Monitoring Tool for Battery Life Calculation. *Eng. Proc.* **2023**, *58*, 17. https://doi.org/10.3390/ecsa-10-16146

Academic Editor: Stefano Mariani

Published: 15 November 2023

1. Introduction

Lithium batteries are the main power source for all electronic components that are widely used on a daily basis, such as portable computers, smartphones and electric vehicles. They consist of four main components: anode, cathode, electrolyte and separator [1]. During discharge, lithium ions flow from the anode to the cathode and vice versa while charging. The electrolyte helps this energy trade and the separator protects the two parts of the battery from internal short circuits [2]. Compared to previous technologies like lead acid they can provide greater energy density with more robust and compact sizing, actively enhancing their adoption in every modern device as storage or renewable plants or electric vehicles which is the latest trend towards a greener environment [3,4].

However, all lithium-ion technologies inherit several flaws. The high temperature caused by fast charging or discharging in any situation, creates lithium dendrite, electrolyte decomposition or even cracking inside the cells leading to total breakdown [5]. This aspect inquires about the need for a monitoring system that is constantly collecting important values like its capacity, voltage, the current drawn or applied to the battery, and most importantly temperatures at the electrodes and of the casing externally [6].

In order to determine the status and protect the battery from any damage, there are 4 specific values that can be calculated, which include [7,8]:

- State of Charge (SoC);
- State of Health (SoH);
- State of Power (SoP);
- Depth of Discharge (DoD).

SoC describes the battery's current charge compared to full, SoH reflects the current capacity after any life cycle and SoP is defined as the peak power that the battery can supply at a certain time period. A variable that can be easily monitored and provide information about battery use and user behavior is the Depth of Discharge, which shows the usage percentage in each case, such as, for example, 100% to 40% equals 60% DoD. Previous research concluded that a high depth of discharge with low SoC at the end of each use stresses the cells and can lead to severe damage and a low State of Health [9,10].

It is clear that these four parameters are crucial for the battery status diagnosis. A low State of Charge (below 20%) associated with a high depth of discharge signifies high battery stress. If the temperature of the cells overcomes the safe limit at 40° Celsius, it will cause lithium deposition on the cathode, leading to less available lithium flowing from the anode to the cathode so that the SoH is lowered [11]. Additionally, if battery health (SoH) is limited, it will be strained to cover the energy demand for the nominal time or any peak loads required, turning to safe mode or even total breakdown [12]. High SoC for long periods can also damage the battery as being close to full charge adds excessive pressure to the electrolyte and the separator [13,14]. Hence, a battery management system protecting the battery with the nominal operational conditions is vital for safe and uninterrupted functioning.

The goal of this paper is to suggest a simple, compact and user-friendly setup based on an Arduino-compatible device. A typical 18,650 battery is employed as the key element for testing with the addition of a current and a voltage sensor for the respective values collection. Moreover, a set of NTC thermistors with different ohmic grades (10k and 100k Ohms) collect the battery temperature at the electrode end while the external temperature and humidity are also collected at a 10 s time frame like the other parameters. Then, all values are transferred to an internet application server with Wi-Fi for real-time monitoring and further processing.

2. Materials and Methods

2.1. Experimental Layout

According to the experiment requirements, a battery tester board is developed to monitor and transfer the data to the central application host. The board includes various components, all deployed within an Arduino-compatible device, which can be programmed using the Arduino IDE 2.2.1 user-friendly interface for further adjustments. The TTGO@ESP32, Shenzhen, China) is selected as the main processing unit due to its high affordability and processing power with low power consumption and size. It includes 25 I/O (Inputs/Outputs) and analog ports, with I2C (Inter-integrated Circuit) and SPI (Serial Peripheral Interface) capable of functioning as a communication interface supporting Wi-Fi and LoRa protocols. In addition, the voltage reading of the analog ports is supported by 4095 step positions, resulting in higher measurement accuracy.

The layout includes two current sensors (ACS712-5A) from Allegro microsystems (Shenzhen, China) which can measure up to 5A corresponding to the analog output 185 mV/Amp, while the sensitivity is set and calibrated by the built-in potentiometer. A set of coupled resistors (100 KΩ) is implemented as the voltage divider to monitor both charging and discharging voltage sequences. For the battery temperature value, two thermistors (NTC) with distinct resistances, 10 kΩ and 100 kΩ, were used for further validation regarding the accuracy of the measurement. In addition, the environmental temperature and humidity values were provided by a single and calibrated barometric sensor with product number GY-213V-HTU21D, (Shenzhen, China).

The battery selected is a NiteCore NL1823 Li-Ion (Shenzhen, China) with a capacity of 2.300 mAh and a nominal voltage of 3.7 volts, placed within a socket soldered into the board.

This module can be charged internally via a DC adapter or a photovoltaic (PV) panel with a supply of up to 6 volts. An additional plug for external load is provided for testing the battery performance. At the working state, all measurements from the battery are collected every 10 s as a time frame. Then, the board sends the measurements (voltage, current, temperature and humidity) to the internet application server over Wi-Fi using HTTP POST messages. The application platform includes a database for storing data and a web server (Grafana UI) for data visualization or reception. Figure 1 below displays the layout of the circuit including the CPU, sensors, battery, DC supply plug and the load output.

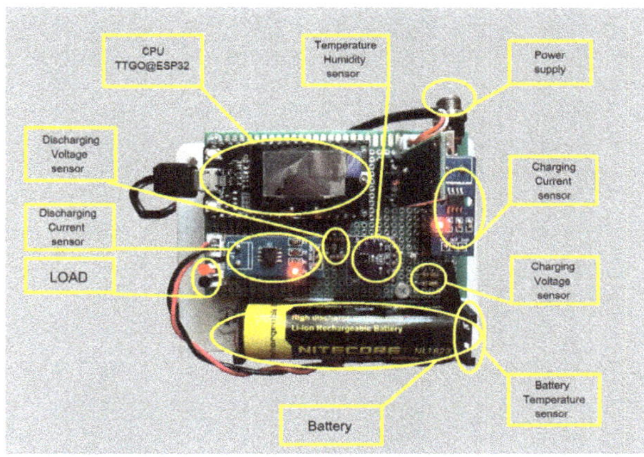

Figure 1. Layout circuit including the processing unit, sensors and lithium battery.

2.2. Calculation of Selected Parameters

The collected values from the circuit are used for the calculation of the previously mentioned parameters (SoC, SoH, SoP and DoD), whose significance is stated previously in the Introduction. According to the literature [10–12], there is a variety of complex equations to obtain these values. For this experiment, the following rules are required:

- SoC equals the ratio of current capacity divided by the nominal value;
- SoH is defined as the maximum voltage of the battery divided by the nominal;
- SoP refers to the maximum current that can be drawn from the battery;
- DoD is calculated manually depending on the SoC before and after charging.

Capacity can be identified by subtracting the current drawn from the nominal and multiplying it with hours to convert it into Ampere hours. To simplify the calculations, the State of Voltage (SoV) term is utilized as a SoC alternative that is identical for battery status monitoring [13]. Therefore, the following equations are utilized to calculate these parameters:

For State of Charge:

$$\text{SoC} = \frac{C}{C_{RATED}} \quad (1)$$

For State of Health:

$$\text{SoH} = \frac{V_{MAX}}{V_{RATED}} \quad (2)$$

For State of Power:

$$\text{SoP} = I_{MAX} \times \frac{dT}{dt} \quad (3)$$

For Depth of Discharge:

$$\text{DoD} = \text{SoC}' - \text{SoC} \quad (4)$$

3. Results and Discussion

The values monitored were collected and presented in Figure 2 at Grafana Interface. Constant monitoring is available through this free web application and data can be extracted for any specific time frame, adding better control and direct data comparison.

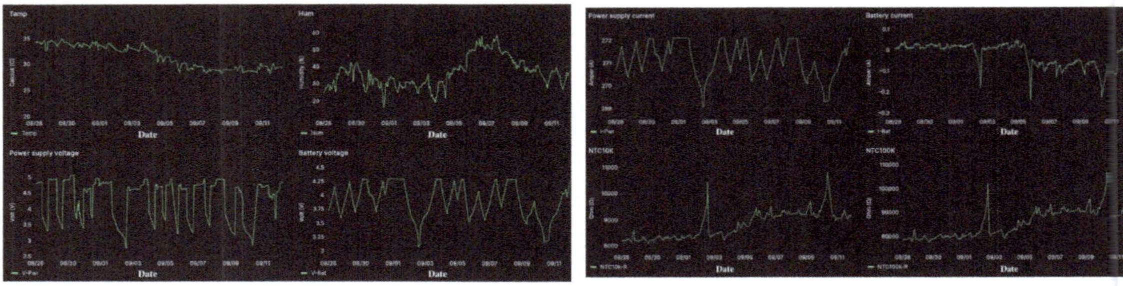

(**a**) (**b**)

Figure 2. The eight values monitored and imprinted through Grafana Interface. Specifically, the following parameters are available: (**a**) humidity, ambient temperature, supply voltage and current, and (**b**) battery voltage and amperage with the output of the NTC thermistors.

Total measurements gathered are 30,000, spread over a month-long period. The comparison between the NTC thermistors to validate the accuracy was fulfilled and presented in Figure 3. The results show that both NTCs, 10k and 100k (after value correction for the first one) provide the same sensitivity in temperature changes and can be utilized for this work.

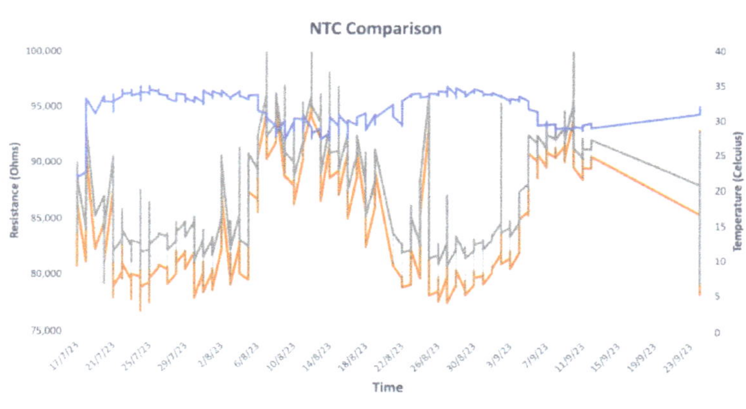

Figure 3. The comparison between the thermistors reveals identical accuracy after value correction.

Testing is based on the experiment ambient temperature, located in the province of Peristeri in Athens, Greece. As mentioned before, temperature is a key factor in battery performance. As the power voltage increases, battery voltage is enhanced but not in a linear way, due to the increased temperature, affecting its internal resistance, which can be seen in the second graph very clearly. The temperature of the battery is highly affected by the voltage applied and is calculated via the 100k NTC datasheet [15]. As resistance increases, lithium-ion flow is limited both at charging or discharging phases, and overheating due to high current or ambient temperature causes dendrites to form at the electrodes so fewer lithium-ions are available to flow and capacity is limited [5–7]. In this testing, the temperature remains within the operation limits of the cells. After measurement 145, the supply voltage was limited to provide further protection. In the graph, a linear decrease can be seen in both voltages as the source is disconnected, with a simultaneous increase in both temperatures. It can then reach the maximum point when the source voltage is reapplied. Figure 4 below imprints this sequence.

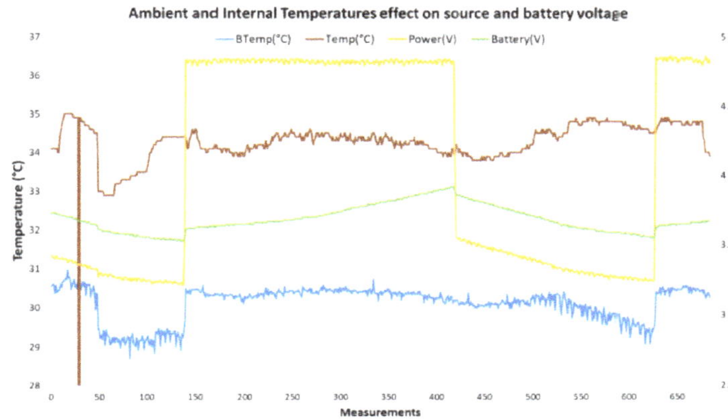

Figure 4. Temperature's role in battery and supply voltage variation influenced by power voltage.

The variation of battery current corresponding to external and internal temperatures is displayed in Figure 5. The current and voltage fluctuations are similar while the ambient temperature and current variations are too modest to provide any major outcomes.

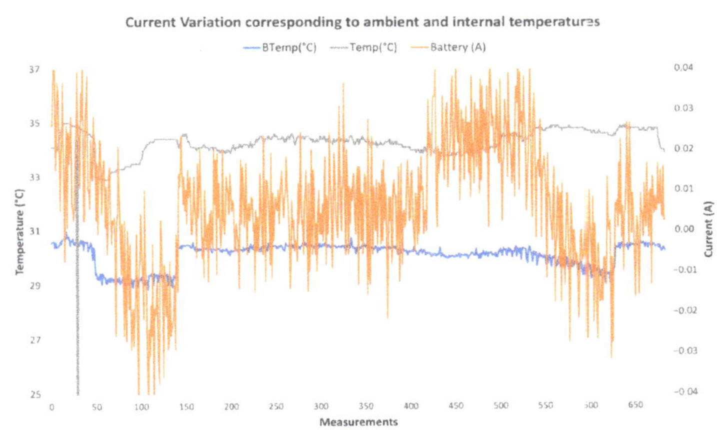

Figure 5. Current fluctuations due to variations of Ambient and internal temperatures.

The next step is to calculate the battery parameters mentioned in Section 2.2, hence, the results are summarized in Table 1. The State of Charge was maintained at 40–80%, hence, a DoD of 40%. If DoD is increased and the battery is dropped below 20% charge, it is susceptible to lithium deposition, causing a reduction in SoH [9]. By not keeping the battery at 100% charge, the State of Health decreased only by 0.1% after 50 cycles. Batteries are manufactured with over 100% available SoH, typically 102–110%, so the drop is not visible in Table 1. The State of Power can reach 6x the battery capacity, with a small decrease after testing, as the battery chemically ages. A significant factor was the high ambient temperature, typically 30–35° Celcius due to summer conditions. So, no major stress on the batteries was applied.

Table 1. Battery parameters calculated before and after testing.

Parameter	Before Testing	After Testing
SoC	90%	60%
SoH	100%	100%
SoP	15 A	14.8 A
DoD [1]	40%	40%

[1] Typical DoD range achievable due to layout limitations.

4. Conclusions

This work presented a compact, affordable and customizable layout of lithium battery monitoring through a web app monitoring platform. Certain values like ambient and internal temperatures were gathered to calculate important values for the battery like the State of Charge and State of Health. The results show that battery stress is mainly affected by battery current and supply voltage that raises internal temperature causing dendrites at the battery electrodes, reducing total capacity. In this work, it is suggested that SoC should be maintained with an optimal range (20–80%), while low DoD and operating current lead to major cell longevity, as SoH is decreased by only 0.1% after 50 cycles while maintained at safe operating conditions. Future work involves an advanced software tool for calculating more parameters like the State of Life and a bigger setup to be tested in EV modules, mainly focused on battery stress verification with the addition of more sensors and the ability to handle increased currents required within the safe temperature zone.

Author Contributions: Conceptualization, D.R. and I.C.; methodology, D.R.; software, V.A.O.; validation, I.C. and V.A.O.; formal analysis, V.A.O.; investigation, P.C. and V.A.O.; resources, I.C.; data curation, D.R.; writing—original draft preparation, D.R.; writing—review and editing, D.R. and I.C.; visualization, I.C.; supervision, V.A.O.; project administration, I.C.; funding acquisition, D.R. All authors have read and agreed to the published version of the manuscript.

Funding: This research received no external funding.

Institutional Review Board Statement: Not applicable.

Informed Consent Statement: Not applicable.

Data Availability Statement: The data that support the findings of this study are available from the corresponding author upon reasonable request.

Conflicts of Interest: The authors declare no conflicts of interest.

References

1. Scrosati, B.; Garche, J. Lithium Batteries: Status, Prospects and Future. *J. Power Sources* **2010**, *195*, 2419–2430. [CrossRef]
2. Manthiram, A. A Reflection on Lithium-Ion Battery Cathode Chemistry. *Nat. Commun.* **2020**, *11*, 1550. [CrossRef] [PubMed]
3. Rahimi, M. Lithium-Ion Batteries: Latest Advances and Prospects. *Batteries* **2021**, *7*, 8. [CrossRef]
4. Rimpas, D.; Kaminaris, S.D.; Aldarraji, I.; Piromalis, D.; Vokas, G.; Papageorgas, P.G.; Tsaramirsis, G. Energy Management and Storage Systems on Electric Vehicles: A Comprehensive Review. *Mater. Today Proc.* **2022**, *61*, 813–819. [CrossRef]
5. Yao, L.; Xu, S.; Tang, A.; Zhou, F.; Hou, J.; Xiao, Y.; Fu, Z. A Review of Lithium-Ion Battery State of Health Estimation and Prediction Methods. *World Electr. Veh. J.* **2021**, *12*, 113. [CrossRef]
6. Zhou, X.; Han, X.; Wang, Y.; Lu, L.; Ouyang, M. A Data-Driven LiFePO4 Battery Capacity Estimation Method Based on Cloud Charging Data from Electric Vehicles. *Batteries* **2023**, *9*, 181. [CrossRef]
7. Gauthier, R.; Luscombe, A.; Bond, T.; Bauer, M.; Johnson, M.; Harlow, J.; Louli, A.; Dahn, J.R. How Do Depth of Discharge, C-Rate and Calendar Age Affect Capacity Retention, Impedance Growth, the Electrodes, and the Electrolyte in Li-Ion Cells? *J. Electrochem. Soc.* **2022**, *169*, 020518. [CrossRef]
8. Reshma, P.; Manohar, V.J. Collaborative Evaluation of SoC, SoP and SoH of Lithium-Ion Battery in an Electric Bus through Improved Remora Optimization Algorithm and Dual Adaptive Kalman Filtering Algorithm. *J. Energy Storage* **2023**, *68*, 107573. [CrossRef]
9. Rimpas, D.; Kiatipis, A. Charging Strategy Effect on Lithium Polymer Battery Capacity: A Case Study. *Int. J. Energy Environ.* **2020**, *3*, 107–118.
10. Andrenacci, N.; Vellucci, F.; Sglavo, V. The Battery Life Estimation of a Battery under Different Stress Conditions. *Batteries* **2021**, *7*, 88. [CrossRef]

11. Zia, M.F.; Elbouchikhi, E.; Benbouzid, M. Optimal Operational Planning of Scalable DC Microgrid with Demand Response, Islanding, and Battery Degradation Cost Considerations. *Appl. Energy* **2019**, *237*, 695–707. [CrossRef]
12. Kostopoulos, E.D.; Spyropoulos, G.C.; Kaldellis, J.K. Real-World Study for the Optimal Charging of Electric Vehicles. *Energy Rep.* **2020**, *6*, 418–426. [CrossRef]
13. Wikner, E.; Thiringer, T. Extending Battery Lifetime by Avoiding High SOC. *Appl. Sci.* **2018**, *8*, 1825. [CrossRef]
14. How, D.N.T.; Hannan, M.A.; Hossain Lipu, M.S.; Ker, P.J. State of Charge Estimation for Lithium-Ion Batteries Using Model-Based and Data-Driven Methods: A Review. *IEEE Access* **2019**, *7*, 136116–136136. [CrossRef]
15. NTC100K Datasheet. Available online: https://datasheetspdf.com/pdf-file/944190/Danfoss/NTC100K/1 (accessed on 27 September 2023).

engineering proceedings

MDPI

Proceeding Paper

Development of a Compact IoT-Enabled Device to Monitor Air Pollution for Environmental Sustainability †

Vijayaraja Loganathan [1,*], Dhanasekar Ravikumar [1], Vidhya Devaraj [1], Uma Mageshwari Kannan [1] and Rupa Kesavan [2]

1 Department of Electrical and Electronics Engineering, Sri Sairam Institute of Technology, Chennai 600044, Tamilnadu, India; dhanasekar.eee@sairamit.edu.in (D.R.); sit21ee010@sairamtap.edu.in (V.D.); sit21ee031@sairamtap.edu.in (U.M.K.)

2 Department of Computer Science and Engineering, Sri Venkateswara College of Engineering, Sriperumbudur 602117, Tamilnadu, India; rupakesavan@svce.ac.in

* Correspondence: vijayaraja.eee@sairamit.edu.in

† Presented at the 10th International Electronic Conference on Sensors and Applications (ECSA-10), 15–30 November 2023; Available online: https://ecsa-10.sciforum.net/.

Abstract: Degrading air quality is a matter of concern nowadays, and monitoring air quality helps us keep an eye on it. Air pollution is a pressing global issue with far-reaching impacts on public health and the environment. The need for effective and real-time monitoring systems has become increasingly apparent to combat this growing concern. Here, an innovative air pollution surveillance system (APSS) that leverages Internet of Things (IoT) technology to enable comprehensive and dynamic air quality assessment is introduced. The proposed APMS employs a network of Io enabled sensors strategically deployed across urban and industrial areas. These sensors are equipped to measure various pollutants, including particulate matter (PM2.5 and PM10), nitrogen dioxide (NO_2), sulfur dioxide (SO_2), ozone (O_3), carbon monoxide (CO), and volatile organic compounds (VOCs). Here, a regression model is created to forecast air quality using sensor data while taking into account variables including weather information, traffic patterns, and pollutants. Additionally, air quality categories (such as good, moderate, and harmful) are classified using classification algorithms based on preset thresholds. The IoT architecture facilitates seamless data transmission from these sensors to a centralized cloud-based platform. The developed APSS monitors the air quality using an MQ-135 gas sensor, and the data are shared over a web server using the Internet. An alarm will trigger when the air quality goes below a certain level. Also, the air quality, which is measured in parts per million (PPM), is displayed on the unit connected to it. Furthermore, when the PPM goes beyond a certain level, an alert message is sent to the air pollution control board, which takes preventive measures to control the pollution and also alerts the people, which helps each person in that society save their environment from pollution and have a good air quality environment. Additionally, the APSS offers user-friendly interfaces, accessible through web and mobile applications, to empower citizens with real-time air quality information. The effectiveness of the IoT-based air pollution monitoring system has been validated through successful field trials in urban and industrial environments, and it has the ability to provide real-time data insights and empower stakeholders in promoting environmental sustainability and fostering citizen engagement.

Keywords: APSS; Internet of Things; parts per million; sensors

Citation: Loganathan, V.; Ravikumar, D.; Devaraj, V.; Kannan, U.M.; Kesavan, R. Development of a Compact IoT-Enabled Device to Monitor Air Pollution for Environmental Sustainability. *Eng. Proc.* **2023**, *58*, 18. https://doi.org/10.3390/ecsa-10-15996

Academic Editor: Stefano Mariani

Published: 15 November 2023

1. Introduction

Air pollution is the contamination of the indoor and outdoor environment by any chemical, physical, or biological agent that modifies the natural characteristics of the atmosphere. It has effects on humans, animals, plants, and the environment. It affects long-term health issues like heart disease, lung cancer, and respiratory diseases in humans.

It also affects animals and plants, causing some diseases and causing damage to the environment. A device used to measure this polluted air is an air pollution monitoring system, which helps control air pollution. There are some existing approaches that are discussed here: A wireless air pollution monitoring system is created, and when the level of measured pollutants exceeds the acceptable amount, alarms are sent to the user through the Internet. The MQ135 sensor is used, together with a microcontroller, to track indoor pollution [1]. Finally, a model which combines Arduino UNO hardware and software (Arduino IDE 2.2.1) with MQ135 and MQ7 sensors and a dust sensor is used to sense gases like NO_2, CO, and PM2.5.

This model also monitors the air quality utilizing Thing Speak and hardware connected to the Internet through a Wi-Fi module [2]. The air-sensing system in [3] monitors the quality over a cloud-based platform and sounds an alarm when the quality goes below a certain level, which shows that there are a number of pollutants present in the surrounding air. The AQI scale becomes higher when the air is more hazardous to human health. The model in [4] combines the Arduino UNO hardware and software with gas sensors such as MQ135 and MQ7 and the dust sensor GP2Y1010AU0F, which aid in detecting gases like NO_2, CO, and PM2.5 via accurate measurements. Additionally, the air quality is monitored using an IoT platform. In [5], sensors will collect data on a range and transmit it to a Raspberry Pi, which serves as a base station. On a Raspberry Pi 3-based webserver, the data are collected by sensors. To show these data on websites, a MEAN stack is created. Distribution to all stakeholders is the key component of the proposed task. The systemin [6] forecasts the city's pollution level and also monitors air contaminants to produce alerts. It is made to accommodate gases produced by vehicle pollution.

A sensor-based hardware module in [7] that can be installed along roads is demonstrated. It can be mounted on lamp posts and send data wirelessly to a remote server, so that traffic can be managed using this information. Through a smartphone application, this system also offers information about air quality, enabling users to choose routes with good quality. Ref. [8] identified that the primary source of air pollution is transportation. To reduce air pollution, electric cars and bicycles can replace other types of transportation. This study shows how crucial it is to predict air pollution levels so that individuals can change their travel routes. An inexpensive, alternative IoT-based air quality monitor is described in [9]. It tracks air pollution in real time and communicates data quickly over a low-power wide-area network. A wide network of these monitors may provide a significant quantity of data, which can be analyzed in the cloud in real time and connected with the time of day, month, year, weather, and other aspects. The microcontroller and sensors communicate through the microcontroller, which collects data from the sensors, processes it, and communicates it to an online server using the IoT [10].

The system uses the IoT to identify the most dangerous airborne contaminants as well as louder noise levels. It also provides real-time updates on the degree of pollution. In summary, an air pollution surveillance system is essential for safeguarding the environment, the public's health, and advancing sustainable development. It gives individuals and governments the power to make decisions that improve living standards and the quality of the environment. It can be helpful in reducing traffic pollution. Different sensors can be used for different ranges and accuracies. The solution presented here consists of a single air quality monitoring sensor which can sense ammonia, sulfur, benzene, carbon dioxide, smoke, and other similar harmful gases; a Wi-Fi module; and an alert system. This helps in keeping track of air quality, and in the case of excess pollutants, it helps in informing the pollution control board. An air pollution surveillance system can be used for real-time analysis. The Kalman filter was employed to increase the measuring device's accuracy since it can make predictions even in the event that sensor data are missing. This paper is split into a description of the proposed system, the modeling of the air pollution surveillance system, and experimentation, with validation of the proposed work.

2. Proposed Air Pollution Surveillance System

The proposed air pollution surveillance system is shown in Figure 1. A power supply is connected to the Arduino UNO. The MQ-135 sensor detects ammonia, sulfur, benzene, carbon dioxide, smoke, and other similar harmful gases and sends them to the Arduino UNO. It is connected via the Wi-Fi module, as shown in Figure 1, which sends data when the pollution rate received from the gas sensor is high. Then, the data are sent to the pollution control board through a hybrid cloud that combines both public and private cloud resources, which takes steps and takes action towards reducing pollution in the air. Since the Kalman filter can make predictions even in the case of missing sensor data, it is used to boost the accuracy of the measuring instrument as shown in Figure 2, and the range of pollutants measured is given in Table 1.

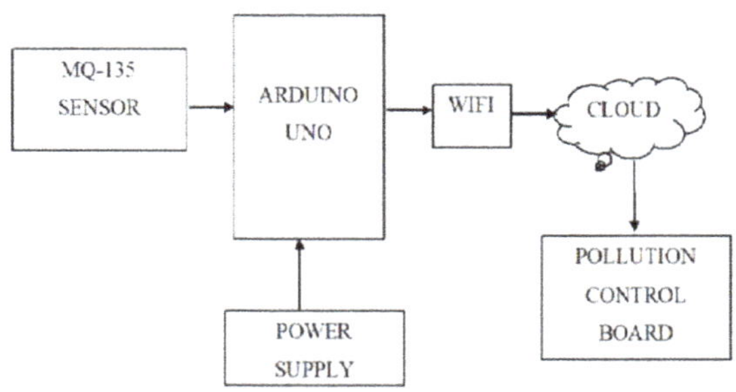

Figure 1. Air pollution surveillance system.

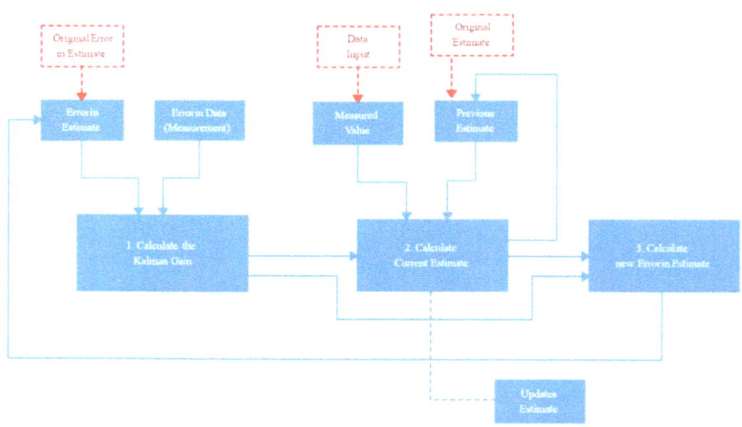

Figure 2. Process of Kalman filter.

Table 1. Pollutants measured and their range.

Measurement Item	Unit	Normal	Bad
CO	ppm	0.0~9.0	9.1~15.0
NH_3	ppm	0~5	6~40
SO_2	ppm	0.0~0.1	0.2~20.0
NO_2	ppm	0.0~0.1	0.2~20.0
O_3	ppm	0.0	0.1~0.15

Alarm system algorithm:
Initialize the hardware components and serial communication.
Continuously read the analog value from the sensor.

If the analog value is greater than the threshold:

- Activate the alarm (red LED and buzzer).
- Display "ALERT" and "EVACUATE" on the LCD.

If the analog value is not greater than the threshold:

- Deactivate the alarm (turn off the red LED and buzzer).
- Display nothing or a different message on the LCD.
- Repeat the process in a loop.

3. Modeling of Air Pollution Surveillance System

As shown in Figure 3, connections are provided in accordance with the circuit. Data are sent from the MQ-135 to the alarm system which sounds and shows an alert message on the LCD when the level of smoke or pollution is excessive. The pollution control board receives data on a continuous basis along with the PPM value. The LCD shows a safe message when the pollution level is below the set threshold. When the circuit is closed, the safe green LED illuminates.

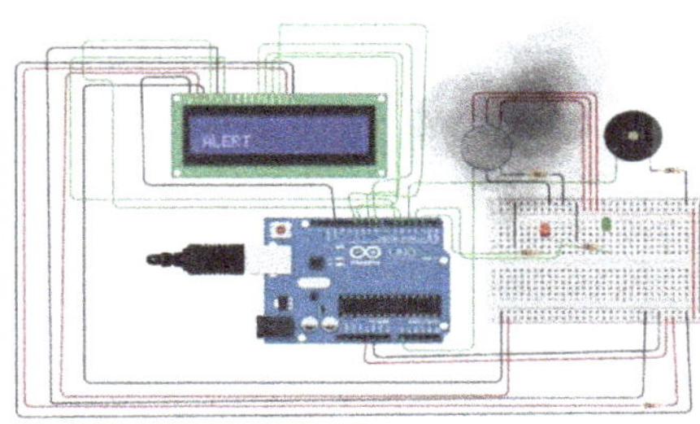

Figure 3. APSS simulation.

The simulation of the proposed air pollution surveillance system is shown in Figure 3. The air pollution surveillance system, which provides the required alert in the form of an LED and LCD, is shown in Figure 4.

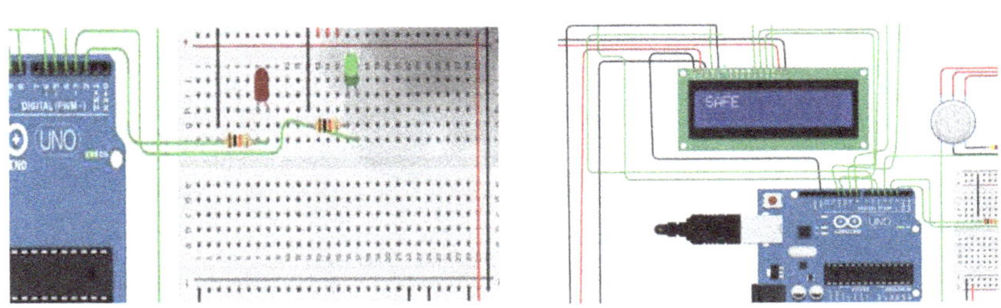

Figure 4. Result of APSS for LED and LCD.

4. Experimentation of Air Pollution Surveillance System

In the developed APSS shown in Figure 5, components such as the MQ-135 gas sensor, Arduino UNO, Wi-Fi module, LCD display, alert system, and LED lights are used. Gases like CO_2, smoke, alcohol, benzene, and NH3 are detected using the MQ-135 gas sensor. The gas sensor continuously monitors the quality of the air at the PPM level. When the PPM is higher than 1000, the LCD shows an alert message; otherwise, the safe message is displayed. The LED is green during the safe message, and the red LED is turned on when the alert message is displayed. The data from the gas sensor are sent through the Wi-Fi module to the pollution control board. The data are shown as PPM to the pollution control board. The alert system gives a sound when the PPM enters the danger level, and an alert message is also sent to the pollution control board. This helps the pollution control board take the required actions to control the pollution.

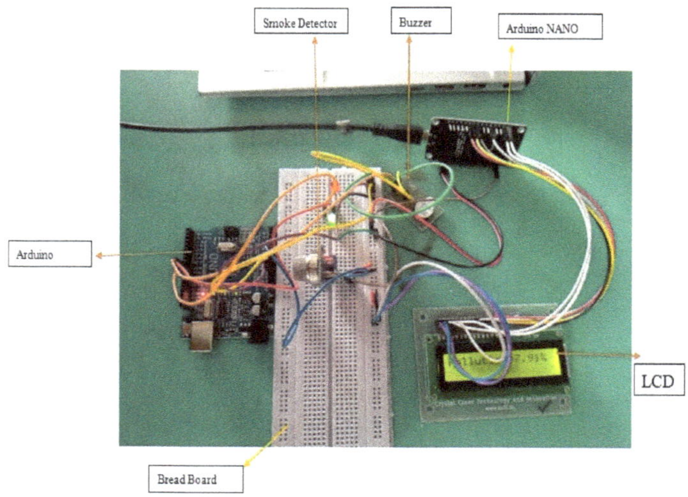

Figure 5. Result of APSS.

Users can also connect to the Wi-Fi module and see the air quality of their locality, which helps them take the necessary measures. The data are displayed continuously with their PPM levels. There cursive mathematical process known as the Kalman filter determines the state of dynamics from a set of noisy measurements taken over time. It can be used to monitor and forecast pollutant concentrations in the context of air pollution monitoring using sensor data. The Arduino is connected to the LCD display, a bread board fixed with LEDs and an MQ-135 gas sensor, a Wi-Fi module, and an alert system.

On observing the range of each pollutant in the cement or petrochemical industry taken as examples, the measured pollutants and their range of PPM levels are shown in Figure 6. To analyze the data observed via the developed APSS, a regression model is created to forecast air quality using sensor data while taking into account variables including weather information, traffic patterns, and pollutants. Additionally, air quality categories (such as good, moderate, and harmful) are classified using classification algorithms based on preset thresholds. By continuously monitoring the PPM levels, the air quality can be detected, a safe environment can be implemented, and diseases influenced by air pollution such as cancer, asthma, kidney, liver, skin diseases, neurological disorders, etc., can be reduced. The people also receive an alert sound when they are connected through the Wi-Fi module when the PPM level becomes high and are shown the level of PPM and a message such as safe or alert. In such a way, a modern gas sensing unit is developed (Figure 5). By analyzing the model in a industry before and after using the Kalman filter, there is at least a 5% increase in accuracy and it showed a better performance. The results gained during the analysis are shown in Figure 6, which shows a bar graph between the gases sensed by the sensor and their ranges. Figure 7 shows the accuracy changes before and after applying the Kalman filter in the gases CO_2 and NO_2, respectively.

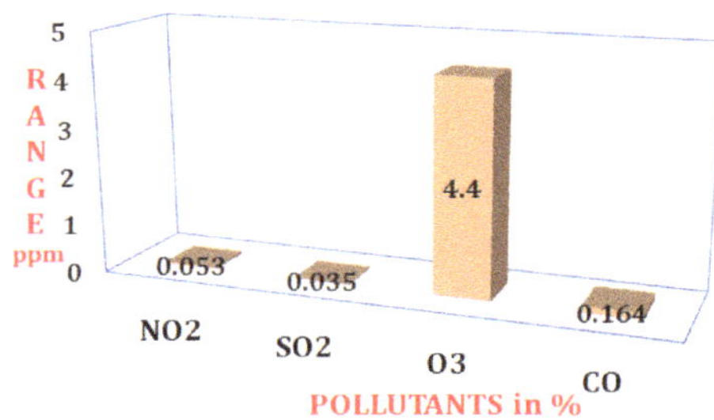

Figure 6. Pollutants' range from the cement or petrochemical industry.

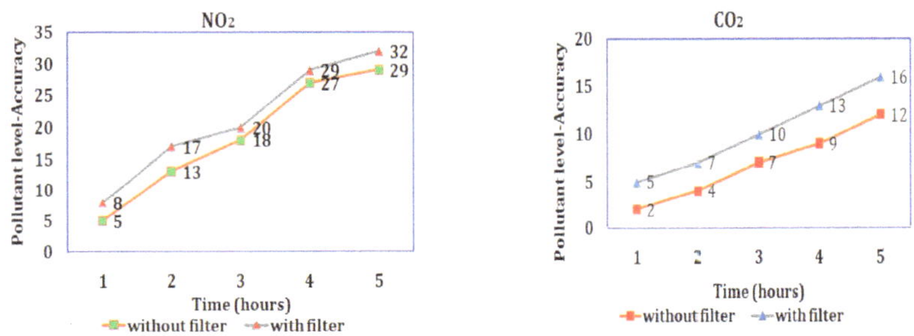

Figure 7. Accuracy of NO_2 and CO_2 measurement before and after filter.

5. Conclusions

The proposed APSS is cost-effective and handy. The fascinating feature of the proposed model is that it reports pollution to the pollution control board, which helps to address polluted areas as soon as they become polluted. The following conclusions are presented:

- A key aspect was to monitor pollution levels to create a profile of a city or region and predict risks at individual levels. This awareness can also lead people to make direct contributions to reduce pollution levels.
- This system can be upgraded by adding more sensors, so it is portable.
- There is no need for any application to be installed; the system sends data directly to the pollution control board.
- The process of the Kalman filter makes the system reliable as it has processing feature based upon previous values.
- It is found that the accuracy is increased after implementing a Kalman filter.
- The proposed system was tested successfully, and it is able to sense gases in a range of 10 to 1000 ppm and process them with a Kalman filter. Making the model more reliable and accurate helps in sending air quality data. This system can be used as a warning system when the air quality becomes severe, and measures could be taken to prevent it.

Author Contributions: Conceptualization, V.L. and D.R.; methodology, V.D.; software, U.M.K.; validation, V.L., V.D. and R.K.; investigation, R.K.; writing—original draft preparation, V.L. and V.D.; writing—review and editing, supervision, V.L. and U.M.K. All authors have read and agreed to the published version of the manuscript.

Funding: This research received no external funding.

Institutional Review Board Statement: Not applicable.

Informed Consent Statement: Not applicable.

Data Availability Statement: Data are available in this manuscript.

Conflicts of Interest: The authors declare no conflict of interest.

References

1. Kodali, R.K.; Pathuri, S.; Raj Narayanan, S.C. Smart Indoor Air Pollution Monitoring Station. In Proceedings of the International Conference on Computer Communication and Informatics (ICCCI), Coimbatore, India, 22–24 January 2020; pp. 1–5.
2. Saikumar, C.V.; Reji, M.; Kishoreraja, P.C. IOT based air quality monitoring system. *Int. J. Pure Appl. Math.* **2017**, *117*, 53–57.
3. Sahoo, B.; Maharana, A.; Murali, M.; Shivani, L.; Suganya, G.; Premalatha, M. Low-Cost Air Sensing System. In Proceedings of the 2019 3rd International Conference on Computing and Communications Technologies (ICCCT), Chennai, India, 21–22 February 2019; pp. 258–267.
4. Jha, R.K. Air Quality Sensing and reporting System Using IoT. In Proceedings of the Second International Conference on Inventive Research in Computing Applications (ICIRCA), Coimbatore, India, 15–17 July 2020; pp. 790–793.
5. Parmar, G.; Lakhani, S.; Chattopadhyay, M.K. An IoT based low-cost air pollution monitoring system. In Proceedings of the International Conference on Recent Innovations in Signal processing and Embedded Systems (RISE), Bhopal, India, 27–29 October 2017; pp. 524–528.
6. Kanpur Rani, V.; Vallikanna, A.L. Air Pollution Monitoring System using Internet of Vehicles and Pollution Sensors. In Proceedings of the 4th International Conference on Electronics, Communication and Aerospace Technology (ICECA), Coimbatore, India, 5–7 November 2020; pp. 249–252.
7. Muthukumar, S.; Sherine Mary, W.; Jayanthi, S.; Kiruthiga, R.; Mahalakshmi, M. IoT Based Air Pollution Monitoring and Control System. In Proceedings of the International Conference on Inventive Research in Computing Applications (ICIRCA), Coimbatore, India, 11–12 July 2018; pp. 1286–1288.
8. Jiyal, S.; Saini, R.K. Prediction and Monitoring of Air Pollution Using Internet of Things (IoT). In Proceedings of the Sixth International Conference on Parallel, Distributed and Grid Computing (PDGC), Waknaghat, India, 6–8 November 2020; pp. 57–60.
9. McGrath, S.; Flanagan, C.; Zeng, L.; O'Leary, C. IoT Personal Air Quality Monitor. In Proceedings of the 31st Irish Signals and Systems Conference (ISSC), Letterkenny, Ireland, 11–12 June 2020; pp. 1–4.
10. Nowshin, N.; Hasan, M.S. Microcontroller Based Environmental Pollution Monitoring System though IoT Implementation. In Proceedings of the 2nd International Conference on Robotics, Electrical and Signal Processing Techniques (ICREST), Dhaka, Bangladesh, 5–7 January 2021; p. 493498.

engineering proceedings

MDPI

Proceeding Paper

LoRa Radius Coverage Map on Urban and Rural Areas: Case Study of Athens' Northern Suburbs and Tinos Island, Greece †

Spyridon Mitropoulos [1,*], Vasilios A. Orfanos [2], Dimitrios Rimpas [2] and Ioannis Christakis [2]

1 Department of Surveying and Geoinformatics Engineering, University of West Attica, 28, Ag. Spyridonos Str., 12243 Egaleo, Greece
2 Department of Electrical and Electronics Engineering, University of West Attica, P. Ralli & Thivon 250, 12244 Egaleo, Greece; vorfanos@uniwa.gr (V.A.O.); drimpas@uniwa.gr (D.R.); jchr@uniwa.gr (I.C.)
* Correspondence: smitro@uniwa.gr
† Presented at the 10th International Electronic Conference on Sensors and Applications (ECSA-10), 15–30 November 2023; Available online: https://ecsa-10.sciforum.net/.

Abstract: As the use and development of Internet of Things is very popular nowadays, one of the most widespread ways of exchanging data from such arrangements is the use of the LoRa network. One of the advantages offered by this technology is the ability to provide low power consumption as well as wide wireless coverage in an area. Although in research, there are references regarding the coverage radius of a geographical area, differences can be detected between urban (cities) and rural (countryside) areas, as in the latter, there are no dense structures nor radio signal noise inside the operating frequency spectrum of LoRa. Thus, results are expected to be better in rural areas than urban areas. Especially in an urban area, apart from the signal noise caused by other LoRa devices (either commercial or private), the coverage varies according to the placement of the LoRa station inside a building, which is related to the height at which the gateway is placed in another building. In this work, the LoRa radio coverage study is presented in a radius of 2 km both in an urban and a rural environment using only one LoRa gateway. To better capture the coverage, LoRa stations are placed on every floor of the selected buildings periodically. The results show the difference in coverage between urban and rural areas which is related to radio signal noise. Furthermore, significant changes in the coverage map in urban areas can be observed, directly related to the installation height of the LoRa station. With the understanding of these variations in LoRa network performance in different environments, informed decisions can be made regarding the deployment of such networks, optimizing their efficiency and ensuring seamless data transmission in both urban and rural settings.

Keywords: IoT LoRa; radio coverage; map coverage

Citation: Mitropoulos, S.; Orfanos, V.A.; Rimpas, D.; Christakis, I. LoRa Radius Coverage Map on Urban and Rural Areas: Case Study of Athens' Northern Suburbs and Tinos Island, Greece. *Eng. Proc.* **2023**, *58*, 19. https://doi.org/10.3390/ecsa-10-16122

Academic Editor: Francisco Falcone

Published: 15 November 2023

1. Introduction

The development of IoT devices in recent years has mainly focused on LoRa protocol for wireless data transmission. LoRa uses wireless data transmission protocols, achieving long-range communication. It is based on Chirp Spread Spectrum (CSS) modulation with a fixed frequency shift step, where it can achieve long-distance data transmission plus low power transmission. It is utilized in a wide range of applications such as smart metering, vehicle and infrastructure monitoring, smart health applications, and industrial production monitoring and control [1–4]. IoT architecture consists of three parts: perception, network, and application. A LoRa network includes three different types of nodes, i.e., a LoRa server, a LoRa gateway (GW), and the end devices [5].

The rapid development of wireless communication technologies in IoT has enabled their widespread use in smart buildings, such as Zigbee, Wi-Fi, LoRa, etc., [6,7]. Smart buildings present an increasing research field because of their ability for real-time dynamic control of various activities and reduction in energy consumption in building operations [8]. In [9], an investigation of communication performance of LoRa wireless technology inside an office building is presented. The experimental measurements show the transmission delay and packet loss rate of LoRa nodes from the same or different floors. The study in [10] concluded that the effect of electromagnetic propagation changes under various environments and cities, as well as rainforests.

The combination of IoT solutions is used for automated control of services such as building access, security systems, lighting, heating systems, ventilation systems, etc. [11,12]. In [13], a wearable LoRa-based wireless sensor network including self-powered environmental sensors was used to monitor harmful environmental conditions. A LoRa-based low-power real-time air quality monitoring system in [14] was implemented to collect both air and gas pollutant values, such as NO_2, CO, PM10, and PM2.5. In the research work of [15], a LoRa-based smart metering system was proposed and applied to smart buildings and smart grid systems. In [16], the monitoring of environmental conditions took place in a university, with LoRa nodes installed indoors and outdoors. Based on measured data the rate of packet loss in the outdoors under line-of-sight nodes was much smaller than the indoor ones. Ref. [17] showed the relation of throughput based on the installation position, presented using RSSI, SNR, and PER (packet error ratio), showing that the transmission signal was attenuated seriously in the basement.

In this work, the innovation of a low-cost water tank measurement device using an ultrasonic sensor is presented, while the data transmission was conducted through a LoRa network. An extensive study of the LoRa network coverage is also presented, using SNR and RSSI measurements. In the urban area, network coverage is studied as a result of LoRa terminal nodes being placed on all floors of buildings. The same experiment is conducted in a rural area without tall and dense buildings or significant radio interference.

2. Materials and Methods

2.1. Experimental Setup

This study presents the implementation of a low-cost water tank monitoring station. It is focused on the reception and network layer of LoRa architecture. The measurement analysis concerns the radio coverage of an area between the LoRa node and LoRa gateway. The low-cost water tank stations are implemented to measure the depth of a water tank. For portability purposes, a small solar panel with a lithium battery pack energizes the nodes. Every station is based on the TTGO@ESP32 (Shenzhen, China) (868 Mhz) [18] module (Figure 1a), meanwhile, LoRa gateway a Mikrotik wAP LR8 kit (Riga, Latvia) [19] (Figure 1b) is used. In addition, custom dipole antennas are made and used for both nodes and the gateway. The antenna performance measurements are shown in Figure 1c. To measure the distance (depth), the ultrasonic JSN-SR04T (Shenzhen, China) sensor is utilized [20].

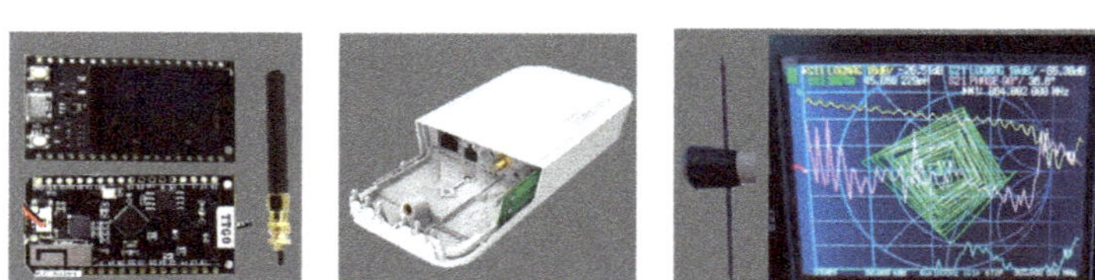

(a) (b) (c)

Figure 1. LoRa node construction: (**a**) TTGO@ESP32. (**b**) LoRa gateway. (**c**) Antenna frequency response.

The final construction of the autonomous LoRa node is presented in Figure 2a. The internal setup parts of the LoRa node are shown in Figure 2b and of the LoRa gateway in Figure 2c. This setup is based on the 868 Mhz frequency band, spread factor 8, bandwidth 125 Khz, and code rate 4/5. Measurements are obtained via the web interface of the LoRa gateway.

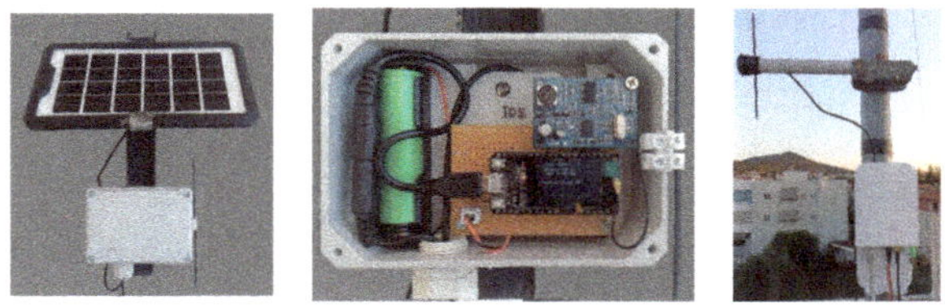

(**a**) (**b**) (**c**)

Figure 2. LoRa node and gateway. (**a**) LoRa node. (**b**) The parts of LoRa node. (**c**) LoRa gateway.

2.2. Experimental Results

The experiment was conducted in two phases. The LoRa nodes and LoRa gateway were at first installed within an urban area in the northeast of Athens in the municipality of Agia Paraskevi, Attica, Greece. The second phase took place in a rural area in the northwest of Tinos Island, in the village of Pirgos. The urban area and the installation points of the LoRa gateway and nodes are shown in Figure 3a. Measurements were obtained in the first two weeks of June 2023 and refer to an area of 1.6 square kilometers. In this area, the surface level comprises two hills, while the average building height is 15 m tall. The visualization of surface elevation (Figure 3b) between the gateway and nodes was retrieved using Google Earth's application.

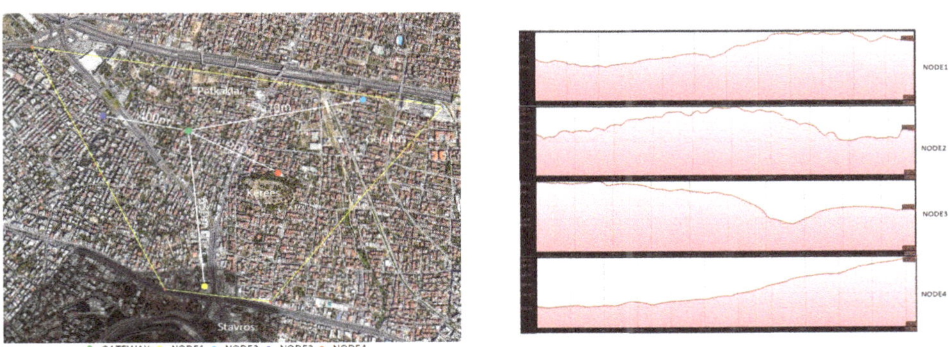

(**a**) (**b**)

Figure 3. Urban area of northeast of Attica. (**a**) Nodes and gateway installation points. (**b**) Elevation graphs between nodes and gateway.

The measurements between the nodes and the gateway are summarized in Table 1. The height of the antenna of the LoRa gateway was 3 m higher than the top roof of the building (total height around 18 m).

Table 1. SNR and RSSI measurements from nodes (urban area).

Node	GW Dist. (m)	Roof (dB)		2nd Floor (dB)		1st Floor (dB)		Ground (dB)	
		SNR	RSSI	SNR	RSSI	SNR	RSSI	SNR	RSSI
1	892	0~2.5	−110~−113	−5.5~−7.5	−112~−114	N/A	N/A	N/A	N/A
2	967	−7.5~−9.5	−113~−115	N/A	N/A	N/A	N/A	N/A	N/A
3	494	−7.0~−10.0	−114~−117	N/A	N/A	N/A	N/A	N/A	N/A
4	565	3~6	−103~−107	−2.0~−1.0	−108~−111	−7.0~−10.0	−115~−118	N/A	N/A

The installation points of the LoRa gateway and LoRa nodes in the rural area are projected in Figure 4a, while the surface elevation between the gateway and the nodes is shown in Figure 4b. For the research purpose, the radio coverage distance extension is presented in Figure 5a,b; measurements were obtained in the first two weeks of July 2023 and refer to an area of 6.0 square kilometers. This area is surrounded by mountains that offer excellent isolation from radio frequency noise. In this area, the average residential building height does not exceed 6 m. Table 2 shows the measurements between the nodes and the gateway.

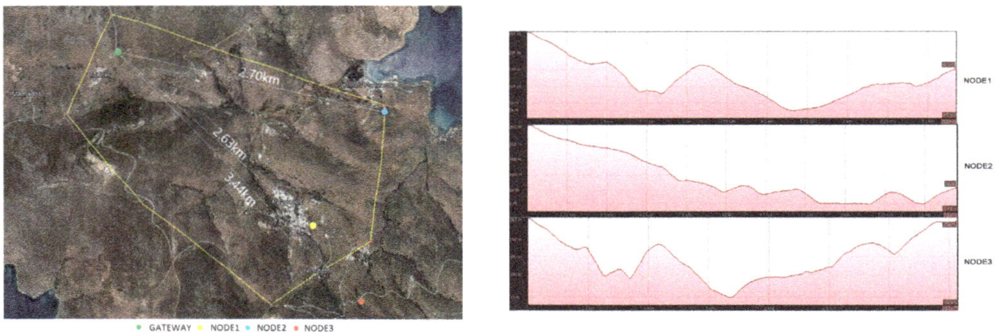

(a) (b)

Figure 4. Rural area of northwest of Tinos Island. (**a**) Nodes and gateway installation points. (**b**) Elevation graphs between nodes and gateway.

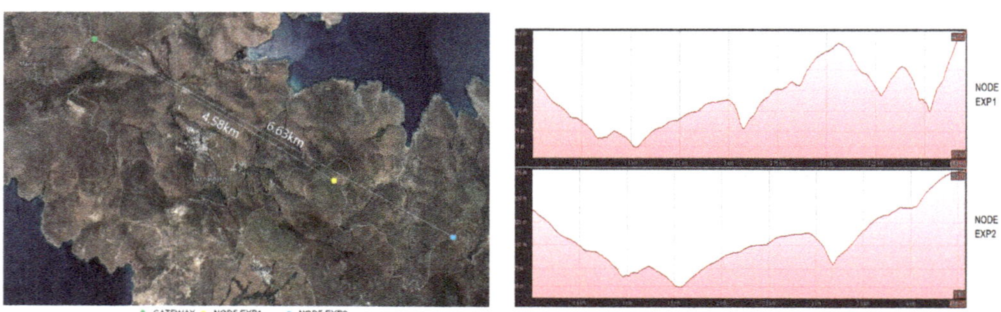

(a) (b)

Figure 5. Rural area of northwest of Tinos Island. (**a**) Experimental nodes and gateway installation points. (**b**) Elevation graphs between nodes and gateway.

Table 2. SNR and RSSI measurements from nodes (rural area).

	Gateway Distance (km)	SNR (dB)	RSSI (dB)
Node 1	2.68	5.5~5.7	−108~−109
Node 2	2.70	4.5~5.1	−109~−110
Node 3	3.57	3.9~4.78	−108~−110
Node EXP1	4.58	9.5~10.5	−95~−91
Node EXP2	6.63	−3.75~5.5	−116~−102

3. Results and Discussion

The visualization of the deepness of each water tank via the ultrasonic sensors is presented dynamically under the Grafana lab environment using a web browser (Figure 6). Figure 6a shows the water tanks' depth measurements, and Figure 6b displays the battery voltage. The reliability of the distance sensor's measurements has previously been verified by calibrating it at different distance lengths with a high accuracy laser measuring device. These measurements prove that the LoRa signal is reliable enough for accurate data acquisition from the stations at long distances.

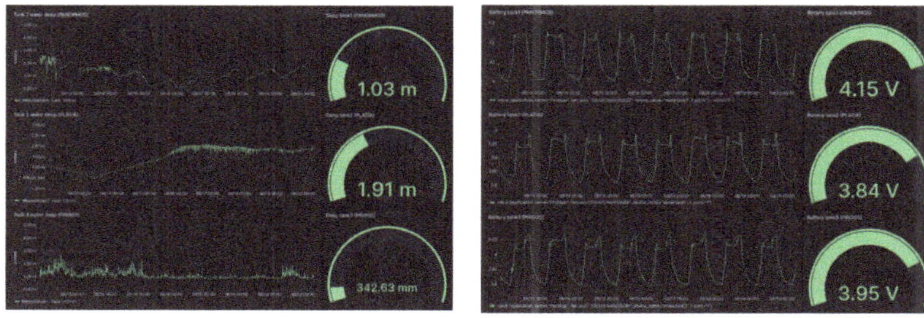

(**a**) (**b**)

Figure 6. Grafana lab visualization interface. (**a**) Water tanks' depth. (**b**) Battery voltage.

According to the measurements in Tables 1 and 2 (SNR; RSSI) of the LoRa radio coverage between the urban and rural area, the coverage graphs are shown in Figure 7 (Figure 7a shows the urban coverage and Figure 7b shows the rural coverage).

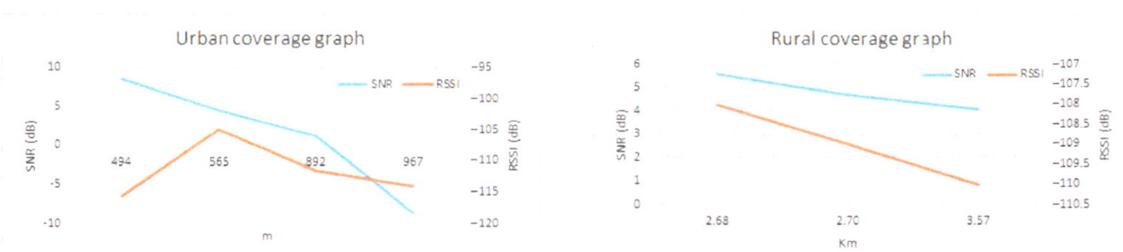

(**a**) (**b**)

Figure 7. SNR and RSSI coverage graphs. (**a**) Urban area. (**b**) Rural area.

From the coverage graphs, it is concluded that the radio coverage of the rural area is greater than the urban area. The reasoning for this behavior is because in the urban area, there is a dense structure of buildings that affects the transmission of the signal. Buildings act as attenuators, according to the SNR measurements. This becomes more evident particularly from the fact that there was no reception when the station was installed on the ground or the first floor. In addition, in urban areas, many forms of radio transmission have been developed, as well as other LoRa networks, that create radio noise in the transmission signal, as derived from the RSSI measurements. On the other hand, small villages and the low height (max 6 m) of their houses do not interfere with the signal as much. Also, radio noise is much weaker as the installations of radio transmitters (television; mobile network) are also weaker in relation to the urban area. According to the above, the results of the LoRa radio coverage in both urban and rural areas are shown; the radio coverage in the urban area refers to an area of 1.6 square kilometers, while in the rural area, it refers to an area of 6.0 square kilometers.

4. Conclusions

In this work, a LoRa-based water tank measuring system was presented. As part of the research study, radio coverage measurements were performed both in urban and rural areas on a similar terrain. The maximum distance of the radio coverage in the urban area was 1 km, while in the rural area, it worked up to 6.5 km. The measurements taken in the city were also conducted for each floor, where in many cases, especially on the ground and first floor, the system could not operate properly. In summary, it was confirmed that in non-line-of-site cases, the LoRa network did not work, as a radio noise affected the radio coverage. In provinces with no radio frequency noise, this system showed excellent radio coverage. On the other hand, the operation of monitoring systems, especially in rural areas, improved the lifestyle of the citizens, hence directly addressing their needs in terms of people's quality of life and livelihood with simple, low-cost but highly important methods.

Author Contributions: Conceptualization, I.C. and S.M.; methodology, I.C. and D.R.; software, I.C.; validation, S.M., D.R. and V.A.O.; formal analysis, S.M.; investigation, I.C.; resources, S.M.; data curation, V.A.O.; writing—original draft preparation, S.M. and I.C.; writing—review and editing, D.R. and V.A.O.; visualization, V.A.O.; supervision, D.R.; project administration, I.C. and D.R. All authors have read and agreed to the published version of the manuscript.

Funding: This research received no external funding.

Institutional Review Board Statement: Not applicable.

Informed Consent Statement: Not applicable.

Data Availability Statement: All of the data created in this study are presented in the context of this article.

Conflicts of Interest: The authors declare no conflicts of interest.

References

1. Adelantado, F.; Vilajosana, X.; Tuset-Peiro, P.; Martinez, B.; Melia-Segui, J.; Watteyne, T. Understanding the Limits of LoRaWAN. *IEEE Commun. Mag.* **2017**, *55*, 34–40. [CrossRef]
2. Orfanos, V.A.; Kaminaris, S.D.; Piromalis, D.; Papageorgas, P. Smart Home Automation in the IoT Era: A Communication Technologies Review. *AIP Conf. Proc.* **2020**, *2307*, 020054. [CrossRef]
3. Petäjäjärvi, J.; Mikhaylov, K.; Pettissalo, M.; Janhunen, J.; Iinatti, J. Performance of a low-power wide-area network based on LoRa technology: Doppler robustness, scalability, and coverage. *Int. J. Distrib. Sens. Netw.* **2017**, *13*, 1550147717699412. [CrossRef]
4. Papadakis, N.; Koukoulas, N.; Christakis, I.; Stavrakas, I.; Kandris, D. An Iot-Based Participatory Antitheft System for Public Safety Enhancement in Smart Cities. *Smart Cities* **2021**, *4*, 919–937. [CrossRef]
5. Augustin, A.; Yi, J.; Clausen, T.; Townsley, W.M. A study of LoRa: Long range & low power networks for the internet of things. *Sensors* **2016**, *16*, 1466. [PubMed]
6. Orfanos, V.; Kaminaris, S.; Papageorgas, P.; Piromalis, D.; Kandris, D. A Comprehensive Review of IoT Networking Technologies for Smart Home Automation Applications. *J. Sens. Actuator Netw.* **2023**, *12*, 30. [CrossRef]
7. Ray, P.P. A Survey on Internet of Things Architectures. *J. King Saud Univ.-Comput. Inf. Sci.* **2018**, *30*, 291–319.
8. Zainordin, N.B.; Abdullah, S.M.B.; Baharum, Z.B.A. Light and Space: Users Perception towards Energy Efficient Buildings. *Procedia-Soc. Behav. Sci.* **2012**, *36*, 51–60. [CrossRef]
9. Liang, R.; Zhao, L.; Wang, P. Performance evaluations of LoRa wireless communication in building environments. *Sensors* **2020**, *20*, 3828. [CrossRef]
10. Ferreira, A.E.; Ortiz, F.M.; Costa, L.H.M.K.; Foubert, B.; Amadou, I.; Mitton, N. A study of the LoRa signal propagation in forest, urban, and suburban environments. *Ann. Telecommun.* **2020**, *75*, 333–351. [CrossRef]
11. Mussab, A.; Zaidan, A.A.; Mohammed, T.; Kiah, M.L.M. A review of smart home applications based on Internet of Things. *J. Netw. Comput. Appl.* **2017**, *97*, 48–65.
12. Yuce, M.R. WE-Safe: A Self-Powered Wearable IoT Sensor Network for Safety Applications Based on LoRa: A scoping review. *Int. J. Med. Inform.* **2017**, *3*, 105–108.
13. Zhang, X.H.; Zhang, M.M.; Meng, F.F.; Qiao, Y.; Xu, S.J.; Senghout, H. A Low-PowerWide-Area Network Information Monitoring System by Combining NB-IoT and LoRa. *IEEE Internet Things J.* **2018**, *6*, 590–598. [CrossRef]
14. Abdelfatteh, H.; Abdelhak, A.; Aziz, D. Performance Evaluation of Low-PowerWide Area based on LoRa Technology for Smart Metering. In Proceedings of the 6th International Conference on Wireless Networks and Mobile Communications, Marrakesh, Morocco, 16–19 October 2018; pp. 97–102.
15. Wang, S.Y.; Chen, Y.R.; Chen, T.Y.; Chang, C.H.; Cheng, Y.H.; Hsu, C.C.; Lin, Y.B. Performance of LoRa-Based IoT Applications on Campus. In Proceedings of the IEEE Vehicular Technology Conference, Toronto, ON, Canada, 24–27 September 2017; Volume 2017, pp. 1–6.
16. Neumann, P.; Montavont, J.; Noel, T. Indoor deployment of low-power wide area networks (LPWAN): A LoRaWAN case study. In Proceedings of the IEEE International Conference on Wireless & Mobile Computing, New York, NY, USA, 17–19 October 2016; p. 8.
17. Ayele, E.D.; Hakkenberg, C.; Meijers, J.P.; Zhang, K.; Meratnia, N.; Havinga, P.J.M. Performance analysis of LoRa radio for an indoor IoT applications. In Proceedings of the 2017 International Conference on Internet of Things for the Global Community (IoTGC), Funchal, Portugal, 10–13 July 2017; p. 8.

18. Santos, S. ESP32 with Built-in SX1276 Lora and SSD1306 OLED Display (Review), Maker Advisor. 2023. Available online: https://makeradvisor.com/esp32-sx1276-lora-ssd1306-oled/ (accessed on 29 July 2023).
19. MikroTik Routers and Wireless. Available online: https://mikrotik.com/product/wap_lr8_kit (accessed on 31 August 2023).
20. JSN-SR04T—Instruction Manual—Openbikesensor. Available online: https://www.openbikesensor.org/docs/hardware/general/collective-order/jsn-sr04t-en.pdf (accessed on 31 August 2023).

engineering proceedings

Proceeding Paper

Assessment of Stress Level with Help of "Smart Clothing" Sensors, Heart Rate Variability-Based Markers and Machine Learning Algorithms †

Liudmila Gerasimova-Meigal [1,*], Alexander Meigal [1], Vyacheslav Dimitrov [2], Maria Gerasimova [1], Anna Sklyarova [1], Nikolai Smirnov [3] and Vasilii Kostyukov [4]

1 Department of Physiology and Pathophysiology, Petrozavodsk State University, 185910 Petrozavodsk, Russia; meigal@petrsu.ru (A.M.)
2 Department of Computer Science and Mathematical Support, Petrozavodsk State University, 185910 Petrozavodsk, Russia; dimitrov@cs.petrsu.ru
3 Department of Probability Theory and Data Analysis, Petrozavodsk State University, 185910 Petrozavodsk, Russia; smirnovn@petrsu.ru
4 Center of Artificial Intelligence, Petrozavodsk State University, 185910 Petrozavodsk, Russia; kostyukov@petrsu.ru
* Correspondence: gerasimova@petrsu.ru; Tel.: +7-911-402-9907

† Presented at the 10th International Electronic Conference on Sensors and Applications (ECSA-10), 15–30 November 2023; Available online: https://ecsa-10.sciforum.net/.

Abstract: Physiological stress in healthy subjects was assessed using heart rate (HR), monitored with the help of Hexoskin smart garments. HRs were collected during daily life activities and in laboratory settings during stress tests. Heart rate variability parameters were computed and referenced with expert levels of stress. The data were processed with the help of machine learning algorithms (Random Forest, CatBoost, XGB, LGBM, SVR). The Random Forest Regressor provided the best rate of correct entries (86%), and the CatBoost Regressor provided the shortest time (2 ms) for the assessment of stress levels.

Keywords: textile wearable sensors; machine learning; artificial intelligence; HRV; stress

Citation: Gerasimova-Meigal, L.; Meigal, A.; Dimitrov, V.; Gerasimova, M.; Sklyarova, A.; Smirnov, N.; Kostyukov, V. Assessment of Stress Level with Help of "Smart Clothing" Sensors, Heart Rate Variability-Based Markers and Machine Learning Algorithms. *Eng. Proc.* **2023**, *58*, 20. https://doi.org/10.3390/ecsa-10-16173

Academic Editor: Francisco Falcone

Published: 15 November 2023

1. Introduction

Stress is a multisystem compensatory response of the body to external and internal stimuli (stressors) [1,2], and it evolved to preserve the constancy of its vital parameters, or homeostasis [1,2]. The response to stressors is represented by the intensification of metabolism due to changes in hormonal, autonomic, nervous and motor functions [1]. The reaction to stress is characterized by stages [1,2] and several levels of its intensity [3]. Despite its originally protective nature, stress can lead to functional "over taxation" [1] and the dysregulation of stress-sensitive systems—nervous, cardiovascular, gastrointestinal and immune [1,4]. Stress can significantly reduce indicators of human well-being, such as working capacity, quality of life, personal capital and social adaptation. Thus, the instrumental assessment, control, prediction and prevention of stress are recognized as critical scientific problems [1,5]. To address this problem, informative biosignal-based markers of stress must be identified [5].

2. Theory

Several groups of stress markers have been proposed, e.g., psychophysiological, autonomic and cognitive, as well as blood and saliva tests for hormones [5–9]. The autonomic markers, e.g., skin sympathetic response, pupil response and heart rate variability (HRV), attract growing attention [6,9–13]. For example, the time- and frequency-domain parameters of HRV are promising for assessing levels of stress [9]. Nonlinear parameters (dimension and entropy) of HRV are increasingly being used to detect disorders of the autonomic nervous system [13], which can be used to assess stress. Textile sensors ("smart clothing" technologies) are suitable for monitoring physiological functions in daily life. Typically, HRV, temperature, respiratory rate, skin sympathetic response and motility are measured with textile sensors. The search for stress markers can be carried out with the help of machine learning (ML) algorithms, which are especially well suited for nonlinear metrics [14]. In this study, we aimed to assess stress levels with HRV markers in healthy individuals in their daily activities using HRV parameters obtained from smart clothing sensors and processed with ML algorithm technologies. The Hexoskin Smart Shirt was chosen for smart clothing because there are reports that it provides reliable HR signals [15,16].

3. Experimental Section

3.1. Subjects and Protocol

Nineteen practically healthy subjects, aged 19 to 55 years, volunteered to participate in the study. All subjects signed their informed consent before the study. The study protocol was approved by the joint ethics committee of the Ministry of Health Care of the Republic of Karelia and Petrozavodsk State University (No. 30, 16 June 2014). The whole study was conducted within the time period November 2021–January 2022 in the city of Petrozavodsk (Republic of Karelia, Russia). The study was conducted in two settings. In the laboratory (15 subjects), three conventional tests were used to induce stress: (1) reaction time tests (simple and choice reaction time, manual target interception) (detailed in [17]), (2) physical exercise (3 min of pedaling on an ergometer with a growing load [7]) and (3) the cold press test (CPT) [7] (3 min immersion of a hand in ice water at 3–4 °C, detailed in [18]). Rest, "without stress", conditions were used for comparisons. A total of 10 different conditions were applied in a mixed order within two or three individual days. The HR data were obtained either with laboratory instruments (VNS-Spectr, Neirosoft, Ivanovo, Russia) or a Hexoskin Smart Shirt.

In field conditions, eight subjects practiced their daily life activities (25 individual measurements), which included walking indoors and outdoors, mental activities (PC- and Internet-based, sitting, phone calling), taking meals (including coffee), sleeping, visiting a fitness center, driving a car, visiting a dental clinic, etc. All subjects kept records of their activity in a form with 15 min intervals. In different trials, measurements lasted 2 to 12 h, depending on the will of the subject. The data were collected only with Hexoskin Smart Shirts.

Momentarily perceived stress was assessed with a visual analogue scale (VAS) [4,7,19]. On the VAS, the left end (value "1") corresponded to the "no stress at all" condition, and the right end (value "10") corresponded to maximal "perceived stress". The VAS was administered at each of the 10 experimental conditions. In field conditions, subjects marked the level of their momentarily perceived stress with the VAS every 15 min.

3.2. Data Acquisition with Smart Clothes

HR was collected by the Hexoskin Smart Shirt (Hexoskin Smart Sensors and AI, Montreal, QC, Canada). It was put on the subject according to the instructions and connected to a logger. After the experiment, data were uploaded into the HxServices (v.4.05) software (Figure 1).

3.3. HR Data Processing

The Hexoskin Smart Shirt provides a wav sound file. This format is convenient for processing in terms of the ready-made libraries of the Python language. To extract signal peaks from a wav file, the ecg_peaks function of the neurokit2 library [20] was used. With a known value of the pulse frequency (for HxS, it is 256 Hz), the value of the time between the peaks was computed, which allowed for calculating the values of the intervals between neighboring R waves (RR interval, RRi). The resulting dataset of RRis was filtered with cutoff HR values < 45 and >180 beats per min. After that, each time series of a trial was subdivided into 3 min serial segments. Then, HRV parameters were calculated for each of the segments with the help of the pyHRV toolbox [21]. Each segment was labeled with the expert level of perceived stress marked by individual subjects with the VAS in one dimension.

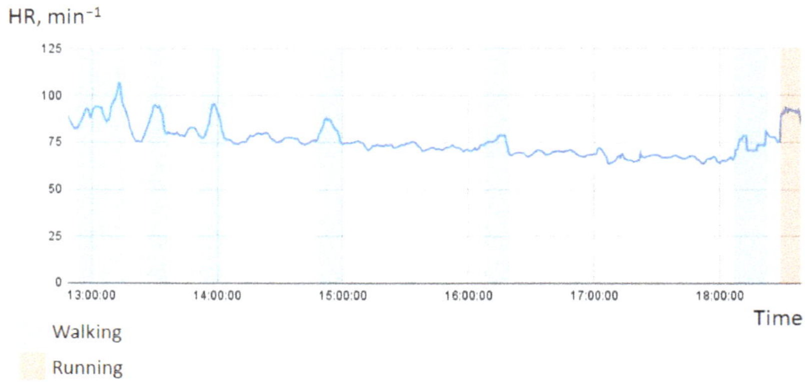

Figure 1. A representative HR signal recorded with a Hexoskin Smart Shirt in field conditions.

The time-domain HRV parameters included the HR (hr_mean), the standard deviation (sdnn), the root-mean-squared difference (rmssd) and the proportion of successive intervals greater than 50 ms (pnn50) in terms of normal RR intervals (nni). The frequency-domain HRV parameters included the total power spectrum of RRis (fft_total); the power spectrum at very low (vlf; <0.04 Hz), low (lf; 0.04–0.15 Hz) and high frequency bands (hf; 0.15–0.40 Hz); and the spectrum structure (vlf_pct, lf_pct, hf_pct, lf_nu, hf_nu). Nonlinear parameters included the sample (sampen), approximate entropy (apen) and sd1 and sd2 of the Poincare plot.

Altogether, 11,570 segments were available in the dataset for the analysis. Table 1 shows the distribution of the number of records corresponding to 10 different levels of stress. Of the 11,570 segments, over 53% had the stress level "2" and 17% had the stress level "3", with further systematic decrement to levels "7–9" (some 1%) and 10 (0%). Of the whole dataset, 75% of the segments were randomly assigned to the "training" set, and the remaining 25% were assigned to the "test" set (Table 1).

Table 1. Distribution of segments among the datasets and their reference to the level of stress.

Stress Level	Number of Segments	Number of Segments in the Training Set	Number of Segments in the Test Set
1	577	433	144
2	6225	4669	1556
3	2051	1538	513
4	1140	855	285
5	584	438	146
6	530	397	133
7	155	116	39
8	184	138	46
9	124	93	31
10	0	0	0
Total number of records	11,570	8677	2893

The calculated parameters were checked for correlation. In the matrix in Figure 2, there is a strong correlation dependence for some pairs of features, which necessitates the selection of significant features. For the regression model for predicting the level of stress, the final set of features was formed from 11 features: pnn50, nni_mean, hr_mean, sd1, sd2, sampen, lf_pct, hf, lf, vlf and hf_pct. The CatBoost Regressor with the random_state = 42 parameter was used as the regression algorithm. The rest of the algorithm's parameters had default values.

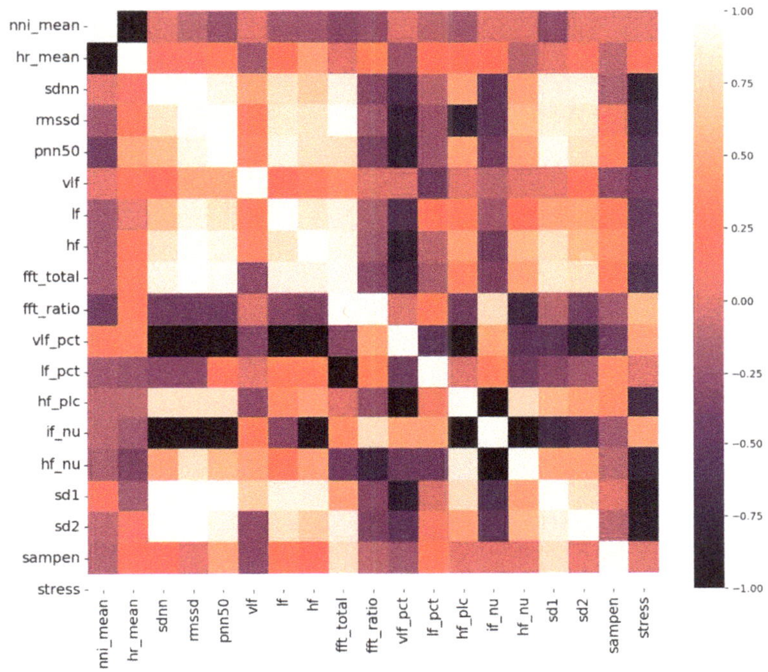

Figure 2. Correlation of features in the dataset with the original set of features.

To reduce the dimension of the input data and the error metrics, the following algorithm for selecting significant features was developed:

- The CatBoost Regressor model was trained on data containing all available features.
- Using the get_feature_importance method from the CatBoost library [22] for the Python language, we obtain an array of feature significance scores for the model; the score values are sorted in descending order.
- An empty current feature set, an empty final feature set and a minimum mean absolute error (MAE) are initialized, which is initially set to infinity.
- The model is iteratively trained and tested on fixed training and test sets.
- The next feature is added to the current set of features in descending order of significance for the model.
- The model is trained on the current set of features, and the MAE is calculated on the test set of records with the current set of features.

If the obtained MAE value is less than the current minimum MAE value, then the considered feature is stored in the final feature set, and the minimum MAE value becomes equal to the current one; otherwise, the considered feature is not stored in the final feature set.

For the regression model for predicting the level of stress, the final set of features was formed from 11 features: pnn50, nni_mean, hr_mean, sd1, sd2, sampen, lf_pct, hf, lf, vlf and hf_pct. The test was carried out for five types of modes: the Random Forestv Regressor, CatBoost Regressor, XGBoost Regressor, LightGBM Regressor and Support Vector Regressor.

4. Results

The final outcome is presented in Table 2. As can be seen from Table 2, the Random Forest Regressor algorithm provided the highest % of correct entries (86.3%). On the other hand, the shortest execution time was produced by the CatBoost Regressor algorithm, which was the second best in terms of correct entries.

Table 2. Stress level regression quality metrics provided by different algorithms.

Method	MAE	R^2	Target Variable Prediction Time for One Record, ms	Number of Correct Entries, %
RandomForestRegressor (random_state = 42)	0.0018	0.9992	8.48	86.3
CatBoostRegressor (random_state = 42	0.0822	0.975	2.02	85.3
XGBRegressor (random_state = 42)	0.0837	0.9719	4.63	80.1
LGBMRegressor (random_state = 42)	0.1628	0.9044	1.97	82.9
SVR (kernel = 'rbf')	0.3913	0.4503	29.93	64.1

5. Conclusions

In conclusion, given the quality of the outcome and prediction time value, the CatBoost Regressor algorithm can be regarded as a reliable algorithm for the assessment of stress levels with HRV parameters obtained with a "smart clothing" device. We also conclude that an 85–86% rate of correct entries (correct linking of HRV parameters to the level of stress) looks promising. Still, one sound limitation can be identified in the study. Namely, the diversity of the levels of stress was not high enough in this study, as more than 70% of the segments were assigned to the stress level 2–3, which means "low stress". In future studies, longer periods of HR monitoring with textile sensors would be reliable.

Author Contributions: Conceptualization, L.G.-M., A.M. and V.K.; methodology, L.G.-M., A.M. and V.K.; software, V.D., N.S. and V.K.; validation, L.G.-M., A.M., V.D., N.S. and V.K.; formal analysis, L.G.-M., A.M., V.D., N.S. and V.K.; investigation, L.G.-M., A.M., M.G. and A.S.; resources, A.M. and V.K.; data curation, L.G.-M., A.M., M.G. and A.S.; writing—original draft preparation, L.G.-M., A.M., V.D., M.G., A.S. and V.K.; writing—review and editing, L.G.-M., A.M. and V.D.; visualization, V.D.; supervision, L.G.-M.; project administration, A.M. and V.K.; funding acquisition, V.K. All authors have read and agreed to the published version of the manuscript.

Funding: This research received no external funding.

Institutional Review Board Statement: The study was conducted in accordance with the Declaration of Helsinki and approved by the joint Ethics Committee of the Ministry of Health Care of the Republic of Karelia and Petrozavodsk State University (Statement of approval No. 30, 16 June 2014).

Informed Consent Statement: Informed consent was obtained from all subjects involved in the study.

Data Availability Statement: The datasets generated for this study are available on request to the corresponding author.

Acknowledgments: The authors thank the subjects for their participation.

Conflicts of Interest: The authors declare no conflicts of interest.

References

1. Anisman, H.; Merali, Z. Understanding Stress: Characteristics and Caveats. *Alcohol Res. Health* **1999**, *23*, 241–249. [PubMed]
2. Cohen, S.; Gianaros, P.J.; Manuck, S.B. A Stage Model of Stress and Disease. *Perspect. Psychol. Sci.* **2016**, *11*, 456–463. [CrossRef] [PubMed]
3. Allen, A.P.; Kennedy, P.J.; Cryan, J.F.; Dinan, T.G.; Clarke, G. Biological and psychological markers of stress in humans: Focus on the Trier Social Stress Test. *Neurosci. Biobehav. Rev.* **2014**, *38*, 94–124. [CrossRef]

4. LeMoult, J.; Battaglini, A.M.; Grocott, B.; Jopling, E.; Rnic, K.; Yang, L. Advances in stress and depression research. *Curr. Opin. Psychiatry* **2023**, *36*, 8–13. [CrossRef] [PubMed]
5. Han, H.J.; Labbaf, S.; Borelli, J.L.; Dutt, N.; Rahmani, A.M. Objective stress monitoring based on wearable sensors in everyday settings. *J. Med. Eng. Technol.* **2020**, *44*, 177–189. [CrossRef] [PubMed]
6. Zontone, P.; Affanni, A.; Piras, A.; Rinaldo, R. Exploring Physiological Signal Responses to Traffic-Related Stress in Simulated Driving. *Sensors* **2022**, *22*, 939. [CrossRef] [PubMed]
7. Skoluda, N.; Strahler, J.; Schlotz, W.; Niederberger, L.; Marques, S.; Fischer, S.; Thoma, M.V.; Spoerri, C.; Ehlert, U.; Nater, U.M. Intra-individual psychological and physiological responses to acute laboratory stressors of different intensity. *Psychoneuroendocrinology* **2015**, *51*, 227–236. [CrossRef]
8. Smets, E.; De Raedt, W.; Van Hoof, C. Into the Wild: The Challenges of Physiological Stress Detection in Laboratory and Ambulatory Settings. *IEEE J. Biomed. Health Inform.* **2019**, *23*, 463–473. [CrossRef]
9. Chen, S.; Xu, K.; Zheng, X.; Li, J.; Fan, B.; Yao, X.; Li, Z. Linear and nonlinear analyses of normal and fatigue heart rate variability signals for miners in high-altitude and cold areas. *Comput. Methods Programs Biomed.* **2020**, *196*, 105667. [CrossRef]
10. Manser, P.; Thalmann, M.; Adcock, M.; Knols, R.H.; de Bruin, E.D. Can Reactivity of Heart Rate Variability Be a Potential Biomarker and Monitoring Tool to Promote Healthy Aging? A Systematic Review with Meta-Analyses. *Front. Physiol.* **2021**, *12*, 686129. [CrossRef]
11. Pham, T.; Lau, Z.J.; Chen, S.H.A.; Makowski, D. Heart Rate Variability in Psychology: A Review of HRV Indices and an Analysis Tutorial. *Sensors* **2021**, *21*, 3998. [CrossRef] [PubMed]
12. von Rosenberg, W.; Chanwimalueang, T.; Adjei, T.; Jaffer, U.; Goverdovsky, V.; Mandic, D.P. Resolving Ambiguities in the LF/HF Ratio: LF-HF Scatter Plots for the Categorization of Mental and Physical Stress from HRV. *Front. Physiol.* **2017**, *8*, 360. [CrossRef] [PubMed]
13. Gerasimova-Meigal, L.; Meigal, A.; Sireneva, N.; Saenko, I. Autonomic Function in Parkinson's Disease Subjects Across Repeated Short-Term Dry Immersion: Evidence From Linear and Non-linear HRV Parameters. *Front. Physiol.* **2021**, *12*, 712365. [CrossRef] [PubMed]
14. Hill, J.R.; Caldwell, B.S. Assessment of Physiological Responses During Field Science Task Performance: Feasibility and Future Needs. *Front. Physiol.* **2022**, *13*, 779873. [CrossRef] [PubMed]
15. Haddad, M.; Hermassi, S.; Aganovic, Z.; Dalansi, F.; Kharbach, M.; Mohamed, A.O.; Bibi, K.W. Ecological Validation and Reliability of Hexoskin Wearable Body Metrics Tool in Measuring Pre-exercise and Peak Heart Rate During Shuttle Run Test in Professional Handball Players. *Front. Physiol.* **2020**, *11*, 957. [CrossRef]
16. Keogh, A.; Dorn, J.F.; Walsh, L.; Calvo, F.; Caulfield, B. Comparing the Usability and Acceptability of Wearable Sensors Among Older Irish Adults in a Real-World Context: Observational Study. *JMIR mHealth uHealth* **2020**, *8*, e15704. [CrossRef]
17. Meigal, A.Y.; Tretjakova, O.G.; Gerasimova-Meigal, L.I.; Sayenko, I.V. Program of Seven 45-min Dry Immersion Sessions Improves Choice Reaction Time in Parkinson's Disease. *Front. Physiol.* **2021**, *11*, 621198. [CrossRef]
18. Gerasimova-Meigal, L.I.; Sirenev, I.M. Cold-induced reactivity in multiple sclerosis patients. *Patol. Fiziol. Eksperimental'naya Ter. (Pathol. Physiol. Exp. Ther.)* **2017**, *61*, 56–62. [CrossRef]
19. Lesage, F.X.; Berjot, S. Validity of occupational stress assessment using a visual analogue scale. *Occup. Med.* **2011**, *61*, 434–436. [CrossRef]
20. Makowski, D.; Pham, T.; Lau, Z.J.; Brammer, J.C.; Lespinasse, F.; Pham, H.; Schölzel, C.; Chen, S.H.A. NeuroKit2: A Python toolbox for neurophysiological signal processing. *Behav. Res. Methods* **2021**, *53*, 1689–1696. [CrossRef]
21. Gomes, P.; Margaritoff, P.; da Silva, H.P. pyHRV: Development and evaluation of an open-source python toolbox for heart rate variability (HRV). In Proceedings of the International Conference on Electrical, Electronic and Computing Engineering (IcETRAN), Veliko Gradište, Serbia, 3–6 June 2019; pp. 822–828.
22. Hancock, J.T.; Khoshgoftaar, T.M. CatBoost for big data: An interdisciplinary review. *J. Big Data* **2020**, *7*, 94. [CrossRef] [PubMed]

engineering proceedings

MDPI

Proceeding Paper

Development of a Monitoring System against Illegal Deforestation in the Amazon Rainforest Using Artificial Intelligence Algorithms †

Thiago Almeida Teixeira [1,*], Neilson Luniere Vilaça [1], Andre Luiz Printes [1,2], Raimundo Claúdio Souza Gomes [1,2], Israel Gondres Torné [1,2], Thierry-Yves Alves Araújo [1] and Arlley Gabriel Dias e Dias [1]

1 Embedded Systems Laboratory, State University of Amazonas, Manaus 69050-020, Brazil; neilsonluniere@hub-uea.org (N.L.V.); aprintes@uea.edu.br (A.L.P.); rsgomes@uea.edu.br (R.C.S.G.); itorne@uea.edu.br (I.G.T.); tyalves@hub-uea.org (T.-Y.A.A.); agdd.ele16@uea.edu.br (A.G.D.e.D.)
2 High School Technology, University of the Amazon State, Manaus 69050-025, Brazil
* Correspondence: tateixeira@hub-uea.org; Tel.: +55-92-99400-9520

† Presented at the 10th International Electronic Conference on Sensors and Applications (ECSA-10), 15–30 November 2023; Available online: https://ecsa-10.sciforum.net/.

Abstract: The Amazon Rainforest represents one-third of the world's tropical forest area. This makes it indispensable for maintaining global biodiversity. However, the increasing occurrences of wildfires and deforestation in the region are notorious. In this sense, it is essential to protect forests to ensure quality of life for future generations and prevent damage that affects the entire planet. In this work, a real-time monitoring device is proposed to identify attempts at deforestation through audio signals from tractors and chainsaws, using embedded artificial intelligence (AI). Additionally, it is capable of communicating with a base station, reaching distances close to 1 km in dense forest, through long-range (LoRa) communication. A user interface is also developed, providing daily alerts such as attack identification, occurrence times, device locations, and battery status. The system has an average power consumption of around 300 nA, employing power management methods defined as ultra-low power mode, sleep mode, prediction mode, and transmission mode. Hence, the device has the potential to promote the sustainable preservation of the Amazon Rainforest, helping to prevent large-scale illegal deforestation.

Keywords: illegal deforestation; Amazon Rainforest; artificial intelligence; LoRa communication; low-power system

Citation: Teixeira, T.A.; Vilaça, N.L.; Printes, A.L.; Gomes, R.C.S.; Torné, I.G.; Araújo, T.-Y.A.; Dias, A.G.D.e. Development of a Monitoring System against Illegal Deforestation in the Amazon Rainforest Using Artificial Intelligence Algorithms. *Eng. Proc.* **2023**, *58*, 21. https://doi.org/10.3390/ecsa-10-16188

Academic Editor: Stefan Bosse

Published: 15 November 2023

1. Introduction

The Amazon Rainforest plays an essential role in climate regulation, capable of controlling precipitation cycles, stabilizing temperatures, and acting as a powerful 'sponge', capturing a significant amount of carbon dioxide, the primary gas responsible for exacerbating global warming [1–4]. The increasing deforestation in the Amazon Rainforest represents one of the greatest threats to biodiversity, global climate balance, and the communities that depend directly on this ecosystem [5–7].

The urgent need to contain and reverse this challenging scenario has driven the development of innovative solutions, among which environmental monitoring devices stand out. In this context, the development of monitoring devices has become a promising strategy for the conservation of the Amazon.

This work aims to present a study on a specific monitoring device that aims to significantly contribute to preventing deforestation in the Amazon Rainforest. The device in question is based on the integration of technologies, including an audio acquisition sensor, LoRa communication networks, and embedded artificial intelligence, in low-power hardware. This information is processed in real time by advanced algorithms capable of detecting patterns and anomalies, providing immediate alerts to the responsible agencies and relevant authorities.

One of the main advantages of this device is its ability to cover vast and hard-to-reach areas, allowing monitoring not only of areas near urban centers but also of remote and border regions. Furthermore, its energy autonomy and real-time data transmission capability make it a valuable tool for making agile and effective decisions.

The environmental impact of this technology is significant because its effectiveness in monitoring the Amazon Rainforest has the potential to deter illegal deforestation activities and, consequently, significantly reduce the rate of forest degradation. However, it is essential to emphasize that the implementation of this device faces challenges, such as the cost of development, operation, and maintenance.

In this context, this work aims to promote a detailed understanding of the proposed monitoring device, and highlighting its benefits and limitations as an important tool for the conservation of the Amazon Rainforest. The discussion on the role of technology in environmental preservation is crucial for guiding policies and practices that promote the sustainability of this ecosystem, which is vital for the planet.

2. Materials and Methods

2.1. Architecture and Hardware Device

The proposed monitoring system should be capable of covering regions of the Amazon Rainforest and identifying deforestation attempts through the acquisition of sounds from trucks and chainsaws. The system has a Central Gateway responsible for the centralized communication of the devices, using LoRa wireless communication. Figure 1a illustrates the architecture of the monitoring system, with devices installed in the forest region, covering areas with a radius of approximately 1 km between the communication nodes and the Central Gateway.

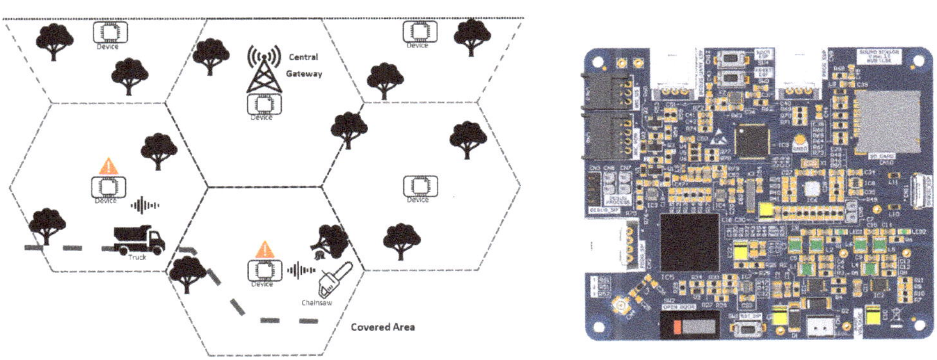

(**a**) (**b**)

Figure 1. Amazon Rainforest monitoring system: (**a**) system architecture; (**b**) developed monitoring device.

The developed hardware device consists of an embedded electronic system with the main components being an SD Card module, digital microphones, and microcontrollers. One microphone is responsible for waking up the system from low-power mode and energizing the main systems, while the second microphone) is responsible for collecting audio data for artificial intelligence processing. A microcontroller is dedicated to managing the system's power and transmitting LoRa packets. Another microcontroller performs embedded artificial intelligence processing. Figure 1b illustrates the developed electronic board.

2.2. Embedded Artificial Intelligence

To perform the processing of the AI model, the TensorFlow Lite 2.14.0 library was used, which can execute machine learning models on microcontrollers using very few kilobytes of memory. The primary execution environment fits within 16 KB on an Arm Cortex M3 and can run many basic models. The library has been tested on various Arm Cortex-M Series processors and has also been ported to other architectures, including ESP32 [8].

For the development of an artificial intelligence model capable of identifying deforestation attempts, audio recordings of chainsaws and tractors were initially collected in a forest environment under various conditions, including different motor acceleration levels, and with varying distances between the tools and the audio collection device.

The collected audio data were categorized into two classes: normalcy and deforestation. Subsequently, the audio data were segmented into 1 s samples, each composed of a vector of 16,000 values. After processing the audio, a dataset was obtained, consisting of 50,404 samples. Among these, 88% were categorized as normalcy samples, and 12% as attack samples, indicating deforestation activity.

With the dataset in hand, the training phase proceeded. In this phase, the dataset was split into two subsets: the training set with 80% of the samples and the test set with 20%. It is worth noting that this split was carried out in a stratified manner to ensure representative proportions of both classes in both subsets. The architecture of the neural network consists of 5 one-dimensional convolutional layers, interspersed with pooling layers. Next, a 1D Global Max Pooling layer was applied to reduce the dimensions, and a dropout layer with a rate of 20% was inserted. In the last layer, a fully connected layer was applied using softmax activation. The Nadam optimizer was used, the cost function adopted was categorical cross entropy, and the evaluation metric chosen was the F1-Score. The learning rate was set to 0.001, and a mini-batch containing 256 samples was applied in the training process.

2.3. LoRa Communication and Web Application

The LoRa communication gateway was installed on a 15 m communication tower in the Ecoforest Adventure Park, located in a rural area of the Amazon region. Its purpose is to establish communication with Sound Sensor boards installed in dense forest areas, with a maximum distance of up to 1100 m between them. Figure 2a shows the LoRa gateway installed on the Ecoforest Adventure telecommunication tower, and Figure 2b presents the device installed on a tree in the Amazon Rainforest, with its respective protective case and antenna operating at a frequency of 915 MHz and a gain of 5dBi.

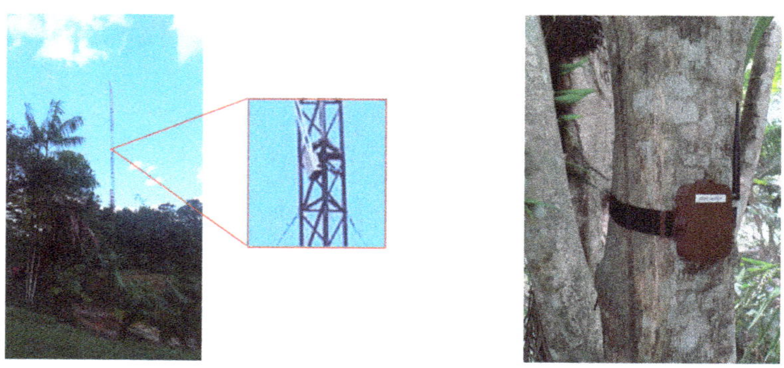

(a) (b)

Figure 2. Installation of the monitoring system: (**a**) communication tower; (**b**) prototype installed on a tree in the Amazon Rainforest.

To visualize the monitoring data from the Amazon Rainforest processed by each Sound Sensor device, a web application was developed using the frameworks NestJS for creating the Sound Sensor API and ReactJS for creating the project's web page. The API comprises two databases, PostgreSQL (a relational database) and InfluxDB (a time-series database). It also integrates with ChirpStack, an open-source LoRaWAN server, which registers the Sound Sensor devices and receives their information via uplink.

3. Results

3.1. Energy Management

With the implementation of the device's operating modes, it was possible to measure the device's energy consumption at each stage. The average time of operation for each of the modes was also established. Table 1 presents these measured parameters.

Table 1. Measured current values of the system's operation modes.

Operation Mode	Period	Consumption
Ultra-Low Power	10 m	300 nA
Sleep	2 s	200 µA
Prediction	1 s	50 mA
Transmission	300 ms	130 mA

3.2. AI Performance

After the completion of training, it was possible to evaluate the performance of the proposed model. The results and performance metrics achieved on the test dataset are described in Table 2, and the confusion matrix is shown in Figure 3.

Table 2. Results of the proposed model on the test dataset.

Metric	Value
Accuracy	99.6%
F1-Score	99%
Loss	0.052

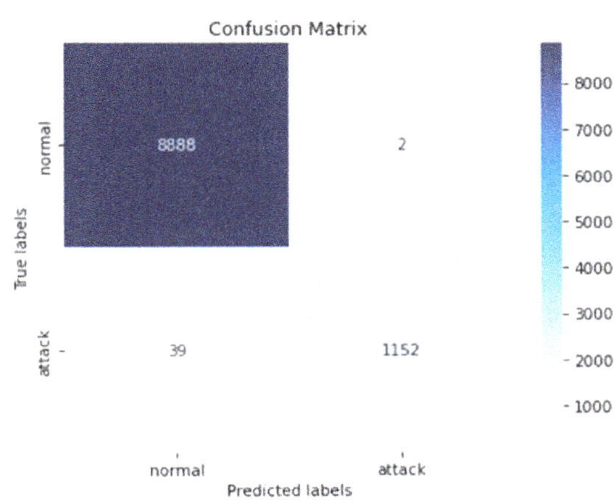

Figure 3. Confusion matrix of the proposed model.

3.3. LoRa Distance

The maximum message reception distance was validated through a route in the Ecoforest Adventure park. The monitoring device was programmed to send LoRa messages with 1 byte of information in the payload every 10 s during the forest journey, with the geographic coordinates of the location being recorded every 1 min. Simultaneously, the uplinks sent by the device were monitored on the developed dashboard, allowing for the real-time observation of the received data. The coordinates were synchronized with the uplink transmission times in a .csv file. This file was loaded into the Google Earth Pro 7.3.6.9345 software, where it was possible to visualize the traveled path as well as the terrain elevation profile of the forest. Figure 4 shows the achieved distance of approximately 1023 m from the gateway installation point to the device and the elevation profile of the terrain along the route in Google Earth Pro.

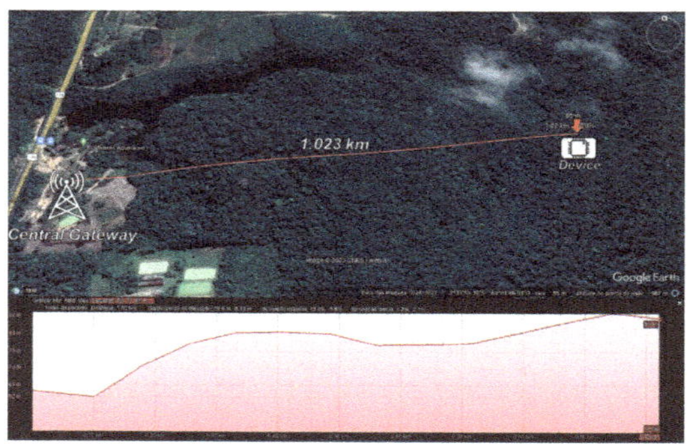

Figure 4. Validation of the LoRa communication route between gateway and device. The terrain elevation profile is also shown.

3.4. Data Visualization

Figure 5 depicts the web application developed, where users can view real-time data from the devices, including installation points on a map, battery levels, last sent sample, reference point, latitude and longitude, city/state information, and predictions from the device indicating whether it is in normal operation or in alert mode. This application enables users to take preventive and protective actions for the Amazon Rainforest against deforestation.

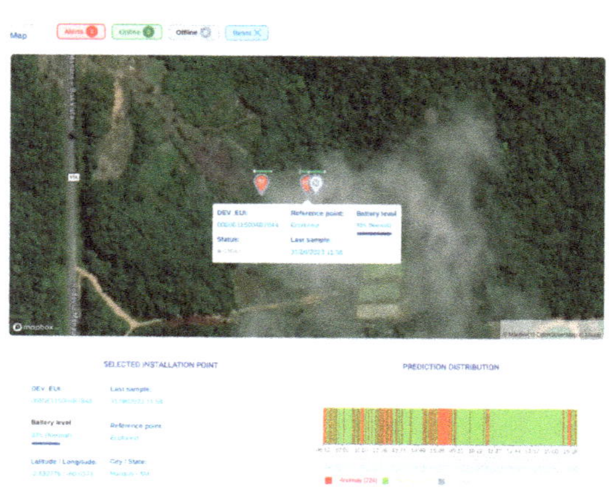

Figure 5. Real−time web application for monitoring the Amazon Rainforest.

4. Conclusions

The illegal deforestation monitoring system for the Amazon involved the development of an embedded electronic device, composed of digital microphones capable of capturing the ambient sounds of the Amazon Rainforest. Through an artificial intelligence model, it identifies the sounds of chainsaws or tractors. The results of this pattern recognition are transmitted via the LoRa network, reaching approximately 1023 m, and can be viewed through a web application. The energy management system achieved ultra-low power consumption, reaching approximately 300 nA of minimum current. The developed neural network was able to accurately identify attacks with an accuracy of 99.6%, with no occurrence of false positives.

In this way, this system has proven to be an efficient means of combating the imminent illegal deforestation of the forest, because it enables the real-time analysis and detection of attacks. It helps maintain biodiversity, the water cycle, and carbon stocks, preserving these natural processes. Furthermore, public policies can be implemented and better managed, even allowing for the recovery of deforested areas. In this sense, this system has the potential to monitor not only the Amazon Rainforest, but also other interconnected ecosystems, promoting environmental preservation in a sustainable manner.

Author Contributions: Conceptualization, R.C.S.G., A.L.P. and I.G.T.; methodology, N.L.V., T.A.T., T.-Y.A.A. and A.G.D.e.D.; software, T.-Y.A.A. and A.G.D.e.D.; validation, N.L.V.; T.A.T., T.-Y.A.A. and A.G.D.e.D.; formal analysis, R.C.S.G. and A.L.P.; investigation, N.L.V., T.A.T., T.-Y.A.A., and A.G.D.e.D.; resources, N.L.V., T.A.T., T.-Y.A.A. and A.G.D.e.D.; data curation, T.-Y.A.A. and A.G.D.e.D.; writing—original draft preparation, N.L.V., T.A.T., T.-Y.A.A. and A.G.D.e.D.; writing—review and editing, R.C.S.G., A.L.P. and I.G.T.; visualization, N.L.V., T.A.T., T.-Y.A.A. and A.G.D.e.D.; supervision, R.C.S.G., A.L.P. and I.G.T.; project administration, R.C.S.G., A.L.P. and I.G.T.; funding acquisition, R.C.S.G., A.L.P. and I.G.T. All authors have read and agreed to the published version of the manuscript.

Funding: This research received no external funding.

Institutional Review Board Statement: Not applicable.

Informed Consent Statement: Not applicable.

Data Availability Statement: The data analyzed during the current study are available from the corresponding author on reasonable request.

Acknowledgments: The authors thank the Embedded Systems Laboratory of the State University of Amazonas for the support provided.

Conflicts of Interest: The authors declare no conflict of interest.

References

1. Boulton, C.A.; Lenton, T.M.; Boers, N. Pronounced loss of Amazon rainforest resilience since the early 2000s. *Nat. Clim. Chang.* **2022**, *12*, 271–278. [CrossRef]
2. Lenton, T.M.; Held, H.; Kriegler, E.; Hall, J.W.; Lucht, W.; Rahmstorf, S.; Schellnhuber, H.J. Tipping elements in the Earth's climate system. *Proc. Natl. Acad. Sci. USA* **2008**, *105*, 1786–1793. [CrossRef] [PubMed]
3. Gatti, L.V.; Basso, L.S.; Miller, J.B.; Gloor, M.; Gatti Domingues, L.; Cassol, H.L.G.; Tejada, G.; Aragão, L.E.O.C.; Nobre, C.; Peters, W.; et al. Amazonia as a carbon source linked to deforestation and climate change. *Nature* **2021**, *595*, 388–393. [CrossRef]
4. Malhi, Y.; Roberts, J.T.; Betts, R.A.; Killeen, T.J.; Li, W.; Nobre, C.A. Climate Change, Deforestation, and the Fate of the Amazon. *Science* **2008**, *319*, 169–172. [CrossRef]
5. Rodrigues, E.; de Almeida, A.B.; Oliveira, A.S.T.; Pacheco, D.L.; Ramos, G.C.; de Sousa, G.O.; de Souza Santos, I.; Bento, J.S.; da Cruz, K.A.S.; Coqueiro, L.F.U. Impactos do desmatamento da floresta Amazônica sobre a saúde pública e as comunidades tradicionais. In *Saúde, História, Ciência e Educação: Perspectivas dos Grupos PET da UNIFESP Durante a pandemia de COVID-19*, 1st ed.; Fontoura: Várzea Paulista, Brazil, 2022; pp. 75–79.
6. Castro, D.M.B.; Haonat, A.I. Fiscalização dos Desmatamentos Ilegais na Floresta Amazônica do Brasil: Paralelo a partir do Livro "Vidas Secas". *Rev. Extensão* **2021**, *5*, 50–58.
7. Fearnside, P.M. *Destruição e Conservação da Floresta Amazônica*; Editora INPA: Manaus, Brazil, 2022.
8. TensorFlow for Microcontrollers. Available online: https://www.tensorflow.org/lite/microcontrollers (accessed on 28 September 2023).

engineering proceedings

MDPI

Proceeding Paper

Advancements in Sensor-Based Technologies for Precision Agriculture: An Exploration of Interoperability, Analytics and Deployment Strategies †

Bishnu Kant Shukla *, Neha Maurya and Manshi Sharma

Department of Civil Engineering, JSS Academy of Technical Education, Noida 201301, India; nehamaurya1223@gmail.com (N.M.); manshi9415@gmail.com (M.S.)

* Correspondence: bishnukantshukla@jssaten.ac.in; Tel.: +91-82849-42192

† Presented at the 10th International Electronic Conference on Sensors and Applications (ECSA-10), 15–30 November 2023; Available online: https://ecsa-10.sciforum.net/.

Abstract: In response to escalating global food demand and growing environmental concerns, the incorporation of advanced sensor technologies in agriculture has become paramount. This paper delves into an in-depth exploration of cutting-edge sensor-based technologies, inclusive of Internet of Things (IoT) applications, machine learning algorithms, and remote sensing, in revolutionizing farming practices for improved productivity, efficiency, and sustainability. The breadth of this exploration encompasses an array of sensors employed in precision agriculture, such as soil, weather, light, humidity, and crop health sensors. Their impact on farming operations and the challenges posed by their implementation are scrutinized. Emphasis is placed on the integral role of IoT-based sensor networks in promoting real-time data acquisition, thereby facilitating efficient decision making. The study examines crucial wireless communication standards like ZigBee, Wi-Fi, Bluetooth, and fifth-generation (5G) and upcoming technologies like NarrowBand Internet of Things (NB-IoT) for sensor data transfer in smart farming. The paper emphasises the necessity of interoperability among various sensor technologies and provides a thorough analysis of data analytics and management techniques appropriate for the substantial data generated by these systems. The robustness of sensor systems, their endurance in difficult environmental settings, and their flexibility in adapting to shifting agricultural contexts are highlighted. The report also explores potential future directions, highlighting the potential of 5G and AI-driven predictive modelling to enhance sensor functions and expedite data processing systems. The challenges encountered in deploying these sensor-based technologies, such as cost, data privacy, system compatibility, and energy management, are discussed in depth with potential solutions and mitigation strategies proposed. This paper, therefore, navigates towards an improved comprehension of the expansive potential of sensor technologies, leading the way to a more sustainable and efficient future for agriculture.

Keywords: smart agriculture sensors; IoT; precision farming; sensor-based technologies; data analytics; interoperability; wireless communication protocols

Citation: Shukla, B.K.; Maurya, N.; Sharma, M. Advancements in Sensor-Based Technologies for Precision Agriculture: An Exploration of Interoperability, Analytics and Deployment Strategies. *Eng. Proc.* **2023**, *58*, 22. https://doi.org/10.3390/ecsa-10-16051

Academic Editor: Stefano Mariani

Published: 15 November 2023

1. Introduction

Agricultural systems will be under tremendous pressure to increase food production in sustainable ways as the world's population is projected to nearly double to 10 billion by 2050 [1]. In order to be prepared for the future, agriculture must find inventive solutions to problems like unstable climatic patterns and dwindling arable lands. Agricultural frameworks in the modern era face a number of difficulties. Unpredictable weather patterns brought on by climate change include more frequent droughts and floods. Such unpredictabilities place a severe pressure on conventional farming techniques, necessitating a change to more robust agricultural practises [2]. Further increasing the need for effective and sustainable agricultural methods is the growing loss of arable land, which is mostly linked to urbanisation and soil deterioration [3].

The potential of technology intervention shines brilliantly in the midst of these difficulties. A ray of hope is provided by precision agriculture, which is supported by cutting-edge sensor technologies. By continuously tracking crop conditions, weather patterns, and soil health, these devices provide actionable data that enable well-informed decisions and prompt treatments. In essence, sensor-based technology integration in farming landscapes resolves many of the current problems and, more importantly, sets the way for a sustainable and productive agricultural future [4].

2. Sensor-Based Technologies: An Insight into Modern Agricultural Practices

The era of traditional farming is evolving towards a tech-centric agricultural paradigm. Sensor-based technologies, leveraging data, promise enhanced production and sustainability. Table 1 highlights key sensor technology advancements and their agricultural applications to illustrate this shift.

Table 1. Advanced sensor technologies being used in modern agriculture.

Sensor Type	Primary Function	Agricultural Application	Reference
Soil Sensor	Measure soil moisture, salinity, and pH levels	Guide irrigation and fertilization strategies	[5]
Weather Sensor	Monitor atmospheric conditions	Optimize planting and harvesting schedules	[6]
Light Sensor	Gauge sunlight exposure	Determine optimal crop growth environments	[7]
Humidity Sensor	Monitor ambient moisture	Assist in microclimate regulation for crops	[7]
Crop Health Sensor	Use spectral imaging to assess crop health	Detect early signs of diseases, pests, or nutrient deficiencies	[8]

By providing real-time information on the qualities of the soil and optimising irrigation and fertilisation, soil sensors have transformed precision farming [5]. Farmers can adjust to changing atmospheric conditions by using weather sensors to optimise planting and harvesting seasons [6]. Crop health and yield are improved by the fine-tuning of microclimatic conditions by light and humidity sensors [7]. Spectral imaging-based crop health monitors notify farmers to problems like infections or nutritional deficiencies and allow for pre-emptive interventions [8].

3. The Convergence of IoT and Wireless Communication in Precision Agriculture

Through the incorporation of sensor technology and cutting-edge wireless communication systems, the Internet of Things (IoT) has transformed agriculture. Transmitting enormous amounts of real-time data over great distances, especially in difficult circumstances, becomes crucial as farms transform into digital ecosystems.

3.1. Sensor Networks in IoT-Based Agriculture

Interconnected sensor networks, which collect various agronomic metrics, are agriculture's IoT [9]. Their true value comes from their ability to interface with centralised databases and decision-making tools, giving farmers the power to make quick decisions based on current data and promoting a precision culture [10,11].

3.2. Communication Technologies: Bridging the Gap

Numerous communication protocols, ranging from well-established ones like ZigBee to more recent ones like 5G (fifth generation) and NB-IoT (NarrowBand Internet of Things) are available in the tech world for smart farming. ZigBee provides balanced power and range, making it perfect for isolated farms [12], while NB-IoT provides extensive coverage with low energy requirements for large-scale agricultural endeavours [13]. With its speed, low latency, and dependability, 5G is poised to redefine data transfer [14].

4. The Interplay of Sensor Technologies and Data Management

Modern precision agriculture is built on a foundation of seamless sensor technology and data management, creating a synergy that offers increased agricultural productivity and better resource management.

4.1. Need for Interoperability among Diverse Technologies

Different sensor systems used in smart agriculture have unique data formats and protocols [15]. Interoperability becomes more important as systems become more complex. Enhanced interoperability optimises yield and resource consumption by consolidating and coherently analysing data.

4.2. Data Analytics, Management Techniques, and Challenges

Artificial intelligence and smart algorithms help with the rigorous management and analysis of the massive amounts of data from sensors [16]. Making the transition from raw data to usable insights presents difficulties, such as managing storage and protecting data privacy. To solve this, you need to have a solid grasp of agriculture and be an expert data manager.

4.3. Techniques for Sensor Data Analytics and Management

To analyse huge amounts of sensor-generated data, advanced analytics, from classical statistical techniques to contemporary machine learning, are required. An overview of these analytics and management methods for agricultural sensor data is provided in Table 2.

Table 2. Comprehensive overview of sensor data analytics and management techniques in precision agriculture.

Technique	Description	Application in Agriculture	Reference
Statistical Analysis	Traditional data interpretation method, using standard deviation, mean, etc.	Analysing soil nutrient variations, crop yield predictions	[17]
Machine Learning	Algorithms that improve through experience	Predictive modelling for weather, pest infestation forecasts	[18]
Neural Networks	Systems modelled on the human brain, capable of recognizing patterns	Early detection of plant diseases through image recognition	[19]
Edge Computing	Data processing at the edge, or source, of data generation	Real-time soil health monitoring, immediate irrigation decisions	[20]
Decentralized Data Storage	Distributed data storage systems	Securely storing vast amounts of data from multiple farm sources	[21]

5. Future Perspectives and Current Challenges in Sensor-Based Agriculture

The nexus of sensor development, telecommunications, and artificial intelligence is the next frontier in agricultural technology. These cutting-edge fields, while they hold the potential for paradigm-shifting effects, are also rife with complex problems that each call for in-depth investigation.

5.1. Potential of 5G Technology and AI Predictive Modelling

5G's ultra-reliable low latency promises more than fast data transfer, facilitating real-time analysis in distant agricultural fields [22]. AI (artificial intelligence) predictive modelling, enhanced by real-time data, offers precise forecasts, from weather to pest detection [23].

5.2. Challenges: Cost, Data Privacy, System Compatibility, Energy Management

Economic difficulties arise with the adoption of cutting-edge technologies, particularly for agrarians who are resource-constrained [24]. Increased encryption and security measures are required due to the increase in data [25]. In order to combine several digital platforms in agriculture, seamless system compatibility is essential [26]. Energy conservation, notably the use of renewable energy for sensors, must be prioritised [27]. For a thorough summary of the obstacles and future directions for sensor-based agriculture, see Table 3.

Table 3. An analytical synopsis of prospective avenues and challenges in sensor-based agriculture.

Parameter	Description	Potential Benefits	Associated Challenges	Reference
5G Connectivity	Ultra-reliable, low-latency communication.	Enhanced real-time monitoring and feedback.	Infrastructure costs and regional network disparities.	[22]
AI Predictive Modelling	AI-driven data analysis for prediction.	Accurate forecasting and proactive strategies.	Complex model training and requisite data volumes.	[23]
Cost Barriers	Economic implications of advanced technology integration.	Improved long-term ROI with optimized farming.	Initial high costs, especially for small-scale farmers.	[24]
Data Privacy	Safeguarding vast datasets generated from sensors.	Secure, robust data storage and transmission.	Potential data breaches and misuse.	[25]
System Compatibility	Ensuring cohesive integration of diverse technological solutions within the agricultural framework.	Seamless, unified farming operations.	Interoperability issues and legacy system challenges.	[26]
Energy Management	Strategies to ensure continuous sensor operations with minimal energy expenditure.	Sustainable, uninterrupted monitoring.	Dependency on non-renewable energy sources.	[27]

6. Conclusions

With the combination of cutting-edge sensor technology and data-driven innovations, agriculture's historical evolution is ready for yet another important revolution. This discussion went deep into the technical details of these technologies and highlighted how they might affect present and future agricultural endeavours. The analysis of our research yields the following key conclusions:

- **Primacy of Sensor Integration**: Sensor technologies that track factors like soil quality, weather erraticness, light, humidity, and crop health are largely responsible for the advancements in precision agriculture. They have fuelled data-centric farming methods by offering granular insights, assuring the best use of resources and raising yields.
- **Sensor Technologies and Data Management Confluence**: Different sensor technologies add different data formats and communication protocols to the agricultural tapestry as they are woven in. Thus, interoperability is essential for achieving a seamless connection. A sophisticated analytics and management solution is also required to transform the massive amount of collected data into a useful agricultural strategy.
- **Prospective Technological Boons:** The impending use of 5G technology and the skill of AI predictive modelling are positive signs for the future of agriculture. These tools could revolutionise agricultural decision making since they promise accurate forecasting and real-time data transfer.
- **Inherent Challenges:** Even with advances in technology, significant problems still exist. These cover issues including the price tag associated with adopting new technology, protecting data privacy in an era of information overload, establishing system interoperability among various tech solutions, and dealing with energy management for remote sensor operations.
- **Implications for Upcoming Research:**
  1. *Focused Investigation:* Future studies may benefit from focusing on particular sensor types as technology continues to progress. This would enable a deeper investigation of the problems and applications that are unique to them.

2. *Holistic System Design:* The creation of complete systems that integrate various sensor technologies should be prioritised in order to streamline data collection, processing, and the extraction of insightful information.
3. *Broader Dimensions:* Prospective studies ought to extend beyond the strictly technical to include ethical considerations of data privacy and the larger socio-economic effects of technology adoption in agriculture.

Fundamentally, even while the fusion of sensor-based technology with agriculture ushers in a new era of productivity and sustainability, the road ahead is not without obstacles. However, the advantages they offer vastly outweigh these challenges, prompting a concerted effort by academics, technologists, and agricultural stakeholders to fully realise this enormous potential.

Author Contributions: Conceptualization, B.K.S. and N.M.; methodology, B.K.S.; software, M.S.; validation, B.K.S., N.M., and M.S.; formal analysis, N.M.; investigation, M.S.; resources, N.M.; data curation, M.S.; writing—original draft preparation, B.K.S.; writing—review and editing, N.M.; visualization, M.S.; supervision, B.K.S.; project administration, N.M.; funding acquisition, N.M. All authors have read and agreed to the published version of the manuscript.

Funding: This research received no external funding.

Institutional Review Board Statement: Not applicable.

Informed Consent Statement: Not Applicable.

Data Availability Statement: Data will be made available upon request.

Conflicts of Interest: The authors declare no conflict of interest.

References

1. Guo, H.; Jiang, J.; Li, Y.; Long, X.; Han, J. An aging giant at the center of global warming: Population dynamics and its effect on CO_2 emissions in China. *J. Environ. Manag.* **2023**, *327*, 116906. [CrossRef]
2. Nadeem, F.; Jacobs, B.; Cordell, D. Mapping agricultural vulnerability to impacts of climate events of Punjab, Pakistan. *Reg. Environ. Chang.* **2022**, *22*, 66. [CrossRef]
3. Liu, C.; Deng, C.; Li, Z.; Liu, Y. Response characteristics of Soil Erosion to Spatial Conflict in the Production-Living-Ecological Space and Their Drivingmechanism: A Case Study of Dongting Lake Basin in China. *Land* **2022**, *11*, 1794. [CrossRef]
4. Abu, N.S.; Bukhari, W.M.; Ong, C.H.; Kassim, A.M.; Izzuddin, T.A.; Sukhaimie, M.N.; Norasikin, M.A.; Rasid, A.F.A. Internet of things applications in precision agriculture: A review. *JRC* **2022**, *3*, 338–347. [CrossRef]
5. Moursy, A.R.; Hassan, M.N.; Elhefny, T.M. Sampling and analysis of soil and water: A review. *Int. J. Geogr. Geol. Environ.* **2022**, *4*, 34–41.
6. Hewage, P.; Behera, A.; Trovati, M.; Pereira, E.; Ghahremani, M.; Palmieri, F.; Liu, Y. Temporal convolutional neural (TCN) network for an effective weather forecasting using time-series data from the local weather station. *Soft Comput.* **2020**, *24*, 16453–16482. [CrossRef]
7. Achour, Y.; Ouammi, A.; Zejli, D. Technological progresses in modern sustainable greenhouses cultivation as the path towards precision agriculture. *Renew. Sust. Energ. Rev.* **2021**, *147*, 111251. [CrossRef]
8. Benelli, A.; Cevoli, C.; Fabbri, A. In-field hyperspectral imaging: An overview on the ground-based applications in agriculture. *J. Agric. Eng.* **2020**, *51*, 129–139. [CrossRef]
9. Ploennigs, J.; Cohn, J.; Stanford-Clark, A. The future of IoT. *IEEE Internet Things Mag.* **2018**, *1*, 28–33. [CrossRef]
10. Neethirajan, S. SOLARIA-SensOr-driven resiLient and adaptive monitoRIng of farm Animals. *Agriculture* **2023**, *13*, 436. [CrossRef]
11. Vangala, A.; Das, A.K.; Chamola, V.; Korotaev, V.; Rodrigues, J.J. Security in IoT-enabled smart agriculture: Architecture, security solutions and challenges. *Clust. Comput.* **2023**, *26*, 879–902. [CrossRef]
12. Alex, N.; Sobin, C.C.; Ali, J. A comprehensive study on smart agriculture applications in India. *Wirel. Pers. Commun.* **2023**, *129*, 2345–2385. [CrossRef]
13. Majumdar, P.; Bhattacharya, D.; Mitra, S.; Bhushan, B. Application of Green IoT in Agriculture 4.0 and Beyond: Requirements, Challenges and Research Trends in the Era of 5G, LPWANs and Internet of UAV Things. *Wirel. Pers. Commun.* **2023**, *131*, 1767–1816. [CrossRef]
14. Garg, D.; Alam, M. Smart agriculture: A literature review. *J. Manag. Anal.* **2023**, *10*, 359–415. [CrossRef]
15. Adesipo, A.; Fadeyi, O.; Kuca, K.; Krejcar, O.; Maresova, P.; Selamat, A.; Adenola, M. Smart and climate-smart agricultural trends as core aspects of smart village functions. *Sensors* **2020**, *20*, 5977. [CrossRef]
16. Bourechak, A.; Zedadra, O.; Kouahla, M.N.; Guerrieri, A.; Seridi, H.; Fortino, G. At the Confluence of Artificial Intelligence and Edge Computing in IoT-Based Applications: A Review and New Perspectives. *Sensors* **2023**, *23*, 1639. [CrossRef]

17. De Alwis, S.; Hou, Z.; Zhang, Y.; Na, M.H.; Ofoghi, B.; Sajjanhar, A. A survey on smart farming data, applications and techniques. *Comput. Ind.* **2022**, *138*, 103624. [CrossRef]
18. Amani, M.A.; Marinello, F. A deep learning-based model to reduce costs and increase productivity in the case of small datasets: A case study in cotton cultivation. *Agriculture* **2022**, *12*, 267. [CrossRef]
19. Ramachandran, V.; Ramalakshmi, R.; Kavin, B.P.; Hussain, I.; Almaliki, A.H.; Almaliki, A.A.; Elnaggar, A.Y.; Hussein, E.E. Exploiting IoT and its enabled technologies for irrigation needs in agriculture. *Water* **2022**, *14*, 719. [CrossRef]
20. Iftikhar, S.; Gill, S.S.; Song, C.; Xu, M.; Aslanpour, M.S.; Toosi, A.N.; Du, J.; Wu, H.; Ghosh, S.; Chowdhury, D.; et al. AI-based fog and edge computing: A systematic review, taxonomy and future directions. *Internet Things* **2023**, *21*, 100674. [CrossRef]
21. Saba, T.; Rehman, A.; Haseeb, K.; Bahaj, S.A.; Lloret, J. Trust-based decentralized blockchain system with machine learning using Internet of agriculture things. *Comput. Electr. Eng.* **2023**, *108*, 108674. [CrossRef]
22. Eswaran, S.; Honnavalli, P. Private 5G networks: A survey on enabling technologies, deployment models, use cases and research directions. *Telecommun. Syst.* **2023**, *82*, 3–26. [CrossRef] [PubMed]
23. Malhotra, K.; Firdaus, M. Application of Artificial Intelligence in IoT Security for Crop Yield Prediction. *J. Sci. Technol.* **2022**, *2*, 136–157.
24. Khan, N.; Ray, R.L.; Kassem, H.S.; Zhang, S. Mobile Internet Technology Adoption for Sustainable Agriculture: Evidence from Wheat Farmers. *Appl. Sci.* **2022**, *12*, 4902. [CrossRef]
25. Shaikh, T.A.; Rasool, T.; Lone, F.R. Towards leveraging the role of machine learning and artificial intelligence in precision agriculture and smart farming. *Comput. Electron. Agric.* **2022**, *198*, 107119. [CrossRef]
26. Chen, Q.; Li, L.; Chong, C.; Wang, X. AI-enhanced soil management and smart farming. *Soil Use Manag.* **2022**, *38*, 7–13. [CrossRef]
27. Suanpang, P.; Pothipassa, P.; Jermsittiparsert, K.; Netwong, T. Integration of kouprey-inspired optimization algorithms with smart energy nodes for sustainable energy management of agricultural orchards. *Energies* **2022**, *15*, 2890. [CrossRef]

engineering proceedings

MDPI

Proceeding Paper

Bioengineered Monoclonal Antibody Chitosan–Iron Oxide Bio-Composite for Electrochemical Sensing of *Mycobacterium tuberculosis* Lipoprotein †

Resmond L. Reaño [1,*], Glenson R. Panghulan [1], Clydee Ann T. Hernandez [1] and Jeffrey P. Tamayo [2]

[1] Department of Engineering Science, College of Engineering and Agro-Industrial Technology, University of the Philippines Los Baños, Los Baños 4031, Philippines

[2] Central Analytical Services Laboratory, National Institute of Molecular Biology and Biotechnology, University of the Philippines Los Baños, Los Baños 4031, Philippines; jptamayo1@up.edu.ph

* Correspondence: rlreano@up.edu.ph

† Presented at the 10th International Electronic Conference on Sensors and Applications (ECSA-10), 15–30 November 2023; Available online: https://ecsa-10.sciforum.net/.

Abstract: In this study, an electrochemical immunosensor for the detection of the 19 kDa *Mycobacterium tuberculosis* lipoprotein LpqH was developed using a monoclonal antibody immobilized on a chitosan-coated iron oxide bio-composite. The bio-composite is composed of magnetic iron oxide at its core and a non-magnetic thin film on the surface formed by chitosan, providing the chemistry for monoclonal antibody immobilization. Cyclic voltammetry was used to characterize and test the immunosensor assembly. Electrochemical measurements showed a strong relationship between the LpqH concentration in phosphate-buffered saline solution and the measured anodic peak current. The electrochemical immunosensor showed a limit of detection equal to 40 μg/mL (2 μM) LpqH.

Keywords: immunosensor; chitosan-coated iron oxide; monoclonal anti-LpqH; bio-nanocomposite; point-of-care testing; tuberculosis

Citation: Reaño, R.L.; Panghulan, G.R.; Hernandez, C.A.T.; Tamayo, J.P. Bioengineered Monoclonal Antibody Chitosan–Iron Oxide Bio-Composite for Electrochemical Sensing of *Mycobacterium tuberculosis* Lipoprotein. *Eng. Proc.* **2023**, *58*, 23. https://doi.org/10.3390/ecsa-10-16065

Academic Editor: Francisco Falcone

Published: 15 November 2023

1. Introduction

Mycobacterium tuberculosis (*MTb*) is the cause of tuberculosis (TB), an airborne infectious disease that infects humans as a primary host [1]. *MTb* is an ancient pathogen that has been persistent due to its complex survival strategies in the living host and the environment. It has developed a repertoire of culture filtrate antigens (Ags) that multiply as the disease progresses [2,3]. Among the notable TB antigens and biomarkers are CFP10, ESAT6, LAM, HBHA, HspX, and LpqH, which can be detected using a serologic immunoassay [1].

LpqH is a mycobacterial lipoprotein, a bacterial secretion that is composed of membrane-anchored proteins characterized by a lipobox motif [4]. LpqH, also known as the 19-kDa antigen, is an *MTb* outer membrane-anchored glycol-lipoprotein that has been shown to exhibit immunosuppressive functions. It has been well established that LpqH interacts with TLR1/TLR2 on the macrophage cell surface, thereby regulating the host immune response. Aside from that, LpqH acts as an adhesin that establishes the colonization and infection of *MTb* onto the host cell surface [5].

Antibodies (Abs) have become a popular candidate for biosensor development due to their high affinity and specificity to their targets. Enzyme-linked immuno-sorbent assay (ELISA), the gold standard for all immunoassays, is still popular and used worldwide in different fields of application, particularly in clinical diagnostics. With the advancement in analytical and bioanalytical chemistry, the incorporation of antibodies directly to the signal transducer's surface gave birth to a combined immunoassay and biosensor technology termed an immunosensor [6,7]. An immunosensor is a biosensor that uses an antibody as a capture element, wherein such antibody forms a stable immunocomplex with the antigen, which results in the generation of a measurable signal given by the transducer. In contrast, in an immunoassay, the signal recognition from Ag-Ab interaction takes place elsewhere [8].

This study aimed to develop an electrochemical immunosensor for *MTb* LpqH detection using a bio-composite composed of a monoclonal antibody that is specific to LpqH (anti-LpqH) incorporated on iron oxide nanoparticles via bioengineering techniques. The study also aimed to assess the affinity of the anti-LpqH and the limit of detection of the proposed electrochemical setup. To our knowledge, there is no study available on electrochemical immunosensors for *MTb* LpqH detection with or without clinical trials. The procedure also provides a cheap alternative to handling biological molecules by using magnetic iron oxide nanoparticles and chitosan as crosslinkers. The future direction of the study includes developing a more sensitive protocol for *MTb* LpqH detection by improving the electrode assembly and using electroactive labels.

2. Materials and Methods

The *Mycobacterium tuberculosis* 19 kDa recombinant and conserved form lipoprotein, an antigen precursor (LpqH), and the monoclonal anti-LpqH, IT-54, produced in vitro were obtained through BEI Resources, NIAID, NIH. The lipoprotein and the anti-LpqH were diluted using PBS Buffer (10 mM phosphate, 138 mM NaCl, and 2.7 mM KCl at pH 7.4) purchased from Sigma-Aldrich (Burlington, MA, USA) in powdered form. The tween 20 used in creating an emulsion and for making 0.05% v/v PBS-tween buffer at pH 7.4 for blocking was purchased from Promega (Madison, WI, USA). All other chemicals, including acids and bases used, were ACS grade and purchased from Sigma-Aldrich, particularly those used for iron oxide and chitosan synthesis.

2.1. Synthesis and Characterization of Monoclonal Antibody Immobilized on Chitosan-Coated Iron Oxide (AbMNP)

Iron oxide (chemical formula: Fe_3O_4, abbr. as IONP) particles were prepared using $FeCl_3$ and $FeSO_4$ and following the methodology described by Hierro et al., 2018 [9]. Chitosan was prepared using shrimp shells obtained from the local market and sun-dried for 2–3 days. The shells were demineralized, and chitin was extracted via the process described by Varun et al., 2017 [10].

A 200 g of as-prepared IONP was mixed with 55 mL mineral oil, 1.2 mL tween 20, and 10 mL 1% w/v CS, creating an emulsion. The mixture was sonicated and stirred for 35 min. After recovering the CS-IONP bio-composite and washing with double distilled water, 3 mL of 25% glutaraldehyde solution was added, and the mixture was stirred for 5 hrs. Anti-LpqH (100 μg/mL) was added to the optimum amount of glutaraldehyde-activated CS-IONP, and the solution was incubated overnight at 4 °C. The AbMNP was isolated, PBS-tween was added for blocking, and PBS solution for storage.

The microstructure of the two samples, the bare IONP and the AbMNP, were observed using a Field Emission Scanning Electron Microscope (JSM-IT500HR, JEOL, Tokyo, Japan). The voltage used was 3.0 kV, with each sample observed at 1000× and 15,000× magnification.

The samples were also tested using a Fourier transform infrared spectrophotometer or FTIR (Shimadzu IR Prestige-21, Kyoto, Japan) with attenuated total reflectance accessory.

2.2. Enzyme-Linked Immunosorbent Assay (ELISA) and Calculation of Dissociation Constant

ELISA was performed using the conventional process, with IgG-Alkaline Phosphatase and pNPP tablet (Sigma, Burlington, MA, USA). The LpqH antigen (50 μL at 20 μg/mL) was immobilized on a Corning® Polystyrene High Bind Microplate (Corning Inc., New York, NY, USA) by incubation overnight at 4 °C. PBS-tween (200 μL at 0.05% v/v) solution was added to the well and incubated for 1 h at room temperature to prevent non-specific binding. After blocking, the wells were washed with PBS-tween. Then, 50 μL of anti-LpqH in binding buffer solution at various dilutions were added to each separate well with immobilized LpqH. The microplate was incubated for 1 h at 37 °C with gentle shaking.

Each well was washed thrice with the PBS buffer. Then, the standard ELISA protocol for the addition of IgG-AP and pNPP was followed until a deep yellow color was developed. The absorbance was measured at 405 nm. The apparent dissociation constant or K_d^{app} was calculated following the method described by Orosz and Ovádi (2002) [11].

2.3. Cyclic Voltammetry and the Limit of Detection

A single-use screen-printed carbon electrode (SPCE) (Biogenes Technologies, Kuala Lumpur, Malaysia) was used to perform the electrochemical measurements. Various concentrations of LpqH were prepared and allowed to interact with the monoclonal anti-LpqH on AbMNP. Cyclic voltammetry (CV) was performed using an electrochemical workstation Em4Stat (PalmSens, Utrecht, The Netherlands). The applied potential varied between −1 and 1 V with a scan rate of 50 mV/s. All electrochemical measurements were performed using 1× PBS buffer. The limit of detection was calculated using the linearized curve of the anodic peak current.

All calculations (linear regression, dissociation constant, limit of detection, and statistical analyses) were performed using a Python script.

3. Results

3.1. Synthesis and Characterization of Monoclonal Antibody Immobilized on Chitosan-Coated Iron Oxide (AbMNP)

Figure 1 shows the FTIR transmittance spectra of the as-prepared IONP and the AbMNP. The transmittance spectra within the 4000–400 cm^{-1} range were recorded for the samples. The wide band gap from 3700 to 3000 cm^{-1} and peak at 1645 cm^{-1} is attributed to the stretching vibration of –OH groups on the surface of the particles. The peak at 2937 cm^{-1} corresponds to the –CH bond of CH_2 bending. The peaks at 1455 cm^{-1} and 1366 cm^{-1} represent the symmetric –CH_3 deformation of chitosan. The peaks at 569 cm^{-1} and 459 cm^{-1} are typical of Fe–O that correspond to the vibrations of tetrahedral Fe–O and octahedral Fe–O bonds. The band at 1070 cm^{-1} is attributed to the C–OH and C–C–C vibrations of the chitosan ring.

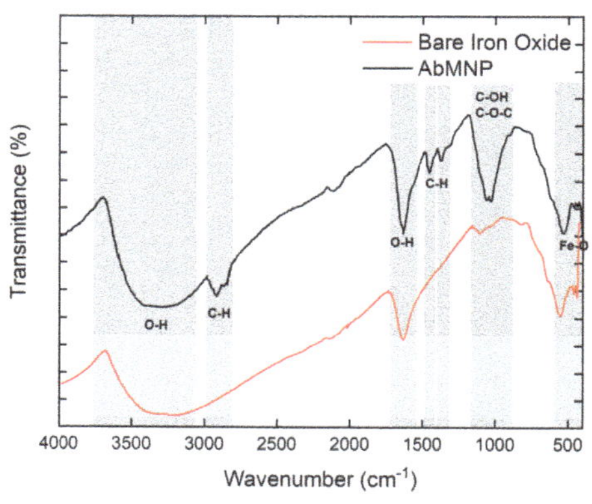

Figure 1. FTIR spectra of bare Iron oxide (Fe_3O_4) and the monoclonal anti-LpqH covalently bonded on magnetic iron oxide via chitosan–glutaraldehyde crosslinking (AbMNP).

The structural morphology of the IONP was investigated by scanning electron microscopy (SEM). Figure 2 depicts the surface images of the particles. Figure 2a,b shows the as-prepared IONP consisting of agglomerated nanoparticles. Figure 2c,d showed that the nanoparticles are coated with chitosan molecules, as shown with smoother lumps.

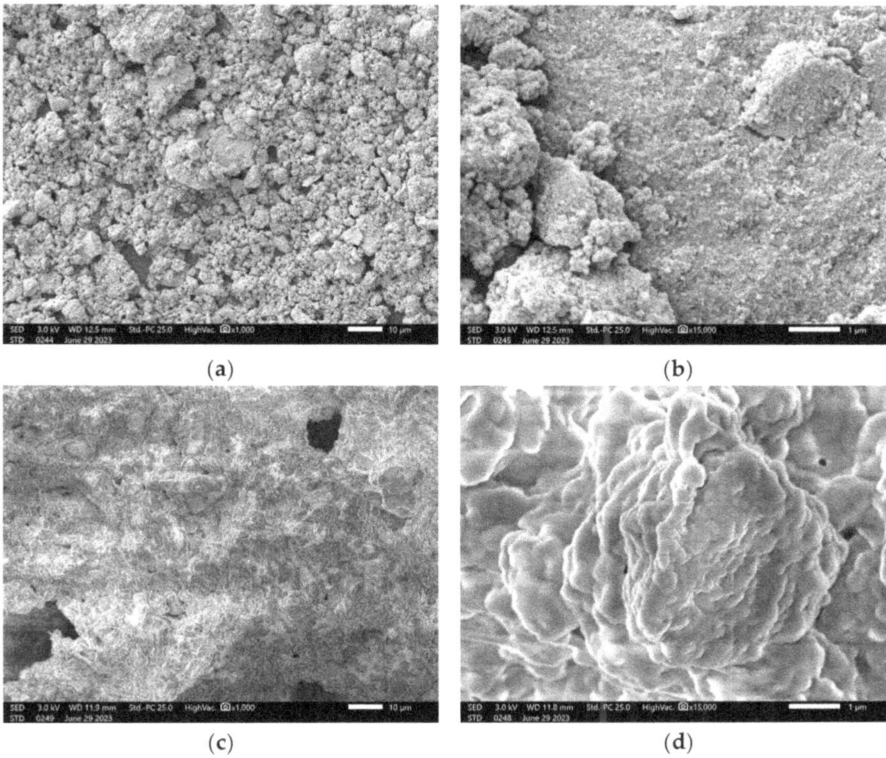

Figure 2. The SEM images of bare IONP at (**a**) 1000× and (**b**) 15,000× magnification, and AbMNP at (**c**) 1000× and (**d**) 15,000× magnification.

3.2. Determination of the Apparent Dissociation Constant (K_d) of the Monoclonal Anti-LpqH Using ELISA

ELISA was performed to provide a quantitative analysis of the interaction of the anti-LpqH and the *MTb.* lipoprotein LpqH by calculating the apparent dissociation constant (K_d^{app}) of the immunocomplex formation. Figure 3a shows the relative absorbance measured for each anti-LpqH dilution titrated with the lipoprotein. The IC_{50} was obtained at 50% relative absorbance ($i = 0.50$) and is equal to 3.140, with the anti-LpqH exact dilution of 7.24×10^{-4}. The IC_{50} was used to calculate the corrected parameters of the linearized curve, with the slope $m = \left(\frac{1}{K_d^{app}}\right) = 100.00$ and the y-intercept $(b) = -1.8550$. The value of $\frac{1}{K_d^{app}}$ implies that with 1000-fold dilution, an effective performance of the anti-LpqH can be anticipated in detecting the LpqH lipoprotein using ELISA. This value was used as the basis for the dilution of the anti-LpqH used in the electrochemical biosensor.

3.3. Cyclic Voltammetry and Determination of Limit of Detection

The resulting voltammogram from testing various *MTb* LpqH concentrations is shown in Figure 4a. The shift in the measured anodic current is highly recognizable as the concentration of the antigen increases. Using the anodic peak current, a linear curve was created as shown in Figure 4b. The limit of detection was computed as 2.0923 µM (39.7545 µg/mL) *Mtb* LpqH.

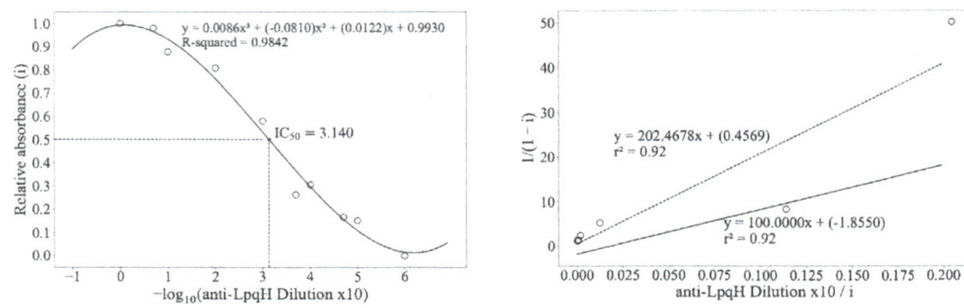

(a) **(b)**

Figure 3. (**a**) Plot of relative absorbance, i vs. $-log_{10}(10{\cdot}dilution)$, data obtained through ELISA. (**b**) The parameters of the linearized curve are shown following the method [11]. The corrected slope is $\left(\frac{1}{K_d^{app}}\right)$ = 100.00, upon using IC_{50} compute for the corrected y-intercept (−1.8550), which was used as linear regression constraint (Pearson correlation = 0.92).

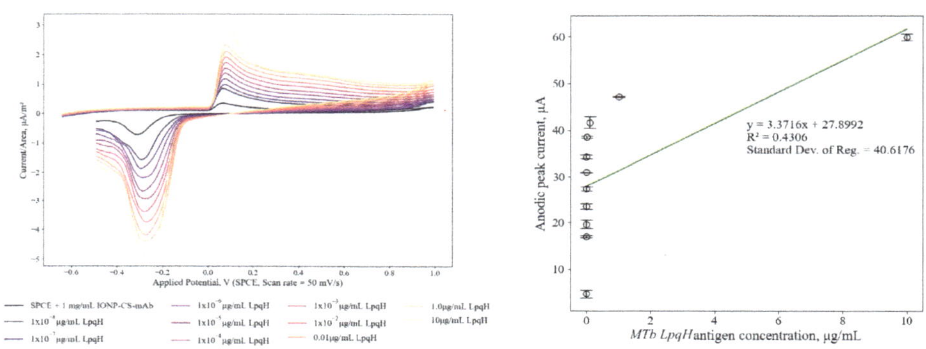

(a) **(b)**

Figure 4. (**a**) Voltammograms of TB lipoprotein LpqH antigen precursor detection using AbMNP as bioreceptor, measured using 50 mV/s scan rate and screen-printed carbon electrode. (**b**) The plot of the anodic peak current with error bars and linearized curve for estimating the limit of detection.

4. Discussion

The AbMNP bio-composite acts as a nanocarrier and captures the target *MTb* LpqH. Using IONP as its magnetic core, AbMNP is easy to collect, wash, and handle using a simple magnet. Reports showed that IONP is electrocatalytic; thus, it can promote the breakdown or release of the anti-LpqH upon interacting with the antigen and/or while current is applied. Biological molecules are non-conductive and their release from the bio-composite's surface supports electric conduction. The LOD can be improved further by optimizing the density of the anti-LpqH or by incorporating an electroactive or electrocatalytic label that can induce a redox reaction with the Ag-Ab interaction.

Author Contributions: Conceptualization, R.L.R. and C.A.T.H.; methodology, R.L.R., G.R.P. and J.P.T.; formal analysis, R.L.R., G.R.P. and C.A.T.H.; investigation, R.L.R. and C.A.T.H.; resources, R.L.R. and J.P.T.; writing—original draft preparation, R.L.R., G.R.P. and C.A.T.H.; writing—review and editing, R.L.R.; supervision, R.L.R. and J.P.T.; project administration, R.L.R., G.R.P., C.A.T.H. and J.P.T.; funding acquisition, R.L.R. and J.P.T. All authors have read and agreed to the published version of the manuscript.

Funding: This research was funded by the Department of Science and Technology (DOST)—Philippine Council for Health Research and Development. Project Title: Development of aptamer based colorimetric and electrochemical point of care type diagnostic test for pulmonary TB and latent TB infection.

Institutional Review Board Statement: Not applicable.

Informed Consent Statement: Not applicable.

Data Availability Statement: Data are contained within the article.

Acknowledgments: The authors would like to acknowledge the BEI Resources, NIAID, NIH, USA for antigens and antibodies used in this study and the NITL, BIOTECH, UPLB, PH for technical support.

Conflicts of Interest: The authors declare no conflict of interest.

References

1. Krishnananthasivam, S.; Li, H.; Bouzeyen, R.; Shunmuganathan, B.; Purushotorman, K.; Liao, X.; Du, F.; Friis, C.G.K.; Crawshay-Williams, F.; Boon, L.H.; et al. An Anti-LpqH Human Monoclonal Antibody from an Asymptomatic Individual Mediates Protection against Mycobacterium Tuberculosis. *NPJ Vaccines* **2023**, *8*, 127. [CrossRef] [PubMed]
2. Samanich, K.M.; Keen, M.A.; Vissa, V.D.; Harder, J.D.; Spencer, J.S.; Belisle, J.T.; Zolla-Pazner, S.; Laal, S. Serodiagnostic Potential of Culture Filtrate Antigens of Mycobacterium Tuberculosis. *Clin. Diagn. Lab. Immunol.* **2000**, *7*, 662–668. [CrossRef] [PubMed]
3. Tucci, P.; Portela, M.; Chetto, C.R.; González-Sapienza, G.; Marín, M. Integrative Proteomic and Glycoproteomic Profiling of Mycobacterium Tuberculosis Culture Filtrate. *PLoS ONE* **2020**, *15*, e0221837. [CrossRef] [PubMed]
4. Becker, K.; Sander, P. *Mycobacterium tuberculosis* Lipoproteins in Virulence and Immunity—Fighting with a Double-Edged Sword. *FEBS Lett.* **2016**, *590*, 3800–3819. [CrossRef] [PubMed]
5. Chatterjee, S.; Kundapura, S.V.; Basak, A.J.; Mukherjee, D.; Dash, S.; Ganguli, N.; Das, A.K.; Mukherjee, G.; Samanta, D.; Ramagopal, U.A. High-Resolution Crystal Structure of LpqH, an Immunomodulatory Surface Lipoprotein of Mycobacterium Tuberculosis Reveals a Distinct Fold and a Conserved Cleft on Its Surface. *Int. J. Biol. Macromol.* **2022**, *210*, 494–503. [CrossRef] [PubMed]
6. Popov, A.; Brasiunas, B.; Kausaite-Minkstimiene, A.; Ramanaviciene, A. Metal Nanoparticle and Quantum Dot Tags for Signal Amplification in Electrochemical Immunosensors for Biomarker Detection. *Chemosensors* **2021**, *9*, 85. [CrossRef]
7. Jafari, M.; Hasanzadeh, M.; Solhi, E.; Hassanpour, S.; Shadjou, N.; Mokhtarzadeh, A.; Jouyban, A.; Mahboob, S. Ultrasensitive Bioassay of Epitope of Mucin-16 Protein (CA 125) in Human Plasma Samples Using a Novel Immunoassay Based on Silver Conductive Nano-Ink: A New Platform in Early Stage Diagnosis of Ovarian Cancer and Efficient Management. *Int. J. Biol. Macromol.* **2019**, *126*, 1255–1265. [CrossRef] [PubMed]
8. Mollarasouli, F.; Kurbanoglu, S.; Ozkan, S.A. The Role of Electrochemical Immunosensors in Clinical Analysis. *Biosensors* **2019**, *9*, 86. [CrossRef] [PubMed]
9. del Hierro, I.; Pérez, Y.; Fajardo, M. Silanization of Iron Oxide Magnetic Nanoparticles with Ionic Liquids Based on Amino Acids and Its Application as Heterogeneous Catalysts for Knoevenagel Condensation Reactions. *Mol. Catal.* **2018**, *450*, 112–120. [CrossRef]
10. Varun, T.K.; Senani, S.; Jayapal, N.; Chikkerur, J.; Roy, S.; Tekulapally, V.B.; Gautam, M.; Kumar, N. Extraction of Chitosan and Its Oligomers from Shrimp Shell Waste, Their Characterization and Antimicrobial Effect. *Vet. World* **2017**, *10*, 170–175. [CrossRef] [PubMed]
11. Orosz, F.; Ovádi, J. A Simple Method for the Determination of Dissociation Constants by Displacement ELISA. *J. Immunol. Methods* **2002**, *270*, 155–162. [CrossRef] [PubMed]

engineering proceedings

Proceeding Paper

Enhanced Pedestrian Dead Reckoning Sensor Fusion for Firefighting †

Tobias Augustin * **and Daniel Ossmann**

Department of Mechanical, Automotive and Aeronautical Engineering, Munich University of Applied Sciences HM, 80335 Munich, Germany
* Correspondence: tobias.augustin@hm.edu

† Presented at the 10th International Electronic Conference on Sensors and Applications (ECSA-10), 15–30 November 2023; Available online: https://ecsa-10.sciforum.net/.

Abstract: Knowing the exact position of firefighters in a building during an indoor firefighting operation is critical to improving the efficiency and safety of firefighters. For the estimation of an individual's position in indoor or Global Positioning System (GPS)-denied environments, Pedestrian Dead Reckoning (PDR) is commonly used. PDR tries to estimate the required position via sensors without external references, for example, using accelerometers and gyroscopes. One of the most common techniques in PDR is step-detection. Applications like firefighting, however, also involve dynamic movements like crouching. Thus, the accuracy of a step-detection algorithm is reduced dramatically. Therefore, this paper presents a novel PDR algorithm that augments the conventional PDR technique with a tracking camera. The position estimates of a zero-crossing step-detection algorithm and tracking camera estimates are fused via a Kalman filter. A system prototype, designed for algorithm validation, is presented. The experimental results confirm that enhancing the system with a secondary sensor also leads to a substantial increase in the position estimation accuracy for dynamic crouching maneuvers compared to conventional step-detection algorithms.

Keywords: Pedestrian Dead Reckoning; Kalman filter; firefighting

Citation: Augustin, T.; Ossmann, D. Enhanced Pedestrian Dead Reckoning Sensor Fusion for Firefighting. *Eng. Proc.* **2023**, *58*, 2459. https://doi.org/10.3390/ecsa-10-16032

Academic Editor: Stefano Mariani

Published: 15 November 2023

1. Introduction

While safety standards in firefighting are continuously improving, indoor operations in burning buildings still present a dangerous task for firefighters. At least 240 injuries and 10 deaths involving firefighters conducting firefighting operations in buildings were reported in the United States in 2022 [1]. To improve safety while performing such a dangerous task, knowing the exact position of firefighters in indoor environments can shorten the rescue time of injured personnel or help firefighters avoid dangerous situations. Such technologies can not only improve the safety of the involved firefighters, but can provide real-time data for so-called internet of emergency services applications, which aim to improve emergency response and disaster management [2]. To determine the position of a person in indoor or GPS-denied environments, a technique called Pedestrian Dead Reckoning (PDR) is used. It relies on sensors such as accelerometers, gyroscopes and magnetometers integrated into wearable devices, smartphones or smartwatches. By continuously tracking a pedestrian's step counts, stride length and heading changes, PDR algorithms can calculate their relative displacement from a known starting point. Other means of PDR include simultaneous locating and mapping [3], magnetic field mapping [4] and magnetic triangulation [5].

For application in firefighting operations, many of the aforementioned PDR methods are not feasible. While radio tracking [6] and magnetic mapping [4] produce accurate results in indoor environments, they are technologies that have to be installed before use. It may be possible to achieve this in large buildings; however, it would not be feasible to do for every building in an area where a fire might occur. For tracking firefighters in any indoor environment, a stand-alone, body-worn device is required. Stand-alone PDR systems often rely on a form of step-detection [7]. Algorithms based on step-detection can accurately estimate position mainly during walking. Movements occurring in firefighting applications, however, also include more dynamic activities like crouching. Those movements are hard to detect by standard step-detection algorithms. Thus, a secondary sensor measuring position or velocity is necessary to improve accuracy in those scenarios. A common sensor chosen for this is a Light Detection and Ranging (LIDAR) sensor [4]. While this approach can yield good results in smoke-free environments, tests show that distance readings of LIDAR systems are heavily influenced by smoke particles and therefore are not usable in a firefighting environment.

Due to these shortcomings in PDR for firefighting applications, in this paper, a novel approach for enhanced PDR is presented. The step-detection algorithm is extended with a stereo tracking camera as a secondary sensor. Despite tracking cameras being readily available and producing accurate tracking results, they are hardly used in Pedestrian Dead Reckoning applications. This tracking camera can visually determine velocity and position relative to a starting point, even in smoky scenarios. A camera providing position and velocity is combined with a step-detection algorithm providing position information. The gathered data are fused using a Kalman filter to robustly estimate the firefighter's position. While in Section 2, the fundamentals of the step-detection and the model for the Kalman filter are presented, in Section 3, the overall PDR system setup, including software and hardware components, is described. Finally, Section 4 discusses the results of a verification campaign in which position data from the proposed algorithm are compared to data generated through step-detection only.

2. Sensor Data Processing Algorithms

PDR relies on an advanced sensor data fusion algorithm combining position data estimated by a step-detection algorithm and the velocity and position data estimates of a secondary sensor.

2.1. Step-Length Estimation

Step-detection describes the process of detecting and counting a person's steps by measuring and analyzing the accelerations of a body-worn inertial measurement unit (IMU). The most common methods of step-detection utilize the vertical acceleration signal and analyze the signal using peak-, zero-crossing or flat zone detection. These simple but accurate methods are considered to be sufficient for this initial study. A zero-crossing detection approach is chosen since flat zone detection only works for foot-mounted sensors and peak-detection accuracy is dependent on a person's walking speed [8]. Zero-crossing detection analyzes the characteristic shape of the vertical acceleration of a torso-mounted sensor [9]. To improve detection accuracy, the high-frequency content of the signal is filtered out using a low-pass filter. A straightforward implementation of a first-order low-pass filter is the so-called exponentially weighted moving average:

$$y(k\Delta_T) = \alpha(k\Delta_T) + (1-\alpha)y((k-1)\Delta_T), \quad (1)$$

where $u(k\Delta_T)$ is the raw signal at time step $k\Delta_T$, and $y(k\Delta_T)$ and $y((k-1)\Delta_T)$ are the filtered signals of the current and the last time steps, respectively [10]. The smoothing factor α lies between 0 and 1 and can be calculated as

$$\alpha = \frac{2\pi\Delta_T f_c}{2\pi\Delta_T f_c + 1}, \quad (2)$$

with Δ_T being the sampling time and f_c being the required cut-off frequency. For a step to be counted as complete, the filtered, vertical acceleration signal a_z has to cross the zero line twice, rising once, i.e.,

$$(a_z > 0) \bigwedge (a_z < 0) \quad (3)$$

and afterwards, falling once, i.e.,

$$(a_z < 0) \bigwedge (a_z > 0). \tag{4}$$

Only if these two conditions are registered in the algorithm can a step finally be counted.

Once a step is registered as complete, the step length has to be added to the current estimated position in the direction of movement. To estimate the step-length d, a method using the relation between hip acceleration and the length of a step following [11] is applied, i.e.,

$$d = \sqrt[4]{a_{\max} - a_{\min}}\, c \tag{5}$$

In Equation (5), $a_{\max}$ is the maximum measured acceleration and $a_{\min}$ is the minimal acceleration, both measured during the last step, and c is a constant for unit conversion. This method produces accurate estimates with low computational effort compared to other algorithms [8,12–15].

2.2. Sensor Data Fusion

Sensor data fusion describes the process of using multiple sensor outputs to estimate the state of a system. A common approach for sensor fusion is complementary filtering, which combines the high-frequency data of one sensor, which is fast but prone to drifting, with low-frequency data from another sensor, which stabilizes the output signal.

In this paper, we use the more advanced method of a Kalman filtering. The idea of the Kalman filter is to use an optimal recursive algorithm for sensor data fusion. The filter operates in two steps: the prediction step, where the system's state is predicted using a prediction model, i.e., a mathematical model of the underlying dynamics, and the update step, where on the one hand, the measurements are used to correct the predicted state via the Kalman gain, and on the other hand, the Kalman gain itself is updated based on the measurements. This gain balances the model's predictions and the actual measurements [16]. The process continually refines the state estimate as new data become available, making it robust against noise and capable of handling real-time applications. The required prediction model is described by the non-linear function $x((k+1)\Delta_T) = f(x(k\Delta_T), u(k\Delta_T))$. As the underlying model in this paper, we define, for each of the three-coordinate axes, the function

$$f(x, u) = \begin{pmatrix} u \\ u\Delta_T \\ x_1\Delta_T + u\Delta_T^2 \end{pmatrix}, \tag{6}$$

with the input u to the model being the measured acceleration a by the inertial measurement unit and the state vector $x = [x_1,\ x_2,\ x_3]^{\mathrm{T}}$. The update step of the Kalman filtering process uses the velocity measured by the tracking camera, the position estimated by the tracking camera and the step-detection to correct the filter estimate. Based on this model the Kalman filter provides estimates of the position and the velocity in the corresponding axis. Note that the dependency of the signals on time, i.e., on $k\Delta_T$, is omitted in Equation (6) for readability reasons. By changing the covariance matrices of the Kalman filter, the accuracy of the predictions and measurement updates is tuned [17].

3. Enhanced Pedestrian Dead Reckoning System

The enhanced PDR makes use of a robust step-detection scheme with which the position of the firefighter is estimated. Additionally, a tracking camera serves as a secondary sensor providing position and speed measurements to back up the step-detection-based position. Finally, all available signals are fused together via a Kalman filter, providing the position and velocity of the firefighter.

3.1. Robust Step-Length Estimation and Secondary Sensor Setup

As the basis, step-detection herein uses the zero-crossing technique, as described in Section 2.1. The vertical acceleration signal used for the step-detection is filtered with a low-pass filter to remove the undesired, high frequency parts of the signal that occur during movement. Since the frequency range of normal human walking is in the range of 1 Hz to 5 Hz, the bandwidth of the low-pass filter is set at $f_c = 10$ Hz in Equation (1) ensuring an adequate roll-off at higher frequencies. Figure 1 shows a comparison of the raw data with the filtered acceleration data. Clearly, sharp peaks and noise are filtered out. To also consider dynamic movements of firefighters such as crouching, the step-detection algorithm's robustness is improved via additional threshold-crossing detection: To initiate the counting process, the acceleration signal has to pass the negative threshold at $\tau_{min} - 2\ \mathrm{m/s^2}$. Afterwards, a step is counted as valid only if between the detection of two subsequent zero-crossings, a rise above the positive threshold τ_{max} $2\ \mathrm{m/s^2}$ is registered. If after initialization via the negative threshold, the described sequence is not completed in a specified time, the step-detection logic is reset and no step is counted. After a step is detected, the length of the step is added to the last known position in the direction of movement that is determined by the heading angle measured by the IMU. The step-length is estimated via Equation (5). It is assumed, that due to the limited field of view and restriction of movement by gear during an indoor operation, firefighters move in the direction with which their body is aligned. Since the IMU is mounted on the air tank of the firefighter, the orientation of the IMU equals the direction of movement. To describe the position in a global reference frame, the coordinate origin of the global reference frame is defined when the device is initialized, where the initial heading defines the x-axis. As the secondary sensor, a stereo tracking camera is used to provide additional position and velocity information. This device has two calibrated cameras that are placed at a certain distance from each other and are horizontally aligned. By measuring the displacement of a tracked object between the two cameras, the distance to the object can be calculated. By doing this for multiple objects and repeating this process for every frame, the position and average velocity are provided [18].

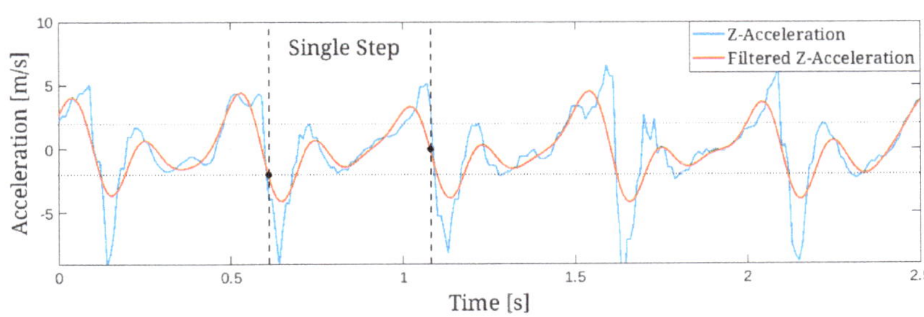

Figure 1. Raw and filtered vertical acceleration during walking, with the illustration of a single step between the two dashed lines.

3.2. Sensor Fusion Algorithm

Based on the model in Equation (6), the Kalman filter estimates the firefighter's velocity and position processing the acceleration signal provided by the IMU, as well as the position and velocity measurements from the tracking camera. The position data from the step-detection and the tracking-camera, however, need to be fused before entering the filter, as the discrete confidence levels of the camera are available, which cannot be handled by the Kalman filter. Thus, the fusion of the two signals is performed via a simple weighting scheme using discrete weights. The tracking camera provides four different confidence level indicators, from the highest confidence to the lowest. For the step-detection, it is assumed that the position accuracy estimated by the step-detection algorithm deteriorates the longer no step is fully registered. To also limit the weighting possibilities to a discrete set for step-detection, three discrete conditions are used to reflect the accuracy after the last detection based on the time passed since the last detection event: if $t_{\text{step}} < 0.5$ s, the best accuracy is assumed; if $0.5\text{ s} \leq t_{\text{step}} < 3$ s, medium accuracy is assumed; and if $t_{\text{step}} \geq 3$ s, the worst accuracy is assumed. The resulting weighting scheme, motivated by the work in [19], includes twelve cases. If the quality of both markers is bad, the step-detection is highly favored, since it provides stable results during walking, even in zero-visibility environments. If the tracking confidence is high, the camera measurements is slightly favored. This is because in theory, the tracking camera's results will be more accurate since it produces continuous position updates and can track the position regardless of the type of movement. A combination of the measurements is performed before they are used in the Kalman Filter. With a weighting gain w, the weighted measurement input for the x- and y-positions is calculated via

$$\begin{bmatrix} x(k\Delta_T) \\ y(k\Delta_T) \end{bmatrix} = \begin{bmatrix} w(k\Delta_T)x_{\text{step}}(k\Delta_T) + (1 - w(k\Delta_T))\, x_{\text{camera}}(k\Delta_T) \\ w(k\Delta_T)y_{\text{step}}(k\Delta_T) + (1 - w(k\Delta_T))\, y_{\text{camera}}(k\Delta_T) \end{bmatrix}, \tag{7}$$

where x_{step} and y_{step} are the position estimates produced by the step-detection, and x_{camera} and y_{camera} are the position estimates from the tracking camera. Since the step-detection cannot estimate the z-position, only the measurement from the tracking camera is used; therefore, no weighting is performed.

3.3. Sensor Hardware Assembly

The IMU used is the Bosch Sensortech BNO055 MEMS absolute orientation sensor. This device measures acceleration on three axes and provides absolute heading data by measuring the earth's magnetic field and fusing gyroscope and magnetometer data. The secondary sensor is a RealSense T265 stereo tracking camera. Its main advantage is its on-chip, online data processing. Thus, no other means of interpreting the data is necessary, and the velocity and position data are directly available for the sensor fusion algorithm presented herein. Note that by using parts of the infrared spectrum, the camera also can produce accurate tracking results in environments with bad lighting.

For validation of the system, a wearable sensor assembly is designed [20]. Both sensors are mounted on the backplate of self-contained breathing apparatus. This design is chosen to imitate application in firefighting settings, where the sensors are placed on the pressurized air tank. For this, a 3D-printed spacer is designed to mount the sensors at the right distance. Weight is added to represent the air tank. The camera and IMU are protected from damage by an enclosure. Figure 2 shows the developed experimental setup.

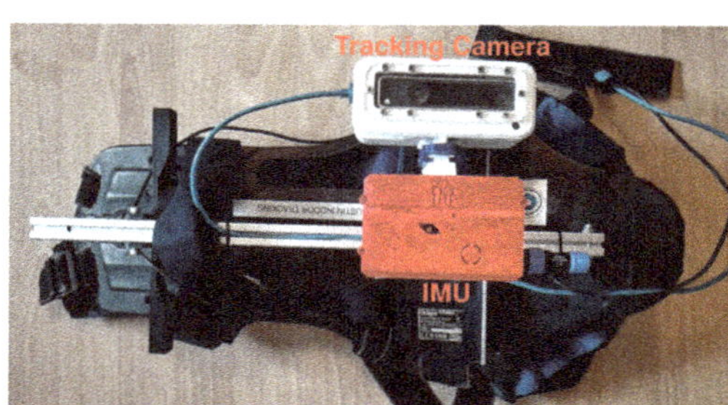

Figure 2. Sensor assembly mounted on the firefighter equipment.

4. Results

For initial validation of the sensor setup and the algorithms, tests on predefined paths were performed. While these tests do not fully mimic the conditions that occur during firefighting operations, they allow an initial feasibility assessment of the setup. Additional experiments in real-life applications are planned in further studies. First, to validate the performance during regular walking scenarios, a 33 m long path was tested. The right diagram in Figure 3 shows the estimated position of the different algorithms compared to the true path. While at the beginning of the test, all three algorithms deliver accurate results, the step-detection deviates strongly after the first heading change. The proposed sensor fusion is able to stay close to the real path and deliver the best results most of the time. The tracking camera alone, however, delivers the best result in the middle of the experiments. This is due to the fact that here, the step-detection shows a big error, also dragging the sensor fusion algorithm away from the real path.

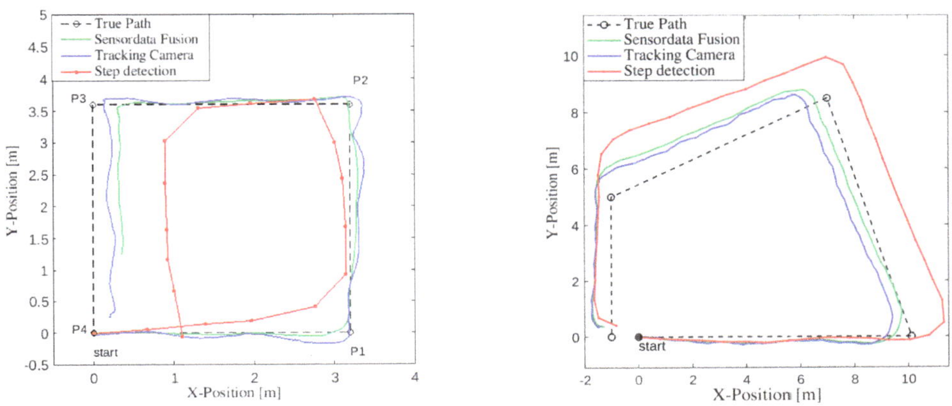

Figure 3. Experimental results comparing step-detection only, tracking camera only, and the proposed sensor data fusion to the true path for crouching (**left**) and walking (**right**).

In the second validation experiment, dynamic crouching, frequently employed in firefighting, was tested. The test path, illustrated in the left diagram of Figure 3, covers a total length of 12 m and includes four 90° turns. The start- and endpoints were identical, and 10 test runs are performed. Clearly, the step-detection alone performs the worst in this scenario, simply because no steps are performed during the movement. The mean values over 10 runs, as listed in Table 1, demonstrate significant improvements in tracking accuracy, with at least a five-fold enhancement at the four corner points (P1 to P4) when utilizing the proposed algorithm compared to relying solely on step-detection. To simulate potential obstructions of the tracking camera caused by dirt or heavy smoke, the tracking confidence is artificially reduced so that step-detection is highly favored. In these scenarios, the mean deviation at each control point is degraded but still lies within the acceptable range of 1 m.

Table 1. Mean deviation from the true path at four corners of the path.

Estimation Method	P1	P2	P3	P4
Step-detection	1.60 m	2.45 m	2.44 m	1.80 m
Sensor data Fusion	0.28 m	0.42 m	0.27 m	0.32 m
Low tracking confidence S.F	0.66 m	0.68 m	0.76 m	0.54 m
Tracking Camera only	0.27 m	0.43 m	0.24 m	0.29 m

The data in Table 1 also indicate that the tracking camera alone performs similarly to the sensor fusion algorithm. In these results, however, the tracking camera confidence was set to its highest possible level. To ensure reliable results, even in scenarios where the camera confidence is degraded, it is essential to incorporate data from step-detection for crouching scenarios. This is crucial because the camera may produce highly inaccurate data in those scenarios. In German firefighting tactics, crouching movement is predominantly used in low-visibility environments (where the camera confidence will be degraded). Thus, the assumption herein is that in these low-visibility situations, reliable step-detection is still possible due to the use of the crouching method. This assumption will be validated in real-life applications in future studies.

5. Conclusions

An enhanced Pedestrian Dead Reckoning method for firefighting applications is presented. Step-detection was successfully upgraded with a secondary sensor to improve position estimates in different moving scenarios. The required sensor fusion algorithm was successfully validated in an experimental validation campaign, showing promising results for the usage of the developed prototype system. To further validate the proposed system, real-world trials with professional firefighters using the equipment will be performed. Such application experiments will provide insight into the limitations of the system in a real-fire scenario and provide feedback to further improve the system setup. This will allow us to define the specific conditions which require the improvement of sensor data fusion algorithm accuracy.

Author Contributions: T.A.: methodology, visualization, analysis, synthesis, testing, validation and original draft preparation. D.O.: supervision, scientific review and proofreading. All authors have read and agreed to the published version of the manuscript.

Funding: This research received no external funding.

Institutional Review Board Statement: Not applicable.

Informed Consent Statement: Not applicable.

Data Availability Statement: The data presented in this study are available on request from the corresponding author.

Conflicts of Interest: The authors declare no conflict of interest.

References

1. Atemschutzunfälle.eu. Unfälle in Amerika. 2023. Available online: https://www.atemschutzunfaelle.de/unfaelle/amerika/ (accessed on 21 December 2023).
2. Damaševicius, R.; Bacanin, N.; Misra, S. From Sensors to Safety: Internet of Emergency Services (IoES) for Emergency Response and Disaster Management. *J. Sens. Actuator Netw.* **2023**, *12*, 41. [CrossRef]
3. Lu, C.; Uchiyama, H.; Thomas, D.; Shimada, A.; Taniguchi, R.-I. Indoor positioning system based on chest-mounted IMU. *Sensors* **2019**, *19*, 420. [CrossRef] [PubMed]
4. Wang, Q.; Luo, H.; Zhao, F.; Shao, W. An indoor self-localization algorithm using the calibration of the online magnetic fingerprints and indoor landmarks. In Proceedings of the 2016 International Conference on Indoor Positioning and Indoor Navigation (IPIN), Alcala de Henares, Spain, 4–7 October 2016.
5. Arumugam, D.D.; Littlewood, P.; Peng, N.; Mishra, D. Long-Range Through-the-Wall Magnetoqua sistatic Coupling and Application to Indoor Position Sensing. *IEEE Antennas Wirel. Propag. Lett.* **2020**, *19*, 507–511. [CrossRef]
6. Cong, L.; Tian, J.; Qin, H. Practical Step Length Estimation Combining FM Radio Signal and Accelerometer. *IEEE Trans. Instrum. Meas.* **2023**, *72*, 1–13. [CrossRef]
7. Hou, X.; Bergmann, J. Pedestrian Dead Reckoning with Wearable Sensors: A Systematic Review. *IEEE Sens. J.* **2021**, *21*, 143–152. [CrossRef]
8. Shin, S.H.; Park, C.G.; Kim, J.W.; Hong, H.S.; Lee, J.M. Adaptive Step Length Estimation Algorithm Using Low-Cost MEMS Inertial Sensors. In Proceedings of the 2007 IEEE Sensors Applications Symposium, San Diego, CA, USA, 6–8 February 2007.
9. Zhao, Y.; Wang, J.; Duan, C. Design and application research of mine underground disaster relief personnel positioning system based on MEMS sensor. In Proceedings of the International Conference on Neural Networks, Information, and Communication Engineering (NNICE 2022), Qingdao, China, 25–27 March 2022; SPIE: Bellingham, WA, USA, 2022; Volume 12258, pp. 695–704.
10. NIST SEMATECH. NIST/SEMATECH e-Handbook of Statistical Methods. 2012. Available online: https://www.itl.nist.gov/div898/handbook/ (accessed on 21 December 2023).
11. Weinberg, H. Using the ADXL202 in Pedometer and Personal Navigation Applications. 2002. Available online: https://www.analog.com/media/en/technical-documentation/application-notes/513772624an602.pdf (accessed on 21 December 2023).
12. Hajati, N.; Rezaeizadeh, A. A Wearable Pedestrian Localization and Gait Identification System Using Kalman Filtered Inertial Data. *IEEE Trans. Instrum. Meas.* **2021**, *70*, 2507908. [CrossRef]

13. Petukhov, N.I.; Zamolodchikov, V.N.; Malyshev, A.P.; Brovko, T.A.; Serov, S.A.; Korogodin, I.V. Synthesis of PDR Algorithm and Experimental Estimation of Accuracy of Step Length Estimation Methods. In Proceedings of the 2022 4th International Youth Conference on Radio Electronics, Electrical and Power Engineering (REEPE), Moscow, Russia, 17–19 March 2022.
14. Zhao, T.; Ahamed, M.J. Pseudo-Zero Velocity Re-Detection Double Threshold Zero-Velocity Update (ZUPT) for Inertial Sensor-Based Pedestrian Navigation. *IEEE Sens. J.* **2021**, *21*, 13772–13785. [CrossRef]
15. Zizzo, G.; Ren, L. Position Tracking During Human Walking Using an IntegratedWearable Sensing System. *Sensors* **2017**, *27*, 2866. [CrossRef] [PubMed]
16. Chui, C.K.; Chen, G. *Kalman Filtering: With Real-Time Applications*, 4th ed.; Springer: Berlin/Heidelberg, Germany, 2009.
17. Welch, G.; Bishop, G. An Introduction to the Kalman Filter. 2006. Available online: https://www.cs.unc.edu/~welch/media/pdf/kalman_intro.pdf (accessed on 21 December 2023).
18. Zaarane, A.; Slimani, I.; Al Okaishi, W.; Atouf, I.; Hamdoun, A. Distance measurement system for autonomous vehicles using stereo camera. *Array* **2020**, *5*, 100016. [CrossRef]
19. Caron, F.; Duflos, E.; Pomorski, D.; Vanheeghe, P. GPS/IMU data fusion using multisensor Kalmanfiltering: Introduction of contextual aspects. *Inf. Fusion* **2006**, *7*, 221–230. [CrossRef]
20. Sadruddin, H.; Mahmoud, A.; Atia, M.M. Enhancing Body-Mounted LiDAR SLAM using an IMU-based Pedestrian Dead 278 Reckoning (PDR) Model. In Proceedings of the 2020 IEEE 63rd International Midwest Symposium on Circuits and Systems (MWSCAS), Springfield, MA, USA, 9–12 August 2020; pp. 901–904.

engineering proceedings

MDPI

Proceeding Paper

Semi-Supervised Adaptation for Skeletal-Data-Based Human Action Recognition †

Haitao Tian * and Pierre Payeur

School of Electrical Engineering and Computer Science, University of Ottawa, Ottawa, ON K1N 6N5, Canada; ppayeur@uottawa.ca
* Correspondence: htian026@uottawa.ca

† Presented at the 10th International Electronic Conference on Sensors and Applications (ECSA-10), 15–30 November 2023; Available online: https://ecsa-10.sciforum.net/.

Abstract: Recent research on human action recognition is largely facilitated by skeletal data, a compact representation composed of key joints of the human body. However, leveraging the capabilities of artificial intelligence on such sensory input imposes the collection and annotation of a large volume of skeleton data, which is extremely time consuming. In this paper, a two-phase semi-supervised learning approach is proposed to surmount the high requirements on labeled skeletal data while training a capable human action recognition model adaptive to a target environment. In the first phase, an unsupervised learning model is trained under a contrastive learning fashion to extract high-level human action semantic representations from an unlabeled source dataset. The resulting pretrained model is then fine-tuned on a small number of properly labeled data of the target environment. Experimentation is conducted on large-scale human action recognition datasets to evaluate the effectiveness of the proposed method.

Keywords: skeletal action recognition; semi-supervised learning; contrastive learning; domain adaptation

1. Introduction

As a notoriously data-driven learning technique, deep learning has demonstrated remarkable effectiveness in human action recognition by involving massive training on large-scale human activity datasets [1,2]. Recently, Graph Convolutional Networks (GCNs), tailored for skeleton data, which is efficiently extracted from 3D imaging sensors and that offers the merit of being robust to variations in the environment, have achieved state-of-the-art performance in human action recognition research [3,4].

Even though it is promising to realize a powerful human action recognition model via leveraging large-scale public datasets, the operation of the resulting model in practice could be challenging. Deploying the model into a target environment, where the distributions of the target data deviate partially from that of the source training data domain due to the particular imaging configuration adopted, refers to the data domain covariate issue in the deep learning community [5]. A common strategy to tackle the problem is to conduct extra fine-tuning rounds with full supervision of the data collected under the target imaging configuration, in order to eliminate the data discrepancy from the source domain to the target domain. However, the collection and annotation of a large volume of skeleton data for fine-tuning is extremely time consuming. Meanwhile, the data collection in real environments could be highly restricted due to privacy considerations.

This work focuses on an important perspective related to the practical deployment of a human action recognition model. It is related to the efficient adaptation of a GCN model from a public dataset domain to a target environment where only a limited number of data will be available for model refining. The semi-supervised adaptation strategy effectively circumvents the overfitting issue in learning with a small number of samples. In turn, it guarantees a target-environment aware action recognition model that is compatible with

Citation: Tian, H.; Payeur, P. Semi-Supervised Adaptation for Skeletal-Data-Based Human Action Recognition. *Eng. Proc.* **2023**, *58*, 25. https://doi.org/10.3390/ecsa-10-16083

Academic Editor: Stefano Mariani

Published: 15 November 2023

state-of-the-art performance. In the *first* stage, an unsupervised learning framework is proposed to discover high-level human action semantic patterns from plenty of unlabeled skeletal data samples of the source data domain. It is inspired by the recent advances in contrastive learning [6] which provides pivotal competency for learning domain invariant representations from unlabeled data. The unsupervised learning phase does not aim to realize a capable action recognition model but to develop an action-pattern aware pretrained model for the next stage. In the *second* stage, the adaptation strategy refines the pretrained model on a small number of labeled skeletal samples to learn a target domain-specific prediction model.

The contribution of the proposed work consists of two aspects. *First*, it investigates the underlying principles of the skeletal data distribution shift issue during the practical action recognition model deployment. *Second*, it investigates domain adaptation strategies to improve the action recognition models' generalizability and robustness by introducing a semi-supervised adaptation strategy which leads to significant reduction of the data requirement in the target domain while achieving convincing performance. The code is available at https://github.com/tht106/SSA (accessed on 14 November 2023).

2. Related Work

The research on human action recognition addresses a variety of downstream computer vision-based tasks, such as human activity analysis, anomaly action detection, and video surveillance in hazardous places. Recently, Graph Convolutional Networks (GCNs) demonstrated the capability of interpreting topological features from multi-dimensional skeleton sequences, thereby dominating the research on skeletal human action recognition [3,4,7]. Yet, the generalization of the action recognition models in real environments is still challenging due to the data distribution shift caused by the variations in sensory configuration, e.g., camera views, heights, orientations, locations, and variations in data collection. Although the current skeletal datasets are devoted to covering the skeletal variations during data collection (e.g., NTU RGB+D [1] involves data variations by configuring three camera angles, 16 differences in height and distance, as well as involving 40 actors into action performance), the expected robustness of the resulting action recognition model remains vulnerable and can be uprooted while facing domain shifts in skeletal data [8].

Contrastive learning formulates unsupervised representation learning by contrasting positive pairs against negative pairs from a pre-defined dynamic dictionary [9,10]. In skeletal action recognition, the contrastive learning scheme regards each skeletal sequence as a unique class represented by a GCN encoder and explcits the skeletal invariances and similarities from the dynamic dictionary formulated by skeleton sequences without using data annotations [6]. Inspired by such advances in contrastive learning, this work is devoted to a semi-supervised adaptation scheme for human action recognition.

3. Method

This section defines the skeletal data distribution shift issue involved in the practical deployment of a human action recognition model. It then proposes a two-phase semi-supervised training framework, depicted in Figure 1, that relies on a significant quantity of public data (unlabeled) for pretraining and then refines a target-adaptive model by using only a small number of data (labeled) of the target environment.

3.1. Data Domain Shift in Skeletal Data

A human action recognition model considers the training dataset $\{X\}$ where $X \in \mathbb{R}^{T \times V \times 3}$ denotes a skeletal sequence composed of T frames in the shape of V human body joints, each one being defined in a calibrated camera reference frame with three-dimensional coordinates. $Y \in \mathbb{R}^{L}$ denotes the action label of X in a range of L categories. The goal of skeletal action recognition is to train a GCN model, composed of a graph convolution encoder **E** and a classification layer **C**, given the input (X_{train}, Y_{train}), where X_{train} is uniformly sampled from the dataset $\{X\}$. Normally, considering the test data sampled from $\{X\}$, i.e., the training and the test data are *i.i.d.* (independent identical distributions), the model achieves a convincing evaluation performance on the test dataset. However, in practical engineering applications, the target deployment environment always presents misaligned data distributions given the fact that the imaging configuration can differ according to the environment, which leads to variations on data distribution such that X_{train} and X_T are not *i.i.d.*, where X_T defines the skeletal data sampled in the target environment $\{X_T, Y_T\}$. The data domain shift corrupts the model performance while the model was well-trained on $\{X_{train}\}$ but evaluated on $\{X_T\}$.

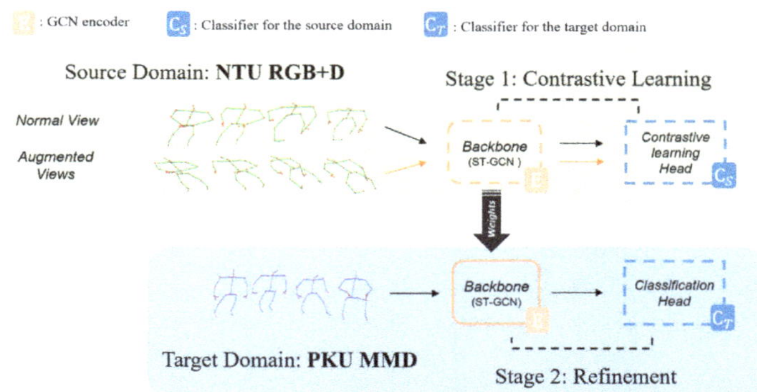

Figure 1. Framework of the proposed semi-supervised adaptation strategy. In the first stage, the training data (unlabeled) from the source domain is utilized to contrastive learning after data augmentation. The learned backbone is recycled in stage 2 for refining over the data samples (labeled) in the target domain. Network embedding is denoted with dotted lines if it is updated during training, and with solid lines otherwise.

3.2. Semi-Supervised Learning

To effectively tackle the underlying issue of skeletal domain shift, a two-stage strategy is proposed which (i) utilizes sufficient public dataset samples to pretrain a GCN encoder **E** with contrastive learning, and then (ii) recycles the pretrained **E** in the second stage to refine a target-specific classification layer, $\mathbf{C}_T$, exclusively on the target domain samples. The learnt **E** and $\mathbf{C}_T$ reassemble the target domain adaptive action recognition model. The details are as follows.

- In the first stage, the Extremely Augmented Skeleton (EAS) scheme [6] is used to augment the training data with $\{X^{aug}_{train}\}$ to enrich the input space in both spatial and temporal dimensions via eight augmentation operations: spatial shear, spatial flip, axis-wise rotate and mask, temporal flip, temporal crop, Gaussian noise and Gaussian blur.

The input query-key pairs (X_{train}, X^{aug}_{train}) compose a dynamic skeletal dictionary upon which the skeleton contrastive learning framework learns underlying topological invariances and semantic similarities from the unlabeled source domain $\{X\}$. The training progress is driven by MoCov2 [11] with the InfoNCE loss:

$$L_{\mathrm{IN}} = -log\frac{\exp(Z_q{\cdot}Z_k)/\tau.}{\exp(Z_q{\cdot}Z_k)/\tau + \sum_{Z_n\in\mathcal{N}}\exp(Z_q{\cdot}Z_n)/\tau} \tag{1}$$

where Z_q denotes the feature representations of the query input X_{train} as $Z_q = \mathbf{C_S}(\mathbf{E}(X_{train}))$, where $\mathbf{C_S}$ denotes the multilayer perception (MLP) projection head for contrastive learning. Likewise, Z_k denotes the feature representation of the key input X^{aug}_{train} obtained by $\overline{\mathbf{C}}_S\left(\overline{\mathbf{E}}\left(X^{aug}_{train}\right)\right)$, where $\overline{\mathbf{C}}_S$ and $\overline{\mathbf{E}}$ are the mean models of $\mathbf{C}_S$ and $\mathbf{E}$, respectively. $\mathcal{N}$ represents negative sample representations to the query input. τ acts as the temperature parameter of the softmax operation in Equation (1).

- In the second stage, the pretrained model is refined on a small number of labeled skeleton data $\{X_T, Y_T\}$ of the target environment. It takes the well-learnt skeletal knowledge-aware GCN encoder $\mathbf{E}$ from the first stage and fine-tunes the reassembled GCN model $\mathbf{C}_T(\mathbf{E}(\cdot))$ over the target domain. This stage is driven by a Cross-Entropy loss:

$$L_{\text{CE}} = -Y_T \cdot \log[\mathbf{C}_T(\mathbf{E}(X_T))] \quad (2)$$

Note that the encoder $\mathbf{E}$ is free of gradient backpropagation in the second stage (as illustrated in Figure 1). As the encoder is pretrained, this fine-tuning process will quickly converge to the target domain data distribution and learn a capable action prediction layer.

4. Experiments

Comprehensive experimentation is conducted to evaluate the effectiveness of the proposed two-stage method in a cross-domain action recognition scenario.

4.1. Datasets and Implementations

This study employs two public datasets to simulate the skeletal distribution shift scenario. In particular, NTU RGB+D [1] is considered as the training dataset and PKU-MMD [2] as the target environment. The former is a popular large-scale skeleton form human action recognition dataset. It presents 56,880 samples covering 60 human daily actions recorded in indoor scenes with three cameras mounted in different locations to support variations in camera views. It is common to utilize such a large-scale dataset in the research community [3–5] for human action recognition. PKU-MMD is another popular public skeletal dataset presenting fewer data samples (20,000 instances over 52 actions), but it involves significant camera view variations in the samples. In this paper, it is employed to mimic practical environments where data collection configurations could be inconsistent. The samples related to 50 actions common to the two datasets are utilized for experimentation. In the first stage, the selected part from the NTU RGB+D dataset is used for unsupervised contrastive learning, while one tenth of the samples (namely 1884) of the PKU-MMD dataset is used for model refining in the second stage. Given the GCN architecture considered, ST-GCN [3] is adopted for the network backbone. The output dimension of $\mathbf{C}_S$ and $\mathbf{C}_T$ is set as 128 and 50, respectively. The parameter τ in Equation (1) is set as 0.07. For model training, it follows the same optimization details as utilized in [6].

4.2. Results

Experimental results are summarized in Table 1 that reports the Top-1 accuracy [1] of the resulting models while tested on the full test set of PKU-MMD. As a comparative, a source-only model is trained with exclusive full supervision using the NTU RGB+D training dataset with annotations. The results in the first row of Table 1 illustrate that the resulting model yields convincing results on the NTU RGB+D benchmark, while the source-only model fails to effectively transfer to the target domain, demonstrating the negative impacts of domain shift on model performance.

Table 1. Comparative performance of the proposed method against full supervised training.

Model	On Benchmark (NTU RGB+D)	On Target Domain (PKU-MMD)
Source only	**69.07%**	57.32%
Adaptation	34.41%	**75.74%**

Next, the model is adapted by utilizing the proposed semi-supervised learning scheme. Results in the second row of Table 1 reflect the significant improvement achieved with the proposed method on the target domain, achieving a gain of 18.42% (from 57.32% to 75.74%). We conjecture that, as the data annotations of the source domain are not utilized in contrastive learning, the learnt encoder $\mathbf{E}$ is action-semantic aware but not overfitting to the source domain, which finally results in a capable human action model for PKU-MMD.

However, a severe performance deterioration is also observed on the benchmark NTU RGB+D evaluation (from 69.07% to 34.41%). It demonstrates that the encoder **E** does not form a competent action recognition stage on the source data domain but rather a reliable intermediate action-semantic aware encoding which is efficiently generated by unsupervised contrastive learning.

With the goal to identify the principle that drives performance increase with respect to the different amounts of target domain samples involved for model refining, Table 2 reports on the results of an experiment where the encoder **E** learnt in the first stage remains fixed but the classification layer $\mathbf{C}_T$ is refined upon different ratios (varying from 5% to 100%) of samples selected from the target domain PKU-MMD. Experimental results demonstrate that the model performance tends to increase monotonically along with the ratio of the refining data samples. Interestingly, even using a very small number (e.g., 5%) of data samples from the target domain, the model still achieves convincing performance compared to the best model (85.06% in Table 2) when using 100% of the PKU-MMD dataset. In conclusion, this study suggests good trade-off conditions between data usage and performance gains while utilizing the proposed two-stage semi-supervised method.

Table 2. Performance gains related to different proportions of the target data samples for $\mathbf{C}_T$ refinement training.

Percentage of data use	5%	20%	30%	50%	70%	100%
Accuracy	71.60%	77.33%	81.03%	82.25%	83.28%	85.06%

4.3. T-SNE Action Clusters Visualization

For better understanding of the effectiveness of the proposed method, closer examination of the embedding features of the two domains is presented using t-SNE [12]. It is expected that a capable classification model can interpret separable and dense action clusters on the feature space. Specifically, Figure 2 visualizes the action clusters of the two respective domains on the last convolutional layer before the classification head as interpreted by the two ST-GCN models ("Source only" on the upper row and "Adaptation" at the bottom). The "Source only" model presents well-separated action clusters when tested on the source domain (left column), which reflects the model's ability to interpret feature representations from the source domain on which the classification head easily determines action-wise classification boundaries. However, under the impacts of domain shift, the source-only model presents less separable action clusters when tested on the target domain (right column). Such a discrepancy on action cluster interpretation leads to the performance difference across the two domains (69.07% vs. 57.32% in Table 1). Conversely, after applying the proposed semi-supervised learning scheme as a refinement stage, the model (bottom row) demonstrates effective adaptation to the target domain whose action clusters are improved in terms of separability. Improvement is observable on both tests considering the source domain (left column) and the target domain (right column).

4.4. Semi-Supervised Learning vs. Supervised Learning

This subsection presents two experiments to evaluate whether utilizing fully supervised learning can reach the same effectiveness as the proposed semi-supervised adaptation method. First, a model trained with full supervision from NTU RGB+D is then fine-tuned also with full supervision on 10% data samples from PKU-MMD. Second, a separate model is trained with full supervision over the combined data samples from NTU RGB+D and PKU-MMD (10%). Both trained models are tested on the test dataset of PKU-MMD. Experimental results in Table 3 demonstrate that the fully supervised learning method presents inferior performance compared to the proposed semi-supervised method. Especially as the fully supervised transfer learning (pretrain on NTU RGB+D and fine-tuned on PKU-MMD) only achieves 45.97% on the target domain, representing a 29.77% gap in accuracy compared to the proposed method.

Figure 2. T-SNE visualization on action clusters on the embedding space of ST-GCN (**upper row**: "Source only"; **bottom row**: "Adaptation"). Clusters are distinguished by colors where twenty actions are randomly selected among fifty actions for clarity. The left column represents action clusters of the source domain (NTU RGB+D) and the right column shows clusters of the target domain (PKU-MMD), respectively.

Table 3. Performance of fully supervised learning vs. semi-supervised learning.

	Full Supervision		**Semi-Supervision**
	NTU RGB+D and 10% PKU-MMD (Fine-Tuning)	**NTU RGB+D and 10% PKU-MMD (Combined)**	**NTU RGB+D and 10% PKU-MMD (Fine-Tuning)**
Accuracy	45.97%	62.61%	75.74%

5. Conclusions

This work proposes a simple but efficient method to deploy a skeleton-data-based human action recognition model to a target environment while requiring only a small amount of labeled data from the latter. The proposed semi-supervised learning strategy utilizes contrastive learning to pretrain a model that learns key skeletal representations from an unlabeled dataset, then fine-tunes the pretrained model on a small number of labeled data samples in the target domain. Experiments are conducted to demonstrate the effectiveness of the proposed strategy. It suggests that the semi-supervised learning method achieves convincing results compared to fully supervised learning that requires voluminous labeled data from both the source and target domains. The research also experimentally characterizes a tradeoff between data usage and model performance, providing reference to develop and deploy future applications.

Author Contributions: Conceptualization, H.T. and P.P.; methodology, software, validation, formal analysis, H.T.; investigation, H.T. and P.P.; resources, P.P.; writing—original draft preparation, H.T.; writing—review and editing, P.P.; supervision, project administration, funding acquisition, P.P. All authors have read and agreed to the published version of the manuscript.

Funding: This research was supported in part by MITACS Accelerate and NSERC Discovery grants.

Institutional Review Board Statement: This research only involved data from public resources and no experiments on human subjects were conducted. Ethical review and approval were waived by the

Office of Research Ethics and Integrity of the University of Ottawa due to project falling under Article 2.5 of the Tri-Council Policy Statement: Ethical Conduct for Research Involving Humans (TCPS 2), and therefore did not require Research Ethics Board review.

Informed Consent Statement: Not applicable.

Data Availability Statement: Not applicable.

Conflicts of Interest: The authors declare no conflicts of interest.

References

1. Shahroudy, A.; Jun, L.; Tian-Tsong, N.; Gang, W. NTU RGB+D: A large scale dataset for 3d human activity analysis. In Proceedings of the IEEE Conference on Computer Vision and Pattern Recognition, Las Vegas, NV, USA, 27–30 June 2016; pp. 1010–1019.
2. Liu, C.; Hu, Y.; Li, Y.; Song, S.; Liu, J. PKU-MMD: A large scale benchmark for skeleton-based human action understanding. In Proceedings of the Workshop on Visual Analysis in Smart and Connected Communities, Mountain View, CA, USA, 23 October 2017; pp. 1–8.
3. Yan, S.; Xiong, Y.; Lin, D. Spatial temporal graph convolutional networks for skeleton-based action recognition. In Proceedings of the AAAI Conference on Artificial Intelligence, New Orleans, LA, USA, 2–7 February 2018; Volume 32.
4. Chen, Y.; Zhang, Z.; Yuan, C.; Li, B.; Deng, Y.; Hu, W. Channel-wise topology refinement graph convolution for skeleton-based action recognition. In Proceedings of the IEEE/CVF International Conference on Computer Vision, Montreal, BC, Canada, 11–17 October 2021; pp. 13359–13368.
5. Ben-David, S.; Blitzer, J.; Crammer, K.; Kulesza, A.; Pereira, F.; Vaughan, J.W. A theory of learning from different domains. *Mach. Learn.* **2010**, *79*, 151–175. [CrossRef]
6. Guo, T.; Liu, H.; Chen, Z.; Liu, M.; Wang, T.; Ding, R. Contrastive learning from extremely augmented skeleton sequences for self-supervised action recognition. In Proceedings of the AAAI Conference on Artificial Intelligence, Virtual, 22 February–1 March 2022; Volume 36, pp. 762–770.
7. Chi, H.G.; Ha, M.H.; Chi, S.; Lee, S.W.; Huang, Q.; Ramani, K. InfoGCN: Representation learning for human skeleton-based action recognition. In Proceedings of the IEEE/CVF Conference on Computer Vision and Pattern Recognition, New Orleans, LA, USA, 18–24 June 2022; pp. 20186–20196.
8. Choi, J.; Sharma, G.; Chandraker, M.; Huang, J.-B. Unsupervised and semi-supervised domain adaptation for action recognition from drones. In Proceedings of the IEEE/CVF Winter Conference on Applications of Computer Vision, Snowmass Village, CO, USA, 1–5 March 2020; pp. 1717–1726.
9. Chen, T.; Kornblith, S.; Norouzi, M.; Hinton, G. A simple framework for contrastive learning of visual representations. In Proceedings of the International Conference on Machine Learning, Virtual, 13–18 July 2020; pp. 1597–1607.
10. He, K.; Fan, H.; Wu, Y.; Xie, S.; Girshick, R. Momentum contrast for unsupervised visual representation learning. In Proceedings of the IEEE/CVF Conference on Computer Vision and Pattern Recognition, Seattle, WA, USA, 13–19 June 2020; pp. 9729–9738.
11. Chen, X.; Fan, H.; Girshick, R.; He, K. Improved baselines with momentum contrastive learning. *arXiv* **2020**, arXiv:2003.04297.
12. Van der Maaten, L.; Hinton, G. Visualizing data using t-SNE. *J. Mach. Learn. Res.* **2008**, *9*, 2579–2605.

engineering proceedings

MDPI

Proceeding Paper

AI-Driven Estimation of Vessel Sailing Times and Underwater Acoustic Pressure for Optimizing Maritime Logistics †

Rosa Martínez *, Jose Antonio García and Ivan Felis

Centro Tecnológico Naval y del Mar, 30320 Fuente Álamo, Spain; joseantoniogarcia@ctnaval.com (J.A.G.); ivanfelis@ctnaval.com (I.F.)

* Correspondence: rosamartinez@ctnaval.com; Tel.: +34-968-19-75-21

† Presented at the 10th International Electronic Conference on Sensors and Applications (ECSA-10), 15–30 November 2023; Available online: https://ecsa-10.sciforum.net/.

Abstract: This paper presents an innovative AI-based approach to estimate vessel sailing times in port surroundings. Leveraging historical vessel data, including ship characteristics and weather conditions, the model employs preprocessing techniques to enhance accuracy. Additionally, an underwater acoustic propagation model is used to study underwater noise pressure, aligning with environmental goals. The dataset, covering January to December 2022 in the Port of Cartagena, Spain, undergoes analysis, revealing intriguing patterns in ship routes. Employing various ML models, the study selects Random Forest as the most accurate, achieving an R2 of 0.85 and MSE of 0.145. The research showcases promising accuracy, aiding port optimization and environmental impact reduction, advancing maritime logistics with AI.

Keywords: port call optimization; artificial intelligence; machine learning; deep learning; maritime logistics; vessel dwell time; environmental impact

Citation: Martínez, R.; García, J.A.; Felis, I. AI-Driven Estimation of Vessel Sailing Times and Underwater Acoustic Pressure for Optimizing Maritime Logistics. *Eng. Proc.* **2023**, *58*, 26. https://doi.org/10.3390/ecsa-10-16091

Academic Editor: Stefano Mariani

Published: 15 November 2023

1. Introduction

Today, a large number of ships and nautical elements are active at sea. According to UNCTAD [1], approximately 80% of world trade is transported by sea, and this number is expected to further increase in the coming years [1]. In addition, shipping companies have been reporting over the years that disruptions and deviations from the initial plan occur frequently, resulting in most cases in delays [2]. These delays contribute to poor port optimization, disruptions in the market chain, and increased pollution, mainly greenhouse gas emissions and underwater radiated noise, due to prolonged idle times of vessels awaiting port calls. In fact, in April 2018, the IMO adopted the Initial Strategy for the reduction of GHG emissions from shipping which sets key ambitions, including cutting annual greenhouse gas emissions from international shipping by at least half by 2050 compared with their level in 2008.

This strategy goes in line with the Zero-Emission Waterborne Transport, the Horizon Europe partnership that aims to deliver and demonstrate zero-emission solutions for all major ship types and services before 2030. Ports over the world are starting to implement R&D tools to optimize their own performance and to be able to partner with these strategies. One example is the Port of Rotterdam, which has developed and implemented tools based on the Just In Time (JIT) arrival criterion, optimizing the speed of each vessel throughout its journey and reducing CO_2 emissions by 14%.

In this paper, we present an innovative approach that incorporates artificial intelligence (AI) models, specifically machine learning (ML), and preprocessing techniques, to estimate the sailing time of vessels in port surroundings. All of this is accomplished by leveraging historical vessel data, such as ship characteristics, movement patterns, weather conditions, and port-specific factors (docks and areas of action). Also, by implementing an underwater acoustic propagation model in each ship in its route, direct aspects related to the underwater noise pressure in the port context are studied. This study aligns with the MSFD, in particular regarding Descriptor 11 [3], searching for a balance between optimizing economical marine activities with good environmental status.

2. Study Area: Cartagena Port

This study encompasses two port docks of Cartagena Port (sited in Murcia, Spain) specialized in different traffic: the Cartagena Dock (sports marinas, cruise ships, and container cargo) and the Escombreras Dock (specialized in liquid and solid bulk traffic and storage activities) which was recently expanded, resulting in the port performing a hegemonic role in the management of this traffic throughout the Spanish Southeast. The port receives important flows that cross the Mediterranean, having dense networks with the Maghreb, the French, and the Italian coasts. The waters of the area are also furrowed by the local professional fishing boats and by the maritime traffic that connects the Atlantic Ocean and the Mediterranean Sea. Moreover, Cartagena receives part of the maritime passenger traffic that connects the Peninsula with the Balearic Islands. Additionally, more and more cruise lines are calling at the port (240,000 cruise tourists on 170 ships in 2019), being one of the national ports that is growing the most in this sense.

3. Methodology

Hence, in the present document, we propose a methodological focus on data analysis, emphasizing preprocessing, to enhance further predictions with machine learning models and underwater acoustic propagation models to assess the underwater radiated noise of ships in the port surroundings with the aim of reducing their impact.

3.1. Data Analysis

The dataset was derived from a Shiplocus (a multi-application platform for port management and maritime traffic exploitation (GMV)) account provided by the APC (Port Authority of Cartagena) through LIFE PortSounds (LIFE PortSounds. Reducing the impact of underwater noise on the marine environment of the Port of Cartagena (LIFE2020)) project. It consisted of 472 MB (1,585,941 $\times$ 32) of vessels and trajectories' relevant parameters of the selected area (Table 1), which corresponds to the Impact Zone (IZ) of the APC. A preliminary data analysis was conducted for computational purposes, and thus undefined and incomplete data were removed, as well as irrelevant columns, maintaining 'Latitude', 'Longitude', 'MMSI', 'Name', 'Date', 'Vessel type', 'SOG', 'COG', 'Length', 'Cargo', and 'Registered Owner' parameters. In addition, 'SOG' (speed over the ground) was used to remove data coming from vessels moving at abnormal speeds, such as very low speeds (1.5 knots) or physically impossible speeds, given by the expression (1). Hence, the dataset obtained after the preliminary analysis consisted of 113 MB (58,632 $\times$ 11).

$$v_{max} = 2.8\sqrt{L} \tag{1}$$

where L refers to the vessel's length.

Table 1. Impact Zone meshgrid coordinates.

Points	Lat	Lon
1	37°37,930′ N	01°10,613′ O
2	37°37,930′ N	00°33,988′ O
3	37°21,783′ N	01°10,613′ O
4	37°21,783′ N	00°33,988′ O

A route was defined as the union of successive AIS messages from a vessel, where successive messages are defined as those between which no more than 5 h have elapsed.

Therefore, AIS messages remaining on the dataset after the data processing steps were used in the crafting of routes.

Given the dataset (consisting of a concatenation of AIS points), routes were transformed into the following features: *'MMSI'*, *'Time spent'*, *'Vessel type'*, *'Length'*, *'Mean SOG'*, *'Cargo'*, *'Owner'*, *'Arrival date'*, *'Start point'*, *'End point'*, *and 'Passing through'*, where the new columns were:

- *'Time spent'*: The duration of the whole route.
- *'Arrival date'*: The timestamp where the route begins.
- *'Start point'*: A sectorization of the area was performed and key areas were defined, so the start point defines the key area where the route begins.
- *'End point'*: As with the start point, the end point is given by the key area where the route ends.
- *'Passing through'*: Coded as 'YES' or 'NO' based on whether a vessel is ending its itinerary in the port area or is just passing through the area but will not end its itinerary.

As can be seen in Figure 1, vessels show great similarities in their routes over the year, except for the tugs, which are always moving close to the docks and show significant uncertainty in the duration of the routes, ranging from 1 h to 22 h. This is due to tugs being vessels that reside in the port and are designed primarily for towing and pushing other vessels in harbors, canals, and other confined waterways. As tugs are vessels from the port, they were excluded from the route's dataset. Also, in Figure 1d, anomalous routes can be observed. These routes were filtered to avoid confusions in the model.

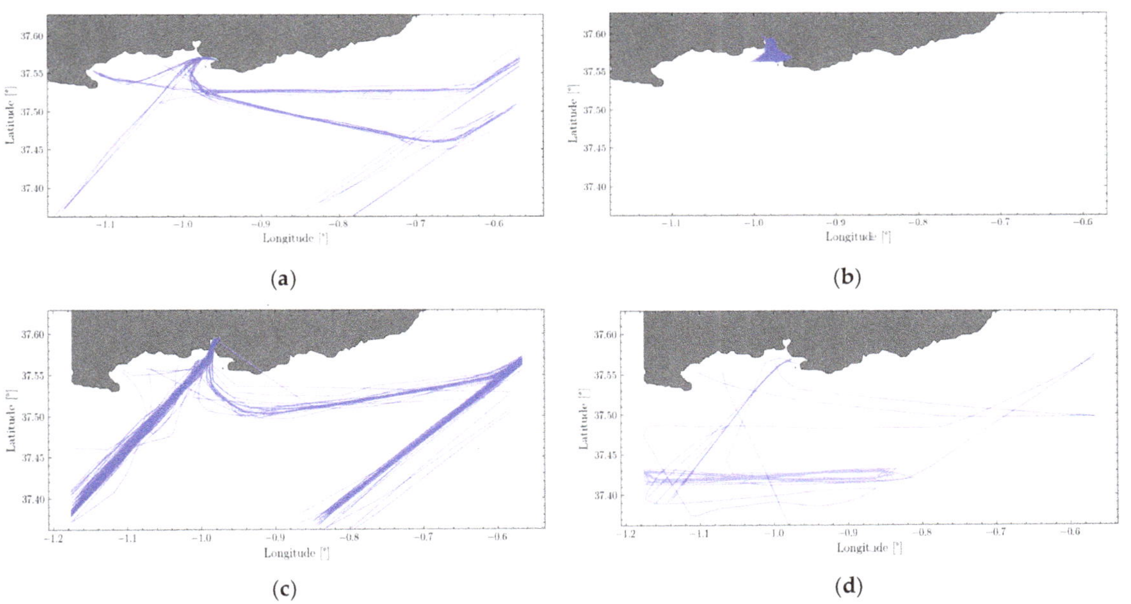

Figure 1. Routes performed by (**a**) LPG tanker vessel; (**b**) tug; (**c**) container ship; (**d**) chemical/oil products tanker.

With this, a thorough exploratory analysis was conducted to analyze route differences and similarities among vessel types to understand their behavior, as well as to sectorize areas depending on the traffic density. Thus, the sectorization made it possible to understand the key areas (areas with the higher density of AIS points) and to filter out those routes that did not start or end in a key area. Finally, a curated dataset (Figure 2) was obtained to be implemented in underwater acoustic pressure and machine learning analysis.

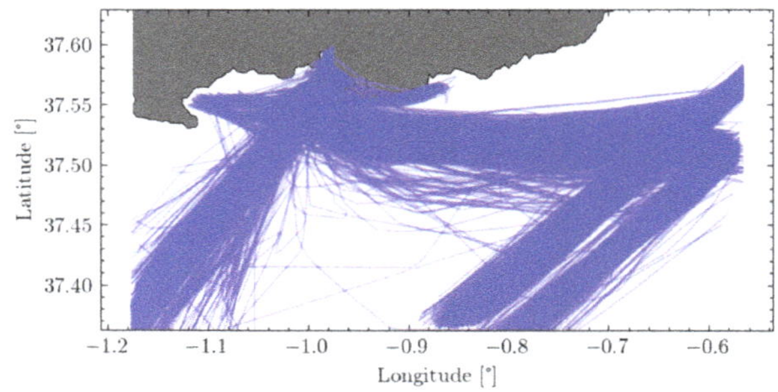

Figure 2. Set of routes after the clustering classification and the filtering of anomalous routes.

3.2. Underwater Acoustic Pressure

To assess the noise emitted by the vessels, the **Ross** model [4] was applied to each route. The **Ross** model considers physical vessel parameters, such as the speed and the length, and also parameters like the frequency to estimate the **Source Level** (SL), which is essentially the Sound Pressure Level (SPL) at 1 m distance of the acoustic source.

After the evaluation of the SL, a spherical loss model dependent only on the distance (for the sake of simplicity) was used to obtain a first approximation of the noise distribution over the area. Two cases were studied: one with a high number of vessels in the impact zone and one with just one vessel in the test site.

For the test with a high number of vessels, four ships (one cruise ship and three cargo ships) were selected from the data set. For these four ships, the SL was obtained at frequencies 62.5 Hz and 125 Hz, which are affected by cavitation noise. Once the SL was obtained, the spherical model was run in the area and the transmission losses were obtained. Finally, the SPL field that would be generated by each of the vessels was calculated and added coherently and 100% additively in linear units to obtain the worst case that could not occur. The physical properties of the 4 ships selected are shown in Table 2.

Table 2. Physical properties of the vessels modelled as sources.

Type	Speed (Knots)	Length
Cruise Ship	15	247
General Cargo Ship 1	10.7	108
General Cargo Ship 2	10.7	108
General Cargo Ship 3	7	90

For the one-vessel case, the cruise ship from the first case was selected as the source to be modelled. The methodology is the same but this time no SPL maps need to be added, as there is only a single source in the area.

3.3. ML Models

Several machine learning (ML) models were tested to predict the time spent for a vessel knowing its arrival date, starting and ending point, and the passing through field. These models were Gradient Boosting and Random Forest Regressor. Gradient Boosting (GB) is a machine learning algorithm that uses an ensemble technique to create a more accurate prediction model from multiple simpler models. The main idea behind GB is to combine several weak models to form one strong model [5]. Random Forest Regressor (RFR) is a meta estimator that fits a number of classifying decision trees on various sub-samples of the dataset and uses averaging to improve the predictive accuracy and control over-fitting [6]. For each model, hyperparameter optimization techniques were conducted and the results were compared to select the best model.

4. Results

4.1. Underwater Acoustic Pressure Assessment

As for the noise assessment, the Source Level of every route was studied and classified for every type of vessel.

As depicted in Figure 3a, the SL generated by the vessels at the 62.5 Hz frequency mostly ranged between 150 and 165 dB. Also, for the 125 Hz component, the SL values were between 110 and 170 dB, where most values were distributed between 140 and 155 dB.

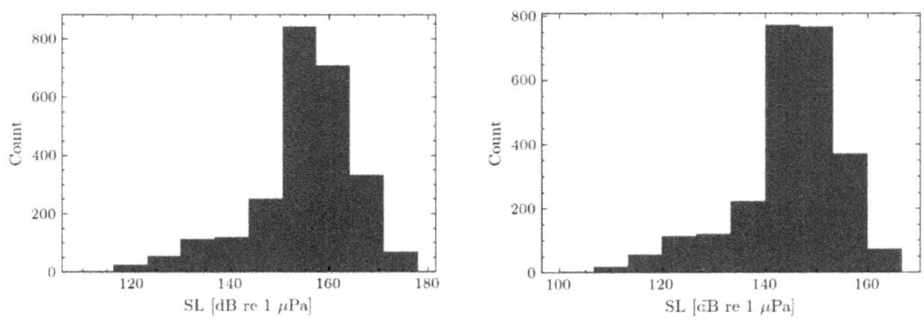

Figure 3. (**a**) SL histogram at 62.5 Hz; (**b**) SL histogram at 125 Hz.

In Figure 4, an SL breakdown per type of vessel and frequency is shown. As can be seen, the lowest values found are from the FSV (Fishing Support Vessel) type. Also, the biggest values are from the cruise ship and the LNG tanker types, whose median values are up to 165 dB in (a) and up to 155 dB in (b).

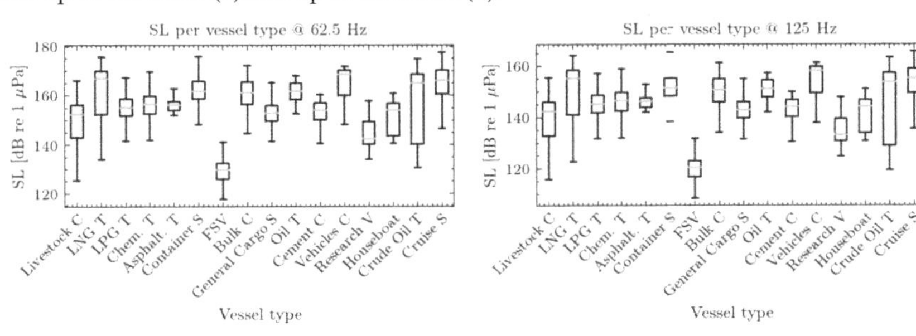

Figure 4. (**a**) Box and Whisker plot of SL values per vessel type at 62.5 Hz component; (**b**) Box and Whisker plot of SL values per vessel type at 125 Hz component.

Using the methodology described in Section 3.2, the SPL maps at 5 m depth for 63 and 125 Hz were obtained, both for the one-vessel case and the four-vessels case in the IZ of the Cartagena port (see Figure 5).

The highest SPL levels were found close to the source and ranged between 140 dB and 130 dB. In Figure 5, a slight difference between the two cases can be appreciated, as the highest SPL in the distance was found for the four-vessel case. For the four-vessel case, mean values of SPL for 62.5 Hz and 125 Hz, are, respectively, 75.8 dB/km^2 and 65.3 dB/km^2. For the one-vessel case, the values found are 72.5 dB/km^2 and 61.7 dB/km^2.

Between the one-vessel case and four-vessel case, the SPL differences found were close to 4 dB/km^2. Also, it should be noted that the SPL levels found for the 125 Hz component are lower than for the 62.5 Hz due to the SL being lower for the 125 Hz.

It should be noted that these models were computed as a first approximation to understand underwater acoustic pressure, and that the 100% coherent and additive addition of the acoustic waves carried out for the four-vessel case will never occur in real terms.

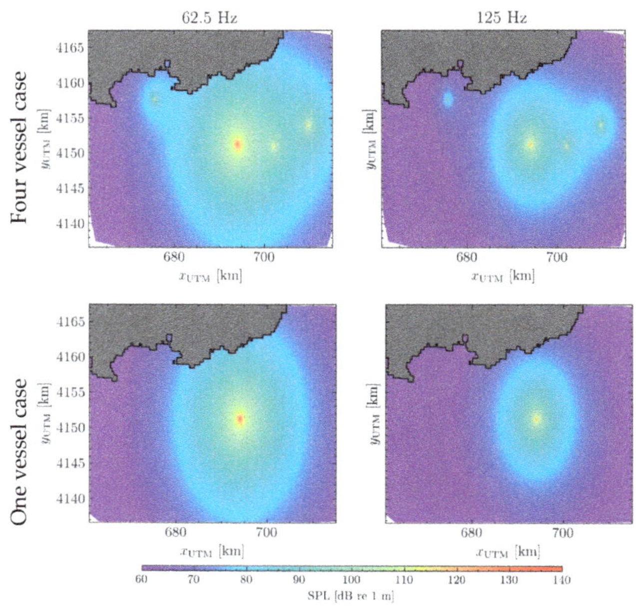

Figure 5. SPL map obtained at 5 m depth for the 4-vessel case (**above**) and for the 1-vessel case (**below**) for 62.5 Hz and 125 Hz on the IZ of the project.

4.2. ML Models

In Table 3 we can see the model metrics after the fine-tuning of the hyperparameters using a GridSearchCV algorithm, where it can be observed that RFR performs the best.

Table 3. Model summary and parameters.

Model	R^2_{train}	R^2_{test}	MSE
Gradient Boosting	0.884	0.8	0.198
Random Forest Regressor	0.974	0.85	0.145

5. Conclusions

This paper presents the study and descriptive analysis of an AIS dataset with the aim of creating an ML-driven tool for the optimization of waiting times in the Port of Cartagena. In addition, a first approach to the visualization of the impact generated by the traffic from an acoustic point of view has been carried out. The machine learning model used was able to predict the transit time of vessels in the defined area with an MSE of 0.145. The acoustic models, although built as a first approximation, showed differences between different frequencies and different numbers of coherent vessels.

Future work will focus on the use of more complex models such as the MMPE (Miami–Monterrey Parabolic Equation) for a better estimation of transmission losses in the area. In addition, these more complex models will be used over a time range of several hours to see the noise signature left by the traffic. On the other hand, more complex models such as NN will be used for a better prediction of the sailing time.

Author Contributions: Conceptualization and methodology, R.M. and I.F.; data curation, J.A.G.; formal analysis, J.A.G. and R.M.; validation, R.M. and I.F.; writing—original draft preparation, J.A.G.; writing—review and editing, R.M. and I.F. All authors have read and agreed to the published version of the manuscript.

Funding: This research was funded by the European Climate, Infrastructure and Environment Executive Agency (CINEA), grant number LIFE20 ENV/ES/000387.

Institutional Review Board Statement: Not applicable.

Informed Consent Statement: Not applicable.

Data Availability Statement: The data presented in this study are not publicly available due to privacy reasons.

Conflicts of Interest: The authors declare no conflict of interest.

References

1. Review of Maritime Transport 2022 | UNCTAD. Available online: https://unctad.org/rmt2022 (accessed on 22 February 2023).
2. Nikghadam, S.; Molkenboer, K.F.; Tavasszy, L.; Rezaei, J. Information sharing to mitigate delays in port: The case of the Port of Rotterdam. *Marit. Econ. Logist.* **2023**, *25*, 576–601. [CrossRef]
3. Vighi, M.; Boschetti, S.T.; Hanke, G. *Marine Strategy Framework Directive Review and Analysis of EU Member States' 2018 Reports*; Publications Office of the European Union: Luxembourg, 2021.
4. McKenna, M.F.; Ross, D.; Wiggins, S.M.; Hildebrand, J.A. Underwater radiated noise from modern commercial ships. *J. Acoust. Soc. Am.* **2012**, *131*, 92–103. [CrossRef] [PubMed]
5. Natekin, A.; Knoll, A. Gradient boosting machines, a tutorial. *Front. Neurorobot.* **2013**, *7*, 21. [CrossRef] [PubMed]
6. El Mrabet, Z.; Sugunaraj, N.; Ranganathan, P.; Abhyankar, S. Random Forest Regressor-Based Approach for Detecting Fault Location and Duration in Power Systems. *Sensors* **2022**, *22*, 458. [CrossRef] [PubMed]

engineering proceedings

MDPI

Proceeding Paper

Application of SERS Spectroscopy for the Study of Biological Molecules †

Pauline Conigliaro [1,2], Stefano Prato [3], Barbara Troian [3], Anton Naumenko [3], Valentina Pisano [1,2] and Ines Delfino [1,*]

1 Dipartimento di Scienze Ecologiche e Biologiche, Università degli Studi della Tuscia, 01100 Viterbo, Italy; pauline.conigliaro@unitus.it (P.C.); valentina.pisano@unitus.it (V.P.)
2 Dipartimento di Economia, Ingegneria, Società e Impresa, Università degli Studi della Tuscia, 01100 Viterbo, Italy
3 A.P.E. Research S.r.l., Area Science Park, Basovizza, 34149 Trieste, Italy; stefano.prato@aperesearch.com (S.P.); barbara.troian@aperesearch.com (B.T.); anton.naumenko@aperesearch.com (A.N.)
* Correspondence: delfino@unitus.it

† Presented at the 10th International Electronic Conference on Sensors and Applications (ECSA-10), 15–30 November 2023; Available online: https://ecsa-10.sciforum.net/.

Abstract: Surface-Enhanced Raman Spectroscopy (SERS) is a specialized spectroscopic technique based on the enhancement of the Raman scattering signals of molecules adsorbed or in close proximity to certain rough or nanostructured metal surfaces. It's an extremely sensitive and powerful analytical tool for the investigation of biological molecules, revolutionizing the field of bioanalytical chemistry. The enhancement of Raman signal is due to various effects, the most important is thought to be the interaction between the electromagnetic wave associated with the laser used and the rough metal substrate (i.e., silver/copper/gold surfaces) on which the molecule is placed. When substrates are used, their characteristics are crucial for the reliability and sensitivity of experiments, as well as the ease of reproducibility of measurements. In the present work, we report on preliminary measurements to investigate the characteristics of two commercial SERS substrates, which have different nanostructures and patterns, properly designed to operate at an excitation wavelength of 785 nm. Aspirin C was used as a representative molecule to evaluate their application for SERS study of biological molecules, thanks to its characteristic fingerprint. Aspirin C is commercially available in the form of effervescent tablets, with acetylsalicylic acid and ascorbic acid as active principles with mainly analgesic and anti-inflammatory properties. The results are discussed also considering future applications for the detection of analytes of environmental interest.

Keywords: SERS; Aspirin C; SERS substrates; biomolecules detection

Citation: Conigliaro, P.; Prato, S.; Troian, B.; Naumenko, A.; Pisano, V.; Delfino, I. Application of SERS Spectroscopy for the Study of Biological Molecules. *Eng. Proc.* **2023**, *58*, 27. https://doi.org/10.3390/ecsa-10-16164

Academic Editor: Francisco Falcone

Published: 15 November 2023

1. Introduction

Surface-enhanced Raman Spectroscopy (SERS) is a variant of Raman scattering that has applications in various fields such as biomedicine, chemistry and biology and has attracted attention because it allows biological molecules of interest with different molecular weights to be identified rapidly, non-destructively and with a certain degree of sensitivity and reproducibility [1,2]. It has been shown to be an extremely sensitive and powerful analytical tool for the investigation of biological molecules [3–5].

The advantage of SERS compared to the Raman technique is the amplification of the signal by several orders of magnitude (up to 10^{10}), using metal nanoparticles or directly using commercial SERS substrates, in which a few microliters of the solutions containing the analyte of interest are deposited in the active area [6,7]. Signal amplification results from the interaction between the electromagnetic wave associated with the laser used and the metal substrate (e.g., silver/copper/gold surfaces) on which the molecule is placed. The efficacy of this amplification is highly dependent on the metal substrate.

In this work, we have presented preliminary evaluation measurements of two commercial substrates, namely Q-SERS™ (Nanova Inc., Columbia, SC, USA) and RAM-SERS-AU (Ocean Insight), that are both optimized to be used with laser excitation at 785 nm and have different characteristics. Aspirin C was used as a representative molecule to evaluate their application for the SERS study of biological molecules, thanks to its characteristic fingerprint. It is commercially available in the form of effervescent tablets with analgesic and anti-inflammatory properties [8]. Its spectral characteristics, in addition to having ascorbic acid and acetylsalicylic acid as active principles, make it a target for study.

2. Materials and Methods

The commercial Aspirin C tablets contain 400 mg of acetylsalicylic acid and 240 mg of ascorbic acid (Vitamin C). To obtain the solution needed for measurements, 100 mg of the tablet was dissolved in 40 mL of water. An appropriate amount of the solution was placed on each of the SERS substrates being investigated and on metal and silicon discs. In all the cases, the solution was left to dry before performing the measurements.

The commercial SERS substrates used were Q-SERS™, supplied by Nanova Inc., characterised by gold nanoparticles with a diameter between 15 and 60 nm, placed on a silicon wafer. The second SERS substrate used was RAM-SERS-AU, manufactured by Ocean Insight, consisting of a layer of gold nanoparticles placed on borosilicate glass. Little information is known about their design, except for the Q-SERS™ substrate which we know is made up of gold nanoparticles of different diameters placed on a silicon layer.

Raman spectra were acquired in the 500–2000 cm^{-1} range using the APE Scanning MicroRaman system (A.P.E. Research S.r.l., Trieste) and APE Raman Mapping software. The laser used has a wavelength of 785 nm and a maximum power of 350 mW. The exposure time was set to 1000 ms for each acquisition.

SERS spectra were analysed using Origin Pro 7.5 software (OriginLab, Northampton, MA, USA).

3. Results and Discussion

To compare the SERS performance of two commercial SERS substrates, Q-SERS™ and RAM-SERS-AU, by using Aspirin C as a representative molecule, the Raman spectrum of Aspirin C was obtained by analysing Aspirin C on a metal disc. A representative Raman spectrum is shown in Figure 1. In the quite noisy spectrum, some of the peaks of acetylsalicylic acid that are observed are visible at 1603 cm^{-1} and 1030 cm^{-1} [9,10], associated with the vibration of the carbonyl group and the presence of the aromatic ring, respectively. Weak signals of aspirin are visible in the region around 840 cm^{-1} (associated with out-of-plane bending C-H), 900 cm^{-1} and 1419 cm^{-1}. As far as ascorbic acid is concerned, a band around 960–980 cm^{-1} associated with the C-H and O-H bending is visible [11].

As stated in the Section 2, the Q-SERS™ is made from nanoparticles on a silicon layer, and the Raman spectrum of Aspirin C on a silicon substrate was recorded and shown in Figure 2.

In the spectrum shown in Figure 2, the silicon signals are visible with greater intensity at 521 cm^{-1} and the band at 1000 cm^{-1} is associated with the second-order resonance peak of the silicon. Conversely, the Aspirin C peaks are not evident.

The spectra of Aspirin C placed on the two commercial substrates were detected and shown in Figure 3.

The two spectra seem to be less noisy than the corresponding ones obtained without SERS substrates. The spectra of Figure 3 show the same bands at 1603 cm^{-1} and 1030 cm^{-1}, associated with Aspirin C. For the spectrum obtained when the Q-SERS™ substrate is used (Figure 3, black line) the band around 521 cm^{-1} associated with the silicon substrate is also visible.

As a first tentative comparison of the enhancement efficacy of the two SERS substrates, the intensity of the Aspirin C-related 1030 cm^{-1} band, present in all the spectra, was used;

the ratio between this intensity that was detected when the solution was placed on the SERS substrate and the corresponding intensity in the Raman spectrum (I_{SERS}/I_{Raman}) was calculated for both the substrates. The preliminary results show that the enhancement in the intensity of the 1030 cm^{-1} band with the RAM-SERS-AU substrate is higher than the corresponding value obtained for the Q-SERS™ substrate.

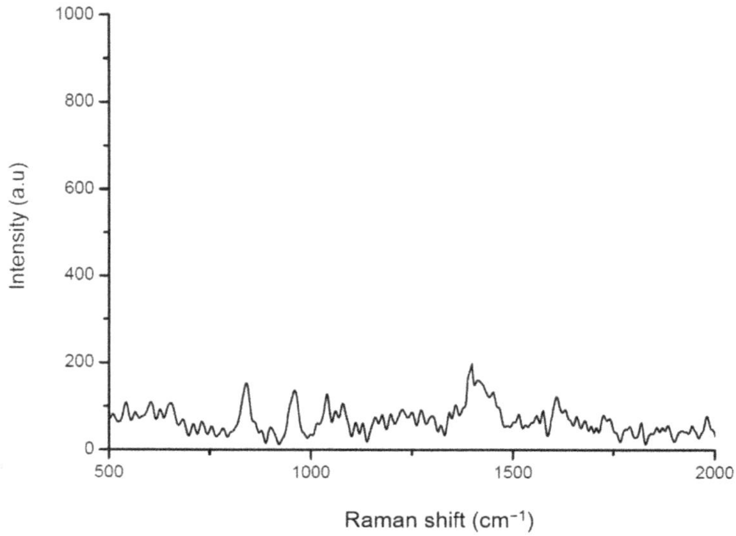

Figure 1. Raman spectrum of Aspirin C solution on a metal disc.

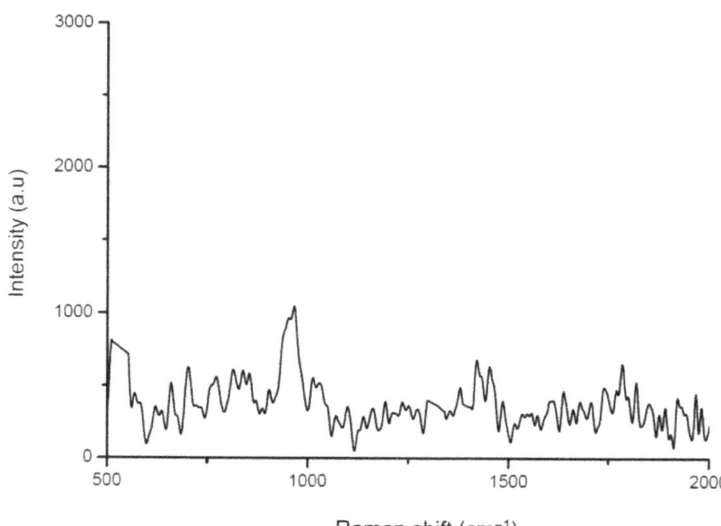

Figure 2. Raman spectrum of Aspirin C solution on silicon.

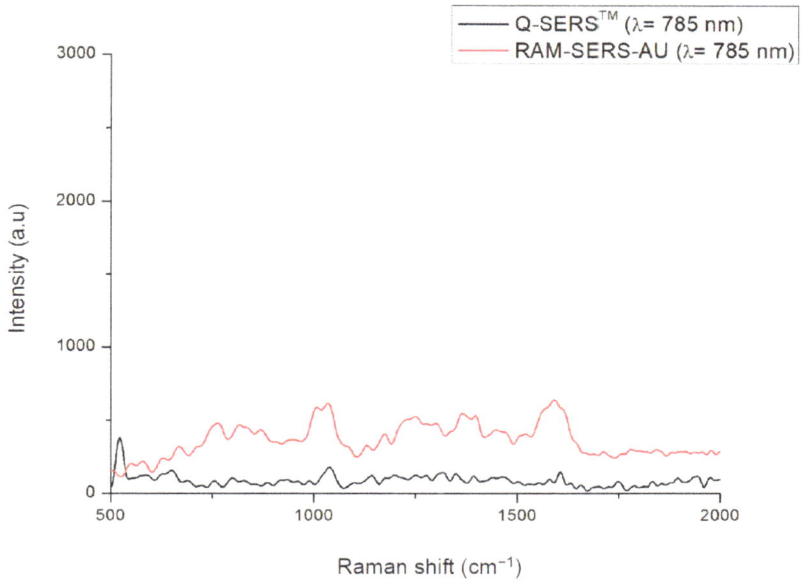

Figure 3. SERS spectra of Aspirin C on Q-SERS™ (black line) and RAM-SERS-AU (red line) substrates by using excitation with a 785 nm laser and optical objective 50x.

4. Conclusions

The results are promising for future applications for monitoring molecules of environmental interest and for designing sensors with advantages in terms of detection limits.

Author Contributions: Conceptualization, S.P. and I.D.; investigation, P.C., S.P., B.T. and A.N.; writing—review and editing, P.C., V.P. and I.D. All authors have read and agreed to the published version of the manuscript.

Funding: This research received no external funding.

Institutional Review Board Statement: Not applicable.

Informed Consent Statement: Not applicable.

Data Availability Statement: The data presented in this study are available upon request from the corresponding author.

Conflicts of Interest: The authors declare no conflict of interest.

References

1. Cialla, D.; März, A.; Böhme, R.; Theil, F.; Weber, K.; Schmitt, M.; Popp, J. Surface-enhanced Raman spectroscopy (SERS): Progress and trends. *Anal. Bioanal. Chem.* **2012**, *403*, 27–54. [CrossRef] [PubMed]
2. Kneipp, K. Surface-enhanced Raman scattering. *Phys. Today* **2007**, *60*, 40–46. [CrossRef]
3. Camerlingo, C.; Portaccio, M.; Delfino, I.; Lepore, M. Surface-Enhanced Raman Spectroscopy for Monitoring Extravirgin Olive Oil Bioactive Components. *J. Chem.* **2019**, *2019*, 9537419. [CrossRef]
4. Mun, J.; Lee, D.; So, S.; Badloe, T.; Rho, J. Surface-enhanced spectroscopy: Toward practical analysis probe. *Appl. Spectrosc. Rev.* **2019**, *54*, 142–175. [CrossRef]
5. Cennamo, G.; Montorio, D.; Morra, V.B.; Criscuolo, C.; Lanzillo, R.; Salvatore, E.; Camerlingo, C.; Lisitskiy, M.; Delfino, I.; Portaccio, M.; et al. Surface-enhanced Raman spectroscopy of tears: Toward a diagnostic tool for neurodegenerative disease identification. *J. Biomed. Opt.* **2020**, *25*, 1. [CrossRef] [PubMed]
6. Liu, Y.; Zhang, Y.; Tardivel, M.; Lequeux, M.; Chen, X.; Liu, W.; Huang, J.; Tian, H.; Liu, Q.; Huang, G.; et al. Evaluation of the Reliability of Six Commercial SERS Substrates. *Plasmonics* **2020**, *15*, 743–752. [CrossRef]
7. Mosier-Boss, P. Review of SERS Substrates for Chemical Sensing. *Nanomaterials* **2017**, *7*, 142. [CrossRef] [PubMed]

8. Khan, M.R.; Alothman, Z.A.; Naushada, M.; Ghfar, A.A.; Wabaidur, S.M. Simultaneous Analysis of Vitamin C and Aspirin in Aspirin C Effervescent Tablets by High Performance Liquid Chromatography–Photodiode Array Detector. *J. Liq. Chrom. Related Technol.* **2012**, *35*, 2454–2461. [CrossRef]
9. Jamur, J.M.S. Raman spectroscopy analysis for monitoring of chemical composition of aspirin after exposure to plasma flame. *Spectrosc. Eur.* **2022**, *34*, 18–22. [CrossRef]
10. Muthuselvi, C.; Dhavachitra, M.; Pandiarajan, S.S. Growth and Characterization of Aspirin Crystal in the Phosphoric acid Medium. *J. Chem. Pharm. Res.* **2016**, *8*, 804–814.
11. Panicker, C.Y.; Varghese, H.T.; Philip, D. FT-IR, FT-Raman and SERS spectra of Vitamin C. *Spectrochim. Acta A Mol Biomol. Spectrosc.* **2006**, *65*, 802–804. [CrossRef] [PubMed]

engineering proceedings

MDPI

Proceeding Paper

Observer Design for Takagi–Sugeno Fuzzy Systems with Unmeasurable Premise Variables Based on Differential Mean Value Theorem †

Wail Hamdi *, Mohamed Yacine Hammoudi and Anouar Boukhlouf

Laboratory of Energy Systems Modeling (LMSE), Department of Electrical Engineering, University of Biskra, BP 145, Biskra 07000, Algeria

* Correspondence: wail.hamdi@univ-biskra.dz

† Presented at the 10th International Electronic Conference on Sensors and Applications (ECSA-10), 15–30 November 2023; Available online: https://ecsa-10.sciforum.net/.

Abstract: In this work, we present the design of an observer for Takagi–Sugeno fuzzy systems with unmeasurable premise variables. Moving away from Lipschitz-based and $\mathcal{L}_2$ attenuation-based methods—which fall short in eliminating the mismatching terms in the estimation error dynamics—we leverage the differential mean value theorem. This approach not only removes these terms but also streamlines the factorization of the estimation error dynamics, making it directly proportional to the estimation error. To ensure the asymptotic convergence of the estimation error, we apply the second Lyapunov theorem, which provides sufficient stability conditions described as linear matrix inequalities. A numerical example applied on a three-tank hydraulic system is presented to demonstrate the observer's effectiveness.

Keywords: Takagi–Sugeno systems; nonlinear systems; fuzzy observer; mean value theorem

1. Introduction

In the industrial sector, cost-effectiveness is paramount. A key strategy to achieve this is using observers to reduce the need for expensive sensors. The Luenberger observer [1] has paved the way for numerous advancements. Takagi–Sugeno (TS) fuzzy systems, which represent nonlinear systems as a weighted sum of linear ones [2], have provided valuable tools for understanding complex dynamics. Using the sector nonlinearity approach, these systems can be accurately described [3]. Based on variables in their weight functions, they're classified into measurable premise variables (MPV) and unmeasurable premise variables (UPVs), with the latter being a primary research focus because it represents the largest category of systems.

Designing observers for systems equipped with UPV tends to be more intricate than their measurable counterparts. These complexities stem predominantly from the mismatching terms in the error dynamics. To address such hurdles, the scientific community has forwarded various techniques. Initially, the Lipschitz-based method emerges as a straightforward solution [4,5], yet stumbles when the nonlinear system's Lipschitz constant exceeds an admissible value, thereby introducing pronounced conservatism in Linear Matrix Inequality (LMI) constraints. An alternative, the $\mathcal{L}_2$-attenuation-based approach [6,7], focuses on minimizing the aforementioned mismatches. Though typically less conservative than the Lipschitz method, there remain instances where the attenuation level is not minimal enough to be accepted, even if the simulation works. This paves the way for the mean value theorem (MVT) method [8,9], allowing the factorization of the estimation error dynamics, which leads to making it proportional completely to the estimation error. Hence, the mismatching terms disappeared from the estimation error dynamics.

The observer based on the MVT, leading to an exact transformation of error dynamics into an LPV system, was introduced in [10]. Subsequent works, such as [11,12], represent

Citation: Hamdi, W.; Hammoudi, M.Y.; Boukhlouf, A. Observer Design for Takagi–Sugeno Fuzzy Systems with Unmeasurable Premise Variables Based on Differential Mean Value Theorem. *Eng. Proc.* **2023**, *58*, 28. https://doi.org/10.3390/ecsa-10-16008

Academic Editor: Francisco Falcone

Published: 15 November 2023

dynamic error using TS representation. This theorem's applications span various studies: real-time motor control in [13], robust H_∞ control for motors in [14], sensorless control for a PMSM in [15], and automotive slip angle estimation in [12]. A notable advancement is in [16], where line integral Lyapunov functions reduce conservatism in MVT observers.

In this paper, we introduce an observer design based on the mean value theorem and validate its efficacy via a simulation on a three-tank hydraulic system. The paper's structure is as follows: Section 2 delves into the TS fuzzy representation and the mean value theorem. Section 3 details the observer design, Section 4 showcases the simulation results, and Section 5 concludes with potential directions for future research.

2. Preliminaries

2.1. Takagi–Sugeno Fuzzy Representation

Let us consider the following nonlinear system:

$$\begin{cases} \dot{x}(t) = f(x(t), u(t)) \\ y(t) = g(x(t)) \end{cases} \tag{1}$$

where $x(t) \in \mathbb{R}^n$ is the state vector, $u(t) \in \mathbb{R}^{n_u}$ is the input vector, and $y(t) \in \mathbb{R}^{n_y}$ represents the output vector.

The TS model reformulates the nonlinear system (1) using a convex combination of linear sub-models. Each ith sub-model follows the given fuzzy rule:

$$\text{If } \xi_1 \text{ is } M_{i1} \text{ and} \ldots \text{ and } \xi_l \text{ is } M_{il} \qquad \text{Then: } \begin{cases} \dot{x}(t) = A_i x(t) + B_i u(t) \\ y(t) = C_i x(t) \end{cases} \tag{2}$$

where ξ is premise variable, l is the number of premise variables, and M_{ij} is the membership function of the ith fuzzy rule corresponding to the jth premise variable. $A_i \in \mathbb{R}^{n \times n}$, $B_i \in \mathbb{R}^{n \times n_u}$, and $C_i \in \mathbb{R}^{n_y \times n}$ are known matrices.

The global TS fuzzy representation of the nonlinear system (1) is described as follows:

$$\begin{cases} \dot{x}(t) = \sum_{i=1}^{r} \mu_i(x(t))(A_i x(t) + B_i u(t)) \\ y(t) = \sum_{i=1}^{r} \mu_i(x(t)) C_i x(t) \end{cases} \tag{3}$$

where $\mu_i(x(t))$ is the weighting functions described by $\mu_i(x(t)) = \prod_{j=1}^{l} M_{ij}(\xi_j)$ and verifies the convex sum property $\sum_{i=1}^{r} \mu_i(x(t)) = 1, \quad 0 \leq \mu_i(x(t)) \leq 1, \quad \forall i = 1, \ldots r.$

2.2. Differential Mean Value Theorem [10]

Let $\Phi(x) : R^n => R^n$ be a differentiable vector function described as follows:

$$\Phi(x) = \sum_{i=1}^{n} e_n(i) \Phi_i(x) \tag{4}$$

where the set E_n is the canonical basis of the vectorial space $\mathbb{R}^n$ for all $n \geq 1$ given by:

$$E_n = \{e_n(i) | e_n(i) = (0, \ldots, 0, \underbrace{1}_{i}, 0, \ldots, 0)^T, i = 1, \ldots, n\} \tag{5}$$

Let $a, b \in \mathbb{R}^n$. Then, there are constant vectors $z_1, \ldots, z_n \in (a, b), z_i \neq a, z_i \neq b$ for $i = 1, \ldots, n$ such that the mean value theorem ensures the following relation:

$$\Phi(a) - \Phi(b) = \sum_{i=1}^{n} \sum_{j=1}^{n} e_n(i) e_n(j)^T \frac{\partial \Phi_i(z_j)}{\partial x_j} (a - b) \tag{6}$$

Applying sector nonlinearity allows for rewriting the above equation using TS transformation:

$$\Phi(a) - \Phi(b) = \sum_{i=1}^{s \leq 2^{n^2}} h_i(z(t))\mathcal{H}_i(a-b) \tag{7}$$

where $\mathcal{H}_i$ represents the sub-models of the nonlinear term $\sum_{i=1}^{n}\sum_{j=1}^{n} e_n(i)e_n(j)^T \frac{\partial \Phi_i(z_j)}{\partial x_j}$, and $h_i(z(t))$ is its weighting function.

3. Mean Value Theorem Observer-Based Design

Let us consider the following observer for fuzzy system (3):

$$\begin{cases} \dot{\hat{x}}(t) = \sum_{i=1}^{r} \mu_i(\hat{x}(t))(A_i\hat{x}(t) + B_iu(t)) + L(y(t) - \hat{y}(t)) \\ y(t) = \sum_{i=1}^{r} \mu_i(\hat{x}(t))C_i\hat{x}(t) \end{cases} \tag{8}$$

The following theorem provides sufficient conditions described as LMI to ensure the asymptotic convergence of the error dynamic:

Theorem 1. *The estimation error converges asymptotically toward zero with decay rate α if there exist matrices $P = P^T \in \mathbb{R}^{n_x \times n_x} > 0$ and $M \in \mathbb{R}^{n_x \times n_y}$ such that the following LMI holds $\forall i = 1, ..., q$:*

$$\mathcal{A}_i{}^T P + P\mathcal{A}_i - MC_i - C_i{}^T M^T + 2\alpha P < 0 \tag{9}$$

The observer gain is given by $L = P^{-1}M$.

Proof. The dynamics of the estimation error $e(t) = x(t) - \hat{x}(t)$ are given as follows:

$$\dot{e}(t) = \underbrace{\sum_{i=1}^{r} \mu_i(x(t))(A_ix(t) + B_iu(t))}_{\Phi_1(x(t))} - \underbrace{\sum_{i=1}^{r} \mu_i(\hat{x}(t))(A_i\hat{x}(t) + B_iu(t))}_{\Phi_1(\hat{x}(t))} - L(\underbrace{\sum_{i=1}^{r} \mu_i(x(t))C_ix(t)}_{\Phi_2(x(t))} - \underbrace{\sum_{i=1}^{r} \mu_i(\hat{x}(t))C_i\hat{x}(t)}_{\Phi_2(\hat{x}(t))}) \tag{10}$$

Using the mean value theorem on the terms Φ_1 and Φ_2, the error dynamics become the following:

$$\dot{e}(t) = \sum_{i=1}^{q} h_i(z(t))(\mathcal{A}_i - LC_i)e(t) \tag{11}$$

To study the stability of the error dynamics, the quadratic Lyapunov function is used:

$$V(t) = e(t)^T Pe(t) \tag{12}$$

The derivative of $V(t)$ with respect to t is as follows:

$$\dot{V}(t) = \sum_{i=1}^{q} h_i(z(t))e^T\Big((\mathcal{A}_i - LC_i)^T P + P(\mathcal{A}_i - LC_i)\Big)e \tag{13}$$

To improve the performance of the estimation, the following decay rate is used:

$$\dot{V}(t) \leq -2\alpha V(t) \tag{14}$$

By substituting (12) and (13) in (14), the following inequality is obtained:

$$\sum_{i=1}^{q} h_i(z(t))e_a^T\Big(\mathcal{A}_i^T P - C_i^T L^T P + P\mathcal{A}_i - PLC_i + 2\alpha P\Big)e_a < 0 \tag{15}$$

The inequality (15) is not linear due to the product of the variables P and L. However, applying the change in variable $M = PL$ provides a solution for achieving the linear stability conditions outlined in Theorem 1. □

In order to solve the inequalities outlined in Theorem 1, we will utilize the Yalmip Toolbox, a well-regarded modeling language in MATLAB designed for formulating optimization problems. This toolbox seamlessly integrates with a variety of solvers, including "Mosek", "SDPT3", and "LMILAB", all recognized for their efficiency in handling LMIs.

In Figure 1, the entire procedure is depicted through a graphical representation:

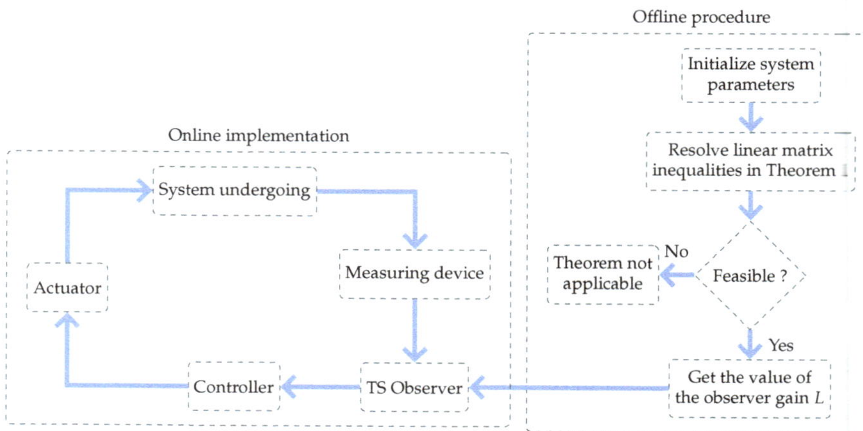

Figure 1. Overall schematic diagram of observer design and implementation.

4. Simulation Results

4.1. Dynamic Model of the System

The three-tank hydraulic system illustrated in Figure 2, based on Guzman's design [17], features three tanks with equal cross-sectional areas S, connected by pipes with areas $S_{p1,2,3}$. Water from a reservoir fills the first and second tanks via pumps P_1 and P_2, with flow rates u_1 and u_2. Valves in each tank manage water release, and the system ensures water levels in the order $x_1 > x_3 > x_2$.

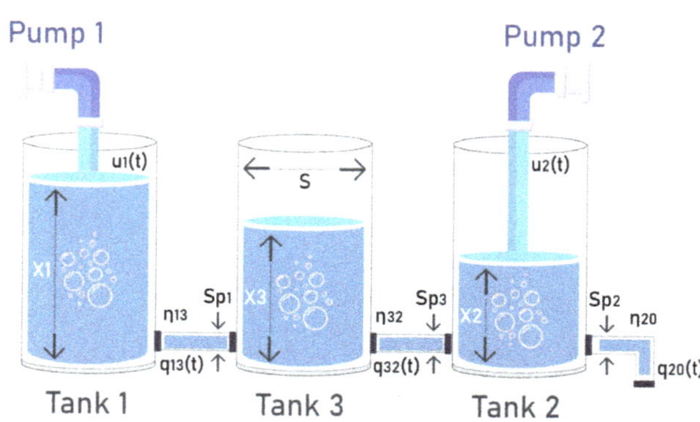

Figure 2. Three-tank hydraulic system.

Let us define $x(t) = \begin{bmatrix} x_1 & x_2 & x_3 \end{bmatrix}^T, u(t) = \begin{bmatrix} u_1 & u_2 \end{bmatrix}^T$ and the premise variables $\xi(x) = \begin{bmatrix} \xi_1(x) & \xi_2(x) & \xi_3(x) \end{bmatrix}^T$. Using these definitions, the system can be represented in the following state-space form:

$$\dot{x}(t) = \frac{1}{S}\begin{bmatrix} -C_1\xi_1(x) & 0 & 0 \\ 0 & -C_2\xi_2(x) & C_3\xi_3(x) \\ C_1\xi_1(x) & 0 & -C_3\xi_3(x) \end{bmatrix} x(t) + \frac{1}{S}\begin{bmatrix} 1 & 0 \\ 0 & 1 \\ 0 & 0 \end{bmatrix} u(t) \tag{16}$$

where

$$\xi_1(x) = \frac{\sqrt{|x_1 - x_3|}}{x_1},\ \xi_2(x) = \frac{\sqrt{x_2}}{x_2},\ \xi_3(x) = \frac{\sqrt{|x_3 - x_2|}}{x_3}$$

$$C_1 = \eta_{13}.S_{p1}.sign(x_1 - x_3).\sqrt{2g}, \quad C_2 = \eta_{20}.S_{p2}.\sqrt{2g}, \quad C_3 = \eta_{32}.S_{p3}.sign(x_3 - x_2).\sqrt{2g}$$

With a gravitational pull of $g = 9.8\ \text{m/s}^2$, the system has discharge coefficients $\eta_{13} = \eta_{32} = 0.456$ and $\eta_{20} = 0.652$. The tubes' cross-sectional areas are $S_{p1} = S_{p3} = 0.5 \times 10^{-4}\ \text{m}^2$ and $S_{p2} = 0.8 \times 10^{-4}\ \text{m}^2$, with all tanks having $S = 154 \times 10^{-4}\ \text{m}^2$. Given the constraint $x_1 > x_3 > x_2$ and these parameters, we derive $C_1 = 1.0094 \times 10^{-4}$, $C_2 = 2.3092 \times 10^{-4}$, and $C_3 = 1.0094 \times 10^{-4}$.

4.2. Observer Design for Three-Tank Hydraulic System

In order to apply Theorem 1, the matrices $\mathcal{A}_i$ and $\mathcal{C}_i$ have to be determined. According to the mean value theorem, $\Phi_1(\xi)$ and its Jacobian $\frac{\partial \Phi_1(\xi)}{\partial x}$ can be defined as follows:

$$\Phi_1(\xi) = \frac{1}{S}\begin{bmatrix} -C_1\xi_1 x_1 \\ -C_2\xi_2 x_2 + C_3\xi_3 x_3 \\ C_1\xi_1 x_1 - C_3\xi_3 x_3 \end{bmatrix}$$

$$\frac{\partial \Phi_1(\xi)}{\partial x} = \frac{1}{S}\begin{bmatrix} \frac{-C_1}{2}\varepsilon_1(x) & 0 & \frac{C_1}{2}\varepsilon_1(x) \\ 0 & \frac{-C_2}{2}\varepsilon_2(x) - \frac{C_3}{2}\varepsilon_3(x) & \frac{C_3}{2}\varepsilon_3(x) \\ \frac{C_1}{2}\varepsilon_1(x) & \frac{C_3}{2}\varepsilon_3(x) & -\frac{C_3}{2}\varepsilon_3(x) - \frac{C_1}{2}\varepsilon_1(x) \end{bmatrix}$$

where the new premise variables $\varepsilon_i(x)$ are given by $\varepsilon_1(x) = \frac{1}{\sqrt{x_1 - x_3}}$; $\varepsilon_2(x) = \frac{1}{\sqrt{x_2}}$; $\varepsilon_3(x) = \frac{1}{\sqrt{x_3 - x_2}}$ and the limits are $L_{\varepsilon 1}(j) = \begin{bmatrix} 2 & 12 \end{bmatrix}$, $L_{\varepsilon 2}(k) = \begin{bmatrix} 2 & 4 \end{bmatrix}$ and $L_{\varepsilon 3}(d) = \begin{bmatrix} 2 & 14 \end{bmatrix}$ for $j, k, d = 1:2$.

By replacing every premise variable by its respective limits in a loop, the matrices $\mathcal{A}_i$ can be obtained as follows for $(i = 1, \ldots, 8)$:

$$\mathcal{A}_i = \frac{1}{S}\begin{bmatrix} \frac{-C_1}{2}L_{\varepsilon 1}(j) & 0 & \frac{C_1}{2}L_{\varepsilon 1}(j) \\ 0 & \frac{-C_2}{2}L_{\varepsilon 2}(k) - \frac{C_3}{2}L_{\varepsilon 3}(d) & \frac{C_3}{2}L_{\varepsilon 3}(d) \\ \frac{C_1}{2}L_{\varepsilon 1}(j) & \frac{C_3}{2}L_{\varepsilon 3}(d) & -\frac{C_3}{2}L_{\varepsilon 3}(d) - \frac{C_1}{2}L_{\varepsilon 1}(j) \end{bmatrix}$$

According to the output equation, which is linear, $\Phi_2(\xi) = Cx(t)$; hence,

$$\frac{\partial \Phi_2(\xi)}{\partial x} = C = \begin{bmatrix} 1 & 0 & 0 \\ 0 & 1 & 0 \end{bmatrix}$$

Therefore, $\mathcal{C}_i = C$.

Solving the LMI in Theorem 1, the following observer matrices are obtained:

$$L = \begin{bmatrix} 0.4840 & 0.9309 \\ -0.9247 & 0.4900 \\ 0.0010 & 0.1682 \end{bmatrix}, P = \begin{bmatrix} 1.1432 & -0.0016 & -0.1489 \\ -0.0016 & 1.1300 & -0.1619 \\ -0.1489 & -0.1619 & 1.8552 \end{bmatrix}$$

4.3. Simulation Validation

The simulation has been validated by considering the initial condition as $x_0 = \begin{bmatrix} 0.08 & 0.06 & 0.07 \end{bmatrix}^T$ and $\hat{x}_0 = \begin{bmatrix} 0.181 & 0.1610 & 0.171 \end{bmatrix}^T$. The system inputs are shown in Figure 3. The tanks levels and their estimation are shown in Figure 4.

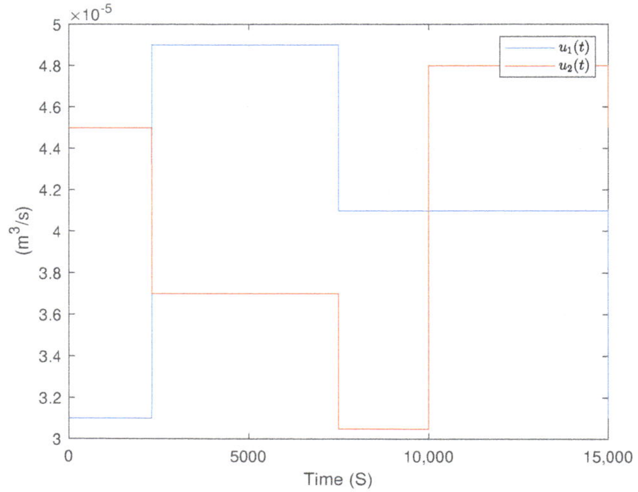

Figure 3. The flow rates of pumps.

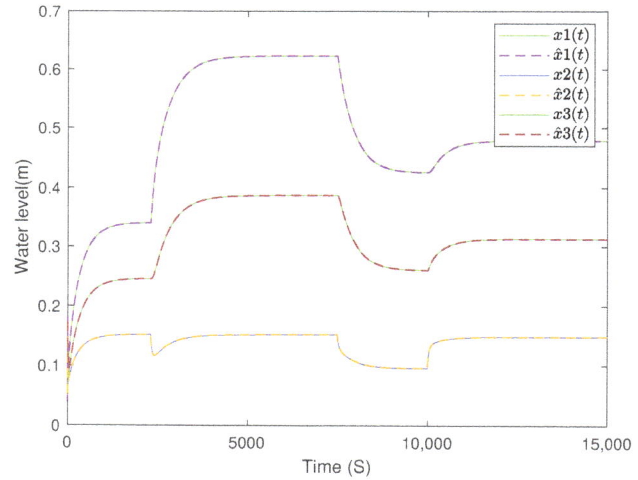

Figure 4. Tank levels and their estimations.

5. Conclusions

This paper introduces a design for the Takagi–Sugeno observer using the mean value theorem, bypassing the commonly used Lipschitz assumption and the $\mathcal{L}_2$ attenuation-based method. While our method adeptly tracks the system's states, it uniquely applies the MVT to systems with nonlinear outputs, contrasting prior works such as [8,9,13,14] that targeted only linear output systems. However, the proposed approach does have limitations, particularly for systems with unknown inputs, directing our future research ambitions. We aim to explore observer designs accommodating unknown inputs and to investigate emerging control system methodologies, like the adaptive fuzzy control for pneumatic active suspensions, as seen in [18]. This paves the way for the Takagi–Sugeno observer's broader applications in complex systems.

Author Contributions: Conceptualization, W.H. and M.Y.H.; methodology, W.H.; software, W.H.; validation, W.H.; investigation, W.H. and M.Y.H.; writing—original draft preparation, W.H.; writing—review and editing, W.H. and A.B.; visualization, W.H. and A.B.; supervision, M.Y.H.; project administration, M.Y.H. All authors have read and agreed to the published version of the manuscript.

Funding: This research received no external funding.

Institutional Review Board Statement: Not applicable.

Informed Consent Statement: Not applicable.

Data Availability Statement: Data are contained within the article.

Conflicts of Interest: The authors declare no conflicts of interest.

References

1. Luenberger, D. An introduction to observers. *IEEE Trans. Autom. Control* **1971**, *16*, 596–602. [CrossRef]
2. Wang, Y.; Zou, L.; Ma, L.; Zhao, Z.; Guo, J. A survey on control for Takagi-Sugeno fuzzy systems subject to engineering-oriented complexities. *Syst. Sci. Control. Eng.* **2021**, *9*, 334–349. [CrossRef]
3. Tanaka, K.; Wang, H.O. *Fuzzy Control Systems Design and Analysis: A Linear Matrix Inequality Approach*; John Wiley & Sons: New York, NY, USA, 2004
4. Louzimi, A.; El Assoudi, A.; Soulami, J.; El Yaagoubi, E. Unknown input observer design for a class of nonlinear descriptor systems: A Takagi-Sugeno approach with Lipschitz constraints. *Nonlinear Anal. Differ. Equ.* **2017**, *5*, 99–116. [CrossRef]
5. Ouzaz, M.; El Assoudi, A. Simultaneous state and fault estimation for Takagi-Sugeno implicit models with Lipschitz constraints. *Int. J. Optim. Control. Theor. Appl.* **2021**, *11*, 100–108. [CrossRef]
6. Boukhlouf, A.; Hammoudi, M.Y.; Saadi, R.; Benbouzid, M.E.H. Hardware-in-the-loop implementation of an unknown input observer for synchronous reluctance motor. *ISA Trans.* **2023**, *133*, 485–494. [CrossRef] [PubMed]
7. Youssef, T.; Chadli, M.; Karimi, H.R.; Wang, R. Actuator and sensor faults estimation based on proportional integral observer for TS fuzzy mod. *El. J. Frankl. Inst.* **2017**, *354*, 2524–2542. [CrossRef]
8. Pan, J.; Nguyen, A.T.; Guerra, T.M.; Sentouh, C.; Wang, S.; Popieul, J.C. Vehicle actuator fault detection with finite-frequency specifications via Takagi-Sugeno fuzzy observers: Theory and experiments. *IEEE Trans. Veh. Technol.* **2022**, *72*, 407–417. [CrossRef]
9. Pan, J.; Nguyen, A.T.; Wang, S.; Deng, H.; Zhang, H. Fuzzy Unknown Input Observer for Estimating Sensor and Actuator Cyber-Attacks in Intelligent Connected Vehicles. *Automot. Innov.* **2023**, *6*, 164–175. [CrossRef]
10. Zemouche, A.; Boutayeb, M.; Bara, G.I. Observers for a class of Lipschitz systems with extension to H_∞ performance analysis. *Syst. Control. Lett.* **2008**, *57*, 18–27. [CrossRef]
11. Ichalal, D.; Arioui, H.; Mammar, S. Observer design for two-wheeled vehicle: A Takagi-Sugeno approach with unmeasurable premise variables. In Proceedings of the 2011 19th Mediterranean Conference on Control & Automation (MED), Corfu, Greece, 20–23 June 2011; pp. 934–939.
12. Phanomchoeng, G.; Rajamani, R.; Piyabongkarn, D. Nonlinear observer for bounded Jacobian systems, with applications to automotive slip angle estimation. *IEEE Trans. Autom. Control* **2011**, *56*, 1163–1170. [CrossRef]
13. Mimoune, K.; Hammoudi, M.Y.; Saadi, R.; Benbouzid, M.; Boukhlouf, A. Real-Time Implementation of Non Linear Observer Based State Feedback Controller for Induction Motor Using Mean Value Theorem. *J. Electr. Eng. Technol.* **2023**, *18*, 615–628. [CrossRef]
14. Allag, M.; Allag, A.; Zeghib, O.; Hamidani, B. Robust h_∞ control based on the mean value theorem for induction motor drive. *J. Control. Autom. Electr. Syst.* **2019**, *30*, 657–665. [CrossRef]
15. Hamidani, B.; Allag, A.; Allag, A.; Zeghib, O. Sensorless non-linear control applied to a PMSM machine based on new extended MVT observer. *J. Electr. Eng. Technol.* **2019**, *14*, 1615–1623. [CrossRef]
16. Mimoune, K.; Hammoudi, M.Y.; Hamdi, W.; Mimoune, S.M. Observer design for Takagi–Sugeno fuzzy systems with unmeasured premise variables: Conservatism reduction using line integral Lyapunov function. *ISA Trans.* **2023**, *142*, 626–634. [CrossRef] [PubMed]
17. Guzman, J.; López-Estrada, F.R.; Estrada-Manzo, V.; Valencia-Palomo, G. Actuator fault estimation based on a proportional-integral observer with nonquadratic Lyapunov functions. *Int. J. Syst. Sci.* **2021**, *52*, 1938–1951. [CrossRef]
18. Ho, C.M.; Ahn, K.K. Adaptive Fuzzy Output Feedback Control Design for Pneumatic Active Suspension with Unknown Dead Zone. *IEEE Access* **2023**, *11*, 66858–66871. [CrossRef]

engineering proceedings

MDPI

Proceeding Paper

Development of a Zigbee-Based Wireless Sensor Network of MEMS Accelerometers for Pavement Monitoring †

Nicky Andre Prabatama [1,*], Pierre Hornych [2], Stefano Mariani [3] and Jean Marc Laheurte [1]

1 Laboratoire Electronique, Systémes de Communication et Microsystémes (ESYCOM), Universite Gustave Eiffel, Centre National de la Recherche Scientifique (CNRS), F-77454 Marne-la-Valle, France; jean-marc.laheurte@univ-eiffel.fr

2 Laboratoire Ausculation, Modélisation, Expérimentation des Infrastructure de Transport (MAST-LAMES), Universite Gustave Eiffel, F-44344 Bouguenais, France; pierre.hornych@univ-eiffel.fr

3 Department of Civil and Environmental Engineering, Politecnico di Milano, 20133 Milano, Italy; stefano.mariani@polimi.it

* Correspondence: nicky-andre.prabatama@univ-eiffel.fr

† Presented at the 10th International Electronic Conference on Sensors and Applications (ECSA-10), 15–30 November 2023; Available online: https://ecsa-10.sciforum.net/.

Abstract: Safety related to pavement ageing is a major issue, as cracks and holes in the road surface can lead to severe accidents. Although pavement maintenance is extremely costly, detecting a deterioration before its surface becomes completely damaged remains a challenge. Current approaches still use wired sensors, which consume a lot of energy and are expensive; further to that, wired sensors may become damaged during installation. To avoid the use of cables, in this work, a prototype of a Zigbee-based wireless sensor network for pavement monitoring was developed and tested in the laboratory. The system consists of a slave sensor and a roadside unit; the slave sensor sends wireless acceleration data to the master, and the master saves the received acceleration dataset in a csv file. Further data processing can be performed in the master on this acceleration dataset. Two laboratory tests were performed for dynamic calibration and simulating five-axle truck pavement displacement. The preliminary results showed that the Zigbee-based wireless sensor network is capable of capturing the required ranges of displacement, acceleration, and frequency. The ADXL354 sensor was found to be the most appropriate accelerometer for this application, with as small as 155 uA power consumption.

Keywords: pavement monitoring; wireless sensor networks; MEMS accelerometers; Zigbee

Citation: Prabatama, N.A.; Hornych, P.; Mariani, S.; Laheurte, J.M. Development of a Zigbee-Based Wireless Sensor Network of MEMS Accelerometers for Pavement Monitoring. *Eng. Proc.* **2023**, *58*, 29. https://doi.org/10.3390/ecsa-10-16236

Academic Editor: Francisco Falcone

Published: 15 November 2023

1. Introduction

Detecting a deterioration in pavements before they become totally damaged remains a costly and very challenging task. Many devices have already been proposed, but most of them still use wired sensors; more and more researchers are currently investigating pavement monitoring systems which are low-cost, low-energy, and wireless. Geophones, accelerometers, and strain sensors are usually used to monitor pavement condition by measuring the strain, displacement, and vertical velocity of the pavement. Many proposed setups still use cables. The monitoring system described in [1] was based on strain gauges, soil pressure gauges used as load cells, thermocouple temperature sensors, and moisture sensors embedded into the pavement. All these sensors are connected to a data logger on the roadside via cables. Geophones are used in the system presented in [2] to measure the vertical velocity at the pavement surface and convert it into vertical displacement (deflection). Other systems are based on MEMS (micro-electromechanical systems) accelerometers [3–9], packaged in a nylon box or covered with resin. These accelerometers are buried into the pavement and then connected to the data logger or master, which is placed next to the road. In the case of the exploitation of data collected with geophones or accelerometers, the measurements are converted into vertical displacements by signal processing using integration [6,8].

On the other hand, thinking of a digital twin of the system, finite element (FE) modeling can be used for the analysis of flexible pavements. In [10], a three-dimensional (3D) FE analysis was carried out on a portion of flexible pavement to predict rut depth under different conditions of temperature, loading, and for different material properties. Another 3D pavement FE model was described in [11] to assess the influence of truck parameters such as wheel set, axle set, vehicle travel speed, and tire pressure on rutting. In this work, a preliminary study was carried out using the state-of-the-art ABAQUS software to model the pavement response and understand the specific effects of moving loads. This analysis enabled us to evaluate the magnitude of the displacement at the surface of the pavement under moving vehicle loads. This was used to define displacement signals for vibrating table tests but also to predict the response in full scale tests carried out on the fatigue Carousel available at Univ. Eiffel in Nantes. The fatigue Carousel is an accelerated pavement testing facility which allows for testing our sensor prototypes under real-life conditions.

In this paper, a novel pavement monitoring system was thus built around a Zigbee-based wireless sensor network prototype and tested in the laboratory. This prototype aims to solve the current problem related to the costs of the monitoring system, getting rid of cables, and heading towards the lowest possible power consumption. MEMS accelerometers were chosen because they are easy to integrate, less costly than other sensors, and consume little energy. Three MEMS accelerometers were selected for a comparative evaluation in the prototype under two laboratory test conditions. The Zigbee communication protocol was chosen primarily because it has the lowest power consumption in both transmit and receive modes. According to the literature, the measured deflection of the pavements should be between about 0.1 mm and 1 mm with a frequency ranging from 0.5 Hz to 20 Hz, resulting in acceleration values from 5 mg to 200 mg. The sensor and system must therefore be able to operate in the low g and low-frequency acceleration ranges.

2. Materials and Methods

2.1. System Architecture and Prototype

The prototype developed was based on the architecture shown in Figure 1a. The on-board unit consisted of an ESP32 Pico-D4 microcontroller(Espressif, Shanghai, China), a Zigbee module (DIGI), and an ADS1115 ADC(Texas Instrument, Dallas, TX, USA). The microcontroller collects the accelerometer data either from the ADS1115 or directly via the I2C communication protocol or SPI for digital accelerometers. It then sends the data directly to the wireless road system using the Zigbee protocol. The road system receives the data from the on-board unit, saves it in a csv file, and then processes it using the digital signal processing algorithm to be implemented. The prototype system, shown in Figure 1b, was developed and tested in the laboratory.

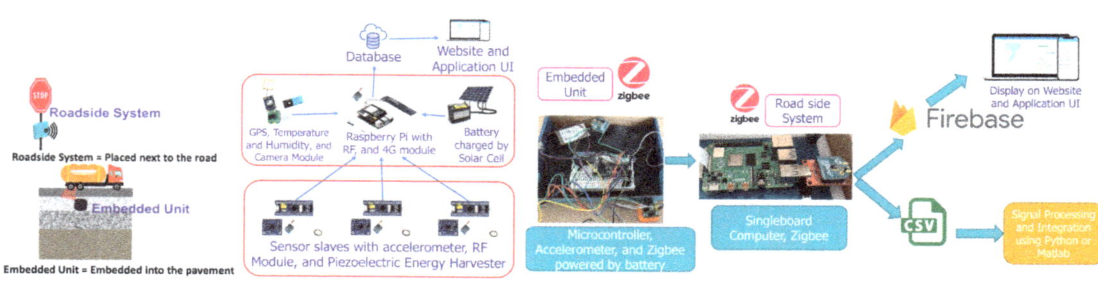

(a) (b)

Figure 1. (**a**) Proposed system architecture; (**b**) first prototype built.

2.2. Finite Element Modeling

A preliminary study using FE modeling was carried out to assess the response of the pavement tested on the fatigue Carousel, subjected to moving loads. The pavement was modeled using the ABAQUS software. All the modeled pavement layers are shown in Figure 2a; the same thicknesses and material properties were adopted as those of the pavement tested on the fatigue Carousel. The pavement section consisted of four layers, from bottom to top, respectively: a 2000 mm thick soil layer; a 750 mm thick granular sub-base layer; a 30 mm thick asphalt base course; and a 50 mm thick asphalt surface course. Infinite elements were used at the side and bottom surfaces of the model to dampen the propagation of waves, avoid spurious reflections, and detect relevant artifacts in terms of high-frequency oscillations.

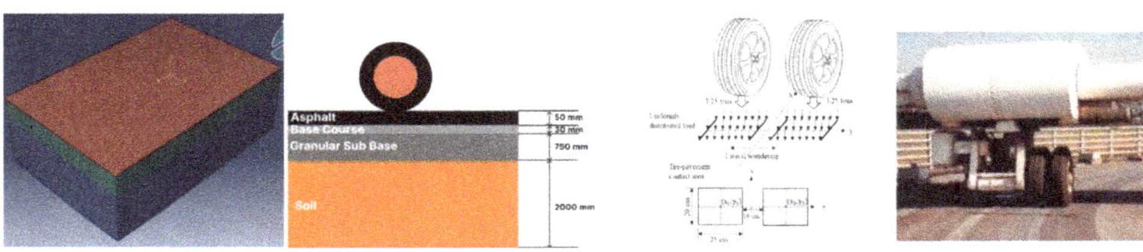

(**a**) (**b**)

Figure 2. (**a**) Three-dimensional structure and layer configuration of the pavement; (**b**) real loading condition.

The pavement of the fatigue Carousel was loaded as depicted in Figure 2b by dual wheels with a total load of 65 kN. The load was assumed to be uniformly distributed over the tire contact areas. A dynamic moving load simulation was carried out to account for the effects of inertia and of the moving tire loads on the pavement. The dynamic simulation was performed using the VDLOAD user subroutine of the FE code to allow the load to move across the structure with a speed of 6 m/s. The adopted values of material properties and layer thicknesses are gathered in Table 1.

Table 1. Values of material properties and layer thickness used in the simulation.

Layer	Young's Modulus *E* (MPa)	Poisson's Ratio	Density (kg/m^3)	Thickness (mm)
Surface course	31,468	0.35	2400	50
Base course	37,554	0.35	2400	30
Granular sub base	160	0.35	2400	750
Soil	95	0.35	2400	2000

2.3. Laboratory Tests

A first laboratory test was carried out using the vibrating pot available in the ESYCOM laboratory. This test enabled the dynamic calibration of each sensor and the collection of important sensor characteristics such as sensitivity, noise, power consumption, and resolution. The shaker used was the LDS V460 equipped with a Bruel & Kjaer PT01 feedback accelerometer (Virum, Denmark) to monitor vibration acceleration. A sinusoidal acceleration signal of low g and different frequencies was used to test three different accelerometers. The test setup is shown in Figure 3b.

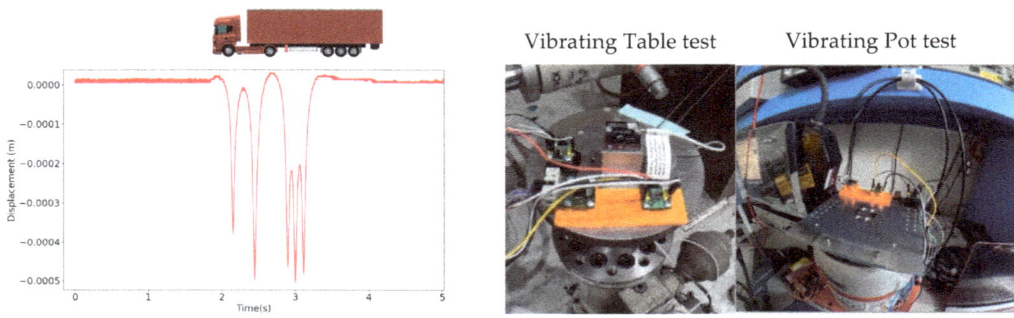

(**a**) (**b**)

Figure 3. (**a**) Typical five-axle truck signal used for the vibrating table test; (**b**) vibrating pot and vibrating table test configurations.

The second laboratory test was carried out using a vibrating table available at the SII laboratory of the Eiffel University in Nantes. The aim of this test was to observe the ability of the MEMS (micro-electromechanical systems) sensors and of the wireless sensor network to measure the signal typical of a five-axle truck passing on the road. The vibrating table, manufactured by Team corporation, can be controlled on the move, as it is equipped with a Messotron LVDT WLC 100 displacement sensor (Seeheim-Jugenheim, Germany). The signal from the displacement transducer was considered to be the reference and used to compare with the displacements resulting from the processing of the data collected by the accelerometers. The test configuration is shown in Figure 3b. A typical signal from a five-axle truck that was used for the laboratory test is shown in Figure 3a. This signal was obtained from a previous experiment carried out by a researcher at Univ. Eiffel in Nantes, which consisted of measuring the pavement response under the loading of a reference truck on a freeway, using geophones and accelerometers, and then converting it into displacement.

While on-board MEMS sensors measure accelerations, it is more relevant, for analyzing the response of the pavement, to determine the displacement of the roadway. Acceleration can be converted into displacement using a two-stage integration. Digital signal processing is therefore required to convert the collected acceleration time history into displacement. Digital signal processing was performed in the master (Raspberry Pi) using python libraries such as numpy, scipy, and matplotlib. The signal processing procedure for converting the raw acceleration signal into displacement was developed with a concept based on [6] which is sketched in Figure 4.

Figure 4. Signal processing method to convert acceleration into displacement.

3. Results and Discussion

3.1. FE Results

Figure 5a shows the variation in the vertical displacement over the top surface of the pavement. The signals correspond to the point where the contour provides the maximum deflection. The corresponding displacement time history at one node along the load path is represented in Figure 5b. The results indicate that a displacement amplitude of approximately 0.35 mm can be expected for the pavement of the Fatigue Carousel under such a load. Thus, during the laboratory tests, a displacement amplitude of this type was adopted to test the system.

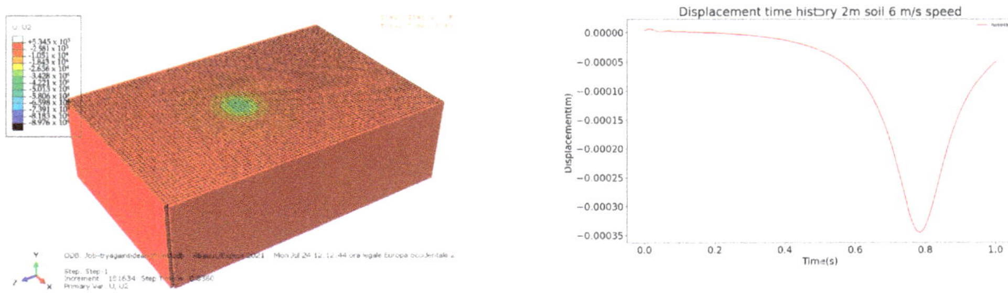

(**a**) (**b**)

Figure 5. (**a**) Contour plot of vertical displacement; (**b**) time history of the vertical displacement at one node.

3.2. Laboratory Test Results

Three MEMS accelerometers were tested: ADXL355, ADXL354, and MS1002. The test results are summarized in Table 2, which shows the output type and sensitivity of each sensor along with the power consumption and noise.

Table 2. Sensor characteristics obtained with the vibrating pot test.

Accelerometer Type	ADXL355	ADXL354	MS1002
Output	Digital Output	Analog Output	Differential Analog Output
Sensitivity	-	384 mV/g	1340 mV/g
Power Consumption	 201.0 μA	 155.2 μA	 12.11 mA
Noise	0.0024 m/s^2	0.022 m/s^2	0.0076 m/s^2

It can be seen that ADXL354 has the lowest power consumption compared with the other two sensors at 155.2 μA, while MS1002 has the highest power consumption at 12.11 mA. ADXL354 has an analog output with a sensitivity of 384 mV/g. Thus, ADXL354 seems to have the best characteristics for the pavement monitoring system as it has the lowest power consumption and good accuracy.

The vibrating table test was carried out at a lower displacement amplitude (0.25 mm) than the signal reported in Figure 5b. This was adopted on purpose to observe whether the accelerometer sensor and the entire system were able to detect very low displacement/acceleration levels, since very low g's are expected to be measured.

The example of raw acceleration signals is shown in Figure 6a. The results of the test with a displacement amplitude of 0.25 mm at a speed of 45 km/h are shown in Figure 6b–d. The result is compared with the signal from the vibration table's integrated displacement sensor, shown in blue. The results show that the proposed digital processing method was capable of converting the measured acceleration value into displacement. However, an error is still visible when comparing the signal-processed displacement with the displacement sensor data.

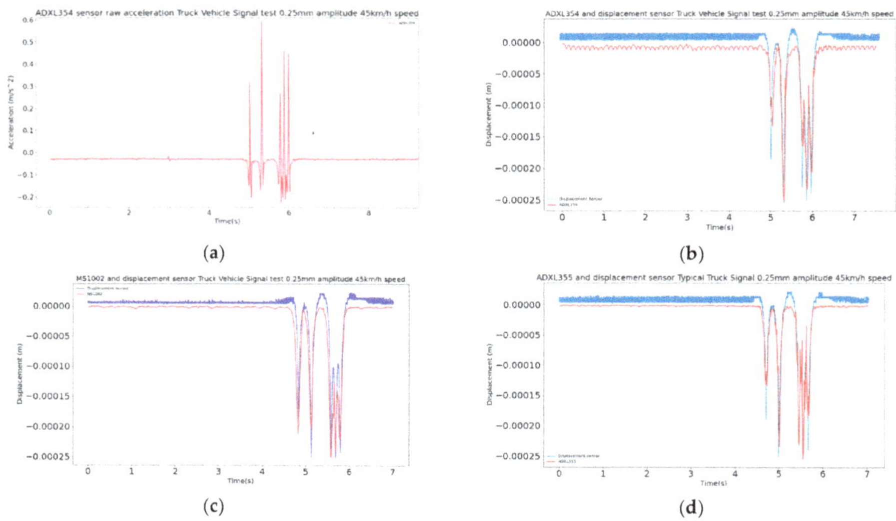

Figure 6. (**a**) Example of raw acceleration signals from the test of the three sensors; (**b**–**d**) displacement conversion for all the sensors.

4. Conclusions

In this work, a prototype Zigbee-based wireless sensor network using a MEMS accelerometer for pavement monitoring was proposed, built, and tested in the laboratory. A preliminary numerical study was carried out to assess the kind of deflection to be expected under the moving load induced by a full-scale fatigue test. In this case, the prototype developed was used to comparatively assess the performance of three MEMS accelerometers by means of two laboratory tests, namely a dynamic calibration and the simulation of the displacement produced by a five-axle truck. A signal processing step was also carried out to convert the accelerations into displacements. The results of the laboratory test showed that the system is capable of monitoring pavement deflections. Also, it has been demonstrated that the ADXL354 is the most suitable accelerometer due to its low power consumption and good accuracy. The system will shortly be tested under real-life conditions at the accelerated pavement testing facility.

Author Contributions: Formal analysis, N.A.P.; writing—original draft preparation, J.M.L. and S.M.; writing—review and editing, N.A.P., J.M.L. and S.M.; visualization, N.A.P.; supervision, J.M.L., S.M. and P.H.; All authors have read and agreed to the published version of the manuscript.

Funding: This project has received funding from the European Union's Horizon 2020 research and innovation program under the Marie Sklodowska-Curie COFUND grant agreement No 101034248.

Institutional Review Board Statement: Not applicable.

Informed Consent Statement: Not applicable.

Data Availability Statement: The data that support the findings of this study are available from the corresponding author upon reasonable request.

Conflicts of Interest: The authors declare no conflicts of interest.

References

1. Xue, W.; Wang, L.; Wang, D.; Druta, C. Pavement Health Monitoring System Based on an Embedded Sensing Network. *J. Mater. Civ. Eng.* **2014**, *26*, 04014072. [CrossRef]
2. Duong, N.S.; Blanc, J.; Hornych, P.; Menant, F.; Lefeuvre, Y.; Bouveret, B. Monitoring of pavement deflections using geophones. *Int. J. Pavement Eng.* **2018**, *21*, 1103–1113. [CrossRef]

3. Ye, Z.; Xiong, H.; Wang, L. Collecting comprehensive traffic information using pavement vibration monitoring data. *Comput. Civ. Infrastruct. Eng.* **2019**, *35*, 134–149. [CrossRef]
4. Bajwa, R.; Coleri, E.; Rajagopal, R.; Varaiya, P.; Flores, C. Pavement performance assessment using a cost-effective wireless accelerometer system. *Comput. Civ. Infrastruct. Eng.* **2020**, *35*, 1009–1022. [CrossRef]
5. Ye, Z.; Yan, G.; Wei, Y.; Zhou, B.; Li, N.; Shen, S.; Wang, L. Real-Time and Efficient Traffic Information Acquisition via Pavement Vibration IoT Monitoring System. *Sensors* **2021**, *21*, 2679. [CrossRef] [PubMed]
6. Bahrani, N.; Blanc, J.; Hornych, P.; Menant, F. Alternate Method of Pavement Assessment Using Geophones and Accelerometers for Measuring the Pavement Response. *Infrastructures* **2020**, *5*, 25. [CrossRef]
7. Levenberg, E. Inferring Pavement Properties using an Embedded Accelerometer. *Int. J. Transp. Sci. Technol.* **2012**, *1*, 229–246. [CrossRef]
8. Arraigada, M.; Partl, M.N.; Angelone, S.M.; Martinez, F. Evaluation of accelerometers to determine pavement deflections under traffic loads. *Mater. Struct.* **2008**, *42*, 779–790. [CrossRef]
9. Ho, C.-H.; Snyder, M.; Zhang, D. Application of Vehicle-Based Sensing Technology in Monitoring Vibration Response of Pavement Conditions. *J. Transp. Eng. Part B Pavements* **2020**, *146*, 04020053. [CrossRef]
10. Asim, M.; Ahmad, M.; Alam, M.; Ullah, S.; Iqbal, M.J.; Ali, S. Prediction of Rutting in Flexible Pavements using Finite Element Method. *Civ. Eng. J.* **2021**, *7*, 1310–1326. [CrossRef]
11. Wang, Y.; Lu, Y.J.; Si, C.D.; Sun, T.C. Finite Element Analysis for Rutting Prediction of Asphalt Concrete Pavement under Moving Wheel Load. *Int. J. Simul. Model.* **2017**, *16*, 229–240. [CrossRef]

engineering proceedings

MDPI

Proceeding Paper

Multi-Modal Human Action Segmentation Using Skeletal Video Ensembles †

James Dickens * and Pierre Payeur

School of Electrical Engineering and Computer Science, University of Ottawa, Ottawa, ON K1N 6N5, Canada; ppayeur@uottawa.ca
* Correspondence: jdick088@uottawa.ca
† Presented at the 10th International Electronic Conference on Sensors and Applications (ECSA-10), 15–30 November 2023; Available online: https://ecsa-10.sciforum.net/.

Abstract: Beyond traditional surveillance applications, sensor-based human action recognition and segmentation responds to a growing demand in the health and safety sector. Recently, skeletal action recognition has largely been dominated by spatio-temporal graph convolutional neural networks (ST-GCN), while video-based action segmentation research successfully employs 3D convolutional neural networks (3D-CNNs), as well as vision transformers. In this paper, we argue that these two inputs are complementary, and we develop an approach that achieves superior performance with a multi-modal ensemble. A multi-task GCN is developed that can predict both frame-wise actions as well as action timestamps, allowing for the use of fine-tuned video classification models to classify action segments and achieve refined predictions. Symmetrically, a multi-task video approach is presented that uses a video action segmentation model to predict framewise labels and timestamps, augmented with a skeletal action classification model. Finally, an ensemble of segmentation methods for each modality (skeletal, RGB, depth, and infrared) is formulated. Experimental results yield 87% accuracy on the PKU-MMD v2 dataset, delivering state-of-the-art performance.

Keywords: action segmentation; skeletal action recognition; video understanding; deep learning; computer vision

1. Introduction

Automated analysis of human behavior from video data has the potential to empower modern health, safety, and surveillance applications. Real-world video data mostly comes in the form of untrimmed streams, hence the motivation for developing action segmentation algorithms, where the goal is to segment a video and classify these segments, providing a dense descriptor of the sequence. Using modern deep learning techniques, human activity understanding has undergone rapid progress, wherein two prominent streams of research and application have emerged. The first such stream is the task of skeletal action understanding, involving both classification and segmentation. Deep learning approaches use large, labelled datasets [1,2], training models on sequences of skeletal frames of a fixed topology to predict action classes. To date, the most prominent approaches for skeletal action recognition employ spatio-temporal graph convolutional networks [3–6].

The second stream is the established research area of video action understanding using RGB videos [7]. Action classification networks employing 3D-CNNs [8–10] as well as vision transformers [11] currently outperform other approaches. For the task of action segmentation, the multi-stage temporal convolutional network (MS-TCN) framework [12,13] has generated impressive results, in which frame-wise features are generated in an offline process by a 3D-CNN's backbone, after which each frame is classified with a temporal convolutional network, making use of dilated convolutions and a refinement stage. Further improvements to the MS-TCN framework can be achieved using timestamp prediction

Citation: Dickens, J.; Payeur, P. Multi-Modal Human Action Segmentation Using Skeletal Video Ensembles. *Eng. Proc.* **2023**, *58*, 30. https://doi.org/10.3390/ecsa-10-16257

Academic Editor: Stefano Mariani

Published: 15 November 2023

with the action segmentation refinement framework (ASRF) model [14], addressing the over-segmentation problem in which too many action segments are predicted.

In this paper, it is argued that these two streams can form complementary features for the goal of action segmentation by using a multi-modal ensemble approach. Skeletal action sequences, while sometimes noisy and dependent on the accuracy of the underlying pose estimation algorithm as well as depth map quality, are invariant to texture, lighting, clothing choice, and the layout of the scene. By contrast, while video understanding algorithms have to deal with the aforementioned factors, they allow for the incorporation of objects and surfaces within the scene to guide predictions. Consider the scenario of an individual interacting with an object, absent in a skeletal sequence, but present in a video. The presence of the object cannot be inferred by skeletal models, only the types of movements associated with the person interacting with said object.

To this end, multi-modal action segmentation approaches are developed in this work to boost segmentation accuracy. The novel contributions of this paper are the following:

- A skeletal segmentation approach that predicts both frame-wise labels and timestamps, using a video classification model to refine predictions.
- A video segmentation approach which predicts both frame-wise labels and timestamps, using a skeletal action recognition model to refine predictions.
- An ensemble of video and skeletal models, each employing their own timestamp-based refinements, to predict frame-wise labels.

Experimental results on the PKU-MMD v2 [2] dataset validate the effectiveness of these multi-modal ensembles, and state-of-the-art results are achieved using a skeletal–video action segmentation ensemble.

2. Methods

In this section, the three proposed approaches are introduced, respectively a skeletal segmentation and video classification ensemble, a video segmentation and skeletal classification ensemble, as well as a skeletal–video segmentation ensemble. To maximize performance, the skeletal predictions in the approaches presented consist of a weighted ensemble of predictions made from joint, bone, joint motion, and bone motion inputs as described by Shi et al. [4] and employed in state-of-the art approaches [6]. Further, video models consist of ensembles of 3 models trained on RGB, infrared (IR), and depth inputs. All ensemble weights are obtained through experimentation. An overview of the first two approaches is given in Figure 1.

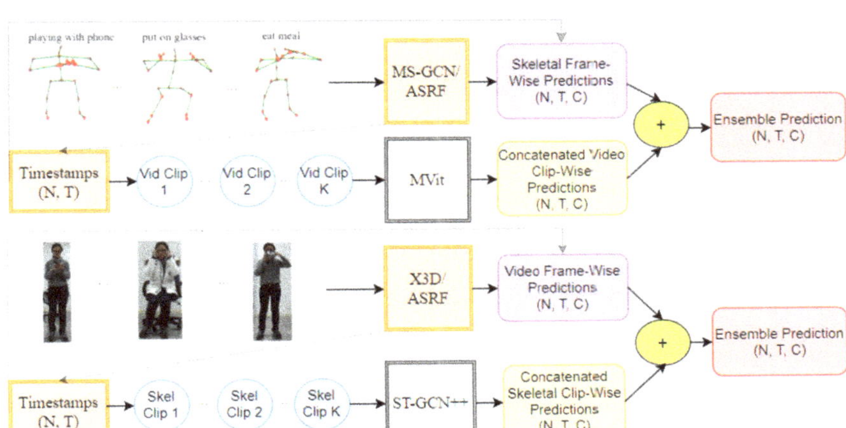

Figure 1. An overview of the proposed methods from Sections 2.1 and 2.2. Top, the skeletal segmentation and video classification ensemble. Bottom, the skeletal classification and video segmentation ensemble.

2.1. Skeletal Segmentation and Video Classification Ensemble

The skeletal segmentation backbone chosen follows the multi-stage spatial-temporal graph convolutional network (MS-GCN) [15], developed by Filtjens et al. It consists of 10 layers of spatio-temporal graph convolutions [3], where the temporal convolutions employ increasingly large dilation factors, originally set to powers of 2^i for i = 1, ... 9, but chosen as non-strictically increasing values determined through experimentation in this work.

The network head follows the ASRF proposed by Ishikawa et al. [14]. The output of the backbone is pooled along the spatial/vertex axis and fed into a softmax layer to obtain a tensor of shape (N, T, C).

Where N is the batch size, T is the number of frames, and C is the number of classes. Next, a series of refinement layers is enacted. Each refinement layer consists of a dimensionality reduction using a 1×1 convolution, followed by a sequence of dilated temporal convolutions of increasing factor (the same as the backbone), followed by a 1x1 convolution with C channels that is then input to a SoftMax layer. Note that the input of each refinement layer is the logits obtained from the previous layer. During training, each refinement stage's output contributes to the segmentation loss, while during inference the output of the last refinement layer yields the model prediction.

A second branch is added that predicts action boundaries, or rather timestamps, a tensor of shape (N, T), where a value larger than 0.5 indicates the presence of an action boundary. The skeletal segmentation predictions can be refined by using a weighted sum of a majority vote of predictions within timestamps with the original frame-wise predictions. In this work, we augment the ASRF framework to be multimodal. Using the predicted timestamps, the corresponding frames in the RGB, depth and IR videos are interpolated to a fixed length and fed to a video classifier to obtain class predictions, which are then expanded to the length of the clip. A weighted sum of predictions yields the final prediction. The loss for training the augmented MS-GCN model is given by

$$L(p, gt) = \sum_{n,i} CE_{seg}\left(p_{n,i}^{seg}, g_{n,i}^{seg}\right) + \lambda_1 \sum_{n,i} CE_{ts}\left(p_{n,i}^{ts}, g_{n,i}^{ts}\right) + \lambda_2 S(p) \quad (1)$$

where p is the model prediction, g is the ground truth, CE_{seg} is the cross-entropy loss applied framewise, n indexes over batch elements, and i indexes over refinement layers. CE_{ts} is a weighted binary cross entropy loss applied to timestamp predictions. S is the smoothing loss given by Abu Farha et al. [12] which is used to prevent over-segmentation by enforcing consistent predictions across local windows of time, and the parameters λ_1, λ_2 weight the timestamp loss and smoothing penalty. The loss is averaged over batch size and refinement layers. During inference, non-maximal suppression with a window size of 5 is applied to timestamps to prevent near-duplicate boundaries.

The video model chosen for classification is the multi-scale vision transformer MVit [11], given its strong performance on the Kinetics-400 dataset [7]. Pre-trained weights are obtained from the PyTorch video (https://pytorchvideo.readthedocs.io/en/latest/model_zoo.html (accessed 14 November 2023)) repository, obtained from training on Kinetics-400. Infrared and depth models are modified such that the weights of the input layer, consisting of a 3D convolution, are averaged across the three input channels to be able to input a single channel for efficiency. The MVit networks are trained on action-specific classification clips rather than long sequences exhibiting various actions, interpolated to 32 frames. Denoting y as MSGCN(x) for a skeletal input x, the overall segmentation prediction is given by

$$\gamma_1\, ASRF(y) + \gamma_2 Concat(MVit(c_1, \ldots c_K)) \quad (2)$$

where ASRF denotes the majority vote refinement within skeletal segmentation predicted timestamp windows, and $Concat(MVit(c_1, \ldots c_K))$ is the concatenation along the temporal axis of ordered window predictions from the video classification model for K windows, and γ_1, γ_2 are weights. The class labels per window are broadcast to the window.

2.2. Skeletal Classification and Video Segmentation Ensemble

For comparison, we experimented with a segmentation ensemble where roles of the skeletal network and video network are swapped from Section 2.1. Following standard protocol [12], video features are computed in an offline process by a 3D-CNN. We employ X3D-large [10] without a network head, using crops of 16 frames of spatial resolution (312, 312), where backbone features are pooled and reshaped to a feature vector of dimension 3888 (432 channels, 3×3 spatial resolution) per frame. The X3D model is pre-trained first on Kinetics-400, then fine-tuned on classification training clips specific to the PKU-MMD v2 dataset split for which features were generated. The network head is the same as the skeletal model in Section 2.1, predicting both frame-wise labels and timestamps. The skeletal classification selected was the ST-GCN++ model developed by Duan et al. [6], which innovates relative to the original ST-GCN by employing a set of parallel convolutions of different kernel sizes and dilation factors in the temporal layer of the graph convolutions, concatenating the result to obtain more discriminative temporal features. The same loss is used as described by equation (1), where the predictions at inference for X3D frame-wise features y is given by

$$\gamma_1 \, ASRF(y) + \gamma_2 Concat(STGCN(c_1, \ldots c_K)) \tag{3}$$

2.3. Skeletal and Video Segmentation Ensemble

An alternative skeletal and video ensemble is proposed as a weighted sum of predictions from the single modality segmentation methods given in Sections 2.1 and 2.2, i.e., pure skeletal segmentation and pure video segmentation models. Since both models predict timestamps, two separate types of window-based refinement were experimented within early trials. With a *single* refinement stage, each modality's timestamp weighted predictions are summed before applying non-maximal suppression, after which refinement occurs. With a *double* refinement scheme, each modality's predicted timestamps are used in separate specific refinements, after which the weighted sum of overall predictions is used as the final network output. Early experiments showed that the double-refinement scheme performed the best by a slight margin, and therefore, it is the approach considered for experiments in this work.

3. Experimental Validation

3.1. Experiments

All experiments were conducted on the PKU-MMD v2 dataset [2], which provides sequences labelled framewise for 42 different action classes in four different modalities: skeletal keypoints for single or two person actions obtained from a Kinect v2 depth camera, RGB videos of resolution 1920×1080, and depth and infrared maps of resolution 512×424. There are 13 subjects and three different camera views, where video lengths are between 1 and 2 min long with an average of seven action instances per video. There are two splits used for testing, the cross subject (cs) in which different subjects are used for testing and training, and the cross view (cv) setup, in which different camera views are used for training versus testing. The cs split has a 773/233 train/test split, while the cv split has a 669/337 split. Due to the small number of training samples, we augmented the training set with examples from the PKU-MMD v1 dataset for the classes that overlap with v2. During training, frames with labels of 0 (the unknown class) were removed for both skeletal and video data types, while during inference, the predictions for these frames were ignored, as is standard practice in semantic segmentation. For all networks trained, the optimizer chosen was SGD with Nesterov momentum, weight decay of 0.0005, and a cosine annealing learning rate scheduler, with all code implemented in PyTorch and trained on a single NVidia RTX 4090 GPU.

For both skeletal segmentation and classification, data augmentation consisted of light rotations about a random axis and small scaling [6]. The batch size for skeletal segmentation was four, trained for 300 epochs, with an initial learning rate of 0.1. The batch size for skeletal classification was 16, with random uniform sampling employed [6] to yield a clip of fixed length of 150, trained for 100 epochs, using an initial learning rate of 0.01. During inference, fixed uniform sampling was used.

With respect to video segmentation, no data augmentation was used, where a batch size of four was used, trained for 250 epochs with an initial learning rate of 0.001. For video classification, clips were resized in an offline process to (224, 224) resolution using bicubic interpolation with anti-aliasing, after a center-wise crop. Depth and IR videos were converted to grayscale videos using min/max normalization. Random right/left flipping was used for all video modalities, and light color jittering applied to entire videos randomly during RGB video training. During training and inference, videos were interpolated along the temporal axis to 32 frames. A batch size of six was used, trained for 30 epochs, with an initial learning rate of 0.001.

3.2. Results and Discussion

The results of all models proposed, including skeletal and video segmentation baselines are shown in Tables 1 and 2 for the cross subject and cross view splits, respectively. On both splits, the skeletal and video segmentation ensemble exhibits the best trade-off between accuracy and segment F1@50 measures; although, skeletal action segmentation exhibits more of an over segmentation issue, affecting the F1@50 score negatively, presumably due to noise in the skeletons causing sudden spurious predictions. The video segmentation models struggled to predict timestamps as successfully and so the corresponding boost of adding a skeletal classifier was not successful on the cv split, and marginal on the cs split. In both splits, the video classification added on top of a skeletal segmentation showed modest improvements in accuracy and segment F1@50 score.

In Table 3, a comparison to the MS-GCN model originally presented by Filtgens et al. [15] is shown. Note that in this original formulation, only the joint representation is used as an input, hence a large performance increase is achieved using the four-stream representation (joints, bones, joint motion, bone motion), before additionally considering the benefit of video predictions. The overall performance increase is quite large with an 18.5% increase in accuracy and a 17.1% increase in the segment F1@50 score.

Table 1. Experimental results on the PKU-MMD v2 cross subject split.

Model	Framewise Acc (%)	Segment F1@50	Timestamp—F1
MS-GCN/ASRF only (Skeletal Segmentation)	84.0	60.7	77.7
X3D/ASRF only (Video Segmentation)	82.2	73.1	64.1
MS-GCN/ASRF + MVit (Skeletal Segmentation + Video Classification)	84.6	60.7	77.7
X3D/ASRF + ST-GCN++ (Video Segmentation + Skeletal Classification)	82.6	**74.4**	64.1
MS-GCN/ASRF + X3D/ASRF (Skeletal Segmentation + Video Segmentation)	**87.0**	68.7	**79.0**

Table 2. Experimental results on the PKU-MMD v2 cross view split.

Model	Framewise Acc (%)	Segment F1@50	Timestamp—F1
MS-GCN/ASRF only (Skeletal Segmentation)	88.4	68.8	77.4
X3D/ASRF only (Video Segmentation)	76.4	64.4	61.9
MS-GCN/ASRF + MVit (Skeletal Segmentation + Video Classification)	**89.2**	70.1	77.4
X3D/ASRF + ST-GCN++ (Video Segmentation + Skeletal Classification)	75.1	65.8	61.9
MS-GCN/ASRF + X3D/ASRF (Skeletal Segmentation + Video Segmentation)	88.4	**74.5**	**79.5**

Table 3. Comparing the best performing model to other work from the literature.

Model	Framewise Acc (%)	Segment F1@50
MS-GCN [15]	68.5	51.6
MS-GCN/ASRF + X3D/ASRF (Skeletal Segmentation + Video Segmentation)	**87.0**	**68.7**

4. Conclusions

In conclusion, in this work we have presented a novel approach to multi-modal human action segmentation, with experimentally validated results showing improved accuracy on the task of action segmentation using both skeletal inputs as well as video inputs (RGB, depth, and infrared). As a potential avenue for future work, the use of lightweight 3D-CNNs as well as compact skeletal graph convolutional networks could be explored for the goal of real-time, or near real-time, multi-modal action segmentation in order to facilitate real-world applications that are time sensitive.

Author Contributions: Conceptualization, J.D. and P.P.; methodology, software, validation, formal analysis, J.D.; investigation, J.D., P.P.; resources, P.P.; writing—original draft preparation, J.D.; writing—review and editing, P.P.; supervision, project administration, funding acquisition, P.P. All authors have read and agreed to the published version of the manuscript.

Funding: This research was supported in part by MITACS Accelerate and NSERC Discovery grants.

Institutional Review Board Statement: This research only involved data from the public domain and no experiments on human subjects were conducted. Ethical review and approval were waived by the Office of Research Ethics and Integrity of the University of Ottawa due to project falling under Article 2.5 of the Tri-Council Policy Statement: Ethical Conduct for Research Involving Humans (TCPS 2), and therefore did not require Research Ethics Board review.

Informed Consent Statement: Not applicable.

Data Availability Statement: Not applicable.

Conflicts of Interest: The authors declare no conflict of interest.

References

1. Liu, J.; Shahroudy, A.; Perez, M.; Wang, G.; Duan, L.-Y.; Kot, A.C. NTU RGB+D 120: A Large-Scale Benchmark for 3D Human Activity Understanding. *IEEE Trans. Pattern Anal. Mach. Intell.* **2019**, *42*, 2684–2701. [CrossRef] [PubMed]
2. Liu, C.; Hu, Y.; Li, Y.; Song, S.; Liu, J. PKU-MMD: A Large Scale Benchmark for Continuous Multi-Modal Human Action Understanding. In Proceedings of the Workshop on Visual Analysis in Smart and Connected Communities, Mountain View, CA, USA, 23 October 2017; pp. 1–8.
3. Yan, S.; Xiong, Y.; Lin, D. Spatial Temporal Graph Convolutional Networks for Skeleton-Based Action Recognition. In Proceedings of the AAAI Conference on Artificial Intelligence, New Orleans, LA, USA, 2–7 February 2018; Volume 32. [CrossRef]

4. Shi, L.; Zhang, Y.; Cheng, J.; Lu, H. Skeleton-Based Action Recognition With Multi-Stream Adaptive Graph Convolutional Networks. *IEEE Trans. Image Process.* **2020**, *29*, 9532–9545. [CrossRef] [PubMed]
5. Chen, Y.; Zhang, Z.; Yuan, C.; Li, B.; Deng, Y.; Hu, W. Channel-wise topology refinement graph convolution for skeleton-based action recognition. In Proceedings of the IEEE/CVF International Conference on Computer Vision, Montreal, QC, Canada, 10–17 October 2021; pp. 13359–13368.
6. Duan, H.; Wang, J.; Chen, K.; Lin, D. PYSKL: Towards Good Practices for Skeletal Action Recognition. In Proceedings of the 30th ACM International Conference on Multimedia, Lisbon, Portugal, 10–14 October 2022; pp. 7351–7354.
7. Carreria, J.; Zisserman, A. Quo Vadis, Action Recognition? A new Model and the Kinetics Dataset. In Proceedings of the 2017 IEEE Conference on Computer Vision and Pattern Recognition (CVPR), Honolulu, HI, USA, 21–26 July 2017; pp. 4724–4733.
8. Tran, D.; Bourdev, L.; Fergus, R.; Torresani, L.; Paluri, M. Learning Spatiotemporal Features with 3D Convolutional Networks. In Proceedings of the 2015 IEEE International Conference on Computer Vision (ICCV), Santiago, Chile, 7–13 December 2015; pp. 4489–4497.
9. Tran, D.; Wang, H.; Torresani, L.; Ray, J.; LeCun, Y.; Paluri, M. A closer look at spatiotemporal convolutions for action recognition. In Proceedings of the 2018 IEEE/CVF Conference on Computer Vision and Pattern Recognition (CVPR), Salt Lake City, UT, USA, 18–23 June 2018; pp. 6450–6459.
10. Feichtonhofer, C. X3D: Expanding Architectures for Efficient Video Recognition. In Proceedings of the 2020 IEEE/CVF Conference on Computer Vision and Pattern Recognition (CVPR), Seattle, WA, USA, 13–19 June 2020.
11. Fan, H.; Xiong, B.; Mangalam, K.; Li, Y.; Yan, Z.; Malik, J.; Feichtenhofer, C. Multiscale vision transformers. In Proceedings of the IEEE/CVF International Conference on Computer Vision, Montreal, QC, Canada, 10–17 October 2021.
12. Abu Farha, Y.; Gall, J. MS-TCN: Multi-Stage Temporal Convolutional Network for Action Segmentation. In Proceedings of the 2019 IEEE/CVF Conference on Computer Vision and Pattern Recognition (CVPR), Long Beach, CA, USA, 15–20 June 2019.
13. Li, S.; Abu Farha, Y.; Liu, Y.; Cheng, M.M.; Gall, J. MS-TCN++ Multi-Stage Temporal Convolutional Network for Action Segmentation. *arXiv* **2020**, arXiv:2006.09220. [CrossRef] [PubMed]
14. Ishikawa, Y.; Kasai, S.; Aoki, Y.; Kataoka, H. Alleviating Over-segmentation Errors by Detection Action Boundaries. In Proceedings of the 2021 IEEE Winter Conference on Applications of Computer Vision (WACV), Waikoloa, HI, USA, 3–8 January 2021; pp. 2321–2330.
15. Filtjens, B.; Vanrumste, B.; Slaets, P. Skeleton-Based Action Segmentation with Multi-Stage Spatial-Temporal Graph Convolutional Neural Networks. *arXiv* **2022**, arXiv:2202.01727. [CrossRef]

engineering proceedings

MDPI

Proceeding Paper

YOLO-NPK: A Lightweight Deep Network for Lettuce Nutrient Deficiency Classification Based on Improved YOLOv8 Nano †

Jordane Sikati [1] and Joseph Christian Nouaze [2,3,*]

1 R&D Center, Guinee Biomedical Maintenance, Bentourayah, Conakry 3137, Guinea; jordanesikati@pusan.ac.kr
2 Department of Electronics Engineering, Pusan National University, Busan 46241, Republic of Korea
3 R&D Center, CAS Corporation, Headquarters, Yangju 11415, Republic of Korea
* Correspondence: krxsange@pusan.ac.kr; Tel.: +82-1094-011-957

† Presented at the 10th International Electronic Conference on Sensors and Applications (ECSA-10), 15–30 November 2023; Available online: https://ecsa-10.sciforum.net/.

Abstract: When it comes to growing lettuce, specific nutrients play vital roles in its growth and development. These essential nutrients include full nutrients (FN), nitrogen (N), phosphorus (P), and potassium (K). Insufficient or excess levels of these nutrients can have negative effects on lettuce plants, resulting in various deficiencies that can be observed in the leaves. To better understand and identify these deficiencies, a deep learning approach is employed to improve these tasks. For this study, YOLOv8 Nano, a lightweight deep network, is chosen to classify the observed deficiencies in lettuce leaves. Several enhancements to the baseline algorithm are made, the backbone is replaced with VGG16 to improve the classification accuracy, and depthwise convolution is incorporated into it to enrich the features while keeping the head unchanged. The proposed network, incorporating these modifications, achieved superior classification results with a top-1 accuracy of 99%. This method outperformed other state-of-the-art classification methods, demonstrating the effectiveness of the approach in identifying lettuce deficiencies. The objective of this research was to improve the baseline algorithm to complete the classification task with a top-1 accuracy above 85%, a FLOP inferior to 10G, and classification latency below 170 ms per image.

Keywords: lettuce nutrient deficiency; classification; deep learning; YOLOV8 Nano

Citation: Sikati, J.; Nouaze, J.C. YOLO-NPK: A Lightweight Deep Network for Lettuce Nutrient Deficiency Classification Based on Improved YOLOv8 Nano. *Eng. Proc.* **2023**, *58*, 31. https://doi.org/10.3390/ecsa-10-16256

Academic Editor: Stefano Mariani

Published: 15 November 2023

1. Introduction

Lettuce (Lactuca sativa) is a widely cultivated leafy vegetable with significant economic and dietary importance. Adequate nutrient supply, particularly of nitrogen (N), phosphorus (P), and potassium (K), is essential for optimal lettuce growth and quality. Nitrogen is a primary component of chlorophyll and essential for photosynthesis. Nitrogen deficiency in lettuce results in stunted growth, pale leaves, and reduced leaf size, affecting the overall yield and nutritional content of lettuce, as well as its susceptibility to diseases [1]. Phosphorus is crucial for energy transfer in plants and plays a key role in root development. Lettuce plants deficient in phosphorus exhibit poor root growth, delayed maturity, and smaller heads. Phosphorus deficiency can also lead to decreased nutrient uptake, negatively impacting overall plant health [2]. Potassium is vital for maintaining plant turgor, enzyme activation, and disease resistance. Lettuce plants with potassium deficiency display wilted leaves, necrosis at the leaf margins, and reduced resistance to pathogens [3]. Potassium deficiency can reduce lettuce's marketability due to decreased visual appeal [4].

This paper is structured as follows: Section 2 discusses previous research on lettuce deficiencies, Section 3 presents the materials and methods, Section 4 discusses the experimental results of the proposed method, and finally, Section 5 provides the conclusions of this article and discusses future work.

2. Related Work

In recent years, there has been growing interest in the development of deep learning-based approaches for the diagnosis and early detection of nutrient deficiencies in lettuce plants. Watchareeruetai et al. introduced, in 2018, an image analysis method for identifying nutrient deficiency in plants based on their leaves using convolutional neural networks [5], setting the stage for subsequent research in this area. In addition, a deep convolutional neural network for the image-based diagnosis of nutrient deficiencies in plants grown through aquaponics is proposed by Taha et al. in 2022 [6]. Furthermore, Lu et al., in 2023, introduced a lettuce pant trace-element-deficiency symptom identification method via machine vision methods [7]. Collectively, these studies represent significant contributions to the field of lettuce NPK deficiency detection and illustrate the increasing reliance on deep learning methodologies for precision agriculture applications. Continued research in this area is crucial to developing sustainable agricultural practices that can meet the increasing demand for high-quality lettuce. In this way, a deep learning approach called YOLO-NPK based on YOLOv8 Nano Classification algorithms [8,9] is employed in this study, to classify those deficiencies. The objective of this research is to enhance the baseline algorithm to achieve a top-1 accuracy above 85%, FLOP inferior to 10G, and classification latency below 170 ms per image.

3. Materials and Methods

3.1. Data Acquisition and Augmentation Strategy

The lettuce NPK dataset [10], sourced from Kaggle, comprises images representing various lettuce deficiency categories alongside Fully Nutritional (FN) lettuce samples. The dataset includes images categorized as FN with 12 images, Nitrogen-deficient (-N) with 58 images, Phosphorus-deficient (-P) with 66 images, and Potassium-deficient (-K) with 72 images. Captured in a controlled environment for hydroponic lettuce deficiency project, the dataset aims to facilitate the development of a system capable of recognizing lettuce deficiencies from images. This system would not only identify deficiencies in hydroponics but also find applications in diverse fields. Figure 1 provide a visual representation of the dataset samples, showcasing a fully nutritional, as well as nitrogen, phosphorus, and potassium deficiency [10].

Figure 1. The dataset samples. (**a**) Fully nutritional lettuce (FN); (**b**) nitrogen deficiency (-N); (**c**) phosphorus deficiency (-P); (**d**) potassium deficiency (-K).

Augmentation techniques were used to increase the training set and the validation set. The following pre-processing was applied to each image: auto-orientation of pixel data (with EXIF-orientation stripping) and resizing to 640 × 640 (Stretch). Furthermore, successive augmentation was applied to create augmented versions of each source image (50% probability of horizontal flip, 50% probability of vertical flip, and equal probability of one of the following 90-degree rotations: none, clockwise, counter-clockwise, upside-down, randomly cropped between 0 and 20 percent of the image, and random shear of between $-15°$ and $+15°$ horizontally and $-15°$ and $+15°$ vertically). In total, 3192 samples were obtained from augmentation, with -K 1175, -N 975, -P 847, and FN 195. Therefore the dataset was split into 70% for the training and 30% for the validation.

3.2. VGG16 (Visual Geometry Group 16) Feature Extractor

VGG16 (Visual Geometry Group 16) is a convolutional neural network (CNN) architecture for deep learning that was developed by the Visual Geometry Group at the University of Oxford [11]. It is part of the VGG family of models and is known for its simplicity and effectiveness in image classification tasks. It consists of 16 weight layers, including 13 convolutional layers and 3 fully connected layers. The architecture uses 3 × 3 convolutional filters with a stride of 1, and 2 × 2 max-pooling layers with a stride of 2. Also, it is characterized by its deep architecture, with small 3 × 3 convolutional filters stacked multiple times. This depth helps the network learn complex hierarchical features from images. The network uses 3 × 3 convolutional filters with a stride of 1 and "same" padding, which means the spatial dimensions of the feature maps do not change after convolutions. Rectified Linear Units (ReLUs) are used as the activation function in the network, helping with the vanishing gradient problem and improving training.

3.3. Depthwise Convolution

Depthwise convolution is a specific type of convolutional operation used in deep learning and convolutional neural networks (CNNs). It is a fundamental building block for various lightweight and efficient neural network architectures, particularly those designed for mobile and edge devices [12]. Depthwise convolution differs from standard convolution in how it processes input channels. In a standard convolution operation, a kernel (also called a filter) slides through the entire input volume, considering all input channels simultaneously. In contrast, in depthwise convolution, each input channel is convolved with a separate kernel. This means that if you have k input channels and k separate kernels, each kernel is responsible for convolving with its corresponding input channel. It significantly reduces the number of parameters in the model compared to standard convolution. This reduction in parameters can lead to models that are more memory-efficient and that compute faster, making them suitable for resource-constrained environments. They are often used in conjunction with pointwise convolution (1 × 1 convolution). This combination is referred to as a depthwise separable convolution. In depthwise separable convolution, the depthwise convolution layer is followed by a 1 × 1 pointwise convolution layer. The pointwise convolution combines the information from the separate channels produced by the depthwise convolution. Lastly, it maintains the spatial dimensions (width and height) of the input, but it can change the number of channels (depth). This contrasts with standard convolution, which can also change spatial dimensions. So, it is particularly efficient when dealing with low-level features in an image, where inter-channel correlations are not as significant. Separating the channels reduces computational complexity.

3.4. YOLOv8 (You Only Look Once Version 8)

The YOLO (You Only Look Once) series [13–18] refers to a family of real-time object detection models that have been widely used in computer vision and deep learning. YOLO was initially introduced by Redmon et al. [9] in 2016 and has since seen several iterations, each with improvements and enhancements [9]. The primary idea behind YOLO is to perform object detection in a single forward pass of a neural network, making it very efficient and suitable for real-time applications. YOLOv8, developed by Ultralytics, represents the most recent iteration of the YOLO series. As an advanced and state-of-the-art model, it expands upon the achievements of its predecessors by introducing novel features and enhancements, resulting in elevated levels of performance, adaptability, and resource efficiency. YOLOv8 boasts comprehensive support for a wide spectrum of vision-based artificial intelligence tasks, encompassing detection, segmentation, pose estimation, tracking, and classification. This versatility empowers users to harness the diverse capabilities of YOLOv8 across a multitude of applications and domains.

3.5. YOLO-NPK

To enhance classification accuracy, a VGG16 feature extractor is integrated into the backbone of YOLOv8n-cls (YOLOv8 Nano Classification). Furthermore, depthwise convolution is introduced within the feature extractor to facilitate feature reuse and empower the deep network to extract more complex and richer features. The diagram below provides an overview of the proposed approach to classifying lettuce deficiencies. The proposed feature extractor receives a 640 × 640 RGB deficient lettuce image as an input and extracts richer features. The classification head fuses the learned feature and performs a classification task, returning the classification result as the output. The schematic representation in Figure 2 delineates the architecture of YOLO-NPK, providing a visual guide to the components and their interactions. In this illustration, Conv signifies the convolutional layer, DW represents depthwise convolution, MP denotes the max-pooling layer, and nc stands for the number of classes. It's important to note that the proposed feature extractor replaces the original backbone of YOLOv8n-cls, while the classification head remains unaltered. This visual scheme aims to elucidate the structural intricacies of our approach, aiding in a comprehensive understanding of the YOLO-NPK framework.

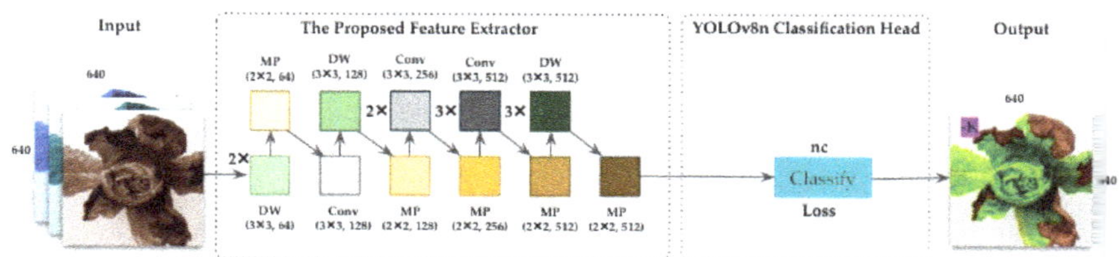

Figure 2. The architecture of YOLO-NPK. Conv, DW, MP, and nc, respectively, stand for convolution, depthwise convolution, max-pooling layer, and number of classes. The original backbone of YOLOv8n-cls was replaced with the proposed feature extractor, and the classification head remained unchanged.

4. Results and Discussion

4.1. Experimental Setup

The experiments were carried out on a computer equipped with the following specifications: an 11th Generation Intel® Core™ i5-11400H processor with 64-bit architecture, equipped with a dual-core CPU, and running at 2.70 GHz. Additionally, the computer was equipped with an NVIDIA GeForce RTX 3050 GPU. The model received input images sized at 640 × 640 pixels. However, due to constraints on GPU memory, the batch size was set to 8 during training. The training process spanned 116 epochs and commenced with an initial learning rate of 0.01, which was later adjusted to a final learning rate of 0.1. Moreover, the following specific hyperparameters were set: a momentum of 0.937 and a weight decay of 0.0005. During the warmup epoch, warmup momentum, and warmup bias learning rate stages, the values were configured at 3.0, 0.8, and 0.1, respectively. The optimizer employed for training the models was Stochastic Gradient Descent (SGD). Data augmentation techniques, such as mosaic, paste-in, and scaling, were used proportionally while training the deep network to avoid unbalanced classes. The early stop mechanism was employed to overcome overfitting.

In the context of classification accuracy, *top*-1 *accuracy* refers to the proportion of correctly classified samples where the model's top prediction matches the true label. It can be mathematically expressed as follows:

$$Top1\ Accuracy = \frac{Number\ of\ Correct\ Predictions}{Total\ Number\ of\ Predictions} \times 100 \quad (1)$$

In this expression, *Number of Correct Predictions* is the count of instances where the model's top prediction matches the true class labels, and *Total Number of Predictions* is the total number of instances or samples in the dataset. The result is typically expressed as a percentage to represent the accuracy rate. *Top-1 accuracy* is a common metric used to evaluate the performance of classification models, where only the highest-confidence prediction is considered for each sample.

4.2. Ablation Study

Several components of the YOLOv8n-cls backbone were modified to obtain the desired results. The overall structure of the backbone was replaced by the VGG16 feature to improve the classification accuracy, and the depthwise convolutional layers were inserted along the feature extractor to allow efficient memory computation and better reuse of features. These operations showed interesting improvement. Table 1 provides details on these diverse modifications.

Table 1. Ablation study on different modifications of YOLO-NPK.

VGG16	Depthwise Convolution	Top-1 Accuracy (%)	FLOPs (G)	CPU Latency (ms)
		93	3.3	19.8
✔		97.5	14.5	68.3
	✔	95.2	2.4	18.2
✔	✔	99	9.2	64.1

4.3. Classification Performance

The performance of YOLO-NPK was measured on the validation set, which represents all the classes. Notably, it shows acceptable results in terms of classification. The model performs efficiently on the FN set and achieves good classification results on other classes (-N, -P, and -K). To gain a comprehensive understanding of the model's overall performance and the intricacies of class-wise classification, refer to Figure 3, which presents a confusion matrix. For a detailed illustration of the model's classification output for each class, consult Figure 4 below.

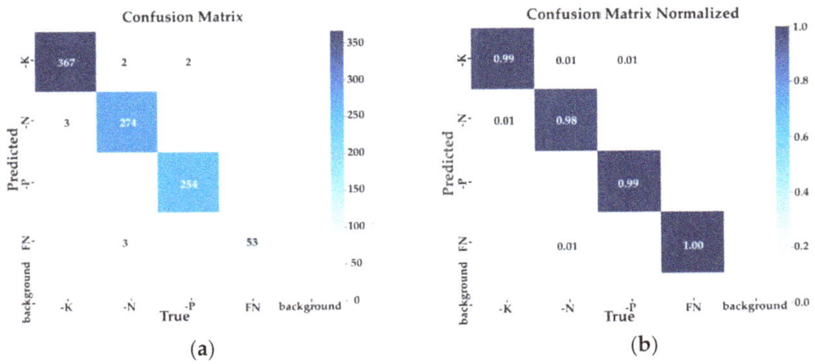

Figure 3. The confusion matrix of YOLO-NPK. (**a**) Confusion matrix. (**b**) Confusion matrix normalized. True represents the ground truth in the dataset, predicted is the classification result, and the background is the images that were missed by the model. This proves the learning capability of the proposed method. More details are provided in Table 2.

(a) (b) (c) (d)

Figure 4. The classification output of YOLO-NPK. (**a**) Fully nutritional lettuce (FN); (**b**) phosphorus deficiency (-P); (**c**) nitrogen deficiency (-N); (**d**) potassium deficiency (-K).

Table 2. Classification performance of YOLO-NPK. FN, -N, -P, and -K, respectively, represent fully nutritional, nitrogen-deficient, phosphorus-deficient, and potassium-deficient lettuce.

Classes	Images	Correctly Classified		Falsely Classified		Missed	
		Count	Rate	Count	Rate	Count	Rate
FN	53	53	100%	0	0%	0	0%
-N	279	274	98.21%	5	1.79%	0	0%
-P	256	254	99.22%	2	0.78%	0	0%
-K	370	367	99.19%	3	0.81%	0	0%

4.4. Comparison of State-of-the-Art Methods

The proposed method, YOLO-NPK, was compared with different state-of-the-art methods. The proposed method shows better classification accuracy. The top-1 accuracy reached 99%, The FLOP 9.2G, and the classification latency per image 64.1 ms. This is in line with the guidelines established before the experiments (top-1 accuracy above 85%, FLOP under 10G, and latency below 170 ms). The other methods satisfied the FLOP and latency conditions but could not fulfil the top-1 accuracy expectation, proving the efficiency and robustness of the proposed model. Table 3 gives details of these comparisons.

Table 3. Comparison of the state-of-the-art method.

Methods	Images Size	Top-1 Accuracy (%)	FLOPs (G)	CPU Latency (ms)
SVM	640	85.3	12	141.6
VGG16	640	87.9	15.2	170.3
MobileNetV2	640	82.5	3.4	41.6
ShuffleNetv2	640	81.6	2.1	30.8
YOLOV8n-cls	640	93	3.3	19.8
YOLO-NPK	640	99	9.2	64.1

5. Conclusions and Future Work

This study introduces YOLO-NPK, a lightweight deep neural network tailored to lettuce deficiency classification, building upon the foundation of YOLOv8 Nano Classification. This research aimed to enhance the baseline algorithm by introducing a custom feature extractor aligned with the study's needs. This goal was successfully met, achieving a top-1 accuracy exceeding 85%, maintaining a FLOP count under 10G, and ensuring a CPU latency below 170 ms per image, meeting the predefined objectives. Future plans involve integrating this solution into more complex systems for smart farming applications.

Author Contributions: Conceptualization, J.S. and J.C.N.; validation, J.S. and J.C.N.; investigation, J.S. and J.C.N.; data curation, J.S. and J.C.N.; formal analysis, J.S. and J.C.N.; methodology, J.S. and J.C.N.; software, J.S. and J.C.N.; visualization, J.S. and J.C.N.; supervision, J.C.N.; writing—original draft preparation, J.S.; writing—review and editing, J.S. and J.C.N.; project administration, J.C.N. All authors have read and agreed to the published version of the manuscript.

Funding: This research received no external funding.

Institutional Review Board Statement: Not applicable.

Data Availability Statement: The Lettuce NPK dataset used in this project was provided by Kaggle. The dataset is available in [10].

Conflicts of Interest: The authors declare no conflicts of interest. CAS Corporation and Guinee Biomedical Maintenance have no conflict of interest.

References

1. Muñoz-Huerta, R.F.; Guevara-Gonzalez, R.G.; Contreras-Medina, L.M.; Torres-Pacheco, I.; Prado-Olivarez, J.; Ocampo-Velazquez, R.V. A Review of Methods for Sensing the Nitrogen Status in Plants: Advantages, Disadvantages and Recent Advances. *Sensors* **2013**, *13*, 10823–10843. [CrossRef] [PubMed]
2. Khan, F.; Siddique, A.B.; Shabala, S.; Zhou, M.; Zhao, C. Phosphorus Plays Key Roles in Regulating Plants' Physiological Responses to Abiotic Stresses. *Plants* **2023**, *12*, 2861. [CrossRef] [PubMed]
3. Nouaze, J.C.; Kim, J.H.; Jeon, G.R.; Kim, J.H. Monitoring of Indoor Farming of Lettuce Leaves for 16 Hours Using Electrical Impedance Spectroscopy (EIS) and Double-Shell Model (DSM). *Sensors* **2022**, *22*, 9671. [CrossRef] [PubMed]
4. Davis, J.L.; Armengaud, P.; Larson, T.R.; Graham, I.A.; White, P.J.; Newton, A.C.; Amtmann, A. Contrasting Nutrient-Disease Relationships: Potassium Gradients in Barley Leaves Have Opposite Effects on Two Fungal Pathogens with Different Sensitivities to Jasmonic Acid. *Plant Cell Environ.* **2018**, *41*, 2357–2372. [CrossRef] [PubMed]
5. Watchareeruetai, U.; Noinongyao, P.; Wattanapaiboonsuk, C.; Khantiviriya, P.; Duangsrisai, S. Identification of Plant Nutrient Deficiencies Using Convolutional Neural Networks. In Proceedings of the 2018 International Electrical Engineering Congress (iEECON), Krabi, Thailand, 7–9 March 2018; pp. 1–4. [CrossRef]
6. Taha, M.F.; Abdalla, A.; ElMasry, G.; Gouda, M.; Zhou, L.; Zhao, N.; Liang, N.; Niu, Z.; Hassanein, A.; Al-Rejaie, S.; et al. Using Deep Convolutional Neural Network for Image-Based Diagnosis of Nutrient Deficiencies in Plants Grown in Aquaponics. *Chemosensors* **2022**, *10*, 45. [CrossRef]
7. Lu, J.; Peng, K.; Wang, Q.; Sun, C. Lettuce Plant Trace-Element-Deficiency Symptom Identification via Machine Vision Methods. *Agriculture* **2023**, *13*, 1614. [CrossRef]
8. Jocher, G.; Chaurasia, A.; Qiu, J. YOLO by Ultralytics 2023. Available online: https://github.com/ultralytics/ultralytics/blob/main/CITATION.cff (accessed on 5 September 2023).
9. Ultralytics Home. Available online: https://docs.ultralytics.com/ (accessed on 5 September 2023).
10. Lettuce NPK Dataset. Available online: https://www.kaggle.com/datasets/baronn/lettuce-npk-dataset (accessed on 5 September 2023).
11. Simonyan, K.; Zisserman, A. Very Deep Convolutional Networks for Large-Scale Image Recognition 2015. *arXiv* **2014**, arXiv:1409.1556.
12. Chollet, F. Xception: Deep Learning with Depthwise Separable Convolutions. In Proceedings of the 2017 IEEE Conference on Computer Vision and Pattern Recognition (CVPR), Honolulu, HI, USA, 21–26 July 2017; pp. 1800–1807.
13. Redmon, J.; Divvala, S.; Girshick, R.; Farhadi, A. *You Only Look Once: Unified, Real-Time Object Detection*; IEEE Computer Society: Washington, DC, USA, 2016; pp. 779–788.
14. Redmon, J.; Farhadi, A. YOLO9000: Better, Faster, Stronger. In Proceedings of the 2017 IEEE Conference on Computer Vision and Pattern Recognition (CVPR), Honolulu, HI, USA, 21–26 July 2017; pp. 6517–6525.
15. Redmon, J.; Farhadi, A. YOLOv3: An Incremental Improvement. *arXiv* **2018**, arXiv:1804.02767.
16. Ultralytics/Yolov5: V2.0 2020. Available online: https://zenodo.org/records/3958273/ (accessed on 5 September 2023).
17. Li, C.; Li, L.; Jiang, H.; Weng, K.; Geng, Y.; Li, L.; Ke, Z.; Li, Q.; Cheng, M.; Nie, W.; et al. YOLOv6: A Single-Stage Object Detection Framework for Industrial Applications. *arXiv* **2022**, arXiv:2209.02976.
18. Wang, C.-Y.; Bochkovskiy, A.; Liao, H.-Y.M. YOLOv7: Trainable Bag-of-Freebies Sets New State-of-the-Art for Real-Time Object Detectors. *arXiv* **2022**, arXiv:2207.02696.

engineering proceedings

MDPI

Proceeding Paper

Time Series Modelling and Predictive Analytics for Sustainable Environmental Management—A Case Study in El Mar Menor (Spain) †

Rosa Martínez *, Ivan Felis, Mercedes Navarro and J. Carlos Sanz-González

Centro Tecnológico Naval y del Mar, 30320 Fuente Álamo, Spain; ivanfelis@ctnaval.com (I.F.); mercedesnavarro@ctnaval.com (M.N.); jcarlossanz@ctnaval.com (J.C.S.-G.)
* Correspondence: rosamartinez@ctnaval.com; Tel.: +34-968-19-75-21

† Presented at the 10th International Electronic Conference on Sensors and Applications (ECSA-10), 15–30 November 2023; Available online: https://ecsa-10.sciforum.net/.

Abstract: In this study on data science and machine learning, time series analysis plays a key role in predicting evolving data patterns. The Mar Menor, located in the Region of Murcia, represents an urgent case due to its unique ecosystem and the challenges it faces. This paper highlights the need to study the environmental parameters of the Mar Menor and to develop accurate predictive models and a standardised methodology for time series analysis. These parameters, which include water quality, temperature, salinity, nutrients, chlorophyll, and others, show complex temporal variations influenced by different activities. Advanced time series models are used to gain insight into their behaviour and project future trends, facilitating effective conservation and sustainable development strategies. Models such as SARIMA and LSTM stand out as valid for predicting the environmental patterns of the Mar Menor.

Keywords: time series; statistical models; machine learning; ARIMA; seasonality; LSTM

Citation: Martínez, R.; Felis, I.; Navarro, M.; Sanz-González, J.C. Time Series Modelling and Predictive Analytics for Sustainable Environmental Management—A Case Study in El Mar Menor (Spain). *Eng. Proc.* **2023**, *58*, 32. https://doi.org/10.3390/ecsa-10-16133

Academic Editor: Stefano Mariani

Published: 15 November 2023

1. Introduction

The Mar Menor is a coastal lagoon in the Region of Murcia (Spain) that faces a series of major environmental and ecological problems, which has generated the need to analyse and understand its evolution, as well as its indicators trend over time. Time series analysis in the context of the Mar Menor provides valuable information on the changes and dynamics of this lagoon. These data provide key information for data collection in the management and conservation of this ecosystem, as well as for the implementation of protection and restoration measures. However, time series analysis present distinctive challenges. They can be complex and influenced by factors as diverse as seasonal cycles, weather events, and human activities. In addition, there may be irregularities, missing data, and noise that make time series difficult to interpret and model. In this context, we aim to address these problems and provide an enhanced understanding of time series dynamics and existing patterns.

In the field of data science and machine learning, time series analysis plays a crucial role in studying and predicting data that evolve over time. The main objective of time series analysis is to understand its performance and predict its evolution, but there are a variety of approaches and algorithms available; different models have different assumptions, characteristics, and capabilities; and their performance can vary significantly.

Therefore, there is a need to identify a standard process for selecting predictive models appropriate to a time series characteristic, allowing the best approach to be identified for each situation, maximising model performance, and minimising prediction error. For this reason, time series characteristics that may have an influence on predictive models' performance were analysed, including trend, seasonality, and time dependence.

Several approaches can be used for this purpose, ranging from classical statistical models such as autoregressive (AR), moving average (MA), autoregressive moving average (ARMA), autoregressive integrated moving average (ARIMA), or autoregressive integrated moving average with seasonality component (SARIMA) models, to models such as the Facebook Prophet or recurrent neural networks (RNNs), concretely, long short-term memory (LSTM) models.

The dataset employed in this paper comes from the Mar Menor Data Web, It consist in data derived from different monitoring stations along the Mar Menor lagoon, with a time interval of five years, from 2017 to 2022.

2. Materials and Methods

2.1. Mar Menor Dataset

The data on Mar Menor's parameters provide essential information for management and conservation decisions in this ecosystem, as well as for implementing appropriate protection and restoration measures. From the Mar Menor Data Web, the downloaded data included a pretreatment as interpolation, which made these data easier to process. The parameters selected to study are: chlorophyll (mg/L), salinity (PSU), oxygen levels (mg/L), phycoerythrin (ppm), water temperature (°C), and transparency (m).

The time interval of these historical series was about 5 years, from 2017 to 2022. Data were extracted from different monitoring stations scattered throughout the Mar Menor and were subsequently standardised by the supplier to a common grid, as shown in Figure 1; OISMA (Oficina de Impulso Socioeconómico del Medio Ambiente) stations are shown in blue, and the Servicio de Pesca stations are shown in red.

Figure 1. OISMA and Servicio de Pesca stations where measurements were taken.

2.2. Time Series and Machine Learning Models

Time series are sequential observations recorded at regular intervals and analysed for patterns or components such as trend or seasonality; in this context, the development of accurate and effective predictive models is essential to obtain reliable results. As mentioned in the Introduction, two approaches of time series analysis were evaluated to study the behaviour of different environmental parameters of the Mar Menor: statistical and machine learning models.

- Statistical models

Autoregressive models, moving average models, and a combination of the two were used. AR(p) models calculate future values with a linear combination of past values (p), which are determined by the partial autocorrelation function, where p is the order of the process indicating the number of previous time steps in the time series that are used to predict the future value of the series. In MA(q) models, the current value of a time series depends only on a small number of past values. This model calculates current values by first determining the average of past errors, achieved by summing them. Subsequently, these averaged values are multiplied by their respective coefficients to obtain the final results. Here, the model order (q) indicates the errors used to obtain the current value. A combination of the properties of the AR and MA processes was considered in which the stationarity of the time series was assumed. The resulting process is stochastic and stationary, called ARMA(p, q) [1]. In addition, there is an "integrated" version of a stationary series, called ARIMA(p,d,q); this model is considered stationary after differentiation. These are the most general classical models in time series forecasting, Where the parameter d represents the differencing order, which is the number of times the data series is differenced to achieve stationarity. Also, SARIMA models consider seasonal patterns and improve forecasting accuracy, and it is necessary to realise a deseasonalisation or seasonal difference (denoted by SARIMA(p,d,q)(P,D,Q), where P, D, and Q represent the seasonal autoregressive, the differencing order, and the moving average order, respectively [2]. Lastly, the Facebook Prophet model, based on the fitting curve technique of the Bayesian model, is appropriate when there is a large seasonality, and it is robust against missing data or trend variations. This model is a non-linear autoregressive additive model, with observations recorded hourly, daily, and monthly over a period of one year or more [3].

- Machine learning models

A recurrent neural network (RNN) is a type of artificial neural network that is specifically used for preprocessing sequential data or time series. These networks are designed to learn from new data and are distinguished by their ability to 'remember' past inputs. This memory informs their decision-making process, affecting both the intake of inputs and the generation of outputs. In fact, RNN results depend on the sequence of past elements, allowing time dependencies in the data to be captured. In this paper, a long short-term memory (LSTM) algorithm, a type of RNN with an input layer, an intermediate layer, and an output layer, was used to introduce different time series as input and to train the network with these data [4].

2.3. *Methodology*

The standardised process was based on a systematic and objective approach. Clear criteria and relevant evaluation metrics were used. A methodology was followed to guide users through the various steps, from initial exploration and data preprocessing to the selection and tuning of appropriate predictive models.

This methodology is structured into five phases: data cleaning and visualization, pattern identification, data transformation for model tuning, model selection and explanation of patterns, and finally, model implementation. In the first phase, a detailed examination of the time series was carried out with the aim of identifying trends and the possible missing data; thus, correction techniques such as interpolation or rolling averaging were used where necessary, and they did not affect the subsequent analysis. Erroneous data were removed, while original data providing useful information were kept. In the pattern identification phase, the Dickey–Fuller test assessed stationarity, assuming the null hypothesis that the series is non-stationary and the alternative hypothesis of stationarity. Partial autocorrelation analyzed seasonality. Transformation techniques, including differentiation and deseasonalization, were applied based on identified characteristics. Following the analysis of all time series, information criteria methods like Akaike (AIC) and Schwarz Bayesian (BIC) guided the selection of the most suitable model for each series, prioritizing those with the lowest AIC and BIC values. Finally, once the best model was selected, predictive models were applied to the series. For this, the dataset for the time series was split into training and testing sets in an 85/15 ratio. Predictions for statistical models were made using a predictive horizon of 7 and the training set was updated at each time step. Meanwhile, both the Facebook Prophet and LSTM models made predictions with a predictive horizon of 7 days on the entire 15% test set. Lastly, in order to assess the fitting of the models, several error metrics were implemented: root mean square error (RMSE), mean absolute error (MAE), and mean absolute percentage error (MAPE).

3. Application

Table 1 presents the results of the Mar Menor dataset. For clarity and relevance, only the phycoerythrin (PE) and water temperature (T) parameters are displayed as examples of the method's implementation. In the data cleaning and visualization step, out of 1462 data points, 4.7% were removed due to being outliers, as shown in Table 1. This removal was part of the data processing which included interpolation, revealing that missing or atypical data did not significantly impact the results.

Table 1. Size, range, and outliers of Mar Menor datasets.

Dataset	Total Data	Range	Outliers
Chlorophyll	975	06/2017–01/2020	47
Salinity	2041	04/2017–10/2022	0
Oxygen	1737	04/2017–12/2021	0
PE	487	07/2021–10/2022	22
Temperature	974	04/2017–11/2019	0
Transparency	2252	09/2016–10/2022	0

After data cleaning, pattern identification involved two key analyses. The first was the Dickey–Fuller test, which treated non-stationarity as the null hypothesis and stationarity as the alternative. The test was conducted with a 95% confidence threshold; a p-value above 0.05 implies rejection of the null hypothesis, indicating non-stationarity of the series. The second analysis focused on seasonality, using partial autocorrelation, where values above 0.5 were deemed significant. The outcomes of both analyses are detailed in Table 2.

Table 2. Seasonality and stationarity for the different datasets.

Dataset	*p*-Value	Correlation Value
Chlorophyll	0.065	0.558
Salinity	0.067	0.824
Oxygen	0.055	0.692
PE	0.232	0.371
Temperature	0.072	0.818
Transparency	0.061	0.807

Based on the established criteria, it can be concluded that none of the datasets are stationary. Moreover, with the exception of the phycoerythrin (PE) data, there is an absence of seasonality in the other datasets. Subsequently, differentiation or deseasonalization techniques were employed to enhance the data's compatibility with the models. This approach aimed to select the most appropriate model, thereby increasing estimation accuracy and reducing error rates.

To determine the best model, i.e., the model with the lowest value of AIC and BIC, we assessed different model fits by making combinations of the hyperparameters p and q, varying them from 0 to 4. Table 3 presents the models obtained for each parameter and the lowest AIC and BIC values calculated. Thereby, predictions were made using these statistical models, in addition to the Facebook Prophet and LSTM models, which were applied to the dataset as well.

Table 3. The most appropriate statistical models for the Mar Menor datasets based on the AIC and BIC.

Dataset	Model	AIC	BIC
Chlorophyll	SARIMA(2,1,1) (0,1,1)	393.866	417.483
Salinity	SARIMA(3,1,2) (0,1,1)	−6482.535	−6443.294
Oxygen	SARIMA(2,1,3) (0,1,1)	−3023.798	−2985.705
PE	ARIMA(2,1,1)	−2418.08	−2401.08
Temperature	SARIMA(2,1,0) (0,1,1)	−1686.45	−1667.05
Transparency	SARIMA(1,1,2) (0,1,1)	−4900.312	−4871.730

4. Results

Table 4 displays the outcomes of the prediction models. As previously noted, these models, including statistical models, Facebook Prophet, and LSTM, were used to make predictions over a 7-day horizon, applying 15% of the data as test data. The error metrics—RMSE, MAE, and MAPE—are detailed in Table 4. Additionally, Figure 2 visualizes these results, highlighting the predictions made using the most effective model for each dataset.

Table 4. Error metrics for different models and datasets.

Dataset	Evaluation	Statistical Model	Facebook Prophet Model		LSTM Model	
		Horizon 7	*Horizon 7*	*15%*	*Horizon 7*	*15%*
Chlorophyll	RMSE	**2.420**	4.377	5.621	9.63	6.431
	MAE	1.465	3.519	4.146	1.608	**1.311**
	MAPE	0.243	0.707	0.504	0.670	**0.102**
Salinity	RMSE	0.180	0.587	1.009	0.475	**0.152**
	MAE	**0.134**	0.488	0.840	0.550	0.359
	MAPE	**0.003**	0.012	0.021	0.028	0.008
Oxygen	RMSE	0.316	0.544	1.157	**0.025**	0.133
	MAE	0.214	0.435	1.014	**0.116**	0.315
	MAPE	**0.036**	0.078	0.184	0.198	0.058
PE	RMSE	0.170	0.334	0.333	0.009	**0.002**
	MAE	0.130	0.305	0.293	0.067	**0.041**
	MAPE	0.317	0.921	0.822	0.904	**0.114**
Temperature	RMSE	0.996	0.919	0.932	**0.130**	0.313
	MAE	0.697	0.743	0.758	**0.277**	0.499
	MAPE	0.032	0.036	0.035	0.256	**0.019**
Transparency	RMSE	0.219	1.153	3.060	0.037	**0.018**
	MAE	0.124	0.983	2.860	0.153	**0.121**
	MAPE	0.045	0.307	0.772	0.286	**0.035**

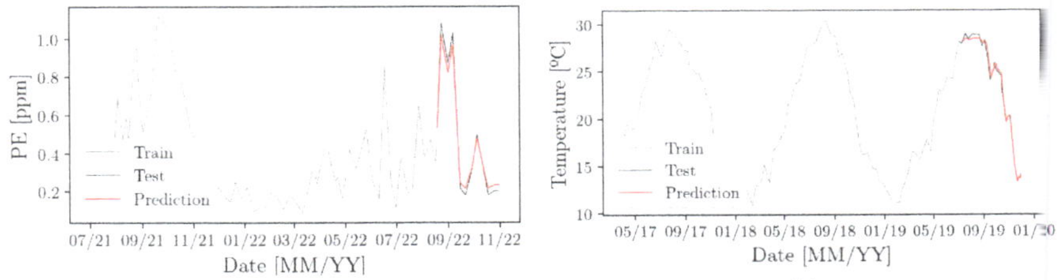

Figure 2. (**a**) Prediction of PE with LSTM (15%); (**b**) prediction of WTª with LSTM (horizon of 7 days).

5. Conclusions

Firstly, this study highlights the significance of thoroughly comprehending the time series and clearly defining the analysis objectives. It identifies two crucial characteristics of time series analysis, emphasizing their essential role in understanding the data, minimizing errors, enhancing the accuracy of predictions, and preparing the data for deeper analysis. Furthermore, it's vital to have a comprehensive understanding of both key performance and error metrics, alongside ensuring data cleanliness, to facilitate the selection of the most appropriate model.

In conclusion, the models generally yielded favorable prediction results, with the statistical and LSTM models emerging as the most effective for this data. Specifically, in the case of the PE data, the LSTM model achieved the lowest error, demonstrating an RMSE of 0.002 over a 7-day predictive horizon.

6. Discussion

The overall results and predictions of this study are positive, suggesting the applied methodology is effective. However, there were limitations regarding the forecasting horizons. Predictions beyond the selected horizon were not feasible with both statistical and machine learning models. While long-term estimations were unattainable with statistical models, machine learning models showed more promise in this regard. Additionally, the data preprocessing was relatively straightforward, benefiting from earlier processing and standardization via the server (L4 level).

One notable issue was the high error rates in chlorophyll predictions, attributable to the training data's significant variance from the prediction data, starting with low values and increasing markedly towards the series' end.

Author Contributions: Conceptualisation and methodology, R.M. and I.F.; data curation, M.N.; formal analysis, M.N. and J.C.S.-G.; validation, R.M. and I.F., writing—original draft preparation, M.N. and J.C.S.-G.; writing—review and editing, R.M. and I.F. All authors have read and agreed to the published version of the manuscript.

Funding: This research was funded by the Instituto de Fomento de la Región de Murcia (INFO) under the Program of grants aimed at Technological Centres of the Region of Murcia for the realisation of non-economic R&D activities. Modality 1: Independent R&D Projects, with File No.: 2022.08.CT01.000040.

Institutional Review Board Statement: Not applicable.

Informed Consent Statement: Not applicable.

Data Availability Statement: The data presented in this study are openly available at https://marmenor.upct.es/thredds/catalog/L4/catalog.html.

Conflicts of Interest: The authors declare no conflict of interest.

References

1. Box, G.; Jenkins, G.; Reinsel, G.; Ljung, G. *Time Series Analysis: Forecasting and Control*, 5th ed.; Wiley: Hoboken, NJ, USA, 2016.
2. Peña, D. *Análisis De Series Temporales*, 2nd ed.; Alianza Edityorial, SA.: Madrid, Spain, 2010.

3. Jha, B.K.; Pande, S. Time Series Forecasting Model for Supermarket Sales using FB-Prophet. In Proceedings of the 5th International Conference on Computing Methodologies and Communication, ICCMC 2021, Institute of Electrical and Electronics Engineers Inc., Erode, India, 8–10 April 2021; pp. 547–554. [CrossRef]
4. Bagnato, J.I. Pronóstico de Series Temporales con Redes Neuronales en Python | Aprende Machine Learning. Available online: https://www.aprendemachinelearning.com/pronostico-de-series-temporales-con-redes-neuronales-en-python/ (accessed on 21 February 2023).

engineering proceedings

MDPI

Proceeding Paper

Design and Implementation of an IoT Based Smart Digestive Health Monitoring Device for Identification of Digestive Conditions †

Rajesh Kumar Dhanaraj [1], Alagumariappan Paramasivam [2,*], Sankaran Vijayalakshmi [3], Cyril Emmanuel [4], Pittu Pallavi [2], Pravin Satyanarayan Metkewar [1] and Manoj Ashwin [2]

1 Symbiosis Institute of Computer Studies and Research (SICSR), Symbiosis International (Deemed University), Pune 411016, India; sangeraje@gmail.com (R.K.D.); pravin.metkewar@gmail.com (P.S.M.)
2 Department of Biomedical Engineering, Vel Tech Rangarajan Dr. Sagunthala R&D Institute of Science and Technology, Chennai 600062, India; vtu19829@veltech.edu.in (P.P.); vtu23548@veltech.edu.in (M.A.)
3 Department of Electronics and Communication Engineering, Vel Tech Rangarajan Dr. Sagunthala R&D Institute of Science and Technology, Chennai 600062, India; drvijayalakshmis@veltech.edu.in
4 Gleneagles Global Health City, Chennai 600100, India; emmanuel69@rediffmail.com
* Correspondence: parama.ice@gmail.com or draparamasivam@veltech.edu.in; Tel.: +91-984-378-0801

† Presented at the 10th International Electronic Conference on Sensors and Applications (ECSA-10), 15–30 November 2023; Available online: https://ecsa-10.sciforum.net/.

Abstract: Over the past few decades, there has been a significant rise in wearable healthcare technologies that have been playing a major role all over the world in monitoring health, alerting individuals during deviations from their normal health conditions and assisting them to stay fit and healthy. Due to the modern lifestyle and consumption of unhealthy food products, there has been an adverse effect on digestive health standards. In this work, a wearable device with textile electrodes is designed and developed to analyze the digestive conditions, namely, pre-prandial and post-prandial, using Electrogastrogram (EGG) signals. Further, the proposed device is comprised of textile electrodes as a sensor, an Analog-to-Digital Converter (ADC) with a Programmable Gain Amplifier (PGA), a Microcontroller with an inbuilt WirelessFidelity (WiFi) module, and an Internet of Things (IoT) cloud platform. Also, the EGG signals are acquired under two different conditions, namely, pre-prandial and post-prandial conditions, and then the Long Short Term Memory (LSTM) deep learning model is utilized to classify pre-prandial and post-prandial EGG signals to identify the eating habits of normal individuals. Results demonstrate that the proposed approach is capable of classifying the pre-prandial and post-prandial EGG signals, which, in turn, identify the fasting or ingestion state of normal individuals.

Keywords: bio-signal acquisition; electrogastrograms; Fast Fourier Transform; Internet of Things; pre-prandial and post-prandial conditions

Citation: Dhanaraj, R.K.; Paramasivam, A.; Vijayalakshmi, S.; Emmanuel, C.; Pallavi, P.; Metkewar, P.S.; Ashwin, M. Design and Implementation of an IoT Based Smart Digestive Health Monitoring Device for Identification of Digestive Conditions. *Eng. Proc.* **2023**, *58*, 33. https://doi.org/10.3390/ecsa-10-16253

Academic Editor: Stefano Mariani

Published: 15 November 2023

1. Introduction

In general, indulging in food on a regular basis is essential for a healthy digestive system. However, people in different age groups are more subjected to a sedentary lifestyle. Further, they prefer to consume highly processed foods, spicy foods, or excessive meals without hunger, skipping meals, eating at unusual times, and eating before sleep [1]. Also, problems with sleep are associated with digestive disorders, either as a form of symptom or a cause [2]. Digestive disorders, otherwise known as gastrointestinal (GI) tract disorders, are impairments and diseases linked to the Gastrointestinal (GI) tract of the human body. These GI tract disorders are linked to abnormal eating habits and unhealthy lifestyles. Overall, this leads to most of the digestive disorders such as Gastroesophageal Reflux Disease (GERD) [1], chronic diarrhea, and major gastrointestinal cancers, namely Colorectal Cancer (CRC), Esophageal Cancer (EC), Pancreatic Cancer (PC) [3], functional dyspepsia, etc. These are some of the serious digestive disorders observed mostly in people worldwide and, thus, can lead to complications as time progresses.

According to a survey report in 2019 [4], of the 6174 participants, 8.2% had GERD. The disease is more prevalent in urban areas (11.1%) than in rural areas (5.1%). Furthermore, GERD is a very common chronic disease in India and can lead to serious chronic diseases like adenocarcinoma if left unnoticed. The common methods of diagnosing digestive disorders include lab tests, imaging tests, endoscopic methods, etc. Also, the imaging tests include a colorectal transit study, a Computerized Tomography (CT) scan, defecography, a Magnetic Resonance Imaging (MRI) scan, ultrasound scanning, etc. The common imaging method used in digestive diagnostics is computerized tomography. In CT, cross-sectional imagery using the X-ray method is used to detect an accurate reading of the organ's histology. According to the study conducted by Herbert L. Fred [5], the method is very expensive, and the amount of radiation dose during a test is extremely high. The prolonged use of the device is extremely dangerous for the patient. Endoscopic procedures involve colonoscopy, endoscopic retrograde cholangiopancreatography, and sigmoidoscopy. Endoscopy is the visual examination of the digestive tract using a camera placed at the tip of a flexible tube, and the tube is inserted through the mouth or the anal orifice. Therefore, it can cause extreme discomfort for the patient during the procedure. Blockages due to the flexibility and maneuverability limitations of the tube can also lead to complications and extreme discomfort for the patient. Moreover, it can be used for histological tests of the gut lining, as it only gives imagery as results [6]. These limitations can affect the accuracy of the diagnostics. To prevent such problems, the non-invasive method of electrogastrogram (EGG) is used.

An Electrogastrogram (EGG) is an electrical signal that gives electrical activity to the stomach and can be recorded with the help of a non-invasive technique called Electrogastrography by placing electrodes over the surface of the stomach. With the use of EGG signals, digestive disorders can be detected in an effective manner [7]. In this era of smart technology, health parameters related to the Electrocardiogram (ECG), blood pressure, heart rate, and oxygen saturation are measured [8]. While surfing various literature, it is evident that the EGG shall be used for the purpose of scientific research, and only a few studies are carried out from a practical point of view. So, the need for EGG tracking as a health parameter with the use of real-time Internet of Things technology is highly beneficial and helps individuals to be alert and able to take care of people with digestive disorders.

EGG involves the measurement of the electrical rhythm of the stomach. In a healthy person, the cycle would be 3 cycles per minute. This frequency can either increase or decrease based on the disorders associated with the system. Further, the increase in frequency is known as Tachygastria, and the decrease in frequency is known as Bradygastria. Also, the increase and decrease in frequency can be measured and detected by using Artificial Intelligence systems, especially learning methods. In a study conducted by Raihan et al. [4], the use of various AI algorithms, including support vector machines (SVM), K-Nearest Neighbor (KNN), and Logistic Regression (LR), was utilized. In machine learning, the computer possesses the ability to learn without any background programming, and the ML algorithm is set to find patterns in the data. Furthermore, Electrogastrogram (EGG) signals can be abnormal by being slower or faster than the normal rate. Therefore, the machine learning algorithm can be used to detect changes in the pattern and provide an accurate result that can differentiate between normal and abnormal EGG signals.

2. Literature Survey

Due to poor electrical activity in the stomach, the majority of people suffer from digestive disorders. Gopu et al. (2008) have proposed a method of recording electrogastrogram signals to avoid the use of endoscopy by using cutaneous Ag-AgCl electrodes and a Signal Conditioning Unit to improve signal quality. The signals are filtered to remove noise, and the converted digital output is sent to a microcontroller [9]. A method presented by Haddab et al. (2009) represents the EGG signal acquisition, using neural networks for noise filtering of motion artifacts with the actual signal and then transmits to the medical care unit through GSM communication [10]. In contrast to passive electrode systems for the acquisition of electrogastrogram signals, an active electrode setup was proposed by Gopu et al. (2010), which showed a higher sensitivity and reliability that helped in the diagnosis of gastric disorders such as ulcers and dyspepsia through preprocessing using Principal Component Analysis (PCA) with the support of a wavelet transform for analysis [11].

Gharibans et al. (2018) discussed that EGG signals can be recorded using multi-channel systems of wearable type and also used some signal processing methods for the removal of artifacts, overtaking the limitations of single-channel measurement and the presence of signal artifacts that resulted in inconsistency in data reliability. This proposed approach shows an increased scope for diagnosing and treating GI disorders in an effective manner [12]. The development of both two- and three-electrode systems and a comparison made by Alagumariappan et al. (2018) for recording the electrical activity of the stomach showed that the three-electrode system showed a higher information content, ensuring progression in the accurate diagnosis of any abnormalities related to electrogastrogram signals [13]. Gharibans et al. (2019) revealed that abnormality in gastrointestinal function is a multifactorial and potential cause of gastroparesis and functional dyspepsia, for which non-invasive cutaneous high-resolution recording of EGG helps in identification of those above-mentioned symptoms [14].

Several researchers have extracted features from acquired EGG signals and classified various digestive diseases [15–20]. Alagumariappan et al. (2020) have discussed the role of electrogastrograms in prior detection of digestion abnormalities in diagnosing Type 2 Diabetes with the help of extracting features by pre-processing the recorded EGG signals using Empirical Mode Decomposition and generic algorithms in picking up good features and relating all these features with the digestive system's mobility [15]. A user-friendly, non-invasive, and wearable approach for monitoring gastrointestinal problems is proposed by Kumar et al. (2020). They used LabVIEW software for analyzing signals and the moving average algorithm in MATLAB for accurate electro-gastrographic extremities [16]. A study conducted by Paramasivam et al. (2021) explores that there is a positive impact of yoga asanas on the digestion process. This is identified through Fast Fourier Transform (FFT) by recording EEG signals before and after yoga, for which the normal frequency range of EGG signals is aligned with post-yoga recorded signals [17].

The objective of this work is to design and develop an Internet of Things-based smart wearable device to alert/notify people once they have skipped their food habits on time. In this paper, the proposed work is organized into four different sections. The first section deals with a brief introduction to the digestion process, its associated electrical signals, and the techniques used to assess the progress of digestion. Further, Section 2 deals with literature relevant to EGG techniques, objectives, and the organization of the research paper. Section 3 explains the proposed methodology, and Section 4 focuses on the results and their analyses. The conclusions reached through analysis are presented in Section 5.

3. Methodology

In this work, a wearable device fabricated with three textile electrodes is designed and developed. Further, the digestive conditions, namely pre-prandial and post-prandial, using EGG signals are analyzed. Figure 1 shows an overall block diagram of the proposed approach.

In the proposed approach, participants without any previous history of digestive health complications are selected. Furthermore, two different things, namely consent forms and questionnaires, are obtained from the participants. The experimental procedures are explained clearly, and informed consent is obtained, whereas the questionnaires are obtained to ensure that the participants are not having any medical complications. After obtaining informed consent and questionnaires, the participants are selected for EGG signal acquisition. Also, the EGG signals are acquired from selected participants for two different conditions, namely pre-prandial and post-prandial conditions These acquired EGG signals are preprocessed, and the unwanted frequency components are removed. Further, the preprocessed pre-prandial and post-prandial EGG signals are given to a deep learning model for the learning process. Once the deep learning model is trained, it provides decision support about digestive habits, namely pre-prandial and post-prandial conditions.

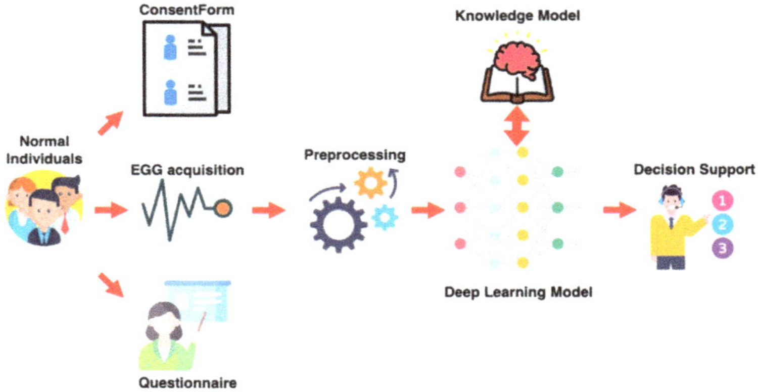

Figure 1. Overall block diagram of the proposed approach.

3.1. Proposed EGG Device

Figure 2 shows the block diagram for the proposed device. The proposed device is a wearable device that is used to monitor digestive habits, namely pre-prandial and post-prandial conditions, effectively, which leads to a healthy life.

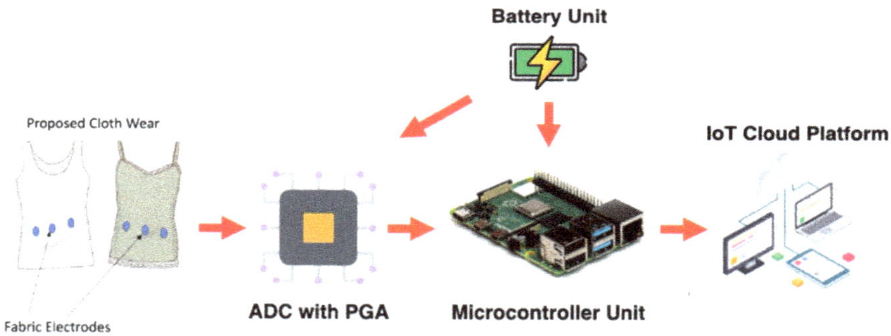

Figure 2. Block diagram of the proposed wearable device.

The proposed device consists of components such as Textile Electrodes, an ADC with programmable gain amplifier, a Microcontroller unit, a battery unit, and an Internet of Things cloud platform.

3.1.1. Textile Electrodes

A conductive thread made up of stainless-steel material is utilized in this work to fabricate electrodes for EGG signal acquisition. Also, the thread is stitched in three distinct places of the innerwear, which forms three fabric electrodes, and its position is determined according to the three-electrode placement protocol suggested by [21]. Furthermore, the wire is tapped from three fabric electrodes, and once the innerwear is worn by the individuals, the fabric electrodes pick up the EGG signals. These acquired EGG signals are given to the Analog to Digital Converter (ADC) for further processing.

3.1.2. ADC with PGA

In this work, an ADS1115-based ADC with a PGA module is used to convert and amplify the EGG signals with less power consumption. The ADS1115 consumes 150 micro-amps of current. Also, the utilized ADS1115 operates from 2 volts to 5 volts and has 4 ADC channels, which can perform two differential input operations. In general, the range of the EGG signals is in micro-volts, and due to this, it is important to amplify the acquired EGG signals to perform further processing. The adopted ADC converter with PGA module performs two different operations, namely amplification and AD conversion. Firstly, the acquired EGG signals are amplified from the micro-volt range to the volt range by setting the gain of the amplifier through programming. Further, the acquired EGG signals are amplified and converted from analog to digital, and they are fed to the microcontroller unit through the Inter-Integrated Circuits (I2C) protocol.

3.1.3. Microcontroller Unit

A Raspberry Pi 3 Model B+ based Microcontroller is used to perform computing operations that are portable, consume less power, and can be connected to the cloud with the help of Internet connectivity. The preprocessing and deep learning algorithms are coded inside the Pi controller using open-source Python programming software. The acquired EGG signals are amplified and preprocessed to remove noise. Further, these preprocessed signals are given to a deep LSTM-based deep learning model for training and testing. Also, the Pi controller updates the decision support produced by the trained deep learning model to the Internet of Things (IoT) cloud platform.

3.1.4. ThingSpeak Internet of Things Cloud Platform

In general, Internet of Things cloud platforms, namely ThingSpeak, are used to visualize and analyze dynamic data remotely. A user account is created on the open-source ThingSpeak Internet of Things cloud platform, and the individual's food habits, namely pre-prandial and post-prandial conditions, are monitored. The microcontroller unit accesses the ThingSpeak user account with the help of an Application Programming Interface (API) key and stores the decision support, which can be viewed by the individual or doctor personnel remotely at any time. Also, the day-wise individual's pre-prandial and post-prandial conditions are logged, which helps the individual lead a healthy life.

3.2. *Data Acquisition and Analysis*

In this work, the Empirical Mode Decomposition (EMD) technique is used to decompose the acquired pre-prandial and post-prandial EGG signals into multiple components called Intrinsic Mode Functions (IMFs) [15]. The frequency of these IMFs is derived using the Fast Fourier Transform (FFT). By analyzing the frequency of all the IMFs, the unwanted IMFs are eliminated and the remaining IMFs are concatenated, which produces a resultant noise-free EGG signal. The frequency components of the sampled EGG signal can be extracted using the FFT algorithm. Also, in this work, the FFT analysis is used to represent the acquired EGG signal in the frequency domain. Further, the FFT of the acquired EGG signal can be computed by the expression (1).

$$f(x) = \sum_{n=0}^{M-1} e^{-2\pi j \frac{xn}{M}} y(n) \tag{1}$$

where $f(x)$ requires a sum of M terms. Also, the frequency with the maximum amplitude will be considered the dominant frequency, which is the fundamental frequency of the particular EGG signal.

LSTM is a type of Recurrent Neural Network (RNN) architecture used in deep learning. Recurrent Neural Network (RNN) tends to have the problem of exploding and vanishing gradients; therefore, it is much more difficult to train. This problem occurs when the gradient either becomes too small or too large during back propagation. This happens in the RNN because they have a recurrent connection that allows them to store information from previous time steps. The LSTM is designed specifically to prevent the problem of an exploding and vanishing gradient. The LSTM architecture uses two types of nonlinear activation functions, namely the logistic sigmoid function and the hyperbolic tangent function.

The sigmoid activation function converts any x coordinate into a y coordinate between 0 and 1. This is used as a gate activation function. Further, the sigmoid activation function is given by the expression (2):

$$\sigma(x) = \frac{e^x}{e^x + 1} \tag{2}$$

The hyperbolic tangent function converts any x coordinate value and converts it into a y coordinate value between −1 and 1. Further, the function is given by the expression (3):

$$(x) = \frac{e^x - e^{-x}}{e^x + e^{-x}} = \tanh(x) \tag{3}$$

As the name suggests, there are two types of memory in the LSTM architecture, namely long-term memory and short-term memory. The long-term memory is also called the cell state, and it can be modified by arithmetic functions, but there are no weights or biases that can modify the function. The short-term memory is also known as the hidden state. Unlike long-term memory, short-term memory has weights and biases that can modify its function. Also, the LSTM architecture consists of memory blocks, and these memory blocks are a set of recurrently connected sub-networks. Furthermore, the memory block maintains its state over time and regulates the flow of information. A vanilla LSTM unit is composed of four components, such as a cell, an input gate, an output gate, and a forget gate [22]

3.2.1. Forget Gate

This step determines how much of the information must be deleted from its previous cell state. This gate uses the sigmoid activation function for its operation, and this is the first block in the LSTM algorithm.

3.2.2. Input Gate

This step is used to update the value of the previous LSTM cell by combining the input value with its biases and weights and the last LSTM output. This gate also uses the sigmoid activation function for its operations.

3.2.3. Cell

This step combines the values of the input value, the input gate value, the forget gate value, and the previous cell value.

3.2.4. Output Gate

This step combines the current input value, the output of the LSTM unit, and the cell value of the last LSTM unit.

The input and output values of the LSTM are combined with the cell values of the last unit and the current unit to obtain the block input value and the block output value, respectively. The values are introduced into the LSTM unit in the form of block input, and the output value is received in the form of block output. In this work, a total of 200 EGG signals are acquired from normal individuals, out of which 100 EGG signals are acquired under the pre-prandial condition and 100 EGG signals are acquired under the post-prandial condition. Furthermore, 80% of the total EGG signals are utilized to train the LSTM deep learning model, and the remaining 20% of the EGG signals are utilized to test the proposed LSTM deep learning model. Also, the proposed deep learning model is incorporated into the developed wearable device, and the device updates the decision support in the Internet of Things cloud.

4. Results and Discussion

Figure 3a,b show a typical EGG signal acquired from normal individuals under pre-prandial and post-prandial conditions. It is observed that the x-axis shows the amplitude of the acquired EGG signal in volts, and the y-axis shows the sample data points acquired at different points in time. Also, it is seen in Figure 3 that the typical EGG signal acquired from normal individuals under pre-prandial and post-prandial conditions has no significant variation by visual examination except for the change in amplitude of both EGG signals.

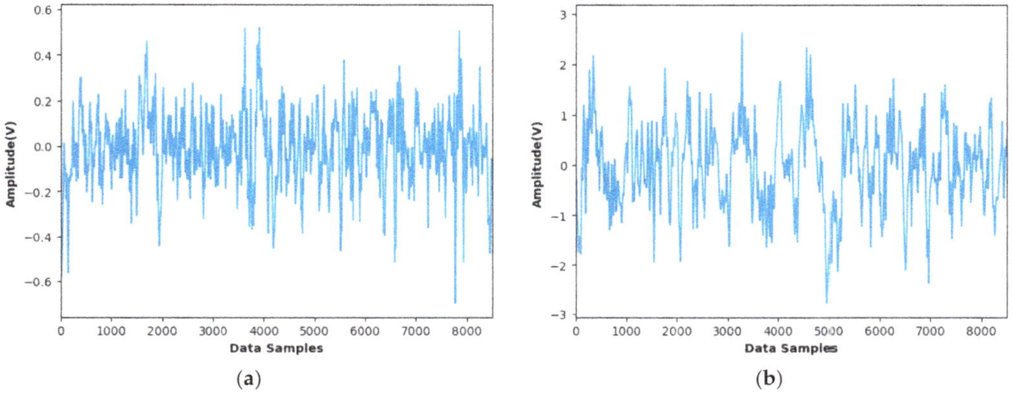

Figure 3. Typical EGG signal acquired under (**a**) pre-prandial conditions; (**b**) post-prandial conditions.

This study is conducted in accordance with the Declaration of Helsinki and approved by the Institutional Ethics Committee of Gleneagles Global Health City, Chennai, India. (Reference number: BMHR/2023/0055). For this proposed study, a total of 200 EGG signals were acquired under pre-prandial and post-prandial conditions from normal individuals with their proper consent.

The EGG signals are analyzed using FFT, and it is observed that there are no significant changes in frequency of the acquired pre-prandial and post-prandial EGG signals. Moreover, 80 pre-prandial EGG signals and 80 post-prandial EGG signals are given to the proposed LSTM deep learning model for the training process. Also, 20 pre-prandial EGG signals and 20 post-prandial EGG signals are given to the proposed LSTM deep learning model for testing. The confusion matrix generated after the testing process is shown in Figure 4. Figure 4 shows the confusion matrix of binary classes, namely pre-prandial and post-prandial, generated by the proposed LSTM deep learning model. Also, it is observed that parametric values such as True Positive (TP), False Positive (FP), True Negative (TN), and False Negative (FN) are given appropriately in terms of true values versus predicted values. By using the above-discussed parametric values, the four different performance metrics are calculated, and the four different performance metrics, such as accuracy, F1_Score, precision, and recall, of the proposed LSTM deep learning model are presented in Table 1.

Table 1. Performance metrics of the LSTM deep learning model.

Performance Metrics	Percentage (%)
Accuracy	96
Precision	94.2
Recall	98
F1_Score	96

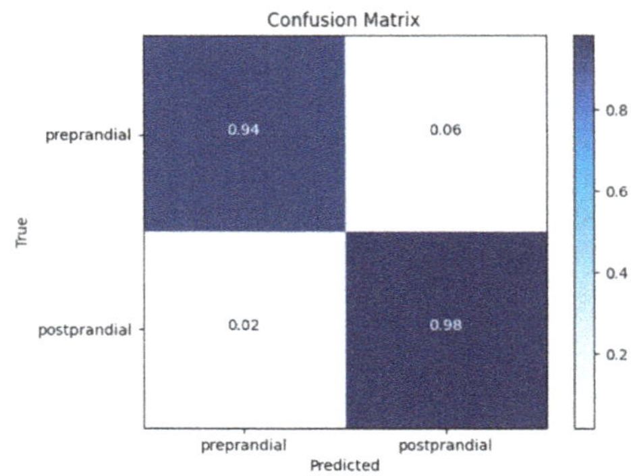

Figure 4. Confusion matrix for pre-prandial and post-prandial conditions.

From Table 1, it is evident that the accuracy of the proposed LSTM model is 96% and the recall of the proposed LSTM deep learning model is 98%. Further, the precision and F1_Score of the proposed LSTM deep learning model are 94.2% and 96%, respectively. Also, it is evident that the proposed LSTM deep learning model is capable of identifying an individual's food habits. Figure 5 shows the ThingSpeak user account for monitoring pre-prandial and post-prandial conditions. Further, the food habits can be monitored by oneself or another person, especially a doctor, by giving proper access. It is shown that the location of the person can also be visualized on the user page. The predicted output of the proposed LSTM deep learning model is logged to the field 2 Chart of the user account using the API key. Furthermore, it is also seen that in the field 2 Chart, the individual food habits are logged with respect to date and time. From the literature, it is evident that the skipping/late consumption of food leads to various digestive abnormalities; however, it is evident that the proposed Internet of Things-based smart digestive health monitoring device is highly efficient at identifying the individual's food habits/consumption results in maintaining a healthy life since digestion plays a vital role in every human's life.

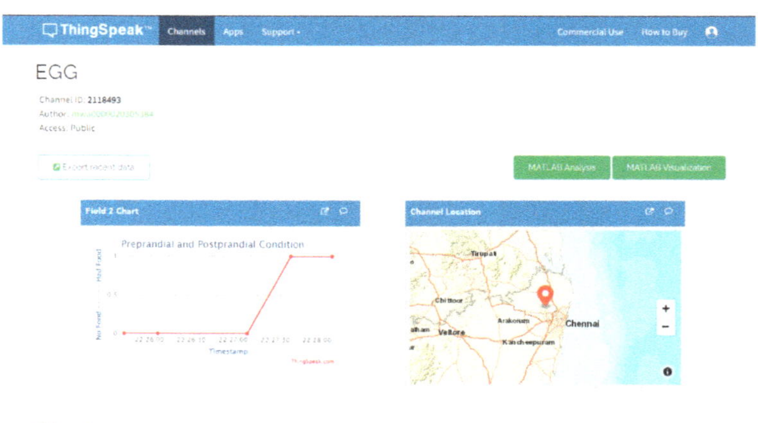

Figure 5. ThingSpeak user account for monitoring pre-prandial and post-prandial conditions.

5. Conclusions

In general, the acid produced by the stomach can sometimes leak into the esophagus because of improper closure of the cardiac sphincter, causing a burning sensation in the esophagus with symptoms such as regurgitations, belching, and coughing. The main cause of this disease is basically linked to the patient's lifestyle and eating habits. In this work, a wearable device was designed and developed to monitor the food intake habits of normal individuals to maintain a healthy lifestyle. Further, the EGG signals were acquired from normal individuals for pre-prandial and post-prandial conditions, and the LSTM deep learning model was utilized to identify the food intake habits of the normal individuals. Results demonstrate that the proposed LSTM model is good at classifying pre-prandial and post-prandial conditions and exhibits an accuracy of 96%. Also, the ThingSpeak Internet of Things cloud platform helps normal individuals monitor their food intake habits day-to-day in a remote manner anytime since the data are being logged regularly to the ThingSpeak Internet of Things user account. Since the proposed device is compact and can be integrated into usual clothing, it shall be used to effectively monitor digestive habits, namely pre-prandial and post-prandial conditions, which leads to a healthy life.

Author Contributions: R.K.D., A.P., S.V., and C.E. conceptualized the idea for this manuscript. R.K.D. provided the resources. A.P. designed and developed the hardware for data acquisition. P.P. and M.A. carried out the investigation and data curation of the acquired data. S.V. validated the data and results acquired. A.P. prepared the original draft. S.V., C.E., and P.S.M. reviewed and edited the original draft. C.E. and P.S.M. supervised the work, and R.K.D. administered the work. All authors have read and agreed to the published version of the manuscript.

Funding: This research received no external funding.

Institutional Review Board Statement: The study was conducted in accordance with the Declaration of Helsinki and approved by the Institutional Ethics Committee of Gleneagles Global Health City, Chennai, India. (Reference number: BMHR/2023/0055).

Informed Consent Statement: Informed consent was obtained from all subjects involved in this study.

Data Availability Statement: The data presented in this study are available on request from the corresponding author. The data are not publicly available due to ethical restrictions.

Conflicts of Interest: The authors declare no conflicts of interest.

References

1. Sharma, D.C.; Naqvi, A.; Chawla, S.; Gulati, C.; Kumar, S.; Singh, N.; Zehra, S. A Survey on Association Between Lifestyle Related Causes of Gastroesophageal Reflux Disease. *J. Exp. Zool. India* **2022**, *25*, 961–964.
2. Vernia, F.; Di Ruscio, M.; Ciccone, A.; Viscido, A.; Frieri, G.; Stefanelli, G.; Latella, G. Sleep disorders related to nutrition and digestive diseases: A neglected clinical condition. *Int. J. Med. Sci.* **2021**, *18*, 593. [CrossRef] [PubMed]
3. Jardim, S.R.; de Souza, L.M.P.; de Souza, H.S.P. The Rise of Gastrointestinal Cancers as a Global Phenomenon: Unhealthy Behavior or Progress? *Int. J. Environ. Res. Public Health* **2023**, *20*, 3640. [CrossRef] [PubMed]
4. Raihan, M.M.S.; Shams, A.B.; Preo, R.B. Multi-Class Electrogastrogram (EGG) Signal Classification Using Machine Learning Algorithms. In Proceedings of the 2020 23rd International Conference on Computer and Information Technology (ICCIT), Dhaka, Bangladesh, 19–21 December 2020; pp. 1–6. [CrossRef]
5. Fred, H.L. Drawbacks and limitations of computed tomography: Views from a medical educator. *Tex. Heart Inst. J.* **2004**, *31*, 345–348. [PubMed]
6. Atar, M.; Kadayifci, A. Transnasal endoscopy: Technical considerations, advantages and limitations. *World J. Gastrointest. Endosc.* **2014**, *6*, 41–48. [CrossRef] [PubMed]
7. Alagumariappan, P.; Krishnamurthy, K.; Jawahar, P.M. Selection of surface electrodes for electrogastrography and analysis of normal and abnormal electrogastrograms using Hjorth information. *Int. J. Biomed. Eng. Technol.* **2020**, *32*, 317–330. [CrossRef]
8. Sangeethalakshmi, K.; Preethi, U.; Pavithra, S. Patient health monitoring system using Internet of Things. *Mater. Today Proc.* **2023**, *80*, 2228–2231.
9. Gopu, G.; Neelaveni, R.; Porkumaran, K. Acquisition and analysis of electrogastrogram for digestive system disorders using a novel approach. In Proceedings of the 2008 International Conference on Electrical and Computer Engineering, Dhaka, Bangladesh, 20–22 December 2008; pp. 65–69.
10. Haddab, S.; Laghrouche, M. Microcontroller-based system for electrogastrography monitoring through wireless transmission. *Meas. Sci. Rev.* **2009**, *9*, 122. [CrossRef]
11. Gopu, G.; Neelaveni, R.; Pokumaran, K.; Shekar, M.G. An enhanced technique for recording and analysis of electrogastrogram using active electrodes. *Sri Lanka J. Bio-Med. Inform.* **2010**, *1*, 21. [CrossRef]
12. Gharibans, A.A.; Smarr, B.L.; Kunkel, D.C.; Kriegsfeld, L.J.; Mousa, H.M.; Coleman, T.P. Artifact rejection methodology enables continuous, noninvasive measurement of gastric myoelectric activity in ambulatory subjects. *Sci. Rep.* **2018**, *8*, 5019. [CrossRef] [PubMed]

13. Alagumariappan, P.; Krishnamurthy, K. An approach based on information theory for selection of systems for efficient recording of electrogastrograms. In Proceedings of the International Conference on Computing and Communication Systems: I3CS 2016, NEHU, Shillong, India, 11–13 November 2018; Springer: Singapore, 2018; pp. 463–471.
14. Gharibans, A.A.; Coleman, T.P.; Mousa, H.; Kunkel, D.C. Spatial patterns from high-resolution electrogastrography correlate with severity of symptoms in patients with functional dyspepsia and gastroparesis. *Clin. Gastroenterol. Hepatol.* **2019**, *17*, 2668–2677. [CrossRef] [PubMed]
15. Alagumariappan, P.; Krishnamurthy, K.; Kandiah, S.; Cyril, E.; Rajinikanth, V. Diagnosis of Type 2 Diabetes Using Electrogastrograms: Extraction and Genetic Algorithm–Based Selection of Informative Features. *JMIR Biomed. Eng.* **2020**, *5*, e20932. [CrossRef]
16. Kumar, G.P.; Prakash, S.O.; Sangeetha, B.; Asha, R.; Suganthi, L.; Divya, B. Wireless Real-Time Electrogastrography Monitoring System. *J. Comput. Theor. Nanosci.* **2020**, *17*, 3724–3732. [CrossRef]
17. Paramasivam, A.; Najumnissa Jamal, D.; Emmanuel, C.; Bhaskar, K.B.; Mohit Jaisingh, M.; Kannan, R. Analysis of Influence of Yoga-Asana on the Digestive Process Using Electrogastrograms. In Proceedings of the International Conference on Computing and Communication Systems: I3CS 2020, NEHU, Shillong, India, 28–30 April 2020; Springer: Singapore, 2021; pp. 423–429.
18. Paramasivam, A.; Kamalanand, K.; Emmanuel, C.; Mahadevan, B.; Sundravadivelu, K.; Raman, J.; Jawahar, P.M. Influence of electrode surface area on the fractal dimensions of electrogastrograms and fractal analysis of normal and abnormal digestion process. In Proceedings of the 2018 International Conference on Recent Trends in Electrical, Control and Communication (RTECC), Chennai, India, 20–22 March 2018; pp. 245–250.
19. Rajagopal, A.; Alagumariappan, P.; Krishnamurthy, K. Development of an automated decision support system for diagnosis of digestive disorders using electrogastrograms: An approach based on empirical mode decomposition and K-means algorithm. In *Disruptive Technology: Concepts, Methodologies, Tools, and Applications*; IGI Global: Hershey, PA, USA, 2020; pp. 661–678.
20. Chowdhury, S.D.; George, G.; Ramakrishna, K.; Ramadass, B.; Pugazhendhi, S.; Mechenro, J.; Jeyaseelan, L.; Ramakrishna B.S. Prevalence and factors associated with gastroesophageal reflux disease in southern India: A community-based study. *Indian J. Gastroenterol.* **2019**, *38*, 77–82. [CrossRef] [PubMed]
21. Parkman, H.P.; Hasler, W.L.; Barnett, J.L.; Eaker, E.Y. Electrogastrography: A document prepared by the gastric section of the American Motility Society Clinical GI Motility Testing Task Force. *Neurogastroenterol. Motil.* **2003**, *15*, 89–102. [CrossRef] [PubMed]
22. Van Houdt, G.; Mosquera, C.; Nápoles, G. A review on the long short-term memory model. *Artif. Intell. Rev.* **2020**, *53*, 5929–5955. [CrossRef]

engineering proceedings

Proceeding Paper

AI-Driven Blade Alignment for Aerial Vehicles' Rotary Systems Using the A* Algorithm and Statistical Heuristics †

Samarth Godara [1,*], Madhur Behl [2], Gaurab Dutta [2], Rajender Parsad [1] and Sudeep Marwaha [1]

1 ICAR-Indian Agricultural Statistics Research Institute, New Delhi 110012, India; rajender.parsad@icar.gov.in (R.P.); sudeep@icar.gov.in (S.M.)
2 Indian Institute of Technology Kharagpur, Kharagpur 721302, India; madhurbehl22@gmail.com (M.B.); gdutta@ece.iitkgp.ac.in (G.D.)
* Correspondence: sgodara@cs.iitr.ac.in
† Presented at the 10th International Electronic Conference on Sensors and Applications (ECSA-10), 15–30 November 2023; Available online: https://ecsa-10.sciforum.net/.

Abstract: In aviation, precise alignment of helicopter blades is paramount for ensuring optimal performance and safety during flight operations. Manual methods for blade alignment often demand extensive calculations and experienced technicians, resulting in time-consuming processes. This research proposes an innovative AI-based algorithm to optimize blade alignment in helicopter rotary systems, integrating the A* algorithm and a statistical heuristic function. The algorithm seeks to minimize the standard deviation of blade distances from the ground, captured using high-speed distance sensors. Firstly, the initial blade positions, along with the swash plate turns' limitations, are given to the algorithm. Later, by exploring all potential adjustments and selecting the most promising sequence to minimize the standard deviation of blade distances (considering the allowable pitch limits), the algorithm achieves precise blade alignment, enhancing helicopter performance and safety. Subsequently, the algorithm outputs the recommended sequence of adjustments to be made in the swash plate. We conducted comprehensive case studies using Mi 17 helicopters as a testbed to validate the algorithm's efficacy. The algorithm was assessed under varying scenarios: near-perfect alignment, single-blade misalignment in upward and downward directions, and multiple blades in asymmetric positions. The results demonstrate the algorithm's capability to swiftly recommend the precise sequence of adjustments for each control rod nut, effectively minimizing blade misalignment and reducing the standard deviation. The implications of this research are far-reaching, with them promising enhanced helicopter performance and safety across diverse application domains. By automating and streamlining the blade alignment process, the algorithm minimizes reliance on human expertise and manual calculations, ensuring consistent and accurate blade alignment in real-world scenarios.

Keywords: AI-driven alignment; A* algorithm; rotor blades; squash plate; statistical heuristics

Citation: Godara, S.; Behl, M.; Dutta, G.; Parsad, R.; Marwaha, S. AI-Driven Blade Alignment for Aerial Vehicles' Rotary Systems Using the A* Algorithm and Statistical Heuristics. *Eng. Proc.* **2023**, *58*, 34. https://doi.org/10.3390/ecsa-10-16137

Academic Editor: Stefano Mariani

Published: 15 November 2023

1. Introduction

The main rotor's set of blades assumes a crucial role in generating lift and controlling an aircraft's movement. However, the conventional flag-tracking system used to monitor the position and motion of these main rotor blades presents certain limitations in terms of accuracy, reliability, and safety [1]. To address this, a track check is conducted to ensure uniform blade paths during rotation and minimize helicopter vibrations, with a permissible deviation limit of 20 mm.

Presently, the established method for track checking involves coloring all blade tips using a glass marking pencil, followed by extending a sheet of white dense paper from a pole at least 500 mm away. By bringing the pole close to the rotating blades, markings are made on the paper, which are then measured using a scale. However, this approach suffers from inaccuracies and consumes time, while the preparatory steps involved are laborious and potentially hazardous, risking blade damage and crew fatigue [2]. Thus the ongoing research aims to evaluate the feasibility of replacing the flag-tracking approach with more efficient, automated alternatives such as artificial intelligence (AI) to overcome these limitations. Furthermore, an advanced process of obtaining the blades' pitch using a high-speed sensor involves the following steps [3]:

1. Sensor Placement and Calibration: A high-speed sensor is strategically positioned to ensure accurate readings. Technicians carefully calibrate the sensor to account for any potential offsets or deviations and ensure that its measurements are reliable.
2. Data Collection during Rotation: With the engine running and the blades in motion, the high-speed sensor continuously collects data on the distance between each blade's tip and the ground. The rotating blades provide a dynamic set of distance measurements as they pass through their rotation cycle.
3. Pitch Calculation: Using the collected distance measurements, technicians calculate the pitch of each blade. The pitch represents the angle at which the blades' rotational plane deviates from the horizontal plane.

Once the pitch of each blade is determined, technicians must decide on the appropriate course of action to bring the blades within the specified range of pitch differences. This task typically involves manual calculations and relies heavily on the expertise and experience of the technicians. The technicians evaluate the pitch data, analyze the current blade alignment, and calculate the necessary adjustments to be made to each control rod of the swash plate to achieve the desired blade alignment. The manual process of determining the course of action can be time-consuming and may lead to suboptimal results due to the complexity of aligning multiple blades accurately. As a result, there is a need for a more efficient and automated approach to this process, which is where the proposed AI-based algorithm utilizing the A* algorithm comes into play. By leveraging AI, the algorithm aims to streamline the course of action determination, reducing the need for manual calculations and human intervention while ensuring swift and accurate blade alignment. This advancement is crucial for enhancing helicopter performance, safety, and maintenance efficiency in various application domains.

2. Methodology

The focal point of the proposed algorithm lies in the utilization of three key inputs to effectively enhance blade alignment within helicopter rotary systems; these inputs are the initial condition of the blades, the turn steps (minimum possible control rod turn), and the pitch and turn limits. As the initial condition, the algorithm takes into account the initial distances of each blade from the ground, along with the specific number of turns applied to the control rod. Furthermore, the algorithm also integrates the concept of pitch and turn limits for each control rod. These limits outline the range within which the control rod can be turned to affect changes in the blade's orientation. Moreover, the algorithm considers the acceptable range between the minimum and maximum pitch values.

Blade Alignment Recommendation System Utilizing the A Algorithm*

The output of the proposed algorithm is the recommended sequence of adjustments for each control rod of the swash plate (in terms of turns), minimizing the standard deviation and achieving optimal blade alignment within the desired pitch difference range. The proposed methodology uses the A* algorithm, which is a widely used search algorithm that efficiently finds the shortest path from a starting node (the initial blade conditions in our case) to a goal node (acceptable solution) in a graph representing the operational space [4]. The technique combines the advantages of Dijkstra's algorithm and greedy best-first search [5]. It intelligently evaluates potential paths by considering both the cost to reach a node from the start and the estimated cost to reach the goal from that node using a heuristic function, ensuring optimal and informed exploration. In our case, the standard deviation of the blades' distance from the ground is used as the heuristic function to be minimized during the A* algorithm, guiding the optimal alignment of helicopter blades for enhanced precision and efficiency. The stepwise, detailed explanation of the proposed algorithm (Figure 1) is as follows.

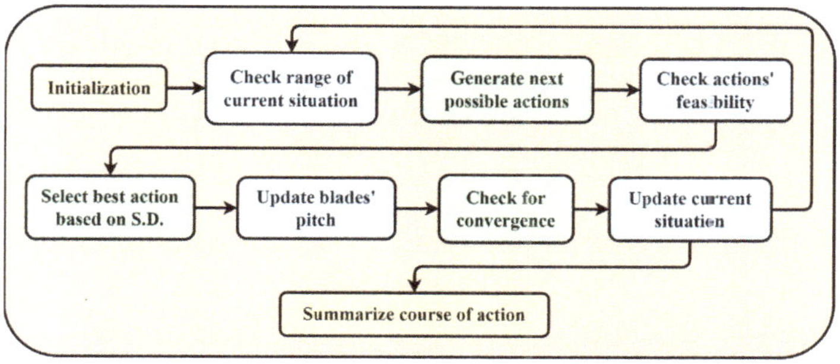

Figure 1. Workflow of the proposed blade alignment-recommender system.

1. Initialization: The algorithm starts by initializing various parameters, including the initial conditions of blade distances and swash plate turns, the minimum and maximum pitch limits, the acceptable range of pitch differences, and the minimum step sizes for distance and turns.
2. Iteration for searching the subsequent operation: The algorithm enters a loop that iterates until all blades are within the desired pitch difference range. It first checks if all blades are within the range; if so, it prints a message and breaks out of the loop.
3. Possible Actions: For each blade, the algorithm considers two possible actions: increasing or decreasing the distance from the ground and the corresponding swash plate turn adjustment.
4. Feasibility Check: The algorithm checks the feasibility of each action by calculating the blades' condition after each action. It also checks if the resulting swashplate turns are within the specified limits.
5. Valid Actions List: If an action is feasible, it is added to the list, along with relevant information such as the resulting blade distances, standard deviation [6] (Equation (1)), and updated swash plate turns.

$$\sigma = \sqrt{\frac{\sum(X-\mu)^2}{N}} \quad (1)$$

Here, X represents the data point in the distribution, μ is the mean of all of the blades, and N represents the total number of blades.

6. Selecting the Best Action: The algorithm stores the possible actions and their corresponding data. It selects the action with the lowest standard deviation from the set, indicating the most promising adjustment.
7. Updating the Blade Alignment: The algorithm checks if the best action leads to a reduction in the initial standard deviation. If so, it updates the blade distances and swashplate turns with the new adjustments. It then prints the details of the recommended operation, including the blade number, action taken, resulting blade distances, updated swash plate turns, and the standard deviation.
8. Iterative Refinement: The loop continues with further iterations until no further improvement is possible in terms of the standard deviation.
9. Final Summary: Once the blade alignment process is complete, the algorithm prints a summary of the recommended operations, displaying the number of turns to be given to each blade's nut to bring the blades within the desired range of pitch differences.

3. Results and Discussion

To assess the effectiveness of the proposed algorithm, researchers focused on the specifications of Mi 17 helicopters equipped with five rotor blades [7]. The algorithm's performance was evaluated using a set of specific input parameters. One of these parameters was the minimum rotation possible for a control rod nut, amounting to 0.1666, which is equivalent to a sixth of a full turn. This small rotation led to a vertical elevation of the blade by 10 mm, showcasing a clear relationship between control adjustments and blade positioning. During algorithm testing, the initial positions of the control rods were standardized at 380 turns. Additionally, the minimum limit for control rod turns was established as 375 turns, while the upper limit was set at 385 turns. These limits played a pivotal role in guiding the algorithm's adjustments and ensuring the controlled nature of the blade alignment process. Furthermore, the researchers considered a permissible track limit of 20 mm, indicating the acceptable extent of deviation in the blade's positioning. The evaluation of the algorithm's efficiency involved the execution of four distinct case studies, each designed to simulate different instances of blade misalignment within Mi 17 helicopters. These scenarios encompassed a range of potential misalignment situations, serving as comprehensive tests for the algorithm's capabilities.

In the first case study, the focus was on a scenario where nearly all blades were appropriately aligned, with each blade's distance from the ground being measured at 352, 353, 350, 349, and 351 cm. Moving to the second case study, the algorithm was tested against a situation of a single blade being misaligned in an upward direction. The corresponding distances for this scenario were 350, 352, 351, 351, and 365 cm. The third case study introduced a more complex misalignment pattern, involving one blade misaligned upward and another misaligned downward. The distances measured for this scenario were 350, 352, 351, 340, and 365 cm. The fourth and final case study presented a scenario in which two blades were misaligned downward while the remaining blades were aligned upward. In this context, the respective distances were recorded as 345, 340, 375, 377, and 385 cm.

For each of these case studies, the algorithm was implemented to optimize blade alignment, with it providing recommendations for necessary adjustments to the control rod nuts. The results of these tests effectively showcased the algorithm's capacity to achieve precise and efficient blade alignment, within the desired pitch difference range. The output of the proposed algorithm corresponding to each of the undertaken scenarios is as follows.

In the first case study, where most blades were already well-aligned, the algorithm's analysis led to two recommended operations for achieving optimal blade alignment (Table 1). Throughout its execution, the algorithm explored two levels within the search space (Figure 2). The initial standard deviation of blade distances was 1.58, which was effectively reduced to the minimum value of 1 after the alignment process was carried out (Figure 3). Moving on to the second case study, which focused on addressing the misalignment of a single blade in an upward direction, the algorithm's outcome indicated the need for a single corrective operation. During its execution, the algorithm traversed 13 distinct levels within the search space. From the outset, the standard deviation was measured at 6.3, which was impressively minimized to 0.83 following the alignment procedure.

Table 1. The recommended actions (turns) corresponding to each blade for each case study.

Case Study	Blade 1	Blade 2	Blade 3	Blade 4	Blade 5
1	0	−0.167	0	0.167	0
2	0	0	0	0	−2.166
3	0	0	0	1.666	−2.166
4	4.498	4.498	−0.5	−0.833	−2.166

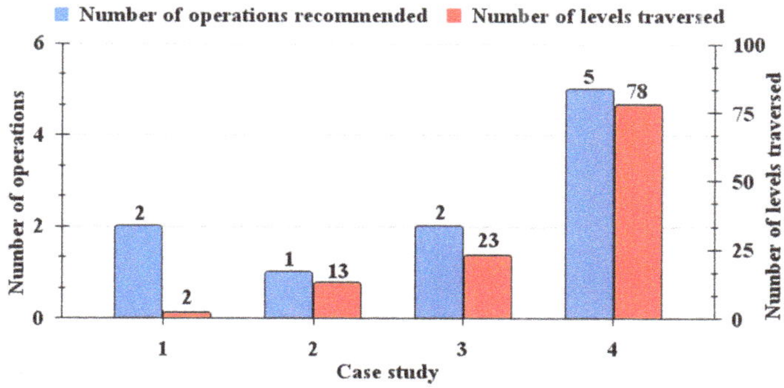

Figure 2. The number of operations recommended along with the levels traversed during the algorithm's execution corresponding to each case study.

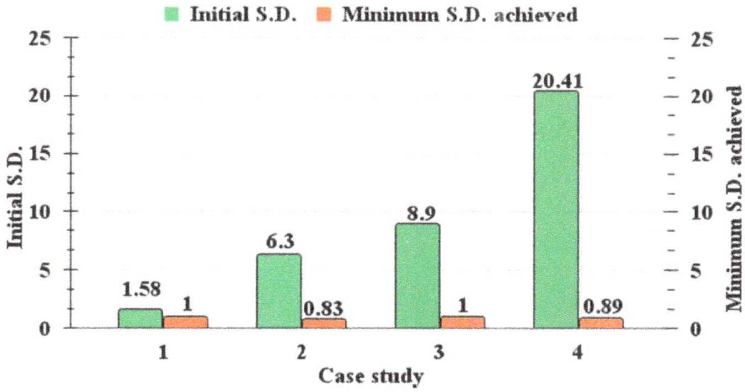

Figure 3. The initial and final standard deviation corresponding to each case study.

In the third case study, where one blade was misaligned upward and another was misaligned downward, the algorithm's optimization process involved suggesting two operations for attaining the best possible blade alignment. As part of its execution, the algorithm navigated through 23 different levels within the search space. The initial standard deviation stood at 8.9, which was substantially reduced to a minimum of 1 after the alignment process was executed. Lastly, in the fourth case study (3.2.4), where two blades were misaligned downward and the remainder were aligned upward, the algorithm's assessment revealed the necessity for five corrective operations to achieve optimal alignment. Throughout its execution, the algorithm traversed a significant 78 levels within the search space. Initially characterized by a standard deviation of 20.41, the alignment process led to a noteworthy reduction, with the minimum standard deviation reaching 0.89. These detailed case studies underscore the algorithm's capability to efficiently analyze and provide practical solutions for a range of blade misalignment scenarios, showcasing its potential for enhancing alignment precision across diverse helicopter configurations.

4. Conclusions

In conclusion, the presented research marks a significant stride towards advancing helicopter maintenance practices through innovative AI-driven solutions. The proposed algorithm's successful application in optimizing blade alignment within Mi 17 helicopters showcases its potential to revolutionize rotorcraft operations. By automating and refining the alignment process, the algorithm not only enhances performance and safety but also reduces reliance on manual intervention. The algorithm's adaptability to diverse helicopter models and potential integration with real-time data acquisition hold promising avenues for further advancements. As this research bridges the gap between cutting-edge AI technologies and rotorcraft operations, its implications extend beyond the laboratory, offering tangible benefits to the aviation industry's reliability, efficiency, and safety.

Author Contributions: Conceptualization, methodology, and writing—original draft preparation, S.G. and M.B.; validation and writing—review and editing, G.D. and S.M.; investigation, resources, and supervision, R.P. All authors have read and agreed to the published version of the manuscript.

Funding: This research received no external funding.

Institutional Review Board Statement: Not applicable.

Informed Consent Statement: Not applicable.

Data Availability Statement: The data and the presented results of the study can be obtained at the website: https://aibharosa.com/.

Conflicts of Interest: The authors declare no conflicts of interest.

References

1. Johnson, L. History: Helicopter Rotor Smoothing. 2005. Available online: http://acversailles.free.fr/documentation/08~Documentation_Generale_M_Suire/Vibrations/Dynamic_Solution_Systeme/History~Helicopter_rotor_smoothing.pdf (accessed on 1 November 2023).
2. Nagy, P.B.; Greguss, P. Helicopter blade tracking by laser light. *Opt. Laser Technol.* **1982**, *14*, 299–302. [CrossRef]
3. Hanif, A.; Yousaf, M.H.; Fazil, A.; Akhtar, S. Inflight helicopter blade track measurement using computer vision. In Proceedings of the 2014 IEEE Region 10 Symposium, Kuala Lumpur, Malaysia, 14–16 April 2014; pp. 56–61.
4. Candra, A.; Budiman, M.A.; Hartanto, K. Dijkstra's and a-star in finding the shortest path: A tutorial. In Proceedings of the 2020 International Conference on Data Science, Artificial Intelligence, and Business Analytics (DATABIA), Online, 16–17 October 2020; pp. 28–32.
5. Wayahdi, M.R.; Ginting, S.H.; Syahputra, D. Greedy, A-Star, and Dijkstra's algorithms in finding shortest path. *Int. J. Adv. Data Inf. Syst.* **2021**, *2*, 45–52. [CrossRef]
6. Lee, D.K.; In, J.; Lee, S. Standard deviation and standard error of the mean. *Korean J. Anesthesiol.* **2015**, *68*, 220–223. [CrossRef]
7. Kal'avský, P.; Petríček, P.; Kelemen, M.; Rozenberg, R.; Jevčák, J.; Tomaško, R.; Mikula, B. The efficiency of aerial firefighting in varying flying conditions. In Proceedings of the 2019 International Conference on Military Technologies (ICMT), Brno, Czech Republic, 30–31 May 2019; pp. 1–5.

engineering proceedings

MDPI

Proceeding Paper

Extended Object Tracking Performance Comparison for Autonomous Driving Applications †

Tolga Bodrumlu *, Mehmet Murat Gozum and Abdurrahim Semiz

AVL Research and Engineering, Istanbul 34885, Turkey; murat.gozum@avl.com (M.M.G.); abdurrahim.semiz@avl.com (A.S.)

* Correspondence: tolga.bodrumlu@avl.com

† Presented at the 10th International Electronic Conference on Sensors and Applications (ECSA-10), 15–30 November 2023; Available online: https://ecsa-10.sciforum.net/.

Abstract: Extended object tracking is crucial for autonomous driving, as it enables vehicles to perceive and respond to their environment accurately by considering an object's shape, size, and motion over time. Two commonly used methods for extended object tracking, Joint Probabilistic Data Association (JPDA) and Gaussian Mixture Probability Hypothesis Density (GM-PHD), were compared in autonomous vehicles using radar data. Both JPDA and GM-PHD perform well in tracking multiple extended objects, but GM-PHD demonstrates a performance advantage, especially in terms of the Generalized Optimal Sub-Pattern Assignment (GOSPA) metric, which measures the accuracy of tracked object positions in comparison to their actual positions.

Keywords: extended object tracking; radar; GM-PHD; JPDA

Citation: Bodrumlu, T.; Gozum, M.M.; Semiz, A. Extended Object Tracking Performance Comparison for Autonomous Driving Applications. *Eng. Proc.* **2023**, *58*, 35. https://doi.org/10.3390/ecsa-10-16201

Academic Editor: Stefano Mariani

Published: 15 November 2023

1. Introduction

Radar technology has revolutionized the way that information is gathered about the surrounding environment, providing valuable insights into the movement and position of objects. One of the most important applications of radar technology is extended object trackers (EOTs), which are sophisticated systems that enable the detection and tracking of multiple objects simultaneously.

An EOT is a critical component of many modern systems, including air traffic control, weather monitoring, and autonomous driving systems. By using advanced algorithms and sophisticated hardware, these systems can accurately identify and track the position, speed, and direction of objects in real time, even in challenging weather conditions. The main aims of an EOT are to track the kinematics and estimate the size of the objects nearby the ego vehicle [1]. As per [2], the shape is often assumed to be rigid, i.e., non-changing. Tracking an extended object is in general a highly complex problem due to the non-linearity of the resulting estimation problem.

Various tracking algorithms have been developed for EOTs. One popular approach is the multi-hypothesis tracking (MHT) algorithm [3], which generates multiple hypotheses about the object's position and shape based on measurements. Another approach is the JPDA algorithm [4], which assigns probabilities to each measurement–object association and updates the probabilities over time. More recently, deep-learning-based object detectors have been proposed, such as the CenterNet [5] and YOLO [6] algorithms, which use convolutional neural networks to predict the object's position and size.

One notable work in this area is the method proposed by Granström et al. [7], who proposed a probabilistic approach to extended object tracking based on the concept of random matrices. Their method also uses a Bayesian filter to estimate the object's state and shape, and it was shown to be effective in tracking vehicles in cluttered environments. Another approach was presented by Ba-Ngu Vo and Wing-Kin Ma [8], who propose a GM-PHD filter for extended object tracking. Their method uses the probability hypothesis density to represent the object's state and shape and was shown to be effective in tracking multiple objects with varying shapes and sizes.

In this paper, two methods are compared for the purpose of extended object tracking, which are JPDA and GM-PHD. The main reason for comparing these two algorithms here is to see which algorithm's results are at the desired level and to determine the algorithm we will use in real-time applications in the future. Both algorithms are coded in the MATLAB environment. The Driving Scenario Designer Toolbox [9] in MATLAB is used to test the implemented algorithms. The created scenarios include an ego vehicle and three surrounding vehicles. Measurements are taken from the radar on the ego vehicle, and the designed algorithms are executed and compared with ground truth data.

The remainder of this paper is structured as follows. Firstly, details of both algorithms are explained, then the Generalized Optimal Sub-Pattern Assignment metric, which is defined for comparison, is given. Finally, the simulation system and results are presented and future work is discussed.

2. Extended Object Tracking

2.1. Joint Probabilistic Data Association Tracker

The JPDA algorithm is a statistical approach used to solve the problem of data association in multiple target tracking [4]. The problem of data association involves assigning observations to a set of targets, where the observations are potentially noisy.

The JPDA algorithm considers all possible combinations of the observations and targets and computes the probability of each association based on the likelihood of the observations given the targets, as well as the prior probability of each target being present. These probabilities are then combined using Bayes' rule to obtain the posterior probability of each target being associated with each observation.

The JPDA tracker is an implementation of the JPDA algorithm for multiple target tracking. The main steps of the algorithm are summarized as follows:

- Initialization: Initialize the target states and their associated covariances and set the prior probabilities of each target being present.
- Prediction: Predict the state and covariance of each target based on a motion model and update the prior probabilities of each target being present.
- Data Association: Compute the likelihood of each observation given each target, and use the JPDA algorithm to compute the posterior probability of each target being associated with each observation
- Track Update: Update the state and covariance of each target based on the associated observations and compute the likelihood of each track.
- Track Management: Decide whether to create a new track for an unassociated observation or terminate a track if its likelihood falls below a certain threshold.

The general structure of the JPDA filter flowchart is shown in Figure 1.

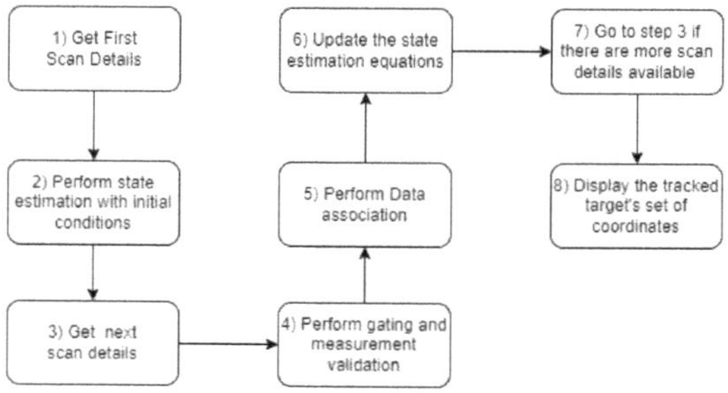

Figure 1. JPDA tracker flowchart.

2.2. Gaussian Mixture Probability Hypothesis Density Filter

The GM-PHD filter is a Bayesian filtering algorithm that is commonly used for multiple target tracking. It is an extension of the probability hypothesis density (PHD) filter, which is a non-parametric approach to multi-target tracking that represents the state of the targets using a single probability density function.

The GM-PHD filter extends the PHD filter by modeling the probability density function of the targets as a mixture of Gaussian components. Each Gaussian component corresponds to a potential target, and its mean and covariance represent the state and uncertainty of the target.

The GM-PHD filter consists of two main steps: prediction and update.

- Prediction: The means and covariances of the Gaussian components are propagated using a motion model. The weights of the Gaussian components are also updated based on the predicted probability of each target being present.
- Update: The filter incorporates the measurement information by calculating the likelihood of each measurement given each Gaussian component. The weights of the Gaussian components are then updated based on the likelihood and the predicted probability of each target being present.

The GM-PHD filter is efficient and scalable, as it can handle a large number of targets and clutter measurements. It also provides a probabilistic representation of the estimated target states, which can be used for decision making and sensor fusion. The general structure of the GM-PHD filter flowchart is shown Figure 2.

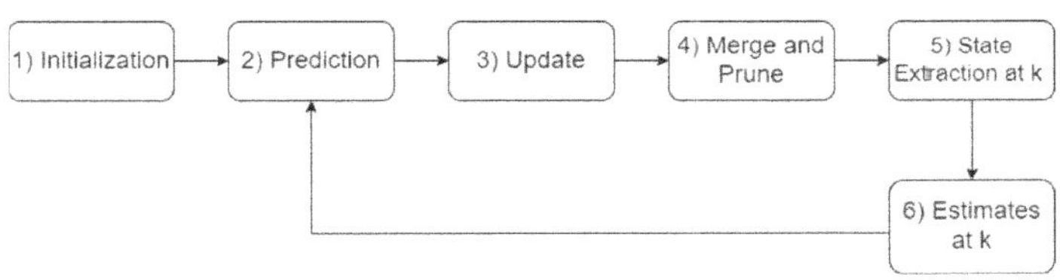

Figure 2. GM-PHD flowchart.

3. Comparison Metric

The GOSPA metric was proposed by Rahmathullah et al. [10] as a way to evaluate the performance of extended object tracking algorithms. The metric compares the estimated and true object trajectories and provides a quantitative measure of the tracking accuracy. It is a generalization of the Optimal Sub-Pattern Assignment (OSPA) metric that considers not only the distances between individual detections/tracks, but also the cardinality differences between the sets being compared.

The GOSPA metric measures the similarity between two sets of detections/tracks based on the distances between their individual elements, as well as the number of false alarms and missed detections. The metric is defined by four parameters: p, c, α, and β. The parameter p determines the order of the distance metric used to compare individual detections/tracks (e.g., $p = 2$ corresponds to the Euclidean distance). The parameters c, α, and β control the trade-off between the cardinality differences and the distances between the sets.

The GOSPA metric is a useful tool for evaluating the performance of multiple object tracking algorithms, as it provides a comprehensive measure of their accuracy and robustness. The code structure of the GOSPA algorithm is detailed in Algorithm 1 briefly.

Algorithm 1: GOSPA metric.

Input: Set of predicted tracks P and set of true tracks T
Output: GOSPA distance

$g = (c, p, fp, m)$; **for** *each true track* $t \in T$ **do**
  $D_t =$; **for** *each predicted track* $p \in P$ **do**
    $D_t(p) =$ calculate distance between t and p;
  $D_t = \text{sort}(D_t)$; **if** $|D_t| > m$ **then**
    $D_t = D_t(1 : m)$;
  $c_t = \sum_{d \in D_t} d$; $m_t = |D_t|$; $c = c + c_t$; $m = m + m_t$;
$p = |P|$; $fp = p - m$; $g = (c, p, fp, m)$; $GOSPA = \sqrt{(g_c + \alpha^2 g_p + \beta^2 g_{fp} + \gamma^2 g_m)}$;
return $GOSPA$;

4. Simulation System and Results

4.1. Simulation System

The MATLAB environment was used as the simulation system. Both the JPDA tracker and GM-PHD algorithm were implemented in the MATLAB environment. The implemented algorithms were tested with radar data obtained from Driving Scenario Toolbox. The studies were carried out on a single radar dataset on the ego vehicle. In the tested scenarios, the number of vehicles in the environment is fixed.

In order to provide more reliable measurement values from surrounding vehicles to the algorithm and to make more accurate decisions about which measurement belongs to which vehicle, a clustering algorithm was applied to the obtained radar measurements. The clustering algorithm was selected as Density-Based Spatial Clustering of Applications with Noise (DBSCAN) [11]. This clustering algorithm takes two parameters as input along with the incoming measurement values. These two parameters are a radius value ϵ and the minimum number of points required to form a dense region (minPts). In the simulations, ϵ was taken as 20 and minPts was taken as 3.

The simulation scenario was created in Driving Scenario Designer Toolbox [9]. In this scenario, the vehicles around the ego vehicle do not only move on a straight road. To increase the complexity of the scenario, lane changes were introduced and the vehicles were brought closer together. In the next section, the implementation of both JPDA and GM-PHD algorithms to the simulation system and the results obtained are shown.

4.1.1. JPDA Tracker

In this section, the JPDA tracker algorithm, which is explained in Section 2.1, is implemented in the MATLAB environment. Studies are carried out on a single radar dataset on the ego vehicle.

A covariance ellipse was created by using the velocity and position of the ego vehicle with the radar data. These created ellipses are also shown in Figure 3. The parameters used in the JPDA tracker algorithm and ellipse generation and their definitions are shown in Table 1. Initial values of these parameters are given as follows.

$$C_v = \text{diag}(10,10), \quad C_{rw} = \text{diag}(10,10,1,1),$$
$$A_p = \text{diag}(1,1,1), \quad C_r = \text{diag}(0.3,0.3,0.1,0.1)$$
$$A_r = \begin{bmatrix} 1 & 0 & 0.05 & 0 \\ 0 & 1 & 0 & 0.05 \\ 0 & 0 & 1 & 0 \\ 0 & 0 & 0 & 1 \end{bmatrix},$$
$$C_p = \text{diag}(0.2,0.1,0.1)$$

Table 1. JPDA parameters.

Symbol	Definition
C_v	Measurement Noise Covariance
C_{rw}	Error Covariance for Kinematic State
A_r	Transition for Kinematic State
A_p	Transition for Shape Parameters
C_r	Pos x,y and Vel x,y Covariance Matrix
C_p	Major and Minor Axes and Azimuth Angle of Ellipse

When observing Figure 3, it can be seen that the position estimations of the vehicles surrounding the ego vehicle are quite close to the ground truth data. However, it is observed that the errors in position estimation begin to increase as the vehicles move closer to each other. When the vehicles are close to each other, the data association part can be a problem due to the distinguishing part.

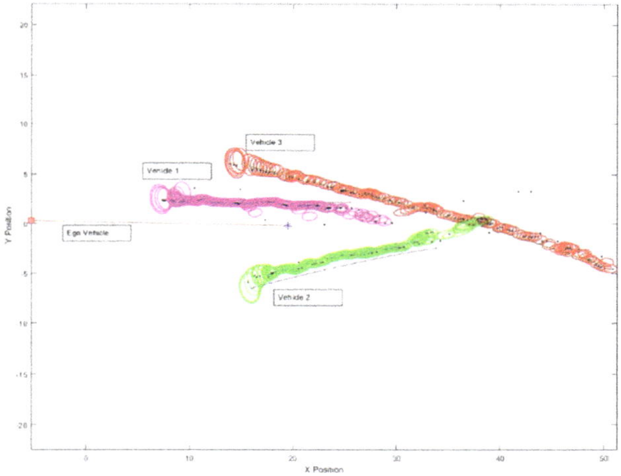

Figure 3. Tracking results obtained for JPDA algorithm.

4.1.2. GM-PHD

In this section, the GM-PHD algorithm, which is explained in Section 2.2, is implemented in the MATLAB environment. Studies are carried out on a single radar dataset on the ego vehicle. The data used here are the same as the data used in the JPDA algorithm. Some of the parameters and their definitions are shown in Table 2. The initial values of these parameters are also given as follows.

Table 2. GM-PHD parameters.

Symbol	Definition
v_{sigma}	Q noise matrix amplitude
P_s	Survival probability
P_d	Probability of detection
w_{birth}	Weighting of birth terms
m_{birth}	Mean of birth terms
P_{birth}	Covariance of birth terms

$$v_{sigma} = 1, \quad P_s{=}0.99, \quad P_d = 0.99$$
$$w_{birth} = 0.03, \quad P_{birth} = \text{diag}(1,1,1,1)$$

$$m_{birth} = \begin{bmatrix} 14.400 & 15.800 & 7.200 \\ 5.832 & 2.901 & 2.999 \\ 6.100 & -6.500 & 2.500 \\ -1.407 & 0.7636 & -0.057 \end{bmatrix}$$

When observing Figure 4, it can be seen that the position estimations of the vehicles surrounding the ego vehicle are quite close to the ground truth data. It can be observed that even when vehicles approach each other, the predictions do not shift to other vehicles. It can easily be said that the data association part works more effectively in the GM-PHD algorithm when compared with Figures 3 and 4.

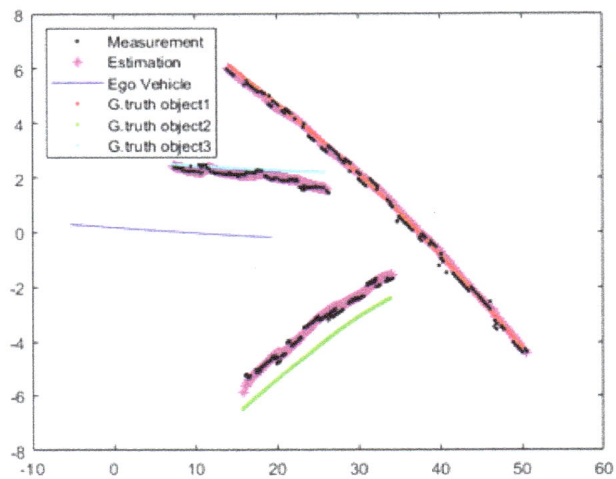

Figure 4. Tracking results obtained for GM-PHD algorithm.

4.2. Performance Comparison of JPDA and GM-PHD Algorithms

The results obtained in Sections 4.1.1 and 4.1.2 qualitatively illustrate the good performance of both algorithms. However, we need a quantitative metric which shows the accuracy of the performance of each algorithm. The GOSPA metric defined in Section 3 is used to illustrate the tracking performance of the JPDA and GM-PHD algorithms. Based on their values (the performance improves as the GOSPA value decreases), the performances of the algorithms are compared. The parameters p and c were chosen to be 2 and 100, respectively. Figure 5 illustrates the comparison of GOSPA values for the tracking performance of JPDA and GM-PHD algorithms.

The GOSPA value comparison illustrates that both algorithms are good at keeping track of the surrounding objects. For both of them, the GOSPA values are quite low. It can be seen that the GOSPA values obtained with GM-PHD are much lower than those obtained with JPDA. The GOSPA values oscillate between 1 and 1.5 for GM-PHD, whereas the GOSPA values obtained with JPDA vary between 1.5 and 5. In brief, the GOSPA results quantitatively illustrate that the tracking performance of GM-PHD is superior to JPDA.

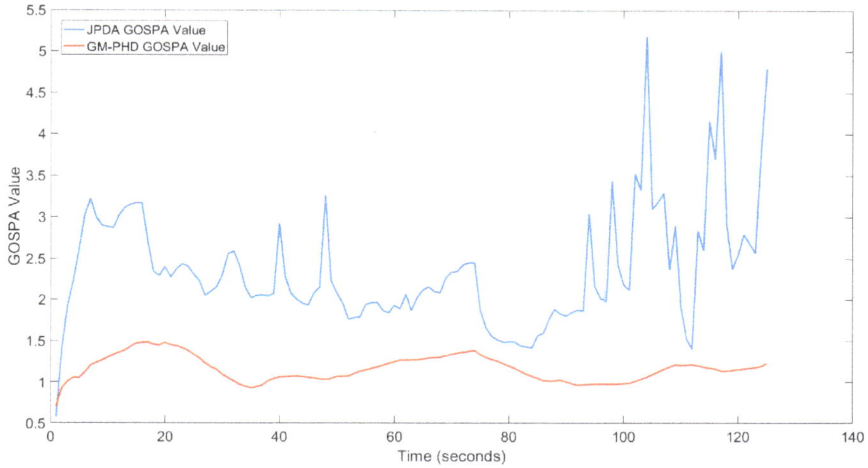

Figure 5. GOSPA value comparison of JPDA and GM-PHD algorithms for tracking performance.

5. Conclusions

In this paper, JPDA and GM-PHD methods are compared with each other for the purpose of extended object tracking. Both methods have advantages and disadvantages with respect to each other. In autonomous driving applications, there is no study which compares the performance of both methods based only on radar data. The GOSPA metric was chosen to compare the performance of GM-PHD and JPDA. The simulation platform was chosen to be the MATLAB Driving Toolbox.

The simulation results show that both algorithms are good at keeping track of the surrounding vehicles, which is understood from the low GOSPA values at all time stamps of the simulation. It is also clearly observed that the GOSPA values obtained with GM-PHD are much lower than the GOSPA values obtained with JPDA, which shows the superiority of GM-PHD over JPDA based on performance. Although it is known that GM-PHD is able to keep track of the objects when the number of objects is varied, unlike JPDA, GM-PHD is not optimal to estimate the cardinality of the objects when the newborn object is modeled as static and not measurement-driven. In the future, we aim to keep track of the surrounding objects by using the adaptive-birth GM-PHD algorithm in real-time implementations for autonomous driving applications.

Author Contributions: Conceptualization, T.B., M.M.G. and A.S.; Data gathering, T.B., M.M.G. and A.S.; Formal analysis, T.B. and M.M.G.; Investigation, T.B., M.M.G. and A.S.; Methodology, T.B., M.M.G. and A.S.; Project administration, T.B.; Resources, T.B.; Supervision, T.B., M.M.G.; Validation, T.B., A.S. and M.M.G.; Visualization, T.B. and A.S.; Writing—original draft, T.B., M.M.G. and A.S.; Writing—review and editing, T.B., and M.M.G. All authors have read and agreed to the published version of the manuscript.

Funding: This research received no external funding.

Institutional Review Board Statement: Not applicable.

Informed Consent Statement: Not applicable.

Data Availability Statement: Data sharing Does not apply to this paper.

Conflicts of Interest: The authors and AVL Research and Engineering declare no conflicts of interest.

References

1. Granstrom, K.; Baum, M. A tutorial on multiple extended object tracking. *techrxiv* **2022**. [CrossRef]
2. Granstrom, K.; Baum, M.; Reuter, S. Extended object tracking: Introduction, overview and applications. *J. Adv. Inf. Fusion* **2017**, *12* , 139–174. [CrossRef]

3. Reid, D.B. An algorithm for tracking multiple targets. *IEEE Trans. Autom. Control* **1979**, *24* , 843–854. [CrossRef]
4. Fortmann, T.; Bar-Shalom, Y.; Scheffe, M. Sonar tracking of multiple targets using joint probabilistic data association. *IEEE J. Ocean. Eng.* **1983**, *8* , 173–184. [CrossRef]
5. Zhou, X.; Wang, D.; Krahenbuhl, P. Objects as points. *arXiv* **2019**, arXiv:1904.07850. [CrossRef].
6. Redmon, J.; Divvala, S.; Girshick, R.; Farhadi, A. You only look once: Unified, real-time object detection. In Proceedings of the IEEE Conference on Computer Vision and Pattern Recognition, Las Vegas, NV, USA, 27–30 June 2016. [CrossRef]
7. Granstrom, K.; Lundquist, C.; Orguner, U. Extended Target Tracking using a Gaussian-Mixture PHD filter. *IEEE Trans. Aerosp. Electron. Syst.* **2012**, *48*, 3268–3286. [CrossRef]
8. Vo, B.-N.; Ma, W.-K. Gaussian mixture probability hypothesis density filter for multi-target tracking. *IEEE Trans. Signal Process.* **2006**, *54*, 4091–4104. [CrossRef]
9. MathWorks. *Automated Driving Toolbox*™; MathWorks: Natick, MA, USA, 2021.
10. Rahmathullah, A.S.; García-Fernández, Á.F.; Svensson, L. Generalized optimal sub-pattern assignment metric. In Proceedings of the 20th International IEEE Conference on Information Fusion, Xi'an, China, 10–13 July 2017. [CrossRef]
11. Ester, M.; Kriegel, H.-P.; Sander, J.; Xu, X. A density-based algorithm for discovering clusters in large spatial databases with noise. In Proceedings of the 2nd International Conference on Knowledge Discovery and Data Mining, Portland, OR, USA, 2–4 August 1996.

engineering proceedings

MDPI

Proceeding Paper

A Prototype to Prevent Fruits from Spoilage: An Approach Using Sensors with Machine Learning †

Uttam Narendra Thakur and Angshuman Khan *

Department of Electronics and Communication Engineering, University of Engineering and Management, Jaipur 303807, India; unthakur08@gmail.com
* Correspondence: angshumankhan2910@gmail.com

† Presented at the 10th International Electronic Conference on Sensors and Applications (ECSA-10), 15–30 November 2023; Available online: https://ecsa-10.sciforum.net/.

Abstract: One of the significant issues facing the world right now is food deterioration. If the freshness or deterioration of a fruit can be determined before it is lost, the fruit waste problem may be mitigated. The goal of this work is to develop a simple model for tracking fruit quality using sensors with a machine learning (ML) approach. This model uses from the gases emitted by fruits to determine the ones that will ripen and require use earlier. Two gas sensors (MQ3 and MQ7) and an Arduino Uno serve as the main processing components of the suggested system. Principal component study (PCA) is a widely employed discriminating approach that has been utilised to differentiate between fresh and rotten apples based on sensed data. The study yielded a cumulative variance of 99.1% over a span of one week. The data were also evaluated using a linear Support vector machine (SVM) classifier, which achieved an accuracy of 99.96%. The distinctive feature of the system is that it evaluates the levels of spoilage based on real-time data and deploys a low-cost, straightforward model that can be used anywhere to preserve any type of fruit.

Keywords: Support vector machine; principal component analysis; classification; machine learning

Citation: Thakur, U.N.; Khan, A. A Prototype to Prevent Fruits from Spoilage: An Approach Using Sensors with Machine Learning. *Eng. Proc.* **2023**, *58*, 36. https://doi.org/10.3390/ecsa-10-16005

Academic Editor: Jean-marc Laheurte

Published: 15 November 2023

1. Introduction

Food spoilage, which causes 40–50% of all losses of root crops, fruits, and vegetables each year, is one of the biggest problems the world is currently experiencing [1]. The development of original fruit offerings relies heavily on smell [2]. Thus, the primary objective of many studies is to characterize the smell or aroma to identify fruits [3]. Fruits smell differently at different stages of ripeness, such as when they are fresh and when they are not. The release mechanism of volatile organic compounds (VOCs) from fruits has to be investigated [4]. This is also influenced by elements like the climate, temperature, humidity, accumulation of carbohydrates, and fungus infections [5]. These kinds of studies would help in the early identification and monitoring of fruit phases, such as raw, ripe, or rotten, protecting sellers and buyers. As an illustration, as a banana is ripening, the flavour chemicals that give it its typical flavour are produced quickly. The aroma's composition also changes at different stages, giving rise to diverse aroma signatures. Several studies have been conducted to investigate the chemical compositions of the various volatile organic compounds (VOCs) emitted by bananas during the ripening process. However, the majority of these studies have mostly focused on the compositional changes occurring in fruits as they mature or undergo processing [6]. The fragrance composition changes during fruit ripening, which helps explain aromatic compound production [7]. Varieties of banana cultivars have distinctive flavours due to differences in their predominant volatile components.

Sensors play an important function in detecting the odours given off by the fruits in various phases of life. A sensory analysis with trained panellists precedes the preparation of the sensory characteristics of food products [3,4]. Composed of an array of non- or partially selective gas sensors, a data processing and pattern recognition system can identify even complex aromatic profiles. The sensor system responds differently to different aromas, and these responses produce a signal pattern that is unique to that aroma [6,7]. Each chemical, however, has unique attributes and traits of its own. Chemical properties are therefore unique and will be used as the input data to create this kind of system [8]. The computerised markings of explicit odours must be gathered into a library (dataset) in order for the programmed recognition process to be successful [9]. The sensor array finds specific compounds in a medium. A unique sensor is responsible for detecting each targeted aroma, or each sensor is in charge of detecting a certain sort of aroma. The primary aim of this study is to present a sensor array system integrated with a machine learning technique for the purpose of identifying distinct stages of fruit ripening. The following are the key contributions of this work:

1. This paper recognizes food deterioration as a critical global issue and seeks to contribute to the mitigation of fruit waste by detecting freshness or spoilage before it occurs.
2. It creates a straightforward model that utilizes sensors and machine learning approaches to monitor and assess the quality of fruits, providing a practical solution to this problem.
3. The paper demonstrates the successful integration of two gas sensors, MQ3 and MQ7, along with an Arduino Uno, as the primary components of the system. These sensors capture emissions from fruits, enabling the assessment of their ripeness and quality.
4. The research employed PCA to classify sensor data with a good accuracy rate over a one-week period. Additionally, an SVM classifier achieved an even higher accuracy when evaluating the collected data.

2. Literature Study

The identification of fruit odours has been the subject of numerous studies; each of them has advantages and disadvantages. In this section, a few pertinent recent works are discussed.

The research in [10] describes the application of a model for the rapid classification of various varieties of Polish honey based on FIGARO semiconductor sensors. It makes use of TGS (FIGARO USA, INC)-type semiconductor sensors, PCA, linear discriminant analysis (LDA), cluster analysis (CA), and LDA algorithms. In [11], a reliable and appropriate non-targeted method is created and evaluated for the differentiation of oranges from three different geographic sources. It uses principal component analysis and Non-linear Iterative Partial Least Squares (NIPALS) techniques, although sensor details are unclear. A strategy for applying an electronic nose for detecting formaldehyde-containing materials and seafood spoilage is described in [12]. In addition, comparisons and evaluations of static and dynamic properties, as well as sensor stability, are made. MOSFET sensors are combined with PCA and deterministic finite automaton (DFA) algorithms. Based on their volatile composition, Moroccan virgin olive oils are explored in [13] with a model to distinguish between distinct geographic origins. It uses six metal oxide semiconductor (MOS) sensors with PCA and LDA algorithms. A bio-inspired neural network, which is based on the olfactory systems of mammals, is given in [14] as a novel data processing method. A set of TGS, MiSC, and SHT75 sensors are used along with a bio-inspired neural network algorithm. In [15], samples are taken from the Pamplona and Toledo DWTPs in the northern Santander region (Colombia) to evaluate a multisensory system. The tests were executed utilising PCA, deterministic finite automaton (DFA), a probabilistic neural network (PNN), and TGS sensors. Membrane filtration (MF) was utilised to validate the detection of E. coli in water.

Various sensors, including semiconductor sensors of the FIGARO and TGS types, MOSFET sensors, and metal oxide semiconductor (MOS) sensors, among others, were used in a number of studies. The literature does not always include thorough sensor specifications and features. The literature also discusses PCA, LDA, NIPALS, DFA, bio-inspired neural networks, and probabilistic neural networks for odour identification and classification. A straightforward approach to determine the best methods for individual applications is lacking.

In this paper, a straightforward ML-based sensor model for fruit quality tracking is presented. The proposed system uses two gas sensors (MQ3 and MQ7) and an Arduino Uno for processing. The experiments use PCA and an SVM classifier. The system analyses rotting levels using real-time data and uses a low-cost, simple model to preserve any fruit.

3. Proposed Model

Figure 1 illustrates a schematic representation of the proposed setup. The proposed configuration is categorically segmented into three components, namely the sensing chamber, data gathering, and processing unit. The sensing chamber comprises a pair of sensors and a sample test platform. The alteration in the environmental conditions within the sensing chamber was measured via sensors and then transmitted to the data acquisition system. The Arduino Uno microcontroller (Arduino, Ivrea, Italy)was employed to acquire data from the sensors and thereafter store them in its memory Subsequently, the response underwent processing by the processing unit in order to facilitate further discrimination and classification.

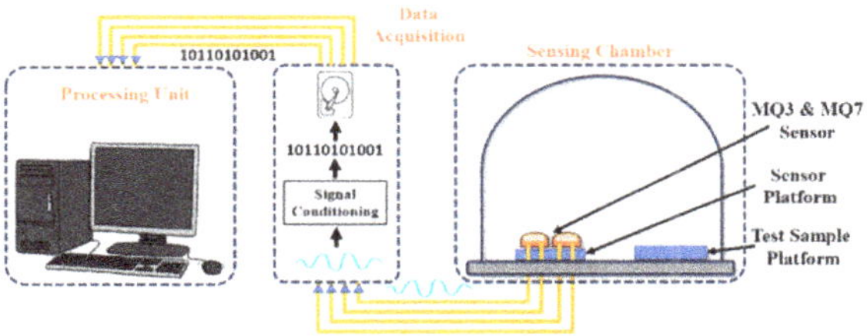

Figure 1. Schematic view of the experimental setup.

The sensors that detect the quantities of volatile chemicals, MQ3 for alcohol and MQ7 for CO, sense the gases released from the samples. The fresh and ripened stages of fruit ripening are perceived separately from the gas concentrations. The ppm levels and associated changes in sensor voltage values obtained when the sensor reacts to each of the many gases that are intended to be monitored are used to calculate the gaseous concentration. The data values are read to a 1 GB secure digital (SD) card in text format with a '.txt' extension. The influence of a change in room temperature on the degree of food rotting is not a concern for the experimental setting; hence, the experiment was only conducted at room temperature. The parallel gas sensor array is installed on the breadboard. The sensor array is supplied with +5 V according to the optimal rating of the MQ series gas sensors, with the analogue pin of the MQ7 sensor wired to pin (A5) and the pin of the MQ3 gas sensor wired to pin (A3) of the Arduino Uno board. The gas sensors are connected to the Arduino Uno board using a breadboard that offers internal connectivity. The ground pin (GND) of the Arduino Uno Board is connected to the gas sensors. The sensors were allowed to read the appropriate gas concentration in the air in order to stabilize the fluctuations in the sensor's voltage level until the sensor was heated up to a particular temperature, ensuring that the sensors detected the particular gas concentration in the aroma of the target food material in this experiment, 'banana.' The data values were transformed from text (.txt) into the comma separated format (.csv), which is the optimum format for using data science applications, to build the dataset in an excel file. Python was chosen as the main programming language for machine learning algorithms (PCA, SVM) and the data science application. PCA and SVM are two popular names in the ML paradigm. PCA is a dimensionality reduction technique widely used in machine learning and statistics. SVMs are a powerful and versatile class of supervised machine learning algorithms primarily used for classification and regression tasks. The dataset was first examined for missing or null values using the Python 'NumPy' module, and then a standard scaling was applied to the data. The PCA, a feature extractor and unsupervised learning clustering tool, was used to analyse the dataset's primary components that determine the categorization between rotten and fresh bananas. The SVM classification, a supervised machine learning approach, was utilised to categorise it in the various groups.

Methodology

A labelled dataset with stipulated accuracy and correctness makes up the current experimental dataset. The specifications are mentioned in the operation of the hardware setup and the environmental circumstances under which the data-gathering technique was carried out. The recorded readings from the MQ7 for carbon monoxide (CO) and the MQ3 for alcohol were grouped together in a single '.csv' file to represent the values seen in the dataset. The dataset includes samples of fresh bananas and examples of damaged bananas. Based on the dataset, it is evident that the MOS gas sensors function as intended because, as the sensor voltage progressively rose, the corresponding ppm levels also increased. This is valid for both of the observed gas concentrations. Since the dataset contains huge null values, data cleaning techniques had to be used before principal component analysis. The dataset's straightforward overview makes it clear that there is a contrast between the two stages of the banana's life cycle, fresh and spoilt, based on the recorded ppm levels and the matching sensor voltage values for each gas concentration. Thus, the dataset is divided into two classes, Class 1 and Class 0, with Class 1 representing the dataset's spoilt characteristics and Class 0 representing its fresh features.

4. Results and Discussion

The dataset in this experiment is split into fresh and rotten categories. The dataset's "sensor_voltage" value for CO and "sensor_voltage" value for alcohol-independent variables served as the basis for the evaluation. A 91.3% percent of the scatter in the data is explained by the first principal component (PC1), while the second PC (PC2) defines the other 7.8%. The principal component analysis plot is displayed in Figure 2a, and it can be seen that there is a clear separation between the two classes of categorization in the two clusters of the eigen values of the two principal components. As a result, a well-fitting hyperplane may be plotted to categorise these two.

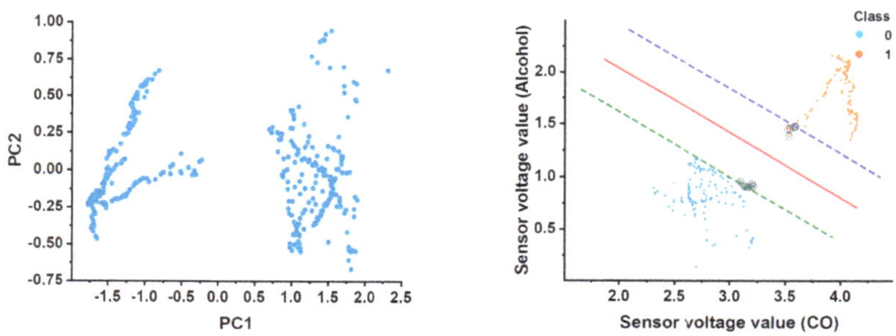

Figure 2. Scatter plots: (**a**) PCA analysis; (**b**) SVM analysis.

In the case of SVM, the assessment was carried out to distinguish between the "sensor_voltage" values of CO and alcohol, respectively. The visualization of these results is provided in Figure 2b. Using 'NumPy' library functions, the dataset was cleaned to eliminate any null or missing values, and it was normalized using 'StandardScaler' from the 'Scikit-learn' framework. Cross-validation, which helped the model avoid overfitting, resulted in an accuracy of 99.96%. The linear SVM model fits a hyperplane with a clear distinction between the two classes of fruit: fresh fruit belongs in class 0, whereas ruined fruit belongs in class 1.

Figure 3 displays one of the sample responses obtained from the sensors. While the reaction magnitude of the MQ7 sensor was somewhat higher than that of the MQ2 sensor, the transient responses of both MQ sensors had nearly comparable response and recovery times. The comparison between our model and earlier research in Table 1 shows that our model has a relatively good accuracy rate.

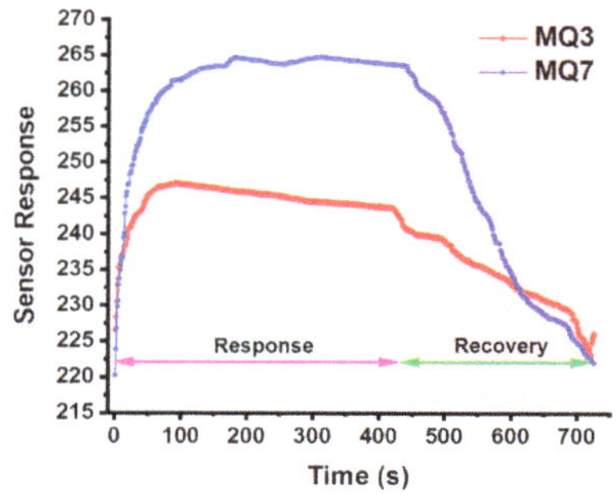

Figure 3. Sensor response of MQ3 and MQ7.

Table 1. Comparison of suggested and existing schemes.

Works	Application	Algorithm Used	No. of Sensors	Type of Sensor	Accuracy
[10]	Classification of Polish honey	PCA, LDA, CA	5	TGS	96.00%
[11]	Discrimination of oranges	PCA, NIPALS (Non-linear Iterative Partial Least Squares)	NR	NR	95.70%
[12]	Spoiling detections in octopus	PCA, DFA	6	MOSFET	93.80%
[13]	Discrimination of geographical origin of virgin olive oil	PCA, LDA	6	MOS	NR
[14]	Chinese liquor classification	Bio-inspired neural network	9	TGS, MiSC, and SHT75	NR
[15]	Detection of the E. coli bacteria in drinking water	PCA, deterministic finite automaton (DFA), PNN (probabilistic neural network)	16	TGS	80.00%
This work	Discrimination and identification of fresh and spoiled fruits	PCA, SVM	2	MQ	99.96%

NR: Not reported.

5. Conclusions

The proposed system uses two gas sensors (MQ3 and MQ7) and an Arduino Uno for processing. Principal component analysis (PCA) is a popular method for distinguishing fresh from bad apples using sensory data. The study generated 99.1% cumulative variance in one week. A linear Support vector machine (SVM) classifier evaluated the data with 99.96% accuracy. The technology uses real-time data to assess deterioration and deploy a low-cost, simple model that can be used anywhere to preserve any fruit. Based on the obtained findings, a low-cost system with the used sensor array and pattern recognition algorithm in this investigation exhibit potential for the detection and classification of under- and over-ripe fruits. Further, with the reported approach, one can distinguish between the different stages in the ripping process.

Author Contributions: Conceptualization, U.N.T.; methodology, U.N.T.; software, U.N.T. and A.K.; validation, U.N.T.; formal analysis, U.N.T.; investigation, U.N.T. and A.K.; resources, A.K.; data curation, U.N.T.; writing—original draft preparation, A.K.; writing—review and editing, A.K.;

visualization, U.N.T. and A.K.; supervision, A.K. All authors have read and agreed to the published version of the manuscript.

Funding: This research received no external funding.

Institutional Review Board Statement: Not applicable.

Informed Consent Statement: Not applicable.

Data Availability Statement: The datasets generated during and/or analysed during the current study are available from the corresponding author on reasonable request.

Conflicts of Interest: The authors declare no conflicts of interest.

References

1. Xue, L.; Liu, G.; Parfitt, J.; Liu, X.; Van Herpen, E.; Stenmarck, Å.; O'Connor, C.; Östergren, K.; Cheng, S. Missing Food, Missing Data? A Critical Review of Global Food Losses and Food Waste Data. *Environ. Sci. Technol.* **2017**, *51*, 6618–6633. [CrossRef] [PubMed]
2. Harvey, J.M. Reduction of Losses in Fresh Market Fruits and Vegetables. *Annu. Rev. Phytopathol.* **1978**, *16*, 321–341. [CrossRef]
3. Stitzel, S.E.; Aernecke, M.J.; Walt, D.R. Artificial Noses. *Annu. Rev. Biomed. Eng.* **2011**, *13*, 607026. [CrossRef] [PubMed]
4. Green, G.C.; Chan, A.D.; Goubran, R.A. Monitoring of food spoilage with electronic nose: Potential applications for smart homes. In Proceedings of the 2009 3rd International Conference on Pervasive Computing Technologies for Healthcare, London, UK, 1–3 April 2009; pp. 1–7. [CrossRef]
5. Aung, M.M.; Chang, Y.S. Temperature management for the quality assurance of a perishable food supply chain. *Food Control.* **2014**, *40*, 198–207. [CrossRef]
6. Loutfi, A.; Coradeschi, S.; Mani, G.K.; Shankar, P.; Rayappan, J.B.B. Electronic noses for food quality: A review. *J. Food Eng.* **2015**, *144*, 103–111. [CrossRef]
7. Berna, A. Metal Oxide Sensors for Electronic Noses and Their Application to Food Analysis. *Sensors* **2010**, *10*, 3882–3910. [CrossRef] [PubMed]
8. Mohamed, R.R.; Yaacob, R.; A Mohamed, M.; Dir, T.A.T.; A Rahim, F. Food Freshness Using Electronic Nose and Its Classification Method: A Review. *Int. J. Eng. Technol.* **2018**, *7*, 49. [CrossRef]
9. Sarno, R.; Wijaya, D.R. Recent development in electronic nose data processing for beef quality assessment. *Telecommun. Comput. Electron. Control* **2019**, *17*, 337–348. [CrossRef]
10. Dymerski, T.; Gębicki, J.; Wardencki, W.; Namieśnik, J. Application of an electronic nose instrument to fast classification of Polish honey types. *Sensors* **2014**, *14*, 10709–10724. [CrossRef] [PubMed]
11. Centonze, V.; Lippolis, V.; Cervellieri, S.; Damascelli, A.; Casiello, G.; Pascale, M.; Logrieco, A.F.; Longobardi, F. Discrimination of geographical origin of oranges (*Citrus sinensis* L. Osbeck) by mass spectrometry-based electronic nose and characterization of volatile compounds. *Food Chem.* **2018**, *277*, 25–30. [CrossRef] [PubMed]
12. Zhang, S.; Xie, C.; Bai, Z.; Hu, M.; Li, H.; Zeng, D. Spoiling and formaldehyde-containing detections in octopus with an E-nose. *Food Chem.* **2009**, *113*, 1346–1350. [CrossRef]
13. Haddi, Z.; Amari, A.; Ali, A.O.; El Bari, N.; Barhoumi, H.; Maaref, A.; Jaffrezic-Renault, N.; Bouchikhi, B. Discrimination and identification of geographical origin virgin olive oil by an e-nose based on MOS sensors and pattern recognition techniques. *Procedia Eng.* **2011**, *25*, 1137–1140. [CrossRef]
14. Liu, Y.-J.; Zeng, M.; Meng, Q.-H. Electronic nose using a bio-inspired neural network modeled on mammalian olfactory system for Chinese liquor classification. *Rev. Sci. Instrum.* **2019**, *90*, 025001. [CrossRef] [PubMed]
15. Carrillo-Gómez, J.; Durán-Acevedo, C.; García-Rico, R. Concentration detection of the E. coli bacteria in drinkingwater treatment plants through an E-nose and a volatiles extraction system (VES). *Water* **2019**, *11*, 774. [CrossRef]

engineering proceedings

MDPI

Proceeding Paper

Enhanced Safety Logic Solver Utilizing 2oo3 Architecture with Memristor Integration †

Chuthong Summatta and Somchat Sonasang *

Faculty of Industrial Technology, Nakhon Phanom University, Nakhon Phanom 48000, Thailand; chuthong@npu.ac.th
* Correspondence: somchat.s@npu.ac.th; Tel.: +66-091-9498-425

† Presented at the 10th International Electronic Conference on Sensors and Applications (ECSA-10), 15–30 November 2023; Available online: https://ecsa-10.sciforum.net/.

Abstract: A safety instrument function (SIF) averts hazardous incidents that may arise due to diverse anomalies within a system. The SIF prevents potential dangers by comprising three integral components—the sensing element, the logic solver, and the final element. The 2oo3 architecture is the optimal configuration for each SIF component, employing both AND and OR logic designs for its voting mechanism. Type A devices, recognized for their passive nature, exemplify robustness and reliability. While these devices are acknowledged as the most dependable, semi-conductor devices or microcontrollers, categorized as Type B, often find application in logic processing. This paper introduces the incorporation of memristors, which are inherently passive devices with memory attributes, into the system. The logic solver, which calculates confidence values, exhibited greater efficacy than Type B devices. Verification was conducted via LTspice circuit simulations. The results of the memristor for Logic Solver in the safety instrumentation function (SIF) IEC 61508/61511 standard are as follows: The voter circuit has the lowest components and failure rate and highest mean time to failure. This is more reliable than the other voter.

Keywords: memristor; 2oo3; logic solver; fail-safe

1. Introduction

The enhanced safety logic solver holds a pivotal role in fail-safe systems, and it is primarily responsible for meticulously processing the monitoring of sensor unit failures within a given system [1–4]. It is the linchpin of safety mechanisms dedicated to averting potential hazards that may ensue from diverse anomalies that can occur for various reasons. Such hazards are especially critical in railway signaling, elevator door mechanisms, and electric motor drive systems. The safety evaluation of the 2oo3 voting logic solver is paramount in achieving a stringent safety level of SIL3. This intricate process involves receiving signals from three Type A sensors and meticulous processing via a static voting component. Subsequently, the signals are directed toward the logic solver and window comparator circuits [5,6], culminating in transmitting a DC signal to the final element. The components within this subsystem encompass both Type A and Type B devices. Type A devices are characterized by their fundamental components, such as resistors, inductors, and capacitors, each with clearly defined failure modes. They consistently operate reliably even under predefined fault conditions, and their data integrity remains intact despite failure. In stark contrast, Type B devices incorporate integrated circuit components or microprocessors, which, while versatile, are less reliable, possess shorter lifespans, and are inherently less verifiable. As a groundbreaking addition to the field, memristors represent a novel category of passive devices [7] and are now recognized as the fourth fundamental electrical element. They exhibit behavior akin to resistors regarding ohmic characteristics while boasting intrinsic memory attributes [7–9]. Memristors have found extensive applications in various circuit domains, including digital circuits [10–14].

Citation: Summatta, C.; Sonasang, S. Enhanced Safety Logic Solver Utilizing 2oo3 Architecture with Memristor Integration. *Eng. Proc.* **2023**, *58*, 37. https://doi.org/10.3390/ecsa-10-16006

Academic Editor: Jean-marc Laheurte

Published: 15 November 2023

In the following sections, this paper presents the design and development of a two-out-of-three voter system that expertly incorporates memristors. These innovative components are harnessed to craft a robust logic circuit that is specifically tailored for deployment within the Safety Instrument Function Level 3 (SIL3) system [15,16]. Performance analysis is carried out via the meticulous examination of the mean time of failure (MTTF) [17], with rigorous testing facilitated by the LTspice simulation program [18].

2. Fundamental Theory and Method

2.1. Safety Instrumentation Function

The safety instrumentation function (SIF), as shown in Figure 1, guided by IEC 61508/61511 standards [15], safeguards operational systems from accidents caused by anomalies. It comprises three key elements. The sensing element converts process variables into standardized electrical signals for monitoring and the control system display. The logic solver processes signals, compares them to set values, and delivers them to the end device. It includes a power supply, central processing, communication, and control components. The final element converts processed signals into a format that is suitable for the system controller, and it is vital for safety considerations. The IEC standards assign a safety integrity level (SIL), ensuring that equipment design aligns with acceptable error levels considering cumulative component failure hazards.

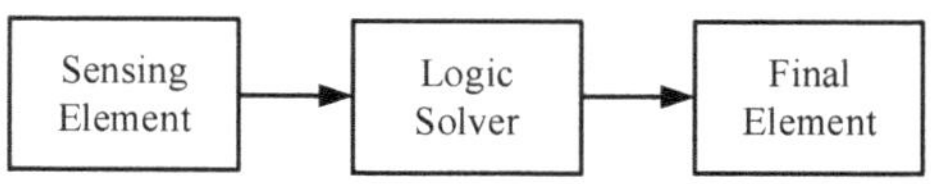

Figure 1. Safety instrumentation function (SIF) [15].

2.2. The 2oo3 Voting

The 2oo3 voting consists of six switches that detect errors in 2 out of 3 detectors, and the logic design has three AND logics and OR logics following Equation (1).

$$Y = AB + BC + CA \quad (1)$$

From past research [1–3], the 2oo3 detection method is very safe. It is a dynamic processing model. Static processing is a process without a contact, like a dynamic one, but uses a static switch instead. The use of the equation from the same logic equation is shown in Figure 2.

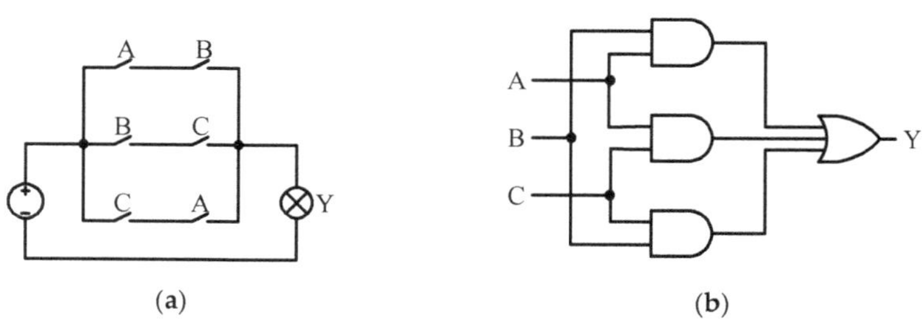

Figure 2. The 2oo3 voting: (**a**) relay circuits; (**b**) logic circuits.

2.3. The Memristor

A memristor is a bipolar passive element with a relationship between charge and flux [7]. A memristor is a combination of the term memory and resistor. Ideally, the resistance of the memristor increases if the charge flows through the memristor in one direction, and it decreases if the current flows in the other direction. When the current stops flowing, the resistance of the memristor remains constant [7–9]. The symbol and current–voltage characteristic curve are shown in Figure 3.

Figure 3. Memristor [7]: (**a**) symbol; (**b**) *I-V* characteristic curve.

The characteristics of the memristor can result in a voltage divider circuit, as presented in [10–14], and the circuit can create OR logic and AND logic gate circuits, as shown in Figure 4.

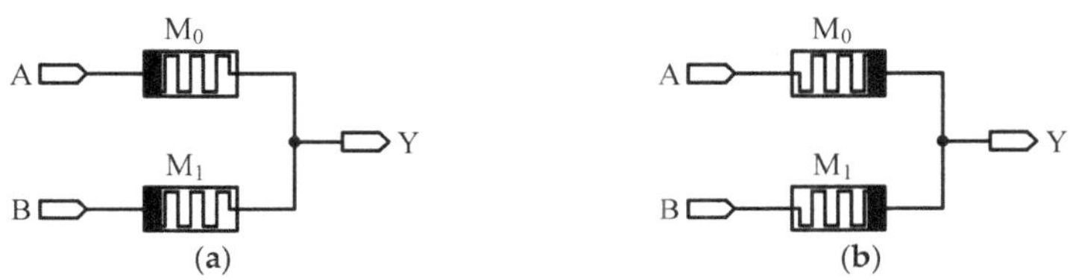

Figure 4. Logical OR and AND with memristor [8,9]: (**a**) logical OR; (**b**) logical AND.

The ideal use of a memristor as a logic circuit is a passive circuit that does not require a power supply, utilizing a logic output signal generated by a voltage divider from the memristor. Moreover, gates and OR gates behave differently due to the property of the memristor in terms of the current flow direction. Input side logic 1 is the supply voltage, and logic 0 is the ground. The output side receives logic 0 or 1 from the voltage provided by the memristor's voltage divider. The operation of AND logic with a memristor is shown in Figure 5.

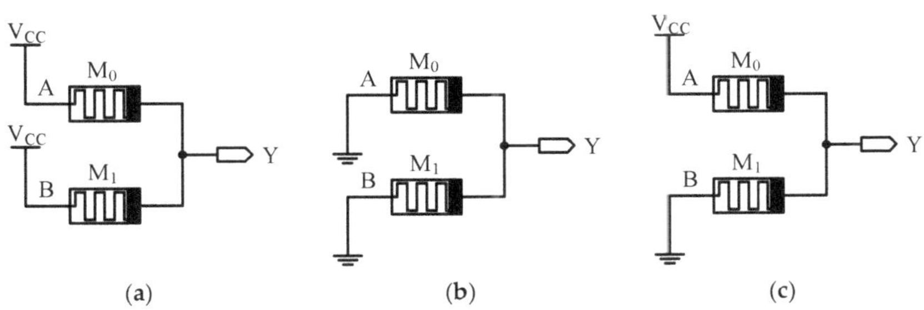

Figure 5. Logical AND with memristor [10,11]: (**a**) $M_0 = M_1 = R_{off}(\mathrm{Y} = \mathrm{V_{CC}})$; (**b**) $M_0 = M_1 = R_{on}(\mathrm{Y} = \mathrm{GND})$; (**c**) $M_0 = R_{off}, M_1 = R_{on}(\mathrm{Y} = logic\text{“}0\text{”})$.

Figure 5a shows that input A is logic 1 and input B is logic 1; no current flows through the circuit, and the output is logic 1 or $\mathrm{V_{CC}}$. Figure 5b shows that input A is logic 0 and B is logic 0; there is no current flow through the circuit, and the output is logic 0 or GND. Figure 5c shows that input A is logic 1, and B is logic 0, the resistance of memristor $\mathrm{M_0}$ increases (R_{off}), the resistance of Memristor $\mathrm{M_1}$ decreases (R_{on}), the current can flow from $\mathrm{V_{CC}}$ to GND, and the output is logic 0.

Figure 6a shows that input A is logic 1, and input B is logic 1. The current can flow through the circuit, and the output is logic 1 or $\mathrm{V_{CC}}$. Figure 6b shows that input A is logic 0 and B is logic 0; there is no current flow through the circuit, and the output is logic 0 or GND. Figure 6c shows that input A is logic 1, and B is logic 0. The resistance of memristor $\mathrm{M_0}$

decreases (R_{on}), the resistance of memristor M_1 increases (R_{off}), the current can flow from V_{CC} to node Y, and the output is logic 1.

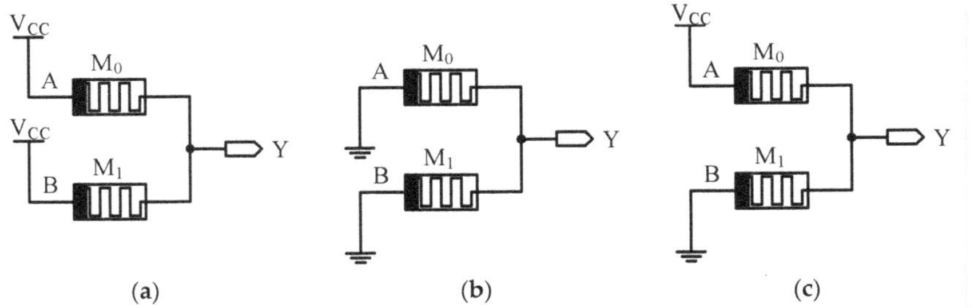

Figure 6. Logical OR with memristor [10,11]: (**a**) $M_0 = M_1 = R_{on}(\mathrm{Y} = \mathrm{V_{CC}})$; (**b**) $M_0 = M = R_{off}(\mathrm{Y} = \mathrm{GND})$; (**c**) $M_0 = R_{on}, M_1 = R_{off}(\mathrm{Y} = logic``1")$.

3. Results and Discussion

3.1. 2oo3 Logic Solver with Memristor

Based on the theories and methods mentioned above, a 2oo3 logic solver with a memristor circuit can be designed, as shown in Figure 7.

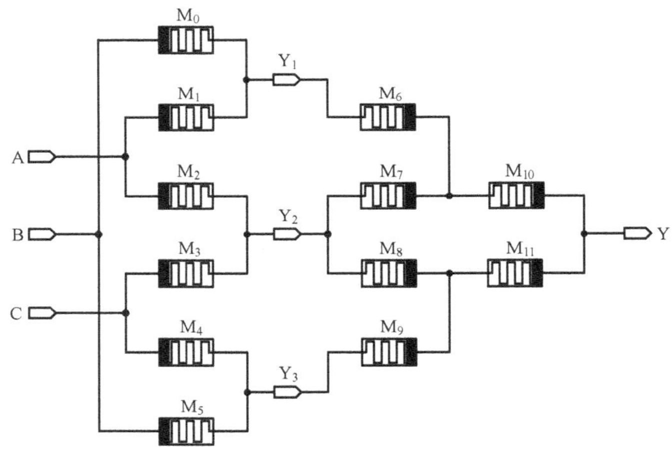

Figure 7. The 2oo3 logic solver with memristor.

Figure 7 shows memristors M_0 to M_5 instead of three two-input AND gates, and it shows M_6 to M_{11} instead of the three-input OR gate. The operation of the circuit was tested with the simulation program LTspice. The results are obtained according to the truth table of 2oo3, as shown in Table 1.

Table 1. The truth table of 2oo3 voting.

A	B	C	Y
0	0	0	0
0	0	1	0
0	1	0	0
0	1	1	1
1	0	0	0
1	0	1	1
1	1	0	1
1	1	1	1

3.2. Reliability Value

The military handbook "Reliability Prediction of Electronic Equipment" (MIL-HDBK-217F) [18] proposed a failure rate for electronic equipment. This work compares 2oo3 voters in different logic IC and memristor devices (in this case, the memristor is replaced with a resistor), as shown in Table 2.

Table 2. The failure of electronic equipment.

No.	Notation	Components	Failure Rate (10^{-6} h)
1	λ_P	Opto-coupler [2]	0.08160
2	λ_D	Diode [2]	0.01213056
3	λ_P	Logic ICs	0.15
4	λ_P	Resistor	0.0069

The failure rate of electronic 2oo3 voter devices can be calculated via the sum of the failure rate, comparing the two devices using the same model: the Chip Carrier (Surface Mount) Package Type Correction Factor. The overall failure rate, the decrease in the number of devices with respect to a lower failure rate, and the mean time of failure [17] are the approximations of reliability with respect to the failure rate. If the failure rate is lower, the average time to failure increases, as shown in Table 3.

Table 3. Comparison of the sum of failure rate and mean time to failure.

Type of 2oo3 Voter Devices	The Sum of Failure Rate (Hour)	The Mean Time to Failure (Hour)
Opto-coupler [2]	0.4896×10^{-6}	2,042,483
Opto-coupler and Diode [2]	0.3385×10^{-6}	2,953,947
Logic ICs	0.6000×10^{-6}	1,666,666
Memristor	0.0414×10^{-6}	24,154,589

Table 3 shows a comparison of the sum of the failure rate and the mean time to failure, showing that the memristor has the lowest failure rate and the highest mean time to failure. Four logic gates and six memristors are used to calculate the sum of the failure rate of logic ICs. The proposed 2oo3 with the memristor has a reliability value that is 14.49 times that of the logic IC model.

4. Conclusions

This paper introduces a novel 2oo3 voter integrated with a memristor for the Enhanced Safety Logic Solver in the safety instrumentation function (SIF) following the IEC 61508/61511 standard. The objective of this study is to showcase an alternative application of memristors. Notably, the proposed voter circuit for the Enhanced Safety Logic Solver exhibits several advantages, including reduced component count, lower failure rates, and superior mean time to failure. These attributes collectively render it a more reliable choice compared to conventional alternatives.

Author Contributions: Conceptualization, C.S.; methodology, C.S. and S.S.; software, C.S. and S.S.; validation, S.S., formal analysis, C.S.; investigation, C.S. and S.S.; writing—original draft preparation, C.S.; writing—review and editing, S.S. All authors have read and agreed to the published version of the manuscript.

Funding: This research received no external funding.

Institutional Review Board Statement: Not applicable.

Informed Consent Statement: Not applicable.

Data Availability Statement: Not applicable.

Acknowledgments: The authors would like to demonstrate gratitude toward the Department of Electronic Technology, Faculty of Industrial Technology of the Nakhon Phanom University for providing research time, the research grant, instrumentation, and financial support.

Conflicts of Interest: The authors declare no conflict of interest.

References

1. Deeon, S.; Hirao, Y.; Futsuhara, K. A Fail-safe Counter and its application to Low-speed Detection. *Trans. Reliab. Eng. Assoc. Jpn.* **2011**, *33*, 137–146.
2. Summatta, C.; Deeon, S. Design and Analysis of 2oo3 Static Voter for SMT function in an Adjustable Speed Electrical Power Drive System. *J. Telecommun. Electron. Comput. Eng. (JTEC)* **2019**, *11*, 1–6.
3. Summatta, C.; Khamsen, W.; Pilikeaw, A.; Deeon, S. Design and Analysis of 2-out-of-3 Voters Sensing in Electrical Power Drive System. In Proceedings of the Conference on Electrical Engineering/Electronics, Computer, Telecommunications and Information Technology 2016, Chiang Mai, Thailand, 28 June–1 July 2016.
4. Summatta, C.; Rattanangam, W. The improvement of a fail-safe counter for low-speed detection. *Przegląd Elektrotechniczny* **2021**, *97*, 23–28. [CrossRef]
5. Summatta, C.; Deeon, S. A Window Comparator Circuit with Digital Switching Levels. In Proceedings of the 9th International Conference on Sciences, Technology and Innovation for Sustainable Well-Being (STISWB 2017), Kunming University of Sciences and Technology, Kunming, China; 2017; pp. 74–78.
6. Summatta, C.; Phurahong, T.; Rattanangam, W.; Chaiyong, W. Low-cost and Compact Window Comparator Circuit with MOSFET-Resistor Voltage References. In Proceedings of the IEEE 2nd International Conference on Power and Energy Applications (ICPEA 2019), Singapore, 27–30 April 2019; pp. 75–78.
7. Chua, L.O. Memristor the missing circuit element. *IEEE Trans. Circuit Theory* **1971**, *18*, 507–519. [CrossRef]
8. Keshmiri, V. *A Study of the Memristor Models and Applications*; Department of Electrical Engineering Linköping Universitet: Linköping, Sweden, 2014.
9. Krestinskaya, O.; Irmanova, A.; James, A.P. Memristors: Properties, models, materials. In *Deep Learning Classifiers with Memristive Networks: Theory and Applications*; Springer: Cham, Switzerland, 2020; Volume 14, pp. 13–40.
10. Borghetti, J.; Snider, G.S.; Kuekes, P.J.; Yang, J.J.; Stewart, D.R.; Williams, R.S. Memristive switches enable stateful logic operations via material implication. *Nature* **2010**, *464*, 873–876. [CrossRef] [PubMed]
11. Singh, T. Hybrid Memristor-CMOS (MeMOS) based Logic Gates and Adder Circuits. *ArXiv* **2015**, arXiv:1506.06735.
12. Huang, Y.; Li, S.; Yang, Y.; Chen, C. Progress on Memristor-Based Analog Logic Operation. *Electronics* **2023**, *12*, 2486. [CrossRef]
13. Fahmy, G.A.; Zorkany, M. Design of a Memristor-Based Digital to Analog Converter (DAC). *Electronics* **2021**, *10*, 622. [CrossRef]
14. Mladenov, V. Application of Metal Oxide Memristor Models in Logic Gates. *Electronics* **2023**, *12*, 381. [CrossRef]
15. *IEC 61508-1:2010*; Functional safety of electrical/electronic/programmable electronic safety-related systems. International Electrotechnical Commission: London, UK, 2010.
16. Krasich, M. How to estimate and use MTTF/MTBF would the real MTBF please stand up. In *Reliability and Maintainability Symposium*; IEEE: Fort Worth, TX, USA, 2009; pp. 353–359.
17. *MIL-HDBK-217F*; Military Handbook Reliability Prediction of Electronic Equipment. US Department of Defense: Washington, DC, USA, 1991.
18. Mladenov, V. A Unified and Open LTSPICE Memristor Model Library. *Electronics* **2021**, *10*, 1594. [CrossRef]

engineering proceedings

MDPI

Proceeding Paper

Traffic Stream Characteristics Estimation Using In-Pavement Sensor Network †

Mu'ath Al-Tarawneh

Department of Civil and Environmental Engineering, Mutah University, Karak 61710, Jordan; muath.altarawneh@mutah.edu.jo

† Presented at the 10th International Electronic Conference on Sensors and Applications (ECSA-10), 15–30 November 2023; Available online: https://ecsa-10.sciforum.net/.

Abstract: The numbers of vehicles on the roads has increased tremendously. Also, the number of roads that are constantly experiencing traffic jams during morning and evening peak hours has increased significantly, which calls for a better understanding of traffic stream characteristics and car-following models. Traffic stream macroscopic parameters (speed, flow, and density) could be estimated through a number of traffic-flow theory models. In order to collect accurate data regarding fundamental of traffic stream parameters, a traffic monitoring system is needed to present the data from different roads. In this study, a real-time traffic monitoring system is introduced for traffic macroscopic parameters estimation. The sensor network has been constructed using a set of linear fiber optic sensors. In order to validate the system for this study, the system was installed at MnROAD facility, Minnesota. Fiber optic sensor detects the propagated strains in highway pavement due to the vehicle movements through the changes of the laser beam characteristics. Traffic flow can be estimated by tracking the peak of each axle passed over the sensor or within the sensitivity area, time mean speed (TMS), and space mean speed (SMS). SMS can be estimated by the different times a vehicle arrived at the sensors. The density can be determined either by using fundamental traffic flow theory model or estimating the time that vehicles occupy the sensor layout. Real traffic was used to validate the sensor layout. The results show the capability of the system to estimate traffic stream characteristics successfully.

Keywords: traffic stream; macroscopic parameters; ITS; in-pavement sensors; traffic data

1. Introduction

The number of vehicles has increased tremendously on roads due to rapid economic growth and urbanization as traffic volume increases on highway networks, contributing to traffic congestion, vehicle accidents, traffic safety, energy consumption, economic impact due to fuel and time waste, and green gas emission [1–3]. To better understand and overcome the traffic volume increment problems, vehicle-to-vehicle (V2V) and vehicle-to-infrastructure (V2I) interactions must necessarily be studied [4]. Traffic flow theory plays a critical role in understanding V2V interactions through a set of macroscopic parameters, including flow (q, vph), density (k, vpkm), and speed (S, kmph). These parameters could be estimated by a process called traffic state estimation (TSE), where an accurate estimation of macroscopic parameters (q, k, and s) plays a central role in the success of the developed traffic management system to mitigate traffic congestion problems [1]. Furthermore, TSE is vital for traffic planning (infrastructure management, maintenance) and traffic management (traffic control, ramp metering) [5].

Currently, the collected traffic data type can be sorted into two categories based on the measurement method: stationary data collected through fixed detectors, including loop detectors, inductive loops, closed-circuit television cameras (CCTV), fiber optic sensors [5–8]; and mobile data, such as on-vehicle GPS probe, OBD2, mounted video cameras, mobile

Citation: Al-Tarawneh, M. Traffic Stream Characteristics Estimation Using In-Pavement Sensor Network. *Eng. Proc.* **2023**, *58*, 38. https://doi.org/10.3390/ecsa-10-16007

Academic Editor: Stefano Mariani

Published: 15 November 2023

phone, connected vehicle [5,9]. Despite expanding the use of mobile-collected data recently, it has problems with sampling and shows the potential for biases [5]. However, stationary collected data through fixed sensors represents every vehicle at their installed locations.

Fiber optic sensors have become widely accepted in traffic management and operation [10,11]. Fiber Bragg Grating (FBG) sensors are preferred among fiber optic sensors due to the distinctive characteristics that give them superiority over counterparts from other fixed sensors, including good linearity in response, accurate output, high resolution, small size, and low cost [12]. However, FBG sensors must be packaged since they are made of fiberglass. Currently, glass fiber reinforced polymers (GFRPs) are used in several fields, such as oil and gas, the pipe industry, structural elements, and aerospace [13], which makes it promising to be used as FBG sensor package material. Thus, this study introduces a stationary system for traffic stream parameter estimation based on GFRP-FBG sensor layout.

2. Sensor Fundamental

FBG sensor detects propagated strains (ε) in highway pavement due to vehicle movements through the changes in the laser beam characteristics, as shown in Equation (1) [14]:

$$\varepsilon = \frac{1}{(1-Op)}\left(\frac{\Delta\gamma_l}{\gamma_l} - \frac{\Delta\gamma_T}{\gamma_T}\right) \tag{1}$$

where Op is the optical elasticity parameter, γ_l is the Bragg wavelength due to mechanical load and temperature, and γ_T is the Bragg wavelength from a compensation temperature sensor.

Figure 1 shows the developed GFRP-FBG sensor by Zhou et al. [15], which will be used in this study for traffic stream characteristics estimation. Figure 1a shows the longitudinal component of the sensor, which is designed to detect the propagated strain in the x-y coordination. Figure 1b illustrates the vertical component of the sensor to monitor vertical strain.

(a) (b)

Figure 1. GFRP-FBG sensor: (**a**) Longitudinal sensor; (**b**) Vertical component sensor. All units in (in.).

3. Methodology

To construct a traffic stream characteristics system, a minimum of two GFRP-FBG sensors must be installed along the road sequentially with a distance of (L) under the proposed vehicle tire track to increase the accuracy of the sensor measurements, with one FBG sensor to compensate for the temperature cross-sensitivity. Figure 2 shows the proposed sensor layout to estimate traffic macroscopic parameters, which consists of three GFRP-FBG sensors: a horizontal sensor (H-S), a vertical sensor(V-S) to track vehicle movement, and a horizontal temperature compensation sensor (H-TS). In this study, the distance L will be 4.87 m, which falls between the optimized distance interval from 2.13 m and 6.1 m to maximize the measurement accuracy for macroscopic parameters estimation.

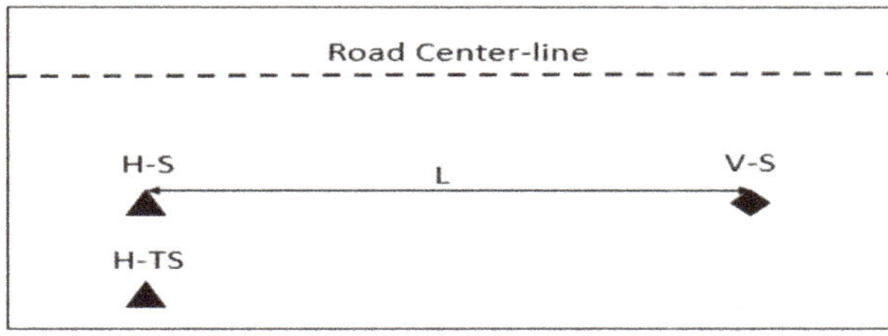

Figure 2. Proposed sensor layout.

Estimating the traveled time for each passing vehicle between the two stationary sensors plays a vital role in deriving all macroscopic parameters (q, k, and s). Figure 3 illustrates the responses from the H-S sensor (positive measurement due to tension in the pavement along the road) and the V-S sensor (negative measurement due to compression at the location of the vertical sensor) for two consecutive vehicles (5 axle leading vehicles and 2 axle following vehicles).

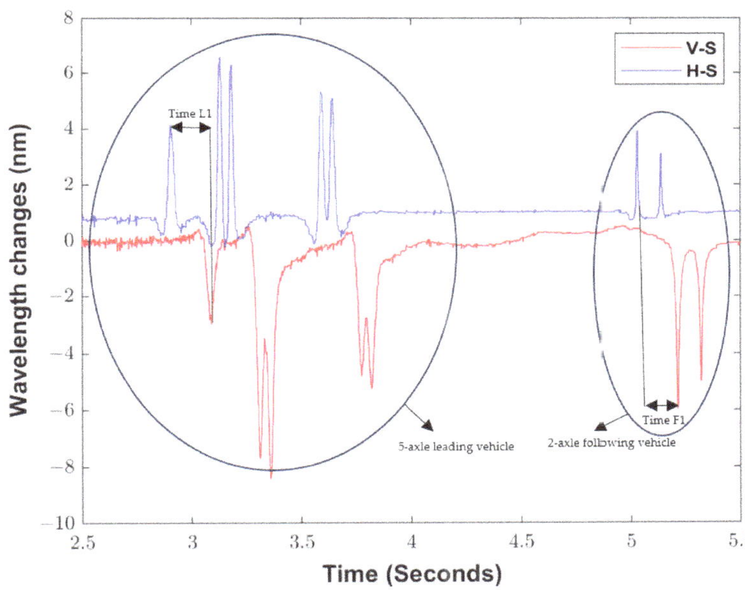

Figure 3. Sensor's response.

As noted in Figure 3, the sensor response will be more likely to have a peak shape due to the compact sensor's size and higher vehicle speed, where each peak indicates the passing of one axle. Thus, the peaks will be tracked and counted accordingly for traffic flow (q) estimation, as stated in Equation (2). The recorded time (Time L1 and Time F1) indicates how long each vehicle travels over the known distance of (L) between the horizontal and the vertical sensors. Then, time mean speed (TMS) and space mean speed (SMS) can be calculated using Equations (3) and (4), respectively.

$$q = \frac{N}{T} \tag{2}$$

$$TMS = \frac{\sum_{i=1}^{N} L/t_i}{N} \tag{3}$$

$$SMS = \frac{N \cdot L}{\sum_{i=1}^{N} t_i} \quad (4)$$

where N is the total number of vehicles in a period of time (T), and t_i is the traveled time for a vehicle (i).

Therefore, density (k) can be calculated using the fundamental flow–density relationship, as follows:

$$k = \frac{q}{SMS} \quad (5)$$

4. Results and Discussion

To validate the system for traffic stream macroscopic estimation, the sensor layout was installed at the Cold Weather Road Research Facility in Minnesota (MnRoad) in Mn, USA. MnROAD consists of 50 test sections distributed on two roadways: a low-volume loop operated by 5 axle semi-truckand a section of (I-94) that contains two lanes operated by live traffic. The proposed sensor layout (Figure 2) was installed inside cell 17, part of I-94 at MnROAD, as shown in Figure 4.

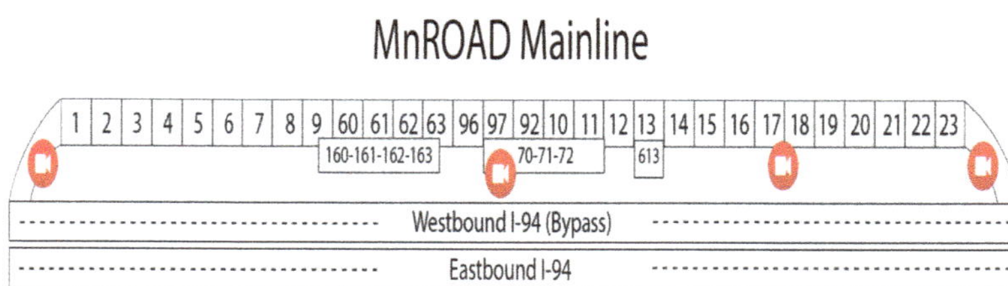

Figure 4. MnROAD test section [16].

After the installation, the sensors were connected to a high-resolution integrator and personal computer to collect data. Figure 5 shows the response of both sensors (vertical and horizontal) when subjected to real traffic for 140 s to validate the system. From Figure 5, a total of 11 vehicles were recorded and tracked by the system. Thus, the flow (q) could be estimated, as equal to 283 vph.

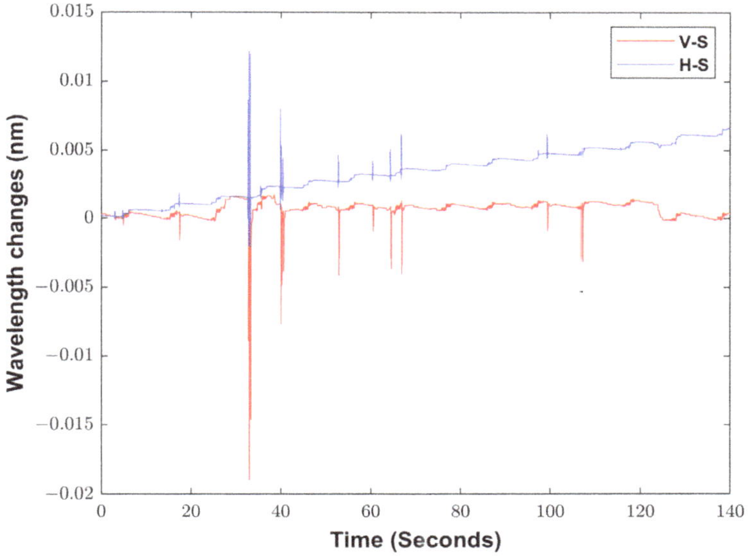

Figure 5. Sensors' response for 140 s.

Table 1 shows the estimated spot speed for the eleven vehicles recorded during 140 s, since the distance between sensors in the layout is defined (4 88 m). The system records the time at which the vehicle enters the sensor layout (Tin) and when the rear wheel exits the layout (Tout) to calculate the travel time for each vehicle, as shown in Table 1. Then, TMS, SMS, and K can be calculated using Equations (2)–(4), respectively. Therefore, TMS equals to 110.00 km\h, SMS equals 108.74 km\h, and k equals to 2.6 v/km.

Table 1. Estimated spot speed.

Vehicle	T_{in}	T_{out}	Travel Time (s)	Spot Speed (km/h)
1	4.923	5.069	0.146	120.25
2	17.44	17.58	0.14	125.40
3	33.23	33.4	0.17	103.27
4	35.61	35.78	0.17	103.27
5	40.52	40.68	0.16	109.73
6	52.82	53	0.18	97.54
7	60.41	60.58	0.17	103.27
8	64.44	64.59	0.15	117.04
9	66.9	67.05	0.15	117.04
10	99.35	99.49	0.14	125.40
11	107.1	107.3	0.2	87.78

5. Conclusions

This study successfully introduces a traffic stream characteristics estimation system based on an in-pavement sensor layout. The developed system will be durable since the sensors are packaged in GFRP material. The number of passed vehicles are tracked and counted for traffic flow estimation; with a higher sampling rate, the travel time of each vehicle is recorded with high accuracy. Then, speed (TMS and SMS) is estimated with previous knowledge of the distance between sensors. Future efforts will be applied to estimate traffic microscopic parameters (headway and spacing).

Funding: This research received no external funding.

Institutional Review Board Statement: Not applicable.

Informed Consent Statement: Not applicable.

Data Availability Statement: Available upon request.

Acknowledgments: The author would like to thank MnROAD technicians who helped during field testing.

Conflicts of Interest: The author declares no conflicts of interest.

References

1. Xing, J.; Wu, W.; Cheng, Q.; Liu, R. Traffic State Estimation of Urban Road Networks by Multi-Source Data Fusion: Review and New Insights. *Phys. A Stat. Mech. Its Appl.* **2022**, *595*, 127079. [CrossRef]
2. Abbas, Z.; Al-Shishtawy, A.; Girdzijauskas, S.; Vlassov, V. Short-Term Traffic Prediction Using Long Short-Term Memory Neural Networks. In Proceedings of the 2018 IEEE International Congress on Big Data (BigData Congress), San Francisco, CA, USA, 2–7 July 2018; pp. 57–65.
3. Zhou, X.; Tanvir, S.; Lei, H.; Taylor, J.; Liu, B.; Rouphail, N.M.; Frey, H.C. Integrating a Simplified Emission Estimation Model and Mesoscopic Dynamic Traffic Simulator to Efficiently Evaluate Emission Impacts of Traffic Management Strategies. *Transp. Res. Part D Transp. Environ.* **2015**, *37*, 123–136. [CrossRef]
4. Asaithambi, G.; Kanagaraj, V.; Srinivasan, K.K.; Sivanandan, R. Study of Traffic Flow Characteristics Using Different Vehicle-Following Models under Mixed Traffic Conditions. *Transp. Lett.* **2018**, *10*, 92–103. [CrossRef]
5. Seo, T.; Bayen, A.M.; Kusakabe, T.; Asakura, Y. Traffic State Estimation on Highway: A Comprehensive Survey. *Annu. Rev. Control.* **2017**, *43*, 128–151. [CrossRef]
6. Kwon, J.; Varaiya, P.; Skabardonis, A. Estimation of Truck Traffic Volume from Single Loop Detectors with Lane-to-Lane Speed Correlation. *Transp. Res. Rec.* **2003**, *1856*, 106–117. [CrossRef]
7. Asmaa, O.; Mokhtar, K.; Abdelaziz, O. Road Traffic Density Estimation Using Microscopic and Macroscopic Parameters. *Image Vis. Comput.* **2013**, *31*, 887–894. [CrossRef]
8. Won, M. Intelligent Traffic Monitoring Systems for Vehicle Classification: A Survey. *IEEE Access* **2020**, *8*, 73340–73358. [CrossRef]
9. Li, H.; Qin, L.; Chang, X.; Rong, J.; Ran, B.; Jia, L. Sensor Layout Strategy and Sensitivity Analysis for Macroscopic Traffic Flow Parameter Acquisition. *IET Intell. Transp. Syst.* **2017**, *11*, 212–221. [CrossRef]
10. Huang, M.-F.; Salemi, M.; Chen, Y.; Zhao, J.; Xia, T.J.; Wellbrock, G.A.; Huang, Y.-K.; Milione, G.; Ip, E.; Ji, P. First Field Trial of Distributed Fiber Optical Sensing and High-Speed Communication over an Operational Telecom Network. *J. Light. Technol.* **2019**, *38*, 75–81. [CrossRef]
11. Liu, H.; Ma, J.; Yan, W.; Liu, W.; Zhang, X.; Li, C. Traffic Flow Detection Using Distributed Fiber Optic Acoustic Sensing. *IEEE Access* **2018**, *6*, 68968–68980. [CrossRef]
12. Guo, H.; Xiao, G.; Mrad, N.; Yao, J. Fiber Optic Sensors for Structural Health Monitoring of Air Platforms. *Sensors* **2011**, *11*, 3687–3705. [CrossRef] [PubMed]
13. Landesmann, A.; Seruti, C.A.; de Miranda Batista, E. Mechanical Properties of Glass Fiber Reinforced Polymers Members for Structural Applications. *Mater. Res.* **2015**, *18*, 1372–1383. [CrossRef]
14. Zhang, Z.; Huang, Y.; Palek, L.; Strommen, R. Glass Fiber–Reinforced Polymer–Packaged Fiber Bragg Grating Sensors for Ultra-Thin Unbonded Concrete Overlay Monitoring. *Struct. Health Monit.* **2015**, *14*, 110–123. [CrossRef]
15. Zhou, Z.; Liu, W.; Huang, Y.; Wang, H.; Jianping, H.; Huang, M.; Jinping, O. Optical Fiber Bragg Grating Sensor Assembly for 3D Strain Monitoring and Its Case Study in Highway Pavement. *Mech. Syst. Signal Process.* **2012**, *28*, 36–49. [CrossRef]
16. MnROAD Mainline Traffic Cameras. Available online: https://www.dot.state.mn.us/mnroad/operations/traffic-cameras.html (accessed on 20 December 2023).

engineering proceedings

MDPI

Proceeding Paper

Tool Wear Estimation in the Milling Process Using Backpropagation-Based Machine Learning Algorithm †

Giovanni Oliveira de Sousa [1,2,*], Pedro Oliveira Conceição Júnior [1], Ivan Nunes da Silva [1], Dennis Brandão [1] and Fábio Romano Lofrano Dotto [1]

1 Department of Electrical and Computer Engineering, São Carlos School of Engineering, University of São Paulo USP, Av. Trab. São Carlense, 400-Pq. Arnold Schimidt, São Carlos 13566-590, SP, Brazil; pedro.oliveiracjr@usp.br (P.O.C.J.); insilva@sc.usp.br (I.N.d.S.); dennis@sc.usp.br (D.B.); fabio.dotto@usp.br (F.R.L.D.)

2 Aranouá Project, Federal Institute of Amazonas IFAM, R. dos Açaizeiros-São José Operário, Manaus 69086-485, AM, Brazil

* Correspondence: giovanni_oliveiradesousa@usp.br; Tel.: +55-92-98114-0320

† Presented at the 10th International Electronic Conference on Sensors and Applications (ECSA-10), 15–30 November 2023; Available online: https://ecsa-10.sciforum.net/.

Abstract: Tool condition monitoring (TCM) systems are essential in milling operations to guarantee the product's quality, and when they are paired with indirect measuring techniques, such as vibration or acoustic emission sensors, the monitoring can happen without sacrificing productivity. Some more advanced techniques in tool wear estimation are based on supervised machine learning algorithms, like several other applications in Industry 4.0's context; however, a satisfactory performance can be obtained with simple techniques and low computational power. This work focuses on an application of tool wear estimation using a simple backpropagation neural network in a milling dataset. Statistical techniques, i.e., the mean, variance, skewness, and kurtosis, were used as features that were extracted from indirect measurements from vibration and acoustic emission sensors' data in a real milling testbench dataset containing multiple experiments with sensor data and a direct measure of the flank wear (VB) in most instances. The data were preprocessed, specifically to acquire clean and normalized values for the neural network training, assuming that the VB measure would be the target variable used to predict tool wear; all incomplete samples without a VB measure, as well as outliers, were removed beforehand. The train and test subsets were chosen randomly after making sure that the maximum values of every variable were represented in the training subset. A multiple topology approach was implemented to test the configurations of multiple backpropagation neural networks to determine the most suitable one based on two performance criteria, i.e., the mean absolute percent error (MAPE) and variance. Although only a simple backpropagation algorithm was used, the results were adequate to demonstrate a balance between accuracy and computational resource usage.

Keywords: tool condition monitoring; backpropagation neural network; tool wear estimation

Citation: de Sousa, G.O.; Júnior, P.O.C.; da Silva, I.N.; Brandão, D.; Dotto, F.R.L. Tool Wear Estimation in the Milling Process Using Backpropagation-Based Machine Learning Algorithm. *Eng. Proc.* **2023**, *58*, 39. https://doi.org/10.3390/ecsa-10-15997

Academic Editor: Stefano Mariani

Published: 15 November 2023

1. Introduction

Machining is a material removal process that is mainly used in computer numerical control (CNC) systems. The system's cutting tool is worn down by the workpiece with each operation, and this can irreversibly affect the surface of the workpiece, so tool condition monitoring (TCM) systems are essential to preventing failures and guaranteeing the quality of the product [1].

The use of TCM systems can be divided into two groups [2]: (1) direct monitoring, which directly measures fault values such as tool wear (TW) and utilizes more expensive lasers and optical sensors, and (2) indirect monitoring, which measures the physical parameters that represent the tool condition parameters indirectly. Some direct monitoring methods that use microscopy or vision systems require the machining process to be interrupted and the tool to be removed from the system, while indirect methods can measure with lower precision without affecting the production line [3]. Various types of sensors can be used for indirect measurements, such as dynamometers to measure cutting forces, accelerometers to measure vibration, acoustic emission (AE) sensors to measure this very parameter in the system, motor current sensors, and even microphones to measure the sound of the process to indirectly measure TW [4].

The next step is to remove signal characteristics (vibration, AE, electric current, etc.) from the pre-processed data. Such techniques can be only in the time domain, e.g., mean and variance, in the frequency domain, e.g., fast Fourier transform (FFT), or in both the time and frequency domain, such as the Wavelet transform (WT) [5].

Several artificial intelligence (AI) tools can be used in the decision-making process for the estimated condition of the tool based on the characteristics of the signals, specifically classical supervised machine learning algorithms such as decision trees and k-nearest neighbors (KNN) for classification and linear regression. However, deep learning algorithms such as convolutional neural networks (CNN) and recurrent neural networks (RNNs) are also used [6].

Indirect measurement can be applied to a specific type of machining operation, such as turning, grinding, or milling. Among these, milling is one of the most common and important, so reducing costs and increasing product quality is essential. The use of TCM systems can provide parameters such as AE, vibration, cutting force, etc., to detect any faults or adverse conditions in the operation [7]. The proposed work focuses on monitoring the milling operation.

Several studies focus on monitoring the TW of milling operations. The authors of [8] propose a method that uses the short-time Fourier transform (STFT) to extract the characteristics of the vibration signal from milling operations. The STFT is used to form a time–frequency map that inputs an image in a supervised convolutional neural network (CNN) to predict the tool flank wear. Another work [1] with the same goal as the previous one uses a comparative approach between statistical techniques and discrete WT with or without the Hölder exponent to analyze milling vibration signals. It also compares the performance of various machine learning algorithms, such as decision trees, k-nearest neighbors (KNN), and multi-layer perceptron (MLP) neural networks. Another method in the literature that predicts flank wear in a milling operation uses the characteristics of a cutting force model and Wavelet packet decomposition (WPD) as the input for a deep MLP neural network [9].

The supervised machine learning methods used to estimate TW using indirect methods, specifically AE and vibration sensor signals, can accurately estimate the wear state of the machining tool. Therefore, this work proposes an MLP artificial neural network (ANN) model to estimate the TW, with the support of statistical techniques for AE and vibration signals. The public dataset used [10] provides real information from milling experiments with several measurements, enough to train a few MLP ANN configurations and test their ability to predict TW.

2. Materials and Methods

This section presents the methods and materials used in the work, specifically the dataset pre-processing and the conditions for implementing, training, and testing the proposed neural networks.

The chosen dataset is made up of experimental data from the milling operation and is called the mill dataset. The pre-processing of the data is necessary to obtain clean and normalized values to serve as inputs for the ANN. The file is a MATLAB-specific. mat file with a mill structure made up of 167 samples and 13 fields with different variables. The explanations of each variable, as well as important information about the dataset, are described in the Readme file next to the data file.

The target variable for the ANN to estimate is called VB, the flank TW. The values of case, a specific test condition, are a combination of the variables DOC, feed, and material, meaning, respectively, the depth of cut, the feed rate of the tool, and the type of material, which can be cast iron (1) or steel (2).

Each case has a unique number of passes in the tool on the workpiece; this variable is called run. The duration of each case is represented by the variable time. Finally, the sensor variables are the root mean square (RMS) values collected in around 9000 samples. Only two of these are used, namely the vibration signals from the variable vib_table and AE from the variable AE_table. Typical raw AE signal values are shown in Figure 1.

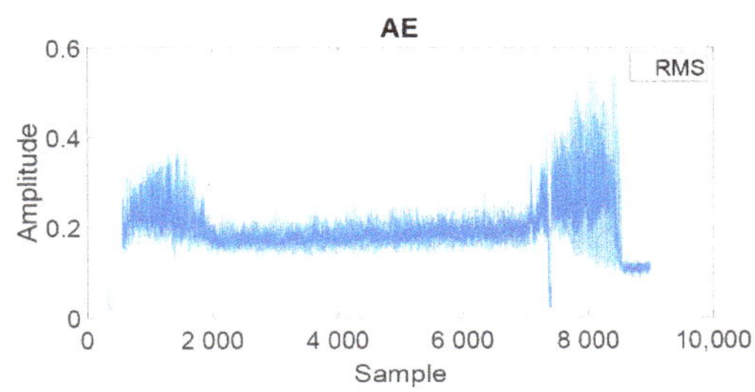

Figure 1. Typical raw RMS values of AE sensor in a run.

Some missing values (NaN) were found in the MATLAB structure. This had already been considered, as the Readme of the dataset mentions that the VB was not measured for each run. Therefore, the samples (run) with missing values were simply removed using MATLAB's own function. The *DOC* and *feed* variables were also removed, considering that they were redundant as the *case* already represents a combination of these variables.

The vib_table and AE_table sensor variables were split into statistical variables, specifically the mean, variance, skewness, and kurtosis for each signal. A simple analysis of the mean values of the statistics highlighted an anomaly in the mean and variance values, leading to a huge difference between the values. The outlier sample value was highlighted (Figure 2); it appears that the signal was corrupted and, consequently, removed from all tables, resulting in much more realistic values. The statistical values replaced the signal values with 8 new variables, specifically, m_AE, m_vib, v_AE, v_vib, s_AE, s_vib, k_AE and k_vib. Each variable's name represents a combination of the first letter of the mean, variance, skewness or kurtosis, and AE or vibration specifying the statistical value and sensor. The resulting table had 145 samples and 13 scalar variables.

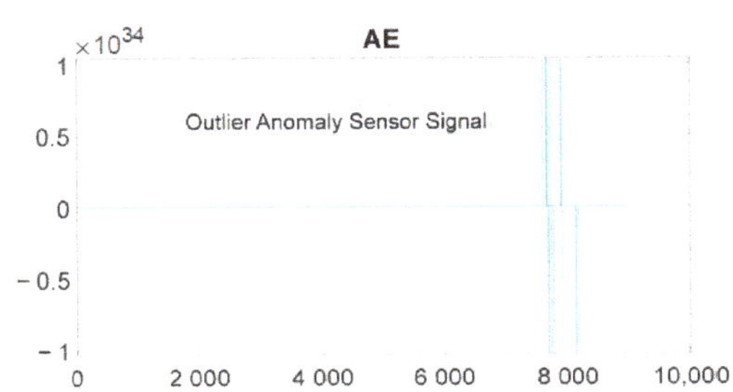

Figure 2. Outlier anomaly AE sensor signal.

The resulting values were normalized between 0 and 1 with a MATLAB function, but the run variable range was dependent on each case; therefore, it needed to be normalized separately, as there are different numbers of runs for each case, especially after removing some samples. The run variable was added to the normalized table after that.

Samples were split into random training and test sets, using 80% for training and 20% for testing, according to the guidelines in [11]. The same author also pointed out that the minimum and maximum values of each variable must be in the training set, so these values were removed from the random selection of the test set, with the rest going into the training set.

2.1. Backpropagation Neural Network

The MLP neural network was used as a universal function approximator, as it estimates a real value between 0 and 1, according to the normalization performed previously. The candidate topologies were 3 when using only one hidden layer, respectively, with 5, 10, and 15 neurons.

The MLP networks were trained using the generalized delta rule three times for each candidate topology. The training set had 116 samples for each training with a maximum number of 1000 epochs, all of which converged for a network accuracy $\epsilon = 0.5 \times 10^{-6}$ and a learning rate $\eta = 0.1$.

The final training results (T1, T2, and T3) are compiled in Table 1, based on the number of epochs, i.e., how many times the algorithm had to be presented with all the training samples until it converged, and the mean squared error (MSE).

Table 1. Training results.

Metric	Training	Network 1	Network 2	Network 3
Epochs	T1	189	199	188
	T2	230	213	203
	T3	177	240	219
MSE	T1	0.0033	0.0032	0.0031
	T2	0.0033	0.0030	0.0030
	T3	0.0032	0.0032	0.0032

The network only converges when the absolute value of the difference between the MSE of the current and previous epoch is less than the ϵ precision. All the training sessions had good results, except for T3 in Network 2; the best results were for T2 in Network 1 and T1 in Network 3, with the lowest MSE and the lowest number of epochs, so it converged faster.

2.2. Tests

The results are based on 9 different tests for each training and topology configuration. The main metrics used are as follows: the mean absolute percentage error (MAPE) and variance (σ^2).

3. Results and Discussion

The plots in Figure 3a–c coincide with the smallest MAPE, all resulting from one of the training sessions (T1, T2, or T3) for each network compiled in Table 2. The output values are very close to the desired values, represented by a black dotted line, but there was no VB result greater than 0.5, contributing to a higher MAPE value. All the networks failed on the first sample. One solution would be to obtain more training data with high values (above 0.5) to balance the distribution, or to even have a better method for separating training and test sets, such as the k-partition cross-validation method, the aim of which would be to evaluate each network topology on different training and test subsets with a specific size (k) [11]. The best results were close to 24% MAPE and 0.43 variance in Network 3.

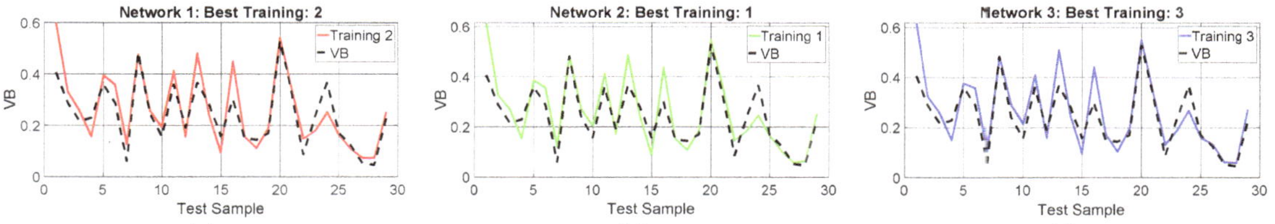

(a) (b) (c)

Figure 3. Best training estimation results in (**a**) Network 1; (**b**) Network 2; (**c**) Network 3.

Table 2. Test results.

Metric	Training	Network 1	Network 2	Network 3
MAPE	T1	27.537	24.820	25.182
	T2	26.791	25.178	25.206
	T3	27.674	25.828	24.407
$\sigma^2(\%)$	T1	0.4469	0.4625	0.5090
	T2	0.4433	0.4423	0.4330
	T3	0.5007	0.4620	0.4710

Figure 4a–c shows the convergence of each network in its best training, i.e., the MSE for each epoch. All networks had similar convergence curves, but both Networks 2 and 3 managed to converge in fewer epochs. Network 2 had the smallest MSE in the test, converging in a similar number of epochs to Network 3 in training; however, it did not perform better in the test phase.

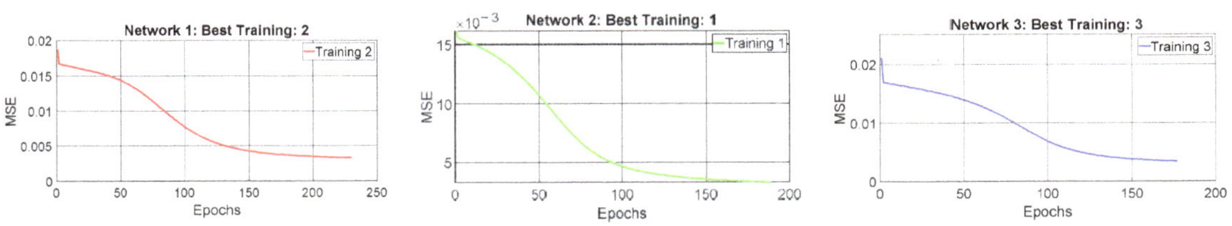

(a) (b) (c)

Figure 4. Best training convergence in (**a**) Network 1; (**b**) Network 2; (**c**) Network 3.

4. Conclusions

This work demonstrated how a relatively simple supervised machine learning algorithm can estimate the values that come from the complex relationships between the milling tool and the workpiece. Simple methods were also used to remove features from the signal and the result was satisfactory as a proof of concept of the power of an ANN for estimating values. Clearly, more complex ANN models such as CNNs and more robust feature extraction methods such as WT would have better results, but they have higher computation costs for training compared to the MLP, which has only one layer and fast training convergences.

Still, it is possible to improve the MLP's performance with more efficient algorithms and better data preparation, including adding two or more hidden layers and being able to reach the deep MLP (DMLP) with at least four hidden layers. Therefore, future studies should compare a more robust MLP with another more advanced ANN, considering both the accuracy and the computational cost of training and implementation. The best algorithm will be the most appropriate considering the problem constraints.

Author Contributions: Conceptualization, data curation, formal analysis, investigation, methodology, validation, writing—original draft, writing—review and editing G.O.d.S.; funding acquisition, project administration, resources, supervision, visualization, writing—review and editing, funding P.O.C.J.; formal analysis, conceptualization, methodology, resources, visualization, writing—review and editing I.N.d.S.; Data curation, methodology, resources, visualization, supervision D.B.; Conceptualization, formal analysis, software, visualization, supervision F.R.L.D. All authors have read and agreed to the published version of the manuscript.

Funding: Pro-Rectory of Research and Innovation of the University of São Paulo under grant: #22.1.09345.01.2.

Institutional Review Board Statement: Not applicable.

Informed Consent Statement: Not applicable.

Data Availability Statement: Publicly available dataset was analyzed in this study. This data can be found here: https://data.nasa.gov/Raw-Data/Milling-Wear/vjv9-9f3x/data (accessed on 1 July 2023).

Acknowledgments: The authors would like to thank the University of São Paulo (USP) and the São Paulo Research Foundation (FAPESP) for the opportunity to carry out and publish the research.

Conflicts of Interest: The authors declare no conflicts of interest.

References

1. Mohanraj, T.; Yerchuru, J.; Krishnan, H.; Aravind, R.S.N.; Yameni, R. Development of Tool Condition Monitoring System in End Milling Process Using Wavelet Features and Hoelder's Exponent with Machine Learning Algorithms. *Measurement* **2021**, *173*, 108671. [CrossRef]
2. Aghazadeh, F.; Tahan, A.; Thomas, M. Tool Condition Monitoring Using Spectral Subtraction and Convolutional Neural Networks in Milling Process. *Int. J. Adv. Manuf. Technol.* **2018**, *98*, 3217–3227. [CrossRef]
3. Mohanraj, T.; Shankar, S.; Rajasekar, R.; Sakthivel, N.R.; Pramanik, A. Tool Condition Monitoring Techniques in Milling Process—A Review. *J. Mater. Res. Technol.* **2020**, *9*, 1032–1042. [CrossRef]
4. Kuntoğlu, M.; Aslan, A.; Pimenov, D.Y.; Usca, Ü.A.; Salur, E.; Gupta, M.K.; Mikolajczyk, T.; Giasin, K.; Kapłonek, W.; Sharma, S. A Review of Indirect Tool Condition Monitoring Systems and Decision-Making Methods in Turning: Critical Analysis and Trends. *Sensors* **2020**, *21*, 108. [CrossRef] [PubMed]
5. Teti, R.; Mourtzis, D.; D'Addona, D.M.; Caggiano, A. Process Monitoring of Machining. *CIRP Ann.* **2022**, *71*, 529–552. [CrossRef]
6. Iliyas Ahmad, M.; Yusof, Y.; Daud, M.E.; Latiff, K.; Abdul Kadir, A.Z.; Saif, Y. Machine Monitoring System: A Decade in Review. *Int. J. Adv. Manuf. Technol.* **2020**, *108*, 3645–3659. [CrossRef]
7. Sener, B.; Serin, G.; Gudelek, M.U.; Ozbayoglu, A.M.; Unver, H.O. Intelligent Chatter Detection in Milling Using Vibration Data Features and Deep Multi-Layer Perceptron. In Proceedings of the 2020 IEEE International Conference on Big Data (Big Data), Atlanta, GA, USA, 10–13 December 2020; pp. 4759–4768.
8. Huang, Z.; Zhu, J.; Lei, J.; Li, X.; Tian, F. Tool Wear Monitoring with Vibration Signals Based on Short-Time Fourier Transform and Deep Convolutional Neural Network in Milling. *Math. Probl. Eng.* **2021**, *2021*, 9976939. [CrossRef]
9. Zhang, X.; Han, C.; Luo, M.; Zhang, D. Tool Wear Monitoring for Complex Part Milling Based on Deep Learning. *Appl. Sci.* **2020**, *10*, 6916. [CrossRef]
10. Agogino, A.; Goebel, K. *Milling Data Set*; Nasa Ames Prognostics Data Repository: Moffett Field, CA, USA, 2007.
11. Nunes, I.S.; Spatti, D.H.; Flauzino, R.A.; Liboni, L.H.B.; Alves, S.F.R. *Artificial Neural Networks: A Practical Course*; Springer: Berlin/Heidelberg, Germany, 2018.

engineering proceedings

MDPI

Proceeding Paper

Enhancing Indoor Position Estimation Accuracy: Integration of Accelerometer, Raw Distance Data, and Extended Kalman Filter in Comparison to Vicon Motion Capture Data †

Tolga Bodrumlu [1,*] and Fikret Çalışkan [2]

1 Mechatronics Engineering Department, Istanbul Technical University, Istanbul 34025, Turkey
2 Control and Automation Engineering Department, Istanbul Technical University, Istanbul 34025, Turkey; caliskanf@itu.edu.tr
* Correspondence: bodrumlu@itu.edu.tr

† Presented at the 10th International Electronic Conference on Sensors and Applications (ECSA-10), 15–30 November 2023; Available online: https://ecsa-10.sciforum.net/.

Abstract: Indoor positioning systems are a significant area of research and development, helping people navigate within buildings where GPS signals are unavailable. These systems have diverse applications, including aiding navigation in places like shopping malls, airports, and hospitals and improving emergency evacuation processes. The purpose of this study is to evaluate various technologies and algorithms used in indoor positioning. This study focuses on using raw distance data and Kalman filters to enhance indoor position accuracy. It employs a trilateration algorithm based on Recursive Least Squares (RLS) for initial position estimation and combines the results with accelerometer data. The designed algorithm using real sensor data collected in an ROS(Robot Operating System) environment was tested, and the results obtained were compared with data obtained from the Vicon Indoor Positioning System. In this comparison, the Root Mean Square Error metric was used. As a result of the comparison, it was observed that the error obtained from the designed algorithm is less than that of the Vicon system.

Keywords: indoor positioning; extended Kalman filter; ROS; sensor fusion

Citation: Bodrumlu, T.; Çalışkan, F. Enhancing Indoor Position Estimation Accuracy: Integration of Accelerometer, Raw Distance Data, and Extended Kalman Filter in Comparison to Vicon Motion Capture Data. *Eng. Proc.* **2023**, *58*, 40. https://doi.org/10.3390/ecsa-10-16089

Academic Editor: Stefano Mariani

Published: 15 November 2023

1. Introduction

In recent years, Unmanned Aerial Vehicles (UAVs) have found extensive applications in various fields, including military, industry, agriculture, and tasks like aerial photography and reconnaissance [1–3]. However, it is worth noting that these applications are primarily designed for the outdoors and rely on robust GPS signals for accurate positioning. In cases where GPS signals are unavailable or weak, the precision of UAV positioning is significantly impacted. Presently, there is a growing need for indoor UAV technology, particularly for inspection purposes, and this area is closely tied to control optimization and precise path tracking.

Ultra-Wideband (UWB) technology has garnered significant attention due to its high precision in indoor positioning. UWB systems leverage short-duration pulses of radio waves spread across a wide spectrum, enabling accurate distance measurements using time-of-flight and trilateration techniques. Aiello and Shalom [4] delved into the principles of UWB and its significance in achieving precise and real-time positioning. Heimovirta, Salantera, and Röning [5], discuss the practical implementation of a real-time indoor localization system based on UWB technology. This work focuses on the utilization of UWB technology for indoor positioning, providing insights into its application in real-world scenarios.

Vicon positioning systems are based on motion capture technology, utilizing cameras and markers to accurately track the movement and position of objects or individuals in a controlled environment. Rhea [6] assessed the accuracy and precision of Vicon motion capture systems for tracking movements in three dimensions (3D). This work focused on evaluating the reliability of Vicon systems in capturing 3D motion data. Gentil, de la Rouviere, and Elton [7] discuss Vicon as a computer vision solution for real-time 3D motion capture. Their work highlights the capabilities of Vicon technology in capturing and analyzing 3D motion data in real-time, particularly in the context of rehabilitation and neuro-engineering. Benn and Martin [8] provide insights into the utilization of Vicon systems for capturing and analyzing motion data in the context of sports and healthcare.

In this paper, an accurate position estimation was made by combining the IMU and the raw distance data with the help of the Extended Kalman Filter (EKF). Initially, position estimation was obtained using the RLS method employed via a trilateration algorithm. This solution was used as a starting point for RLS. Afterward, this position estimation was fused with the acceleration data. As a result, the estimated position obtained with the designed EKF was compared with the Vicon indoor positioning system, and the results are presented both graphically and in tabular form.

2. Position Estimation Algorithm

The position estimation algorithm is based on a geometric approach and EKF algorithms. After these algorithms were designed in a simulation environment, they were integrated into an ROS and tested with real sensor data, and the results were observed.

2.1. Geometric Approach

Herein, a geometric approach is put forward on the basis of research. As shown in Figure 1 below, three reference points are given, $B_1(x_1,y_1,z_1)$, $B_2(x_2,y_2,z_2)$, and $B_3(x_3,y_3,z_3)$, and interval measurements, d_1, d_2, and d_3, up to point A are given. The determination of the coordinates of point A is carried out by solving a system of quadratic equations.

$$\begin{aligned}(x-x_1)^2+(y-y_1)^2+(z-z_1)^2&={d_1}^2\\(x-x_2)^2+(y-y_2)^2+(z-z_2)^2&={d_2}^2\\(x-x_3)^2+(y-y_3)^2+(z-z_3)^2&={d_3}^2\end{aligned} \quad (1)$$

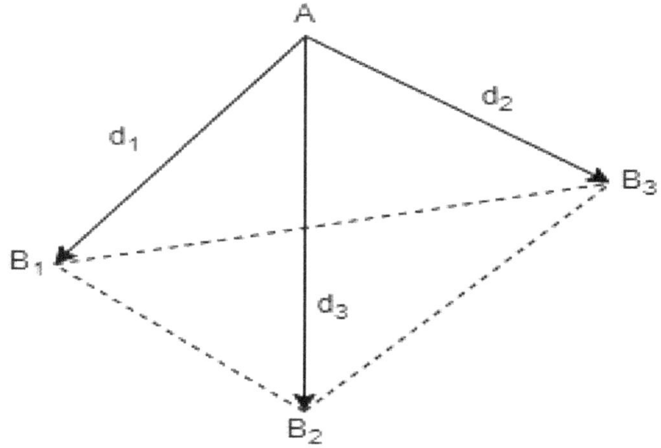

Figure 1. Reference points and interval measurements.

The system of equations given here can be expressed as follows:

$$\begin{aligned}(x^2+y^2+z^2)-2x_1x-2y_1y-2z_1z&=d_1{}^2-x_1^2-y_1^2-z_1^2\\(x^2+y^2+z^2)-2x_2x-2y_2y-2z_2z&=d_2{}^2-x_2^2-y_2^2-z_2^2\\(x^2+y^2+z^2)-2x_3x-2y_3y-2z_3z&=d_3{}^2-x_3^2-y_3^2-z_3^2\end{aligned}\tag{2}$$

In addition, this expression can be shown in matrix form as presented below:

$$\begin{bmatrix}1&-2x_1&-2y_1&-2z_1\\1&-2x_2&-2y_2&-2z_2\\1&-2x_3&-2y_3&-2z_3\end{bmatrix}\begin{bmatrix}x^2+y^2+z^2\\x\\y\\z\end{bmatrix}=\begin{bmatrix}s_1^2-x_1^2-y_1^2-z_1^2\\s_2^2-x_2^2-y_2^2-z_2^2\\s_3^2-x_3^2-y_3^2-z_3^2\end{bmatrix}\tag{3}$$

This matrix form can generally be expressed as follows:

$$\begin{aligned}A_0.x&=b_0 \qquad\qquad \mathrm{x}\in E\\E&=\left\{\ (x_0,x_1,x_2,x_3)^T\epsilon\quad x_0=x_1^2+x_2^2-x_3^2\right\}\end{aligned}\tag{4}$$

When examining the solution set, there are generally two different approaches. These are divided into three-reference-point-based and more-than-three-reference-point-based solutions. The details of these solution sets are elaborated in [9,10]. In brief, for the first scenario, the solution space varies depending on whether three reference points are in the same line. The second scenario involves having more than three reference points, which necessitates the use of the recursive least squares method for the solution set.

Distance data are used together with the recursive least square algorithm to aid in calculating the position. In the next section, the details of a more accurate position estimation with the help of EKF will be explained.

2.2. Sensor Fusion Algorithm

The system's state vector was set as

$$X=\begin{bmatrix}p(w)\\v(w)\\a(w)\end{bmatrix}\tag{5}$$

The vector *p(w)* denotes an object's position in the world coordinate system, specifying its coordinates on the x, y, and z axes as [*px*, *py*, *pz*]. The object's velocity along these axes in the world coordinate system is represented as *v(w)* = [*vx*, *vy*, *vz*], while its accelerometer data in the world coordinate system are denoted as *a(w)* = [*ax*, *ay*, *az*].

To elaborate further, the time interval between measurements is defined as Δt, and $(\Delta t)w(k)$ denotes the process noise related to acceleration. Specifically, $\frac{\Delta t^2}{2}w(k)$ represents the process noise affecting velocity, and $\frac{\Delta t^3}{6}w(k)$ indicates the process noise that influences position.

Equations of motion in the (k + 1) time interval can be expressed as follows:

$$\begin{aligned}p(k+1)&=p(k)+v(k)(\Delta t)+a(k)\tfrac{\Delta t^2}{2}+\tfrac{\Delta t^3}{6}w(k)\\v(k+1)&=v(k)+a(k)(\Delta t)+\tfrac{\Delta t^2}{2}w(k)\\a(k+1)&=a(k)+(\Delta t)w(k)\end{aligned}\tag{6}$$

The state equation can be represented in matrix form as follows:

$$x(k+1)=Ax(k)+Gw(k)\tag{7}$$

The matrices A and G represent the transition matrix and noise process matrix, respectively. The process noise vector and Q covariance can be defined as follows:

$$w(k) = \left[w_x(k)\ w_y(k)\ w_z(k)\right],\ Q = diag\left(\left[\sigma_{ax}^2\ \sigma_{ay}^2\ \sigma_{az}^2\right]\right)$$

$$A = \begin{bmatrix} 1 & 0 & 0 & \Delta t & 0 & 0 & \Delta t^2/2 & 0 & 0 \\ 0 & 1 & 0 & 0 & \Delta t & 1 & 0 & \Delta t^2/2 & 0 \\ 0 & 0 & 1 & 0 & 0 & \Delta t & 0 & 0 & \Delta t^2/2 \\ 0 & 0 & 0 & 1 & 0 & 0 & \Delta t & 0 & 0 \\ 0 & 0 & 0 & 0 & 1 & 0 & 0 & \Delta t & 0 \\ 0 & 0 & 0 & 0 & 0 & 1 & 0 & 0 & \Delta t \\ 0 & 0 & 0 & 0 & 0 & 0 & 1 & 0 & 0 \\ 0 & 0 & 0 & 0 & 0 & 0 & 0 & 1 & 0 \\ 0 & 0 & 0 & 0 & 0 & 0 & 0 & 0 & 1 \end{bmatrix}, G = \begin{bmatrix} \frac{\Delta t^3}{6} & 0 & 0 \\ 0 & \frac{\Delta t^3}{6} & 0 \\ 0 & 0 & \frac{\Delta t^3}{6} \\ \frac{\Delta t^2}{2} & 0 & 0 \\ 0 & \frac{\Delta t^2}{2} & 0 \\ 0 & 0 & \frac{\Delta t^2}{2} \\ \Delta t & 0 & 0 \\ 0 & \Delta t & 0 \\ 0 & 0 & \Delta t \end{bmatrix} \quad (8)$$

The observation vector $z(k)$ contains the distance values used in the geometric approach and an additional noise vector:

$$z(k) = \begin{bmatrix} d_1(k) + n_1(k) \\ d_2(k) + n_2(k) \\ d_3(k) + n_3(k) \\ d_4(k) + n_4(k) \end{bmatrix} = H(k)x(k) + n(k) \quad (9)$$

The observation matrix is represented by $H(k)$, and the vector of noise, with a mean of zero and a covariance matrix, is represented by $n(k)$.

$$R = diag\left(\left[\sigma_{r1}^2 \sigma_{r2}^2 \sigma_{r3}^2 \sigma_{r4}^2\right]\right) \quad (10)$$

Here, $r1$, $r2$, $r3$, and $r4$ are the distances between each of the values of the Marvelmind sensor. The $H(k)$ observation matrix is a Jacobian matrix and is calculated as follows:

$$H(k) = \begin{bmatrix} \frac{\delta d_1(k))}{\delta p_x(k)} & \frac{\delta d_1(k)}{\delta p_x(k)} & \frac{\delta d_1(k)}{\delta p_x(k)} \\ \frac{\delta d_2(k)}{\delta p_x(k)} & \frac{\delta d_2(k)}{\delta p_x(k)} & \frac{\delta d_2(k)}{\delta p_x(k)} \\ \frac{\delta d_3(k)}{\delta p_x(k)} & \frac{\delta d_3(k)}{\delta p_y(k)} & \frac{\delta d_3(k)}{\delta p_z(k)} \\ \frac{\delta d_4(k)}{\delta p_x(k)} & \frac{\delta d_4(k)}{\delta p_x(k)} & \frac{\delta d_4(k)}{\delta p_x(k)} \end{bmatrix} \quad (11)$$

The equations described in the structure of the applied EKF are utilized. The initialization stage of this filter is one of the crucial factors, where the initial values of the covariance matrices are assigned. Subsequently, the prediction and update steps are executed in sequence.

$$\begin{gathered} \overline{\mathrm{x}}_\mathrm{k} = \mathrm{f}\left(\hat{\mathrm{x}}_\mathrm{k}^{-1}, \mathrm{u}_\mathrm{k}\right) \\ \overline{\mathrm{P}}_\mathrm{k} = \mathrm{A}_\mathrm{k}^{-1}\hat{\mathrm{P}}_\mathrm{k}^{-1}\mathrm{A}_\mathrm{k}^{-1\mathrm{T}} + \mathrm{Q}_\mathrm{k}^{-1} \\ K_k = \overline{P}_k H_k^T \left[H_k \overline{P}_k H_k^T + R_k\right]^{-1} \\ \hat{\mathrm{x}}_\mathrm{k} = \overline{\mathrm{x}}_\mathrm{k} + \mathrm{K}_\mathrm{k}(\mathrm{z}_\mathrm{k} - \mathrm{h}(\overline{\mathrm{x}}_\mathrm{k})) \\ \hat{P}_k = (I - K_k H_k)\overline{P}_k \end{gathered} \quad (12)$$

In the following section, the integration of this geometric approach with EKF algorithms into the ROS platform is described, and its implementation with a real dataset is discussed. The obtained results are presented in detail.

3. Implementation and Simulation System Results

After the geometric approach and EKF structure were designed, they were transferred to the simulation environment using C++. The code structure was integrated into the PX4 code, making it compatible with the ROS environment. The acceleration data from UWB sensors are received at approximately 40 Hz, while the raw distance data arrive at around 80 Hz. To address this discrepancy, time synchronization was achieved between the acceleration and distance data. The acceleration data were upsampled to 80 Hz using interpolation. Additionally, since Vicon data are received at a rate of 20 Hz, it was necessary to perform time synchronization with the position data obtained from the EKF results. Therefore, the Vicon data were upsampled to 80 Hz using interpolation to ensure proper time synchronization.

When looking at Figures 2 and 3, it is easily noticeable that the X, Y, and Z positions obtained from the EKF are closer to the reference results than those obtained from the Vicon system. Additionally, Figure 3 depicts the differences between the EKF and Vicon in relation to their respective reference values. Upon examining the position error values in Table 1, it is evident that the average difference between the EKF and the reference is 0.205 m, while for the Vicon, this difference is 0.255 m. Moreover, when looking at the minimum and maximum errors, it is evident that the values obtained from the EKF are lower than those obtained from the Vicon system. In summary, through a numerical comparison of the results, it is apparent that the designed EKF algorithm provides better results compared to the Vicon positioning system.

Table 1. Position error values.

	Minimum Error (m)	Mean Error (m)	Maximum Error (m)
Ref-EKF	0.019	0.205	0.424
Ref-Vicon	0.07	0.255	0.441

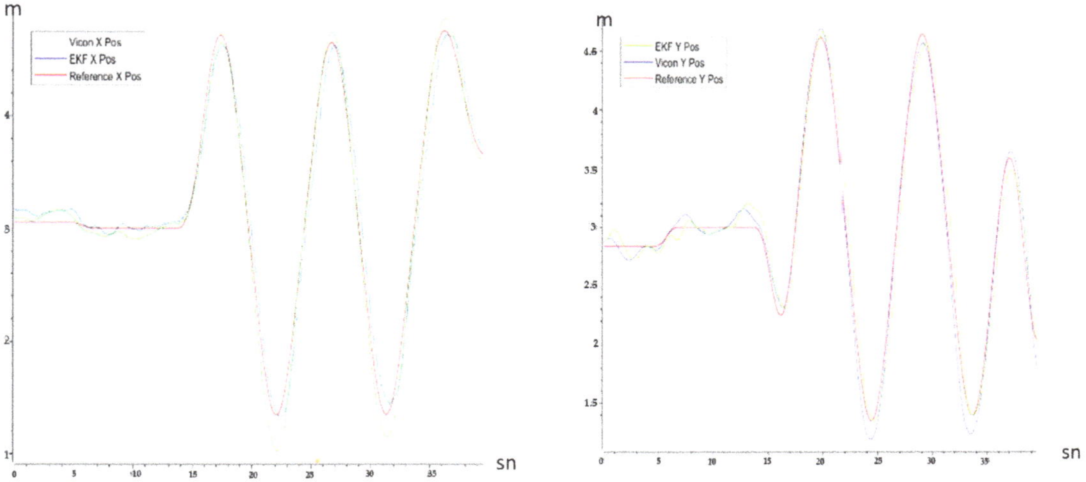

Figure 2. X and Y positions (Vicon-EKF-Reference).

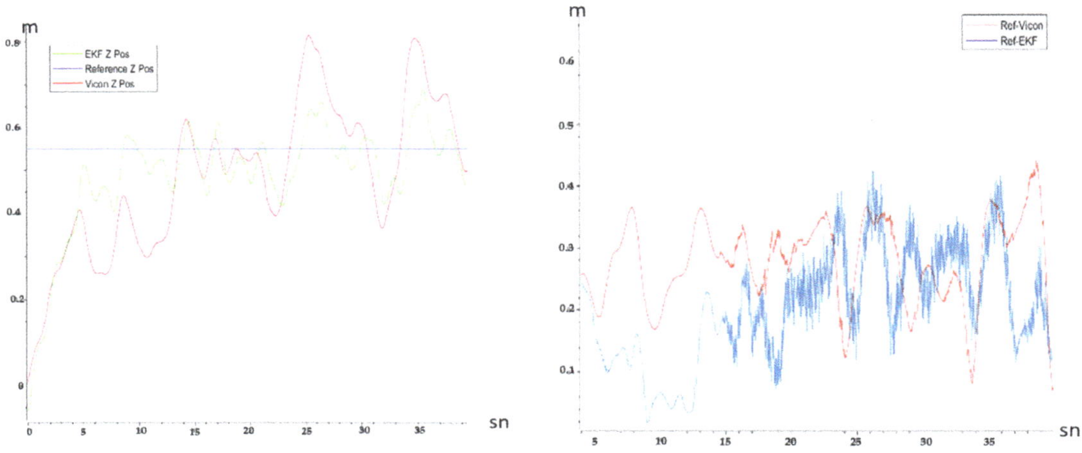

Figure 3. Z position and Vicon and EKF differences from reference.

4. Conclusions

In this study, position estimation was carried out by fusing raw distance data with the IMU using sensor fusion algorithms, specifically via the EFK. The fundamental structure of the designed algorithm incorporates a geometric approach. Firstly, a position calculation was derived from the geometric approach, and then the accuracy of this position was enhanced using accelerometer data and the EKF algorithm. The designed algorithm was transformed for incorporation into the C++ environment and integrated into an ROS. Real sensor data is used during the testing of the algorithm. Sensor data were collected using the ROS platform, and the algorithm was executed within the ROS to observe the obtained results. According to the results obtained, the designed EKF structure yielded more successful outcomes compared to the Vicon position system. These results are supported both graphically and by numerical tables.

Author Contributions: Conceptualization, T.B. and F.Ç.; Methodology, T.B.; Software, T.B.; Validation, T.B.; Data Acquisition, T.B; Formal Analysis, T.B.; investigation, T.B.; writing—original draft preparation, T.B.; writing—review and editing, T.B. and F.Ç; visualization, T.B.; supervision, F.Ç. All authors have read and agreed to the published version of the manuscript.

Funding: This research was partially financially supported by Istanbul Technical University, grant number 42754.

Institutional Review Board Statement: Not applicable.

Informed Consent Statement: Not applicable.

Data Availability Statement: Data are contained within the article.

Conflicts of Interest: The authors declare no conflicts of interest.

References

1. Valavanis, K.P.; Vachtsevanos, G.J. *UAV Applications: Introduction*; Springer: Amsterdam, The Netherlands, 2015.
2. Samad, T.; Bay, J.S.; Godbole, D. Network-centric systems for military operations in urban Terrain: The role of UAVs. *Proc. IEEE* **2007**, *95*, 92–107.
3. Li, Z.; Liu, Y.; Walker, R.; Hayward, R.; Zhang, J. Towards automatic power line detection for a UAV surveillance system using pulse coupled neural filter and an improved Hough transform. *Mach. Vis. Appl.* **2010**, *21*, 677–686.
4. Aiello, R.; Bar-Shalom, Y. Ultra-Wideband Technology for Precise Real-Time Location Systems. *Proc. IEEE* **2000**, *88*, 62–85.
5. Heimovirta, E.; Salantera, S.; Röning, J. Real-Time Indoor Localization Using Ultra-Wideband Technology. In Proceedings of the 11th International Conference on Mobile and Ubiquitous Systems: Computing, Networking and Services, London, UK, 2–5 December 2014; pp. 265–274.

6. Rhea, C.K. Accuracy and Precision of Vicon Motion Capture Systems for Tracking in 3D. *Gait Posture* **2004**, *20*, 332–335.
7. Gentil, C.; de la Rouviere, S.; Elton, C. Vicon: A Computer Vision Solution for Reliable Real-Time 3D Motion Capture. *J. NeuroEng. Rehabil.* **2009**, *6*, 11.
8. Benn, A.; Martin, A. Vicon Revue: Motion Capture Technology for Rehabilitation and Sports Medicine. *Biomech. Med. Swim.* **2004**, *IX*, 122–126.
9. Norrdine, A. An Algebraic Solution to the Multilateration Problem. In Proceedings of the 2012 International Conference on Indoor Positioning and Indoor Navigation, Sydney, Australia, 13–15 November 2012; pp. 1–4.
10. Bodrumlu, T.; Caliskan, F. Indoor Position Estimation Using Ultrasonic Beacon Sensors and Extended Kalman Filter. *Eng. Proc.* **2022**, *27*, 16. [CrossRef]

engineering proceedings

MDPI

Proceeding Paper

Development of a Low-Cost Particulate Matter Optical Sensor for Real-Time Monitoring †

Martín Aarón Sánchez-Barajas [1], Daniel Cuevas-González [1,*], Marco A. Reyna [1,*], Juan C. Delgado-Torres [2], Eladio Altamira-Colado [1] and Roberto López-Avitia [2]

1 Instituto de Ingeniería, Universidad Autónoma de Baja California, Mexicali 21280, Mexico; martin.aaron.sanchez.barajas@uabc.edu.mx (M.A.S.-B.); ealtamira@uabc.edu.mx (E.A.-C.)
2 Facultad de Ingeniería, Universidad Autónoma de Baja California, Mexicali 21280, Mexico
* Correspondence: cuevas.daniel@uabc.edu.mx (D.C.-G.); mreyna@uabc.edu.mx (M.A.R.)
† Presented at the 10th International Electronic Conference on Sensors and Applications (ECSA-10), 15–30 November 2023; Available online: https://ecsa-10.sciforum.net/.

Abstract: Air pollution is a critical public health problem that has increased during the past decades. High levels of air pollution have affected natural environments and people's health, causing significant problems and, in severe cases, premature death. A growing trend called "Personal air monitoring" has become important for prevention of and reduction in exposure to air pollutants. The development of personal particulate matter sensors is still a topic of study among the scientific community. Some important identified challenges are improving the sample rate, precision, stability, dimensions and costs, making personal monitoring of air quality affordable. This work proposes the development of a low-cost particulate matter optical sensor to count the number of particles in real time using the Arduino platform and wireless transmission. Our results demonstrated that using a digital input of the microcontroller instead of the analog–digital converter, after conditioning the sensor signal, allows a very high max particle count, which can be compared to that of expensive sensors. In addition, particulate matter (PM) measurements were compared with a GP2Y1014AU0F dust sensor to validate the accuracy of the sensor.

Keywords: PM counter; personal air monitoring; low-cost sensor; air pollution monitoring

Citation: Sánchez-Barajas, M.A.; Cuevas-González, D.; Reyna, M.A.; Delgado-Torres, J.C.; Altamira-Colado, E.; López-Avitia, R. Development of a Low-Cost Particulate Matter Optical Sensor for Real-Time Monitoring. *Eng. Proc.* **2023**, *58*, 41. https://doi.org/10.3390/ecsa-10-16025

Academic Editor: Stefano Mariani

Published: 15 November 2023

1. Introduction

Air pollution is a critical public health problem that has increased during the last decades. There are natural sources of air pollution such as natural fires, volcanoes and earth storms. However, anthropological activities such as burning fossil fuels, excessive use of transportation and generation of electricity and household pollution have significantly aggravated the problem [1]. The most commonly encountered air pollutants are particulate matter (PM), ozone (O_3), sulfur dioxide (SO_2), nitrogen oxides (NO_x), carbon monoxide (CO) and lead [2]. Exposure to high levels of air pollution has affected natural environments and people's health, causing significant problems and, in severe cases, premature death. In 2019, the World Health Organization (WHO) estimated about 6.9 million deaths associated with air pollution [3]. A growing trend called "Personal air monitoring" has become important for prevention of and reduction in exposure to air pollutants. The development of personal sensors for particulate matter is still a topic of study among the scientific community. There are low-cost, portable and low-power consumption optical particle counter (OPC) devices on the market such as those mentioned in [4–7], as well as low-cost and miniaturized devices developed in the studies of [8,9] that use a laser diode and a photodiode for particle counting through light scattering. Particle counting may, depending on the study, be limited to the sample-throughput capability of the ADC converter or the digital stage of the microcontroller. In [4], the commercial OPC-N2 device from Alphasense has a maximum particle count of 10,000 particles per second, unlike the studies in [5–8] where the particle count per second is not mentioned. In [10], they developed a prototype to monitor the pollution levels in the environment using the optical dust sensor GP2Y1010AU0F as a basis for its development. In [11], the optical dust sensor GP2Y1010AU0F was used to analyze the correlation between the sensor drift and the accumulated production in a steel factory. Finally, in [12,13], they comprehensively analyzed and graphed the optical dust sensor GP2Y1010AU0F.

This work develops a low-cost particulate matter optical sensor that implements a digital counter to measure the number of particles in real time using the Arduino platform and transmit the information wirelessly (see Figure 1).

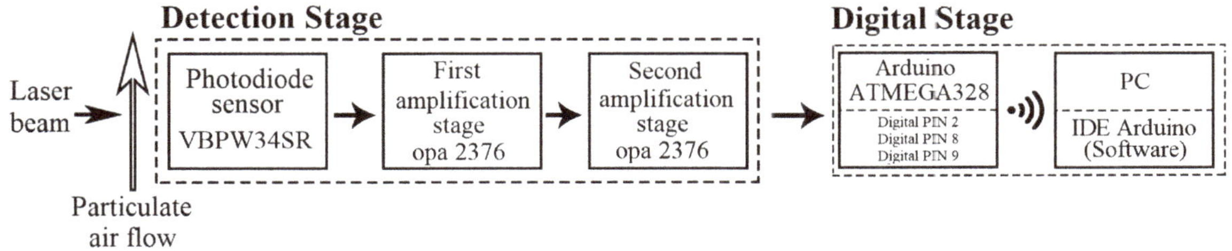

Figure 1. Diagram of the particulate matter optical sensor.

2. Materials and Methods

This paper presents the methodology to develop a low-cost PM sensor to count the number of particles in real time using the Arduino platform and wireless transmission.

2.1. Optical Sensor Operation

The circuit uses the principle of light scattering employing a laser beam and a photodiode that detects the variations that generate the flow of PM. These small variations are amplified and filtered by three stages (See Figure 2). The variations in an analog signal can be measured on an oscilloscope or quantified by a digital system using a microcontroller.

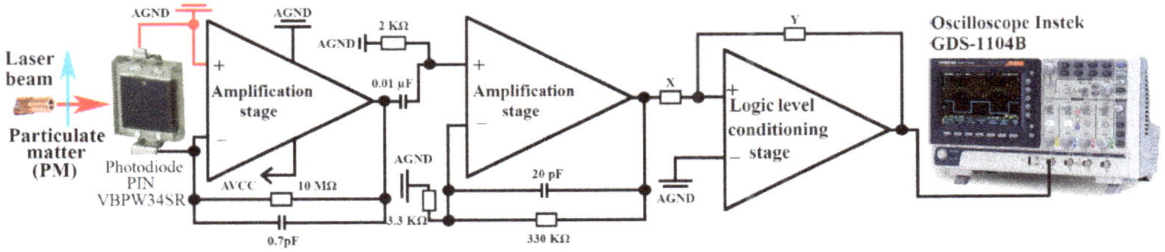

Figure 2. Circuit of the optical stage of the particulate matter sensor.

2.2. Printed Circuit Board (PCB) and Optical Stage Chambers

The circuit shown in Figure 2 was initially implemented on a breadboard; however, because the circuit is very sensitive to noise, a PCB was designed using DipTrace software 4.2.0.1 (see Figure 3A,B) to avoid the noise captured by the breadboard. Subsequently, two chambers were designed using SolidWorks software 2020 SP0.0 to avoid noise produced by light from the environment (see Figure 3C,D), and a chamber with 9 × 9 mm dimensions and another of 3 × 3 mm to identify how the air flow and dimensions affect the system.

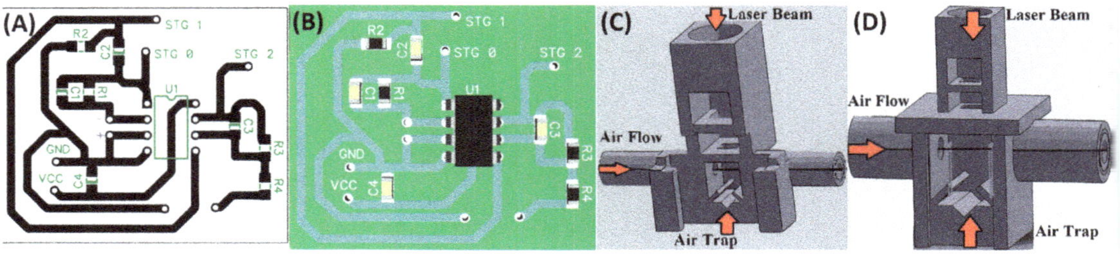

Figure 3. PCB and optical stage chambers designs. (**A**) PCB traces of optical stage and amplification, (**B**) 3D PCB design, (**C**) 3 × 3 mm PM flow chamber and (**D**) 9 × 9 mm PM flow chamber.

2.3. *Digital Stage of the System—Pulse Counter*

This stage presents the circuit design to count the pulses (digitization) of the analog signal due to the scattered light in particulate material, and then transmit via Bluetooth to a PC (See Figure 4). In order to determine the maximum number of particles that our device can count, it was necessary to generate a square wave with a 2.4 Vp amplitude and frequencies from 60 Hz to 150 Hz with the GW Instek AFG-2225 function generator (See Figure 4).

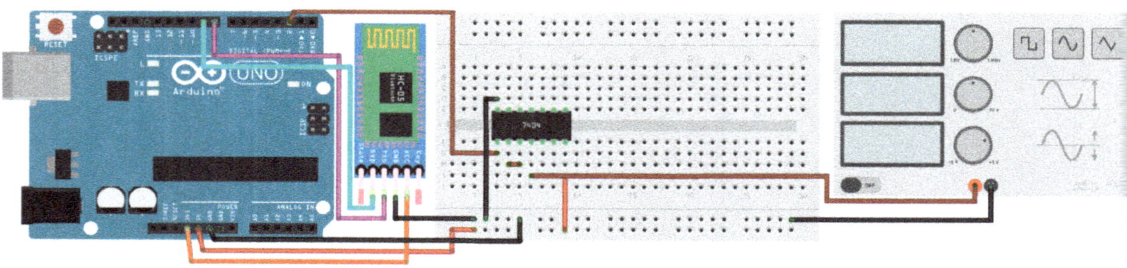

Figure 4. Digital stage validation circuit with Bluetooth transmission.

2.4. *Operation of the Optical Sensor in the Chambers with Particle Counting*

Once the optical circuit PCB has been independently evaluated on the oscilloscope, and validated regarding the maximum operating limits of the pulse counter system, and accuracy, a final test consists of performing these steps in conjunction with an exposure to smoke air flow and observing the particle count and comparing it with a GP2Y1014AU0F dust sensor.

3. Results

3.1. *Optical Sensor Operation*

A visual comparison is shown in Figure 5 between the signal obtained using the GP2Y1014AU0F dust sensor and our developed low-cost PM sensor. Figure 5a is a graph found in the design guide of sensor TIDA-00378 from Texas Instruments that summarizes the relation between pulse height and particle size, as well as pulse spacing and particle concentration. Figure 5b shows the signal obtained from the GP2Y1014AU0F dust sensor, and Figure 5c shows the signal obtained from our developed low-cost PM sensor.

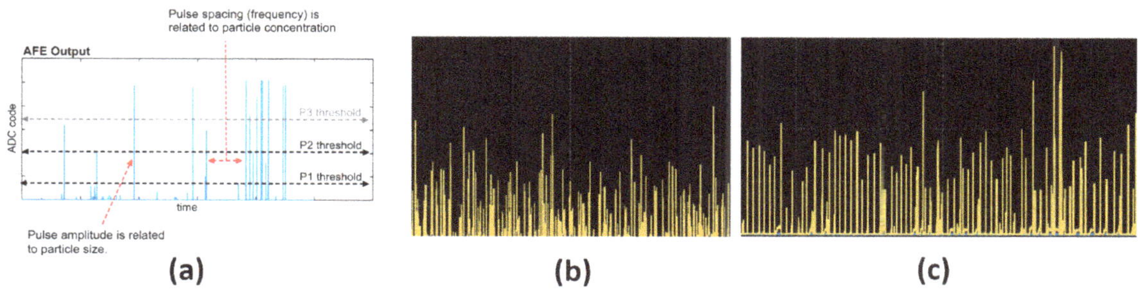

Figure 5. (**a**) Pulse amplitude and spacing relation with particle size and particle concentration. (**b**) GP2Y1014AU0F dust sensor output signal. (**c**) Developed low-cost PM sensor output signal.

3.2. Digital Stage of the System—Pulse Counter

To determine the maximum sampling frequency of the digital counter, tests were performed at different frequencies from 60 Hz to 156.53 kHz with a baud rate of 1MB/s. It was calculated that the error for a frequency of 150 kHz is 0.06% and increases as the frequency increases, as shown in Table 1.

Table 1. Results of the digital pulse counter with its relative error percentage and its efficiency.

Frequency (Hz)	Theoretical Result	Measured Result	Relative Percentage Error (%)	Efficiency (%)
60	216,000	215,999	0.0004	99.99
1000	3,600,000	3,599,972	0.0007	99.99
10,000	36,000,000	35,989,777	0.0283	99.97
100,000	360,000,000	359,847,668	0.0423	99.95
150,000	540,000,000	539,654,597	0.0639	99.93
156,530	563,508,000	540,418,640	4.09	95.90

4. Discussion

A limitation of traditional particle counters is the microcontroller's analog–digital converter or external analog–digital converter, which carries out the particle counting of the analog stage. In our work, we used the Arduino platform's microcontroller, which is limited to a 9.6 kHz frequency setting by a 10 bits analog–digital converter. Therefore, to solve the limit in the sampling frequency of the analog–digital converter, a digital reading alternative was implemented, which registers particles with a maximum frequency of 150 kHz. Our proposal, instead of using the microcontroller's analog–digital converter, introduces the signal through a digital pin of the microcontroller to detect a high-voltage input that corresponds to a detected particle. However, for this method, it is necessary to implement a conditioning stage to ensure an appropriate logic level of the signal for the correct detection of particles as shown in Figure 2.

Another presented problem in the analog stage is that market-available devices have complex hardware and a complicated calibration, and they do not provide user support. Accuracy is a crucial parameter for pollutant measurement; therefore, it is important that sensors are robustly validated and can be easily calibrated. In the work of [10], they found that the GP2Y1010AU0F sensor has calibration issues, which can produce incorrect data while taking measurements, as well as evidence that the sensor does not make homogeneous measurements. Meanwhile, in [11], they mention that the manufacturer does not provide enough information about the sensor, which presents a problem when working with the sensor.

Finally, it was found in the digital stage after comparing our prototype with commercial devices, as shown in Table 2, that our prototype registers 150,000 particles per second with an accuracy of 99.93%, while commercial particle counters with a price of

USD 50 or less can detect up to 1000 to 40,000 particles per second, which represents an improvement of about 275%. Devices with a higher particle count frequency (similar to our prototype) have a cost from USD 200 to 1000 approximately, while our prototype would cost less than USD 15 and considering a mass production the price could be around USD 8–10. The limitations identified in our work are the test with the 9 × 9 mm chamber to identify how the air flow and dimensions affect the system; calibration tests with another robust commercial sensor, optimizing the code to increase the particle count in the digital stage; and performing statistics using the intraclass correlation coefficient (ICC).

Table 2. Specifications comparative of each sensor available on the market.

Sensor	Price (USD)	Release	Frequency (Hz)	Max. Particle Count
AirCasting AirBeam [14].	250	2021	2	10,000
AirViz Speck [15].	150	2022	2	10,000
AlphaSense OPC-N2 [16].	1500	2020	10	100,000
Dylos DC1100/DC1100 Pro [17].	260	2019	2	100,000
GP2Y1010A [18].	10	2015	1	1000
Nova PM2.5 [19].	30	2021	1	10,000
PMS5003 [20].	50	2017	2	20,000
PPD42NS [21].	20	2022	2	10,000
Shinyei PMS-SYS-1 [22].	1000	2021	2	100,000
TSI AirAssure [23].	1000	2022	2	100,000
TZOA PM Research Sensor [24].	90	2022	2	100,000
Our prototype.	14	2023	1	150,000

5. Conclusions

The development of personal monitoring sensors comprises important challenges such as improving sampling rate, precision and stability without increasing the cost.

The analog signal conditioning and filtering stage is susceptible to noise, and variability due to the angle of the laser/light source and the receiver; multiple amplifier stages can also cause variations. Developers of air personal monitoring must be careful to avoid these factors of variation and inaccurate readings. The analog–digital converters with high sampling rates increase the cost of a sensor significantly; therefore, using a digital counter for the digital stage, although it requires a more complex signal conditioning in the analog stage, is better to decrease costs and increase the sampling rate. As shown in the results, we obtained a particle count frequency (150 kHz) with a USD 14 cost.

As future work, some improvements are suggested such as increasing the particle count frequency in the analog stage, optimizing the code to increase the particle count in the digital stage, reducing the number of components, miniaturizing and developing a communication interface for smart devices.

The development of new PM sensors, technology assimilation and cost reduction could make these sensors more accessible to prevent and reduce the exposure to air pollutants.

Author Contributions: Conceptualization, D.C.-G. and M.A.R.; methodology, D.C.-G., M.A.S.-B., J.C.D.-T. and E.A.-C.; software, J.C.D.-T. and M.A.S.-B.; validation, D.C.-G., M.A.S.-B. and J.C.D.-T.; formal analysis, D.C.-G. and M.A.R.; writing—original draft preparation, D.C.-G., M.A.S.-B. and J.C.D.-T.; writing—review and editing, D.C.-G., M.A.S.-B., J.C.D.-T. and E.A.-C.; visualization, R.L.-A.; supervision, D.C.-G., M.A.R. and R.L.-A.; project administration, D.C.-G. and M.A.R. All authors have read and agreed to the published version of the manuscript.

Funding: This research received no external funding.

Institutional Review Board Statement: Not applicable.

Informed Consent Statement: Not applicable.

Data Availability Statement: Data sharing is not applicable.

Conflicts of Interest: The authors declare no conflicts of interest.

References

1. Tiotiu, A.I.; Novakova, P.; Nedeva, D.; Chong-Neto, H.J.; Novakova, S.; Steiropoulos, P.; Kowal, K. Impact of air pollution on asthma outcomes. *IJERPH* **2020**, *17*, 6212. [CrossRef] [PubMed]
2. Types of Pollutants—Air Quality and Health. Available online: https://www.who.int/teams/environment-climate-change-and-health/air-quality-and-health/health-impacts/types-of-pollutants (accessed on 5 August 2023).
3. Ambient (Outdoor) Air Quality and Health—Air Quality and Health. Available online: https://www.who.int/news-room/fact-sheets/detail/ambient-(outdoor)-air-quality-and-health (accessed on 5 August 2023).
4. Crilley, L.R.; Shaw, M.; Pound, R.; Kramer, L.J.; Price, R.; Young, S.; Lewis, A.C.; Pope, F.D. Evaluation of a low-cost optical particle counter (Alphasense OPC-N2) for Ambient Air Monitoring. *Atmos. Meas. Tech.* **2018**, *11*, 709–720. [CrossRef]
5. Kelly, K.E.; Whitaker, J.; Petty, A.; Widmer, C.; Dybwad, A.; Sleeth, D.; Martin, R.; Butterfield, A. Ambient and laboratory evaluation of a low-cost Particulate Matter Sensor. *Environ. Pollut.* **2017**, *221*, 491–500. [CrossRef] [PubMed]
6. Feinberg, S.; Williams, R.; Hagler, G.S.W.; Rickard, J.; Brown, R.; Garver, D.; Harshfield, G.; Stauffer, P.; Mattson, E.; Judge, R.; et al. Long-term evaluation of air sensor technology under ambient conditions in Denver, Colorado. *Atmos. Meas. Tech.* **2018**, *11*, 4605–4615. [CrossRef] [PubMed]
7. Huang, C.-H.; He, J.; Austin, E.; Seto, E.; Novosselov, I. Assessing the value of complex refractive index and particle density for calibration of low-cost particle matter sensor for size-resolved particle count and PM2.5 measurements. *PLoS ONE* **2021**, *16*, e0259745. [CrossRef] [PubMed]
8. Li, X.; Iervolino, E.; Santagata, F.; Wei, J.; Yuan, C.A.; Sarro, P.; Zhang, G. Miniaturized particulate matter sensor for Portable Air Quality Monitoring Devices. *Proc. IEEE Sens.* **2014**. [CrossRef]
9. Tian, R.; Dierk, C.; Myers, C.; Paulos, E. MyPart: Personal, Portable, Accurate, Airborne Particle Counting. In Proceedings of the 2016 CHI Conference on Human Factors in Computing Systems, San Jose, CA, USA, 7–12 May 2016. [CrossRef]
10. Martínez Villalobos, W.D. Desarrollo de un Prototipo Para la Medición de Partículas PM2.5 en la Ciudad de Bogotá. Bachelor Thesis, Universidad Católica de Colombia, Bogotá, DC, USA, November 2023.
11. Winkler, N.P.; Neumann, P.P.; Schaffernicht, E.; Lilienthal, A. Gather Dust and Get Dusted: Long-Term Drift and Cleaning of Sharp GP2Y1010AU0F Dust Sensor in a Steel Factory. In Proceedings of the Danubia-Adria Symposium on Advances in Experimental Mechanics, Poros Island, Greece, 19 December 2022.
12. Wang, Y.; Li, J.; Jing, H.; Zhang, Q.; Jiang, J.; Biswas, P. Laboratory evaluation and calibration of three low-cost particle sensors for particulate matter measurement. *Aerosol Sci. Technol.* **2015**, *49*, 1063–1077. [CrossRef]
13. Air Quality Monitoring. Available online: https://www.howmuchsnow.com/arduino/airquality/ (accessed on 20 September 2023).
14. HabitatMap. Available online: https://www.habitatmap.org/ (accessed on 14 September 2023)
15. Speck Sensor. Available online: https://www.specksensor.com/ (accessed on 14 September 2023).
16. Alpha Sense. Available online: https://www.alphasense.com/products/optical-particle-counter/ (accessed on 14 September 2023).
17. Dylos Corporation. Available online: http://www.dylosproducts.com/ornodcproair.html (accessed on 14 September 2023).
18. Sparkfun. Available online: https://www.sparkfun.com/datasheets/Sensors/gp2y1010au_e.pdf (accessed on 10 September 2023).
19. Reichelt Elektronik Blätterkatalog. Available online: https://cdn-reichelt.de/documents/datenblatt/X200/SDS011-DATASHEET.pdf (accessed on 15 September 2023).
20. South Coast AQMD. Available online: https://www.aqmd.gov/docs/default-source/aq-spec/resources-page/plantower-pms5003-manual_v2-3.pdf (accessed on 14 September 2023).
21. Mouser Electronics. Available online: https://www.mouser.com/datasheet/2/744/Seeed_101020012-1217636.pdf (accessed on 14 September 2023).
22. ALLDATASHEET. Available online: https://pdf1.alldatasheet.com/datasheet-pdf/view/1300723/MSYSTEM/PMS-1-D.html (accessed on 14 September 2023).
23. TSI Knowledge Beyond Measure. Available online: https://tsi.com/products/indoor-air-quality-meters-instruments/continuous-iaq-monitoring-instruments/airassure-8144-6-indoor-air-quality-monitor/ (accessed on 15 September 2023).
24. Karagulian, F.; Gerboles, M.; Barbiere, M.; Kotsev, A.; Lagler, F.; Borowiak, A. *Review of Sensors for Air Quality Monitoring*; EUR 29826 EN; Technical Report by the Joint Research Centre; Publications Office of the European Union: Luxembourg, 2019. [CrossRef]

engineering proceedings

MDPI

Proceeding Paper

Indoor Air Quality Assessment Using a Low-Cost Sensor: A Case Study in Ikere-Ekiti, Nigeria †

Ademola Adamu [1], Kikelomo Mabinuola Arifalo [1] and Francis Olawale Abulude [2,*]

1 Department of Chemical Sciences, Bamidele Olumilua University of Education, Science and Technology, Ikere-Ekiti 361101, Ekiti State, Nigeria; adamu.ademola@coeikere.edu.ng (A.A.); karifalo@yahoo.co.uk (K.M.A.)

2 Environmental and Sustainable Research Group (ESRG), Science and Education Development Institute, Akure 340214, Ondo State, Nigeria

* Correspondence: walefut@gmail.com; Tel.: +234-8034458674

† Presented at the 10th International Electronic Conference on Sensors and Applications (ECSA-10), 15–30 November 2023; Available online: https://ecsa-10.sciforum.net/.

Abstract: Individuals who spend most of their time indoors are especially sensitive to indoor air quality (IAQ), which significantly impacts their general well-being and health. Traditional IAQ measurement techniques, however, are frequently pricy, complicated, and labor-intensive. In this study, we used a low-cost, simple-to-use, and handy sensor system to track the levels of carbon dioxide (CO_2), nitrogen dioxide (NO_2), ozone (O_3), particulate matter ($PM_{1.0}$, $PM_{2.5}$, and PM_{10}), temperature, and relative humidity (RH) in a laboratory at the Bamidele Olomilua University of Education, Science, and Technology in Ikere-Ekiti for a month. We contrasted the outcomes with other benchmarks and WHO recommendations. However, the NO_2 levels (144.00–303.00 ppb) exceeded the suggested levels (National Institute for Occupational Safety and Health (NIOSH)—70 ppb; National Ambient Air Quality Standards (NAAQS)—100 ppb; National Environmental Standards and Regulations Enforcement Agency (NESREA)—120 ppb; and World Health Organization (WHO)—25 ppb), suggesting a possible cause of indoor contaminants. We also noticed that the temperature and humidity varied considerably throughout the day, which impacted the inhabitants' thermal comfort and ventilation. The principal component analysis (PCA) findings indicate that particulate matter, the weather, photochemical reactions, and combustion processes are the key contributors to fluctuation in the air quality measurements. Based on their quantities and relationships, these elements can have a variety of effects on both the natural environment as well as well-being. Our monitoring device can give immediate information and warnings, assisting in locating and reducing indoor airborne pollutant sources and enhancing indoor air quality (IAQ). This work shows that adopting a low-cost sensor system for IAQ measurement in underdeveloped nations, where such data are sparse and frequently erroneous, is both feasible and beneficial.

Keywords: air pollution; particulate matter; principal component analysis; meteorological factors; low-cost sensor; Nigeria

Citation: Adamu, A.; Arifalo, K.M.; Abulude, F.O. Indoor Air Quality Assessment Using a Low-Cost Sensor: A Case Study in Ikere-Ekiti, Nigeria. *Eng. Proc.* **2023**, *58*, 42. https://doi.org/10.3390/ecsa-10-16021

Academic Editor: Stefano Mariani

Published: 15 November 2023

1. Introduction

Indoor air quality (IAQ) issues have received a lot of attention recently because of their considerable effects on human health and well-being [1]. A variety of respiratory and cardiovascular problems can develop as a result of the increase of pollutants like carbon dioxide (CO_2), nitrogen dioxide (NO_2), ozone (O_3), and particulate matter (PM) of different sizes ($PM_{1.0}$, $PM_{2.5}$, and PM_{10}) in indoor environments, which can lower quality of life [1]. IAQ must be rigorously assessed and monitored in order to overcome these issues. While earlier studies have looked at IAQ in a variety of contexts, this study takes a fresh approach by using inexpensive sensors to thoroughly assess the indoor air quality parameters in Ikere-Ekiti, Nigeria.

This study is innovative in that it uses inexpensive sensors to assess a wide range of indoor air quality parameters, including CO_2, NO_2, O_3, $PM_{1.0}$, $PM_{2.5}$, and PM_{10}, in an interior environment and this work is the first of its kind in Ikere-Ekiti. The scope and breadth of IAQ assessments are constrained by traditional research's frequent reliance on pricy monitoring apparatus [1]. Especially in resource-limited places, this work pioneers the use of accessible sensor technologies, enabling extensive data gathering and creating a more inclusive understanding of IAQ dynamics.

Although earlier IAQ studies [2–4] have provided insightful information, they frequently concentrate on certain pollutants or make use of expensive monitoring tools, which limits the breadth and depth of data collection. On the other hand, our study covers a wide range of CO_2, NO_2, O_3, and different PM fractions in addition to single pollutants. Furthermore, the incorporation of low-cost sensors allows for broader spatial coverage and long-term data collection, facilitating a more nuanced analysis of IAQ trends and patterns in Bamidele Olumilua University of Education, Science and Technology, Ikere-Ekiti (BOUESTI).

The primary objectives of this study are to provide a holistic assessment of indoor air quality by measuring CO_2, NO_2, O_3, $PM_{1.0}$, $PM_{2.5}$, and PM_{10} levels in a medical laboratory indoor environment at BOUESTI health center. This multifaceted approach enables a deeper understanding of IAQ variations and potential sources of pollutants and by correlating the IAQ data with established air quality standards and guidelines, the study intends to evaluate the potential health risks posed by indoor pollutants. Lots of anthropogenic activities take place in this laboratory; unfortunately, in the room there is low ventilation (no fan, air conditioner, or fume extractor). This analysis will shed light on the implications for the well-being of its occupants and help formulate recommendations for IAQ improvement strategies.

In conclusion, this paper sets out to advance our understanding of indoor air quality by embracing innovation in sensor technology and adopting a holistic approach. By extending the scope of assessment to encompass multiple IAQ parameters and employing cost-effective sensor solutions, this study seeks to provide actionable insights for policymakers, building managers, and residents to enhance indoor environments and promote public health in BOUESTI and Ikere-Ekiti, Nigeria.

2. Materials and Methods

The study area was located at a medical laboratory at the Bamidele Olomilua University of Education, Science, and Technology Health Center (7.4952° N and 5.1747° E) in Ikere-Ekiti (Latitude: 7.5000° N 5.2333° E), Ekiti State (7.40001° N 5.15000° E), Nigeria, which is situated in the southwest of the nation. In terms of agriculture, transportation, industry, housing, and population, the town and its environs are expanding swiftly. The monitoring was continuous for 24 h during the rainy season period and was performed for a month (19 July–18 August 2023) as a preliminary study using a low-cost sensor (Model: SentinAir S3) developed and designed by a group of researchers from the Italian National Agency for New Technologies, Energy, and the Environment (ENEA), Department of Sustainable Development, Brindisi Research Center, Italy [5]. Following the correct procedures [5], the sensor, which was suspended four meters in the air while fixed on a rack, was able to detect the presence of carbon dioxide (CO_2), nitrogen dioxide (NO_2), ozone (O_3), particulate matter ($PM_{1.0}$, $PM_{2.5}$, and PM_{10}), temperature, and relative humidity (RH). The pollutants' PCA was established. The locations of the sampling sites were located with the aid of a Garmin satellite navigator. The produced data were statistically examined using Minitab and Excel versions.

3. Results and Discussion

The results (Table 1) reveal CO_2 concentrations of 537.48 $\pm$ 46.91 ppm, which are lower than the results (856.9 $\pm$ 400 and 987 $\pm$ 400 ppm) reported by Obisesan and Weli [6], 10,000 ppm NIOSH [7] and 1000 ppm WHO [8] (Figure 1). Though the figures might vary slightly depending on measurement methods and locations, this value aligns with the overall upward trend observed in recent decades. A comparative analysis of these studies provides a comprehensive view of how CO_2 concentrations have changed over time and across regions. NO_2 and O_3 concentrations vary thus: 197.91 ± 34.93 ppb, 2.03 (skewness), 2.61 (kurtosis) and 0.16 ± 14.72 ppb, −1.55 (skewness), 0.83 (kurtosis). The PM_1, $PM_{2.5}$, and PM_{10} levels were measured at 9.59 μg/m^3, 14.14 μg/m^3, and 15.09 μg/m^3, respectively, indicating a gradual increase in particle size. The $PM_{2.5}$ and PM_{10} concentrations are higher, reflecting their greater prevalence and potential health risks. A temperature of 32.76 °C can impact particle dynamics. Warmer temperatures may enhance atmospheric turbulence, leading to particle dispersion and dilution, potentially lowering PM concentrations [9]. With a relative humidity of 58.50%, particles could experience hygroscopic growth, causing them to absorb water and become larger. Higher RH might also aid in particle settling, potentially contributing to lower airborne PM levels. Understanding the intricate interplay between PM sizes, temperature, and relative humidity is crucial for accurate air quality assessments and effective pollution management strategies [10].

Table 1. Description of the results of the pollutants and weather parameters.

Parameter	Mean	StDev	CoefVar	Min	Q1	Q3	Max	skewness	kurtosis
CO_2 (ppm)	537.48	46.91	8.73	57.60	511.67	557.50	1505.90	2.60	37.62
NO_2 (ppb)	197.91	34.93	17.65	144.00	180.00	193.00	303.00	2.03	2.61
O_3 (ppb)	0.16	14.72	22.96	12.00	68.00	72.00	77.00	−1.55	0.83
$PM_{1.0}$ (μg/m^3)	9.59	5.83	60.78	0.00	5.00	14.00	64.00	1.29	5.92
$PM_{2.5}$ (μg/m^3)	14.14	8.88	62.80	0.00	8.00	20.00	124.00	1.93	11.92
PM_{10} (μg/m^3)	15.09	10.19	67.52	0.00	8.00	20.00	196.00	2.82	23.49
Temp (°C)	32.76	1.45	4.44	23.90	31.00	33.80	35.90	−0.73	2.46
RH (%)	58.50	3.93	6.72	49.50	56.50	60.00	91.90	2.74	14.74

Figure 1 shows the NO_2 concentration of 197.91 ppb and contrasts it with daily international guidelines set by NIOSH, NAAQS, NESERA, and the WHO. The observed NO_2 concentration surpasses the 70 ppb [7], 100 ppb [11], 120 NESERA [12] 120 ppb, and 25 ppb WHO [8] limits. This discrepancy is attributed to excessive NO_2 emission due to heavy vehicular movement within the study area. Effective pollution control measures and collaborative efforts are essential to curb NO_2 levels and ensure a healthier and cleaner environment. The recorded O_3 concentration falls below NIOSH's 100 ppb, NAAQS's 70 ppb, NESERA's 100 ppb, and the WHO's 100 ppb limits. While it adheres to most standards, its proximity to these thresholds necessitates vigilant monitoring. Ozone at elevated levels can exacerbate respiratory conditions and harm vegetation. While the observed concentration meets many standards, the closeness to limit values still implies potential health and ecological risks. The impacts of this pollutant can lead to complex health outcomes. The $PM_{1.0}$ concentration of 9.59 μg/m^3 falling below international standards is due to the low volume of vehicular and human activities because the institution is not in session. Maintaining this trend requires efforts to minimize emissions, improve air quality, and safeguard both human health and the environment. The recorded $PM_{2.5}$ concentration is below NAAQS's 35 μg/m^3, and NESERA's 40 μg/m^3, but comparable to the WHO's 15 μg/m^3 standard. The differences in standards emphasize the need for harmonization and stringent measures to curb $PM_{2.5}$ pollution. The recorded PM_{10} concentration is below NAAQS's 150 μg/m^3, NESERA's 150 μg/m^3, and the WHO's 45 μg/m^3 standards. The reasons for these differences could be due to the location of the monitoring station, meteorological parameters, time of day (especially rush hour), and season [13]. Also, the temperature and humidity did not have much effect due to the presence of standing and ceiling fans working and window ventilation.

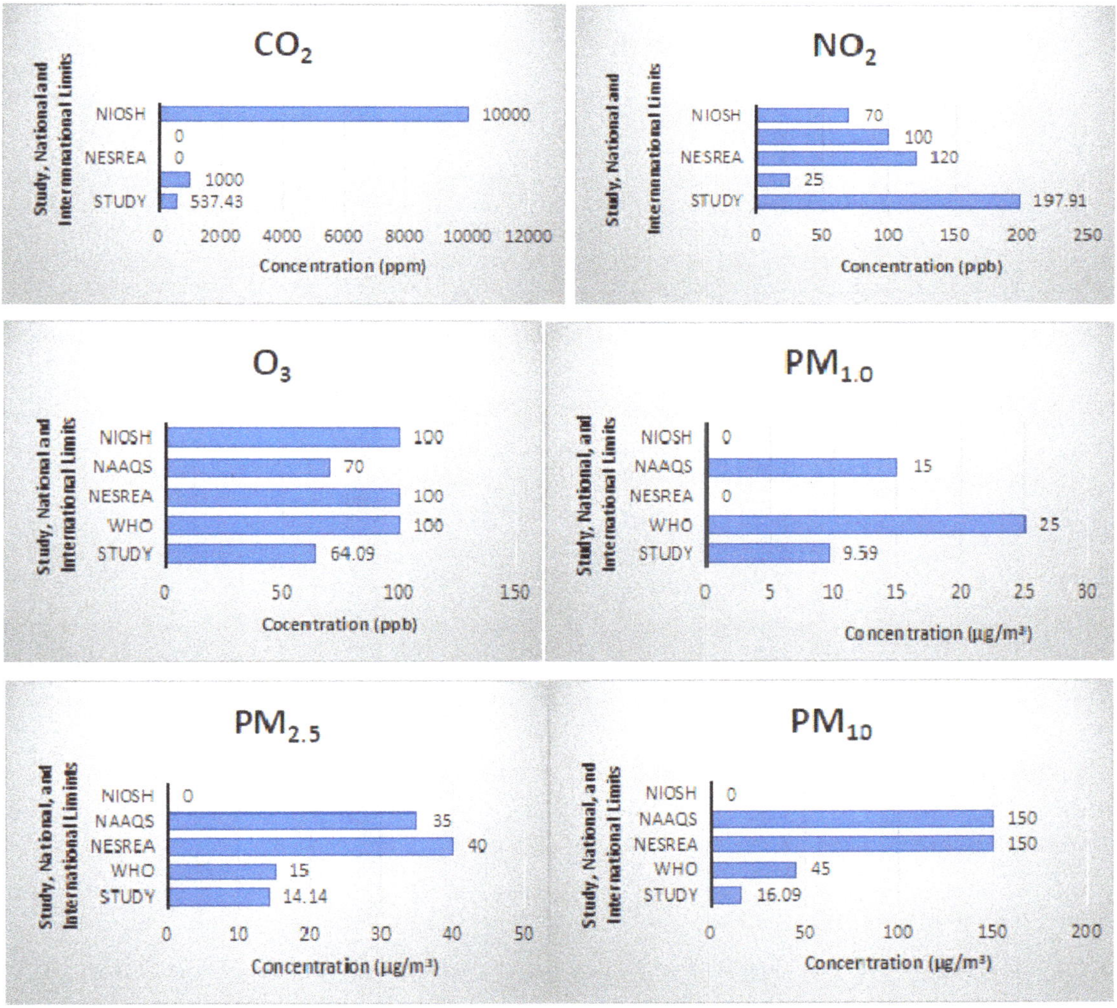

Figure 1. The comparisons of the results with the national and international standards.

According to the study's PCA results, particulate matter (PM_1, $PM_{2.5}$, and PM_{10}) showed substantial loading in PC1 (0.515, 0.524, and 0.522, respectively). High positive loadings show that PM concentrations tend to rise concurrently, which points to a general source of air pollution that affects a range of particle sizes, like vehicle emissions. The primary causes of variation in the air quality measurements are photochemical reactions and combustion processes, which are connected to atmospheric variables (temperature (−0.549) and relative humidity (0.544)) captured via PC2. Nitrogen dioxide (−0.689) and ozone (0.676) concentration fluctuation was recorded via PC3. This is a sign of intricate interactions in atmospheric chemistry or frequent sources of pollution, such as exhaust from moving vehicles.

4. Conclusions

This study assessed the indoor air quality of a BOUSTI by measuring CO_2, NO_2, O_3, $PM_{1.0}$, $PM_{2.5}$, and PM_{10} levels in a laboratory at the BOUESTI health center. The results obtained were compared to international and national IAQ data established standards and guidelines. Based on the recorded PM1 (9.59 μg/m^3), $PM_{2.5}$ (14.14 μg/m^3), and PM_{10} (15.09 μg/m^3) concentrations, the air quality is within the safe limits of NAAQS's 150 μg/m^3, NESERA's 150 μg/m^3, and the WHO's 45 μg/m^3 standards. The simple fact is that the institution is on holiday, so there are minimal vehicular and human movements that could have caused elevated pollutants. The temperature and humidity did not have much effect due to the presence of standing and ceiling fans working and window ventilation.

Author Contributions: Conceptualization, A.A., K.M.A. and F.O.A.; methodology, F.O.A.; software, A.A.; validation, K.M.A. and F.O.A.; formal analysis, A.A.; investigation, A.A. K.M.A.; resources, F.O.A.; data curation, K.M.A.; writing—original draft preparation, F.O.A.; writing—review and editing, A.A. and K.M.A.; visualization, A.A., K.M.A. and F.O.A.; project administration, A.A.; funding acquisition, K.M.A. All authors have read and agreed to the published version of the manuscript.

Funding: This research was funded by the Tertiary Education Trust Fund (TETFUND), Nigeria, grant number TETF/RD&D/UNI/IKERE/IBR/2021/VOL.1.

Institutional Review Board Statement: Not applicable.

Informed Consent Statement: Not applicable.

Data Availability Statement: Data are unavailable due to privacy.

Acknowledgments: The authors are grateful to TETFUND, Nigeria for providing the funds used in this study. Also, the authors thank Adeoluwa, O.V and Faloye, B.O. (Centre for Research and Development (CERAD)), Bamidele Olumilua University of Education, Science, and Technology, Ikere-Ekiti, Ekiti State, Nigeria for approving the funds. Lastly, the authors appreciate the individuals who hosted the sensors deployed in their residents and offices.

Conflicts of Interest: The authors declare no conflicts of interest.

References

1. Abulude, F.O. Right and Cheap Information on Air Quality: The Cases of Real-time Air Quality Index Monitors (Satellite and Phone apps). *Acad. Lett.* **2021**, *2*, 3401. [CrossRef]
2. Schibuola, L.; Tambani, C. Indoor environmental quality classification of school environments by monitoring PM and CO_2 concentration levels. *Atmos. Pollut. Res.* **2020**, *11*, 332–342. [CrossRef]
3. Cordell, R.L.; Panchal, R.; Bernard, E.; Gatari, M.; Waiguru, E.; Ng'ang'a, M.; Nyang'aya, J.; Ogot, M.; Wilde, M.J.; Wyche, K.P.; et al. Volatile Organic Compound Composition of Urban Air in Nairobi, Kenya and Lagos, Nigeria. *Atmosphere* **2021**, *12*, 1329. [CrossRef]
4. Feng, Y.; Yang, C.; Cao, X. Intermediate volatile organic compounds in Canadian residential air in winter: Implication to indoor air quality. *Chemosphere* **2023**, *328*, 138567. [CrossRef] [PubMed]
5. Suriano, D. Users Guide Version (v. 1.3) Refers to the SentinAir system Version 1.3 Available at the Github SentinAir Repository. 2021. Available online: https://github.com/domenico-suriano/SentinAir (accessed on 10 October 2022).
6. Obisesan, A.; Weli, V.E. Assessment of Air Quality Characteristics across Various Land-Uses in Port-Harcourt Metropolis. *J. Environ. Pollut. Manag.* **2019**, *2*, 106.
7. *NIOSH Pocket Guide to Chemical Hazards*; U.S. Department of Health and Human Services, Public Health Service, Centers for Disease Control, National Institute for Occupational Safety and Health: Cincinnati, OH, USA, 1994; DHHS (NIOSH) Publication No. 94–116.
8. WHO. Global air quality guidelines. In *Particulate Matter ($PM_{2.5}$ and PM_{10}), Ozone, Nitrogen Dioxide, Sulfur Dioxide and Carbon Monoxide*; World Health Organization: Geneva, Switzerland, 2021.
9. Wang, J.; Xie, X.; Fang, C. Temporal and Spatial Distribution Characteristics of Atmospheric Particulate Matter (PM_{10} and $PM_{2.5}$) in Changchun and Analysis of Its Influencing Factors. *Atmosphere* **2019**, *10*, 651. [CrossRef]
10. Manisalidis, I.; Stavropoulou, E. Stavropoulos, A.; Bezirtzoglou, E. Environmental and Health Impacts of Air Pollution: A Review. *Front. Public Health.* **2020**, *8*, 14. [CrossRef] [PubMed]
11. NAAQS. *Reviewing National Ambient Air Quality Standards (NAAQS): Scientific and Technical Information*; US EPA: Washington, DC, USA, 22 May 2023.

12. *National Environmental Standards and Regulations Enforcement Agency (Establishment) (Amended) Act, 2018*; National Environmental (Air Quality Control); Federal Government Printer: Lagos, Nigeria, 2021; pp. B 3364–B 3374.
13. Giemsa, E.; Soentgen, J.; Kuschi, T.; Beck, C.; Munkel, C.; Cyrys, J.; Pitz, M. Influence of Local Sources and Meteorological Parameters on the Spatial and Temporal Distribution of Ultrafine Particles in Augsburg, Germany. *Front. Environ. Sci.* **2021**, *8*, 609846. [CrossRef]

engineering proceedings

MDPI

Proceeding Paper

Effect of Glutathione on the Destruction Kinetics of Silver Nanoparticles in Aqueous Solutions: An Optical Study under Neutral and Alkaline Conditions †

Praskoviya Boltovets *, Sergii Kravchenko, Eduard Manoilov and Borys Snopok

V.E. Lashkaryov Institute of Semiconductor Physics, National Academy of Sciences of Ukraine, 03028 Kyiv, Ukraine; kravchenko@isp.kiev.ua (S.K.); manoilov@isp.kiev.ua (E.M.); snopok@isp.kiev.ua (B.S.)
* Correspondence: boltovets@isp.kiev.ua
† Presented at the 10th International Electronic Conference on Sensors and Applications (ECSA-10), 15–30 November 2023; Available online: https://ecsa-10.sciforum.net/.

Abstract: The interaction of nanostructured metal particles with the molecular components of biosystems differs significantly from the processes that take place in the presence of ions of the same metals. This unequivocally indicates the need to take into account not only the course of chemical processes but also implies to discuss certain physical effects that are usually neglected when considering such interactions. In this work, we studied the interaction of silver nanoparticle dispersion (Ag-NP) in ethylene glycol with a particle size less than 100 nm (Sigma-Aldrich 658804) with glutathione in a water and carbonate buffer (pH 10). The choice of glutathione (GSH) is due to the fact that it plays a significant role in intracellular processes, participating in the protection of intracellular components from the toxic effects of heavy metal ions; at the same time, differences in its interaction with silver ions and nanoparticles were experimentally demonstrated. A series of optical studies of the absorption and emission spectra of solutions of silver nanoparticles with GSH was carried out in order to establish the dominant processes in the system. It was shown that the above-mentioned silver nanoparticles in aqueous solutions spontaneously decompose over time, while glutathione differently affects these processes in water and carbonate buffer. It was shown that not only the local surface plasmon resonance bands but also the emission spectra of Ag-NP~GSH solutions in the region of 350–550 nm change with time. The sources of such radiation can be carbon quantum dots (CQD), which, according to published data, can be formed during the synthesis of silver nanoparticles and effectively luminesce in this region of the spectrum. Raman spectroscopy data confirm the presence of CQD in the Ag-NPs solution. The presence of quantum dots in the system makes it possible to indirectly track the presence of silver nanoparticles, which are booster centers, enhancing the emission of CQDs. The studies allow us to state that the interaction of glutathione with silver nanoparticles is a complex topochemical process in which, in addition to chemical reactions, the processes of transformation of silver nanoparticles and changes in the distribution of their sizes and chemical/physical functionality take place.

Keywords: glutathione; silver nanoparticles; carbon quantum dots; plasmon; luminescence

Citation: Boltovets, P.; Kravchenko, S.; Manoilov, E.; Snopok, B. Effect of Glutathione on the Destruction Kinetics of Silver Nanoparticles in Aqueous Solutions: An Optical Study under Neutral and Alkaline Conditions. *Eng. Proc.* **2023**, *58*, 43. https://doi.org/10.3390/ecsa-10-16254

Academic Editor: Francisco Falcone

Published: 15 November 2023

1. Introduction

Glutathione is a major endogenous antioxidant that protects cells against oxidative stress through its ability to bind to and reduce ROS. The production of free radicals induced by nanoparticles leads to a reduction of GSH to an oxidized form, followed by the induction of oxidative stress [1]. Thus, preservation of the GSH-mediated antioxidant defense system is critical for cell survival. However, various studies have indicated that cellular levels of GSH are either increased or decreased after in vitro treatment with AgNPs [2]. The increased levels of GSH observed in some AgNP-treated cells [3] may involve cellular responses to cope with AgNP-mediated oxidative damage. By contrast, the decreased levels of GSH noted in AgNP-treated human skin carcinoma and fibrosarcoma cells [4] suggest an inhibition of GSH-synthesizing enzymes and/or abnormally increased demand for GSH in conjugation with electrophilic molecules.

The interaction of nanostructured metal particles with the molecular components of biosystems differs significantly from the processes that take place in the presence of ions of the same metals [5]. This unequivocally indicates the need to take into account not only the course of chemical processes but also implies to discuss certain physical effects that are usually neglected when considering such interactions.

In this work, we studied the interaction of silver nanoparticle dispersion (Ag-NP) in ethylene glycol with particle size less than 100 nm (Sigma-Aldrich 658804) with glutathione in a water and carbonate buffer (pH 10). The choice of glutathione as described above is due to the fact that GSH plays a significant role in intracellular processes, participating in the protection of intracellular components from the toxic effects of heavy metal ions; at the same time, differences in its interaction with silver ions and nanoparticles were experimentally demonstrated. The thiol group of glutathione has been shown to form an extremely strong bond with silver [6].

It was shown that silver nitrate significantly affects the quantity of glutathione in the blood plasma by oxidizing the reduced glutathione to disulfide 2 GSH <=> $^{-}GSSG^{-}$ $2H^{+}$) or the formation of the Ag-SG complex, which also indicates the protective role of glutathione, thus neutralizing the toxic effect of silver [7]. As for Ag NP, it was demonstrated that the cytotoxicity induced by the nanoparticles was even higher compared to that observed when $AgNO_3$ was used as a silver ion source. Namely, Ag NPs induce reactive oxygen species (ROS) generation [8] (which play important roles in a variety of normal biochemical functions, but abnormality in their functions results in pathological processes, for example, inducing apoptosis [9]) and suppression of reduced glutathione in human Chang liver cells. ROS generated by AgNPs resulted in damage to various cellular components, DNA breaks, lipid membrane peroxidation, and protein carbonylation [10]).

In this work, a series of optical and luminescence studies in the absorption and emission spectra of aqueous silver nanoparticle solutions with GSH were carried out. The obtained results showed that the AgNPs spontaneously decompose over time in aqueous solutions with different mechanisms in water and carbonate buffer when glutathione was added to the solution.

2. Materials and Methods

AgNPs (<100 nm, 10% wt) in ethylene glycol, Glutathione (99%), Sodium Carbonate (ACS reagent $\geq$ 99.5%), and Sodium Bicarbonate (ACS reagent, $\geq$99.7%) were received from Sigma-Aldrich (St. Louis, MI, USA). The optical absorbance was studied by UV-Vis spectroscopy on Umico SQ2800 (UNICO, Dayton, NJ, USA). Luminescence spectra were measured using spectrometer USB2000+ (OCEAN OPTICS INC, Dunedin, FL, USA).

3. Results and Discussion

A 0.01 M carbonate buffer solution (pH~10) was used to prepare the AgNPs buffer solution and the GT one. First, a 5 mM AgNPs buffer solution was prepared by mixing the initial AgNPs (10% wt) in ethylene glycol with a carbonate buffer solution in terms of the silver amount. Then, a GT solution was prepared in carbonate buffer at concentrations of 20 mM. The AgNPs buffer solution and the GT one were mixed in equal volumes [11]. Analogously, the AgNPs solution in water was prepared, and 20 mM GT water solution in equal volumes was added to the AgNPs solution. The absorbance spectra of the prepared solutions were measured at 00, 10, 20, 30, 40, 50 min, 1, 1:30, 2, 2:30, 3, 3:30, 4 h, and 2 days after mixing, while the appropriate luminescence spectra were measured in 30 min and 1 h.

UV-Vis spectra showed that the Ag plasmon peak maximum was characterized by different behavior in the water and the buffer solution of AgNPs after adding the GT (Figures 1a and 2a).

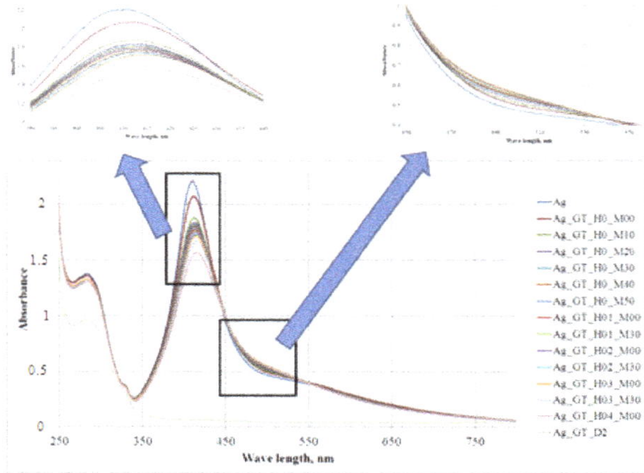

(**a**)

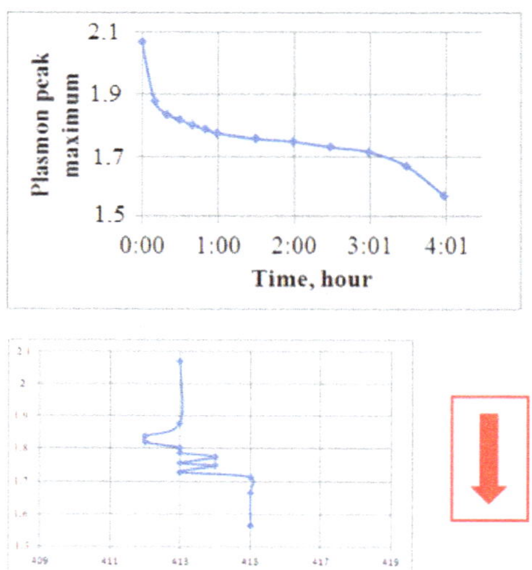

(**b**)

Figure 1. (**a**) UV-Vis spectra of the water solution of AgNPs after adding the GT. (**b**) Dynamics behavior of the plasmon peak maximum in the water solution of AgNPs after adding the GT.

Figure 2a,b shows the change dynamics of the plasmon peak maximum in both the water solution of AgNPs and the buffer solution. The plasmonic band intensity at 413 nm decreased with a slight bathochromic shift in maximum in the water solution of AgNPs, along with decreasing the plasmonic intensity at 520 nm, which indicated the fast AgNPs dispergation (Figure 1a). A quite different process was observed in the buffer solution, where in the beginning the decrease in the plasmon intensity at 415 nm occurred, and then its increasing followed by decreasing was observed. On the other hand, the plasmon band at 550 nm was increasing, which indicated the occurrence of AgNPs agglomeration in the first stage.

Also, the luminescence spectra relieved the different behavior in the water and the buffer solution of AgNPs (Figure 3a,b).

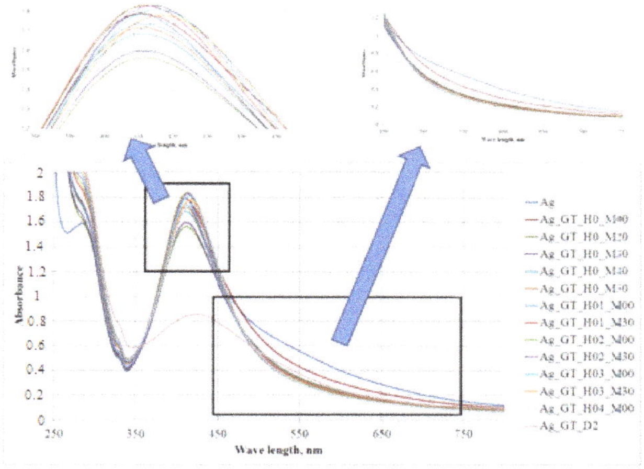

(**a**)

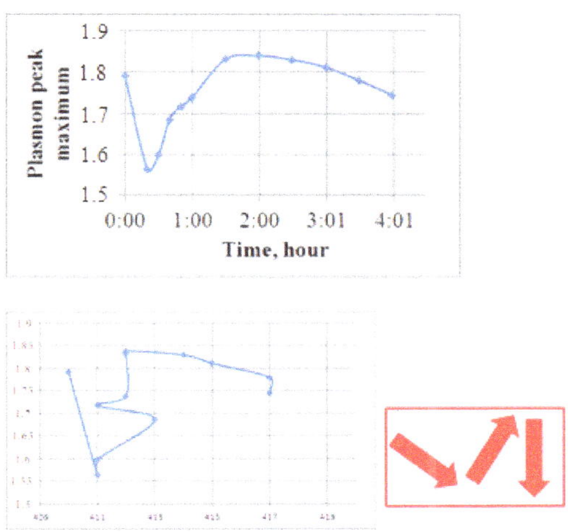

(**b**)

Figure 2. (**a**) UV-Vis spectra of the buffer solution of AgNPs after adding the GT. (**b**) Dynamics behavior of the plasmon peak maximum in the buffer solution of AgNPs after adding the GT.

The spectra of the initial water and buffer AgNPs solution were characterized by the presence of two wide bands at 3.4 and 2.52 eV. The sources of such radiation can be carbon quantum dots (CQD), which can be formed during the AgNPs preparation and have effectively been luminescent in this region of the spectra [12–14]. Raman spectroscopy data confirm the presence of CQD in the used dispersion nanoparticles of silver [11]. The AgNPs defragmentation in the water and buffer solution resulted in the luminescence band diminishing at 3.4 and 2.52 eV. It is interesting to notice that the appearance of the new luminescence band at 2.9 eV diminished during the time in the water AgNPs solution (Figure 3a).

Therefore, the presence of quantum dots in the solution makes it possible to indirectly track the presence of silver nanoparticles, which are booster centers, enhancing the emission of CQDs.

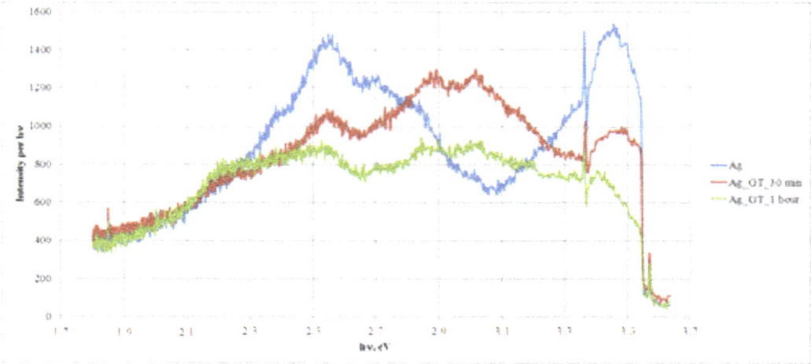

(**a**)

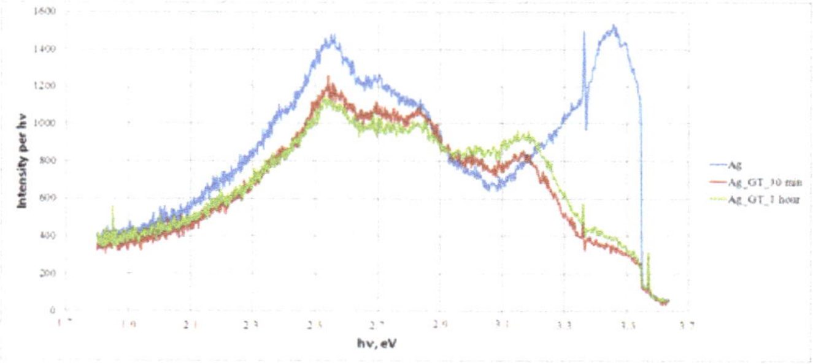

(**b**)

Figure 3. (**a**) Luminescence spectra of the water solution of AgNPs after adding the GT. (**b**) Luminescence spectra of the buffer solution of AgNPs after adding the GT.

4. Conclusions

This study allows us to state that the interaction of glutathione with silver nanoparticles is a complex topochemical process in which, in addition to chemical reactions, the processes of transformation of silver nanoparticles and changes in the distribution of their sizes and chemical/physical functionality take place.

Author Contributions: Conceptualization, B.S. and S.K; methodology, E.M.; software, P.B.; validation, E.M., S.K. and P.B.; formal analysis, S.K.; investigation, E.M.; resources, B.S.; data curation, P.B.; writing—original draft preparation, S.K.; writing—review and editing, P.B.; visualization, S.K.; supervision, B.S.; project administration, E.M.; funding acquisition, B.S. All authors have read and agreed to the published version of the manuscript.

Funding: This research received no external funding.

Institutional Review Board Statement: Not applicable.

Informed Consent Statement: Not applicable.

Data Availability Statement: Data are contained within the article.

Conflicts of Interest: The authors declare no conflicts of interest.

References

1. Talarska, P.; Boruczkowski, M.; Żurawski, J. Current Knowledge of Silver and Gold Nanoparticles in Laboratory Research-Application, Toxicity, Cellular Uptake. *Nanomaterials* **2021**, *11*, 2454. [CrossRef] [PubMed]
2. Kim, S.; Ryu, D.Y. Silver nanoparticle-induced oxidative stress, genotoxicity and apoptosis in cultured cells and animal tissues. *J. Appl. Toxicol.* **2013**, *33*, 78–896. [CrossRef] [PubMed]
3. Farkas, J.; Christian, P.; Gallego-Urrea, J.A.; Roos, N.; Hassellov, M.; Tollefsen, K.E.; Thomas, K.V. Uptake and effects of manufactured silver nanoparticles in rainbow trout (Oncorhynchus mykiss) gill cells. *Aquat. Toxicol.* **2011**, *101*, 117–125. [CrossRef] [PubMed]

4. Arora, S.; Jain, J.; Rajwade, J.M.; Paknikar, K.M. Cellular responses induced by silver nanoparticles: In vitro studies. *Toxicol. Lett.* **2008**, *179*, 93–100. [CrossRef] [PubMed]
5. Snopok, B.A.; Snopok, O.B. Nanoscale–Specific Analytics: How to Push the Analytic Excellence in Express Analysis of CBRN. In *Advanced Nanomaterials for Detection of CBRN. NATO Science for Peace and Security Series A: Chemistry and Biology*; Bonča, J., Kruchinin, S., Eds.; Springer: Dordrecht, The Netherlands, 2020. [CrossRef]
6. Huang, G.G.; Hossain, M.K.; Han, X.X.; Ozaki, Y. A novel reversed reporting agent method for surface-enhanced Raman scattering; highly sensitive detection of glutathione in aqueous solutions. *Analyst* **2009**, *134*, 2468–2474. [CrossRef] [PubMed]
7. Khan, H.; Khan, M.F.; Asim-Ur-Rehman; Jan, S.U.; Ullah, N. The protective role of glutathione in silver induced toxicity in blood components. *Pak. J. Pharm. Sci.* **2011**, *24*, 123–128. [PubMed]
8. Carlson, C.; Hussain, S.M.; Schrand, A.M.; Braydich-Stolle, L.K.; Hess, K.L.; Jones, R.L.; Schlager, J.J. Unique cellular interaction of silver nanoparticles: Size-dependent generation of reactive oxygen species. *J. Phys. Chem. B* **2008**, *112*, 13608–13619. [CrossRef] [PubMed]
9. Martindale, J.L.; Holbrook, N.J. Cellular response to oxidative stress: Signaling for suicide and survival. *J. Cell. Physiol.* **2002**, *192*, 1–15. [CrossRef]
10. Piao, M.J.; Kang, K.A.; Lee, I.K.; Kim, H.S.; Kim, S.; Choi, J.Y.; Choi, J.; Hyun, J.W. Silver nanoparticles induce oxidative cell damage in human liver cells through inhibition of reduced glutathione and induction of mitochondria-involved apoptosis. *Toxicol. Lett.* **2011**, *201*, 92–100. [CrossRef]
11. Kravchenko, S.; Boltovets, P.; Snopok, B. Chemical Transformation of Typical Biological Recognition Elements in Reactions with Nanosized Targets: A Study of Glutathione Coated Silver Nanoparticles. *Eng. Proc.* **2023**, *35*, 31.
12. Liu, J.; Wang, L.; Bao, H. A novel fluorescent probe for ascorbic acid based on seed-mediated growth of silver nanoparticles quenching of carbon dots fluorescence. *Anal. Bioanal. Chem.* **2019**, *411*, 877–883. [CrossRef] [PubMed]
13. Lin, S.; Wang, Z.; Zhang, Y.; Huang, Y.; Yuan, R.; Xiang, W.; Zhou, Y. Easy synthesis of silver nanoparticles-orange emissive carbon dots hybrids exhibiting enhanced fluorescence for white light emitting diodes. *J. Alloys Compd.* **2017**, *700*, 75–82. [CrossRef]
14. Mai, X.-D.; Le, Q.-T.; Thi, L.-A.N.; Thi, P.N.; Thi, P.L.; La, V.-H. Photosynthesis of Silver Nanoparticle—Carbon Quantum Dots Nanocomposites. *Mat. Sci. Res. India* **2019**, *16*, 118–124. [CrossRef]

engineering proceedings

MDPI

Proceeding Paper

An AI-Powered, Low-Cost IoT Node Oriented to Flood Early Warning Systems †

Evangelos Skoubris * and George Hloupis

Department of Surveying and Geoinformatics Engineering, School of Engineering, University of West Attica, Egaleo Park Campus, Ag. Spyridonos Street, 12243 Egaleo, Greece; hloupis@uniwa.gr
* Correspondence: eskoubris@uniwa.gr
† Presented at the 10th International Electronic Conference on Sensors and Applications (ECSA-10), 15–30 November 2023; Available online: https://ecsa-10.sciforum.net/.

Abstract: The present study aims to design a low-cost smart AI-powered node to serve as a flood Early Warning System with a complete solution. The node is designed to predict forthcoming flood events by capturing and combining critical data related to such phenomena. Such data are the water levels at rivers or other water discharge basins, rainfall, soil moisture, and material displacement at river slopes. The node will autonomously monitor the above quantities at a high frequency rate and selectively upload them to a server only when verified conditions for a forthcoming flood will exist. These conditions will be evaluated by the local ML model. This will allow each node to reliably predict flood events and issue local and remote alarms. The combination of several nodes at an area of interest will form a robust and reliable Early Warning System.

Keywords: Early Warning Systems; floods; AI; low cost; IoT; nodes; sensors; cellular network

Citation: Skoubris, E.; Hloupis, G. An AI-Powered, Low-Cost IoT Node Oriented to Flood Early Warning Systems. *Eng. Proc.* **2023**, *58*, 44. https://doi.org/10.3390/ecsa-10-16023

Academic Editor: Stefano Mariani

Published: 15 November 2023

1. Introduction

Natural disaster management presents a challenging field of research and technology applications. Especially during the past decade, climate change has led to several occurrences of extreme weather events, leading to severe natural disasters [1]. Researchers from various scientific fields aim towards achieving viable technological solutions to aid disaster management. One of the most critical kinds of natural disasters is floods. Their extent and magnitude often cause huge social, economic, health, and safety impacts [2].

Flood management via precipitation prediction and hydraulic modeling was proven to be less effective to predict floods [3]. To aid the accuracy of the prediction, several more measurements are required. At this scope, many attempts have been made in the past to employ the Internet of Things (IoT) technology [4], and several IoT-aided Early Warning Systems related to floods have been designed and proposed [2,5–8]. The IoT can be roughly described as a system that incorporates small and remote electronic devices with low power and usually a low cost that samples data and transmits them to an IoT server for further processing. In the case of EWSs, data from the nodes are used as the input data for several models of disaster prediction.

Since the number of IoT nodes presents a high growth, it is reported that both the network communication latencies [9] and the time needed to process the vast amount of data is increased in order to conclude certain alarms and/or actions. Also, the increasing number of nodes stresses the wireless communication channels, and this can also induce further communication latencies. The increasing latencies lead to delays in taking certain actions. Taking into consideration that the data usage can also be translated to cost, the need to locally process the collected data directly at the end devices has emerged.

While some actions can be triggered using simple threshold values of the received data, more complex decisions, such as those related to flood prediction, require sophisticated algorithms. According to the latest studies in the literature [4,9,10], the most popular method of applying such complex algorithms is the use of Artificial Intelligence (AI). When implemented directly on the IoT nodes, this technology is called the Artificial Intelligence of Things (AIoT) [2]. The AIoT requires more system resources, such as computational power and memory, which are offered in modern embedded platforms and Microcontroller Units (MCUs). In [7], the authors claim to have developed the first large-scale IoT-based real-time flood forecast system that has been enabled by AI and deployed in the real world.

In the present study, a node capable of executing TinyML models and performing AI algorithms is designed. The node has all of the specifications to serve as an IoT device. It is power-autonomous, network-connected, and interfaces with various sensors and actuators. The employed sensor suite matches the needs to serve as part of an EWS designed for floods. Since major flood phenomena relate to riverbank collapses, the design is also oriented towards this phenomenon. To the best of the authors' knowledge, this is the first study that utilizes accelerometer data at river-related terrestrial locations to study the relation of riverbank/slope movements to an upcoming flood event. In all other flood-related studies, the combination of accelerometer data mainly relates to civil infrastructure monitoring.

2. Materials and Methods

The proposed system is based on a designed system-level electronic device, referred to as the node. The node is completely custom-designed, both in hardware and in software. The hardware design and the operating sequence of the device are described in the following subsections.

2.1. Hardware

The main parts and modules of the node are the Microcontroller Unit (MCU), the power supply unit (PSU), and the data connectivity unit (modem). All circuitry was hosted in a custom designed printed circuit board (PCB), and within a commercial IP65 rated project box, in order to withstand external environmental conditions at the place of deployment. A short description of each main part of the node is presented at the following paragraphs.

The selected MCU is the Microchip ATmega2560 8-bit AVR controller. This MCU offers enough resources (20 MHz clock, 8 kB of SRAM, 256 kB Flash) to be able to run TinyML models, custom-designed for 8-bit architectures. The MCU also has plenty of on-board peripherals, supporting multiple serial communication protocols and Analog-to-Digital converters. The selected module hosting this MCU is the RobotDyn Mega 2560 PRO [11]. The module also offers a USB-to-TTL converter to easily re-program the MCU, together with a flexible power management circuit and ultra-compact footprint.

The PSU is based on a single 3.7 V, 5 Ah, 26650 form factor Li-ion battery cell, paired with a battery charger and a protection circuit using the TP4056 Integrated Circuit (IC). The power source to charge the battery is a 6 V solar panel able to provide a maximum of 1 W of charge power.

Data connectivity is provided via Simcom's SIM7600G modem [12]. The modem is fitted at an OEM breakout board, which also bears a power management circuit, a SIM card socket, antenna connectors, and all components needed for the modem's recommended operation. The modem provides LTE/GSM (4G/2G) connectivity.

The block diagram of the system is presented at Figure 1. The various peripherals communicate differently to the MCU. The modem occupies a hardware USART port. Regarding the sensors, the accelerometer uses the I2C line of the MCU, the analog output of the soil moisture sensor is read using one 10-bit ADC, while the rain gauge's output pulses are read by the MCU by utilizing one external interrupt pin. The raw battery voltage output is used as the main Vcc powering all electronic modules, since each module has its own power management and voltage regulation circuitry. The primary voltage regulator is omitted.

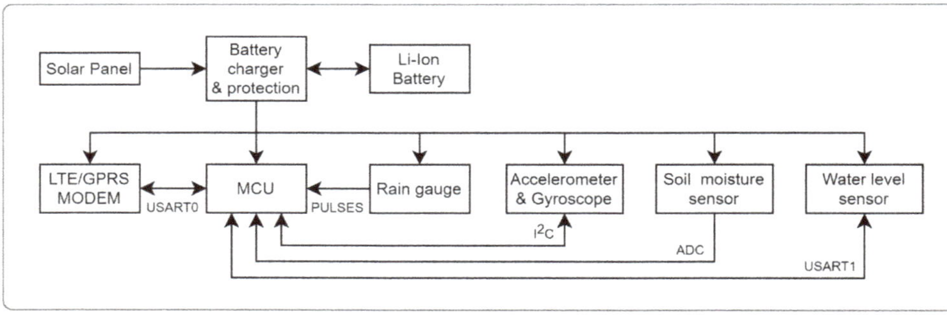

Figure 1. The block diagram of the node.

The various external sensors are connected to the node with a cable. The use of cable glands ensures the IP rating of the node's case and relieves any stress that may be applied to the cables. The sensors selected for this application are a soil moisture sensor (Truebner SMT50 [13]), an ultrasonic ranger used as a water level sensor (MaxBotix MB7066-100 [14]), an accelerometer and gyroscope sensor (Invensense MPU6050) hosted at an OEM breakout board, and a rain gauge sensor (DFRobot SEN0575 [15]). A photograph of a complete prototype and a map of the first deployment location are depicted in Figure 2.

Figure 2. A photograph of the complete prototype (**left**) and the first deployment location (**right**). The red dot shows the exact location of the first system deployment.

2.2. Operating Sequence

The second important part is the firmware (FW) and the operating sequence of the device. The operating logic diagram is presented in Figure 3. When the node is first powered up, an initialization routine is run to check that the system is fully functional. Then, the node proceeds to a periodic data acquisition from the sensors. Inertial measurement unit (IMU) data from the accelerometer are retrieved at a frequency of 4 Hz and stored at a circular buffer that has a total length of 240 measurement sets. This provides the opportunity to recall a detailed IMU history of 60 s at any moment. Data from the soil moisture sensor, the water level meter, and the rain gauge are sampled at a period of 5 min and stored a secondary circular buffer with total length of 288, providing a measurement history of 24 h.

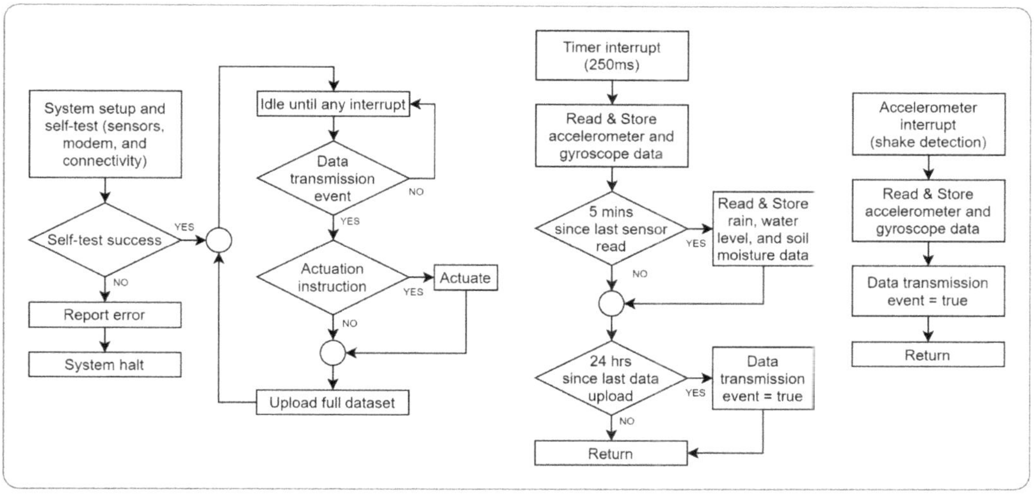

Figure 3. The logic diagram of the firmware.

The secondary buffer is uploaded once per day to the server to keep record of the rain and the soil moisture. On the contrary, the one-minute historical data of the accelerometer are uploaded only at certain occasions. These are the MPU6050 internal programmable interrupt at shake detection or the existence of an event of an angle change calculated according to the following equation [16]:

$$a_F = c \times a_G + (1 - c) \times a_A, \quad (1)$$

where:

a_F: the filtered angle;
c: filter tunable constant ($0 < c < 1$ in order to have a complementary filter that neither overshoots nor attenuates);
a_G: calculated angle using gyroscope data;
a_A: calculated angle using accelerometer data.

This logic ensures that critical short-term historical data related to possible material displacement at river slopes will be retrieved. Each of these datasets will be matched to the actual physical phenomenon of flood or landslide at the riverbank, if apparent, and thus be classified. These data will be used as the main training dataset for the tinyML model to detect floods and landslides at riverbanks. Alongside the accelerometer data, precipitation, temperature/RH, and soil moisture data will also contribute to the model training.

The first version of the system will act mostly as a filtered critical data acquisition device, and when the TinyML models are developed, a new firmware will be uploaded to test the efficiency of the model. The Newton framework will be used to train and produce the TinyML model, customized for the 8-bit architecture of this system. Newton is a trusted high-level framework, perfectly fitted for applications requiring fast development, avoiding an in-depth Neural Network optimization.

After loading the TinyML code to the node, when a node detects the risk of a flood, certain local actions can be taken, such as issuing an alarm sign or actuating electromechanical devices to engage a certain anti-flood infrastructure. Certainly, both the data related to the detection of flood risk and the actions applied by the end device will be uploaded to the IoT server for further processing and consequent management actions.

3. Results and Discussion

The overall system proved to be well designed. Operationally, the node's performance was stable, and the sensor readings were consistent and reliable. Power-wise, the node presented a moderate average current consumption of about 30 mA when acquiring data, and about 50 mA when in the data transmission mode. The overall power autonomy is approx. 5–6 days on a fully charged battery. Although this battery life seems rather short, the real-time and continuous sampling of the sensor data justifies this performance.

The FW logic in terms of data acquisition and upload worked seamlessly. Communication via a cellular network proved to be very consistent, with short network registration and upload/download latencies. The customized data upload protocol reduced data usage to the extent that an inexpensive data plan is sufficient (i.e., 10 MB/month). The prototype node was manually triggered to upload five IMU datasets with a 60 s duration for each. Thus, a 300 s IMU signal was obtained, and it is presented in Figure 4.

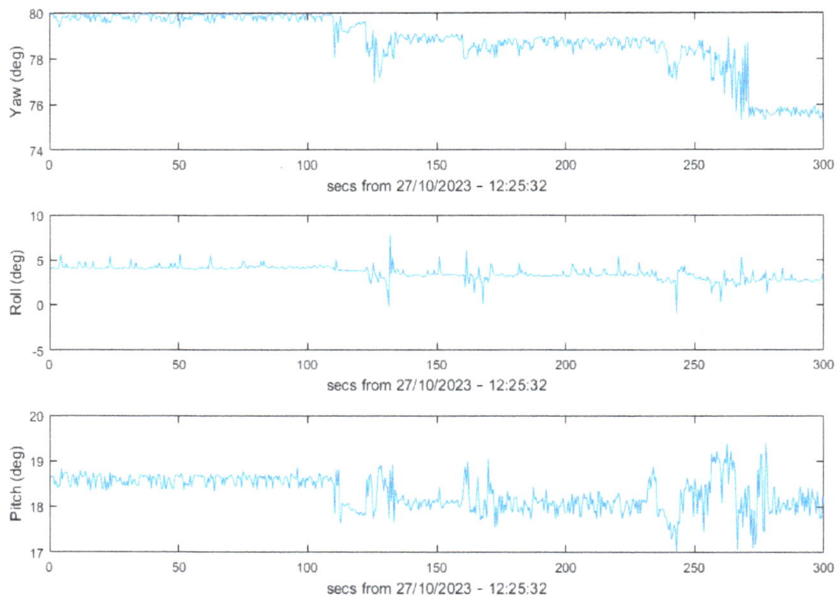

Figure 4. Sample IMU signals (yaw, roll, and pitch, from top to bottom).

According to the project's timeline, it is expected that the first training dataset will be ready by the end of Q2 2024, and that immediately after, the TinyML models will be deployed. This way, the proposed system will be tested in real-world conditions during the wet season of 2024.

4. Conclusions

This research presented the concept, architecture, design, and development of a low-cost, AI-powered IoT node aimed for flood Early Warning Systems. The node interfaces to several sensors, monitoring critical quantities related to floods, such as the water level at a riverbed, the precipitation, the soil moisture, and the material displacement at river slopes. The first stage of the system, i.e., the development of the prototypes, is concluded. The nodes were tested thoroughly in a laboratory environment and proved to be ready for the next stage of this research. The second stage will be the deployment of a certain number of nodes to the area of interest, the collection of the training dataset for the TinyML model, and the actual training of the model. Further on, the third stage will be the upload of the model to the end nodes, the test of the model, and the characterization of the accuracy for flood prevention. New datasets retrieved from the third stage may also be used for the further training and improvement of the model. All findings are scheduled to be published when available.

Author Contributions: Conceptualization, E.S. and G.H.; methodology, E.S. and G.H.; software, E.S.; writing—original draft preparation, E.S.; writing—review and editing, G.H. All authors have read and agreed to the published version of the manuscript.

Funding: This research received no external funding.

Institutional Review Board Statement: Not applicable.

Informed Consent Statement: Not applicable.

Data Availability Statement: Data are unavailable until 31 December 2024 due to privacy agreement restrictions with the Corinth's Canal Authority.

Conflicts of Interest: The authors declare no conflicts of interest.

References

1. Yang, T.H.; Liu, W.C. A general overview of the risk-reduction strategies for floods and droughts. *Sustainability* **2020**, *12*, 2687. [CrossRef]
2. Boulouard, Z.; Ouaissa, M.; Ouaissa, M.; Siddiqui, F.; Almutiq, M.; Krichen, M. An Integrated Artificial Intelligence of Things Environment for River Flood Prevention. *Sensors* **2022**, *22*, 9485. [CrossRef] [PubMed]
3. Aljohani, F.H.; Abi Sen, A.A.; Ramazan, M.S.; Alzahrani, B.; Bahbouh, N.M. A Smart Framework for Managing Natural Disasters Based on the IoT and ML. *Appl. Sci.* **2023**, *13*, 3888. [CrossRef]
4. Esposito, M.; Palma, L.; Belli, A.; Sabbatini, L.; Pierleoni, P. Recent Advances in Internet of Things Solutions for Early Warning Systems: A Review. *Sensors* **2022**, *22*, 2124. [CrossRef] [PubMed]
5. Wang, Q. Machine Learning Model Design for IoT-Based Flooding Forecast. In Proceedings of the 2022 International Conference on Cyber-Enabled Distributed Computing and Knowledge Discovery, Suzhou, China, 14–16 October 2022; pp. 97–103. [CrossRef]
6. Samikwa, E.; Voigt, T.; Eriksson, J. Flood Prediction Using IoT and Artificial Neural Networks with Edge Computing. In Proceedings of the 2020 International Conferences on Internet of Things (iThings) and IEEE Green Computing and Communications (GreenCom) and IEEE Cyber, Physical and Social Computing (CPSCom) and IEEE Smart Data (SmartData) and IEEE Congress on Cybermatics (Cybermatics), Rhodes, Greece, 2–6 November 2020; pp. 234–240. [CrossRef]
7. Wang, Q.; Abdelrahman, W. High-Precision AI-Enabled Flood Prediction Integrating Local Sensor Data and 3rd Party Weather Forecast. *Sensors* **2023**, *23*, 3065. [CrossRef] [PubMed]
8. Wilson, A.J.; Pon Bharathi, A.; Anoop, M.; Angelin Jeba Malar, J. Information system for flood monitoring based on IoT and AI. In Proceedings of the 2023 2nd International Conference on Smart Technologies and Systems for Next Generation Computing (ICSTSN), Villupuram, India, 21–22 April 2023; pp. 1–4. [CrossRef]
9. Hou, K.M.; Diao, X.; Shi, H.; Ding, H.; Zhou, H.; de Vaulx, C. Trends and Challenges in AIoT/IIoT/IoT Implementation. *Sensors* **2023**, *23*, 5074. [CrossRef] [PubMed]
10. Sun, W.; Bocchini, P.; Davison, B.D. Applications of artificial intelligence for disaster management. *Nat. Hazards* **2020**, *103*, 2631–2689. [CrossRef]
11. Mega 2560 PRO (Embed) CH340G/ATmega2560-16AU. Available online: https://robotdyn.com/mega-2560-pro-embed-ch340g-atmega2560-16au.html (accessed on 14 November 2023).
12. SIM7600X Module 4G Wireless Solutions. Available online: https://www.simcom.com/product/SIM7600X.html (accessed on 25 October 2023).
13. Truebner SMT50 Datasheet. Available online: https://www.truebner.de/assets/download/Manual_SMT50.pdf (accessed on 25 October 2023).
14. MaxBotix MB7066-100 Datasheet. Available online: https://maxbotix.com/pages/xl-maxsonar-wr-datasheet (accessed on 25 October 2023).
15. DFRobot Rainfall Sensor (SEN0575). Available online: https://wiki.dfrobot.com/SKU_SEN0575_Gravity_Rainfall_Sensor (accessed on 25 October 2023).
16. Hloupis, G.; Pagounis, V.; Tsakiri, M.; Doxastakis, G.; Zacharis, V. Low-cost warning system for the monitoring of the Corinth Canal. *Appl. Geomat.* **2017**, *9*, 263–277. [CrossRef]

engineering proceedings

MDPI

Proceeding Paper

Remote Embedded System for Agricultural Field Monitoring: Enhancing Resource Allocation in Agriculture †

Ronald J. Contijo, Lucas C. de Camargo, Renan O. A. Takeuchi, André L. S. Moscato, Lafaiete H. R. Leme and Wenderson N. Lopes *

Paraná Federal Institute of Education, Science and Technology, Avenida Dr. Tito, 801, Jardim Panorama, Jacarezinho 86400-000, Brazil; ronaldques1@gmail.com (R.J.C.); lucasxz.our@gmail.com (L.C.d.C.); renan.takeuchi@ifpr.edu.br (R.O.A.T.); andre.moscato@ifpr.edu.br (A.L.S.M.); lafaiete.leme@ifpr.edu.br (L.H.R.L.)
* Correspondence: wenderson.lopes@ifpr.edu.br

† Presented at the 10th International Electronic Conference on Sensors and Applications (ECSA-10), 15–30 November 2023; Available online: https://ecsa-10.sciforum.net/.

Abstract: This research addresses the need to enhance agricultural management due to rapid population growth in the 20th and 21st centuries. It focuses on integrating sensors and embedded systems to collect data on soil and air conditions, including temperature, humidity, and solar radiation. This information is obtained using ESP32 microcontrollers and stored in a centralized database. This system uses JavaScript and the Leaflet library's interpolation algorithm to create interactive maps, allowing the farmers in the northern region of Paraná, Brazil, to monitor their fields and activate irrigation according to the predefined routines. This innovative system provides a data-driven approach to agriculture and automated irrigation, conferring advantages to the farmers and their families.

Keywords: relational database; JavaScript; interpolation; embedded systems; crop cultivation; family farmers

Citation: Contijo, R.J.; de Camargo, L.C.; Takeuchi, R.O.A.; Moscato, A.L.S.; Leme, L.H.R.; Lopes, W.N. Remote Embedded System for Agricultural Field Monitoring: Enhancing Resource Allocation in Agriculture. *Eng. Proc.* **2023**, *58*, 45. https://doi.org/10.3390/ecsa-10-16046

Academic Editor: Jean-marc Laheurte

Published: 15 November 2023

1. Introduction

Since the late 20th century, the agricultural industry has embraced technological advancements catalyzed by the internet revolution to bolster productivity, sustainability, and resource allocation. Notably, the utilization of remote embedded systems with network connectivity has emerged as a dependable instrument, catalyzing the transformation of agricultural field monitoring. In line with the global population projections of 2019, it is anticipated that the world's population will increase by approximately 26% by 2050. With this impending demographic milestone, it is projected that farmers will face the unprecedented challenge of meeting the world's growing food demand, surpassing all the historical production records [1]. This surge in demand necessitates the strategic enhancement of resource allocation. The incorporation of monitoring tools stands as a promising avenue to identify the precise locations, methods, and extent of resource deployment.

Arable land, being a fragile and limited resource, requires careful management to maximize its potential for increased production without the need for expansion. This objective can be accomplished through the judicious management of various components within the agricultural environment, including fertilisers, water, and pesticides. Despite notable technological advancements, the average water usage in fields has shown minimal changes over the last five decades, resulting in water surplus in specific areas of plantation, while causing shortages in the others [2].

This study seeks to design a computerized precision agriculture system employing a microcontroller system with network connectivity. The system will measure the key variables, including air temperature, air humidity, and soil moisture, at multiple locations within an agricultural field and transmit these data to a central server. The data will be interpolated to create maps depicting the agricultural environment's status in relation to the variables under analysis. This system aims to provide an automated irrigation tool.

Hence, the primary objective of this study is to develop a computerized precision agriculture system capable of employing a network-accessible micro-controlled system to collect the measurements of vital variables, such as air temperature, air humidity, and soil moisture, at various locations within an agricultural field These measurements are then sent to a server, and these data are subjected to interpolation te1chniques to create maps of the agricultural environment, illustrating its condition concerning the analyzed variables, both current and historical.

This research has been divided into the following structure: (i) the Materials and Methods, which show the theoretical framework and its selected programming languages for the project, the microcontrollers and sensors used for measurements, and it also introduces the core concepts of network architecture chosen for the project; (ii) the Results and Discussion present the prototype developed in the research; (iii) the Conclusions addresses the final reflections, challenges faced, and suggestions for improvements.

2. Materials and Methods

Structured Query Language (SQL) was chosen for the creation of the database due to its versatility, user-friendliness, and ability to handle extensive sets of structured data. MySQL was employed as the Database Management System (DBMS). A DBMS is a suite of software tools that enables a database to be queried and modified through the use of its proprietary communication sockets [3].

In creating the visual part of the website that delivers information to the user, HyperText Markup Language (HTML) was employed to define the elements, and Cascading Style Sheets (CSS) were used for styling and positioning these elements on the page. Additionally, the Bootstrap framework was utilized for the interface design. Bootstrap is a robust and extensible frontend toolkit, constructed and customized with Sass, allowing for the use of a pre-programmed grid system [4].

The web service responsible for receiving data and executing operations on the database utilized Hypertext Preprocessor (PHP). PHP, a high-level language, is particularly well suited for the creation of dynamic web pages. It offers an array of tools for facilitating user interaction through forms, URL parameters, and links. What sets PHP apart from languages like JavaScript and CSS is its server-side execution, delivering preprocessed, pure HTML. This architectural choice enables interactions with databases and server applications without exposing the primary source code of the system, thereby ensuring data security [5]. In the generation of interpolated temperature and humidity maps, the Leaflet library was employed. Leaflet is an open-source tool for rendering interactive maps [6].

The selected microcontroller was the ESP32, which is manufactured by the company Espressif. This embedded board comes with built-in Wi-Fi and Bluetooth connectivity, a low-power processor, and provides a robust and highly integrable platform capable of meeting the ongoing demand for energy-efficient consumption. It offers essential tools for ensuring data security due to its ability to operate with encrypted network protocols [7].

To conduct temperature and air humidity measurements using the data collection microcontrollers, the DHT11 sensor was employed. This sensor is capable of measuring both parameters and generating them as a digital signal. It can measure temperatures in the range of 0 °C to 50 °C and humidity in the range of 20% to 80%, which typically cover the measurements within the arable fields of northern Paraná, with only occasional exceptions. For soil humidity measurements, a corrosion-resistant resistive hygrometer was used.

The microcontroller network features autonomous power supply through photovoltaic cells. To maintain compactness, prismatic lithium polymer batteries were employed, along with a protection circuit for charging and discharging. Typically, these batteries provide the highest energy density available on the market.

Methods

The development methodology employed for this project was the Waterfall model, which is one of the most traditional approaches to software management. It follows a linear sequence of stages, where each phase of the project must be completed before the next one can begin. In the initial stage, system requirements are defined. In the second stage, the project is developed, with the aid of UML diagrams. Upon completion of this phase, implementation begins, involving the writing of program code and the development of functionalities. Subsequently, the unit testing phase is undertaken to ensure that each component of the system functions correctly.

For system requirements definition, a literature review was conducted within the scientific literature on the use of resources in agricultural fields in the Paraná state's Northern Pioneer region. Subsequently, the Entity-Relationship Diagram (ERD), as depicted in Figure 1, was created to understand how the entities in the system's database should relate to one another. The ERD is one of the key diagrams in UML. MySQL Workbench, a tool for database creation and management, was employed to construct the research diagram.

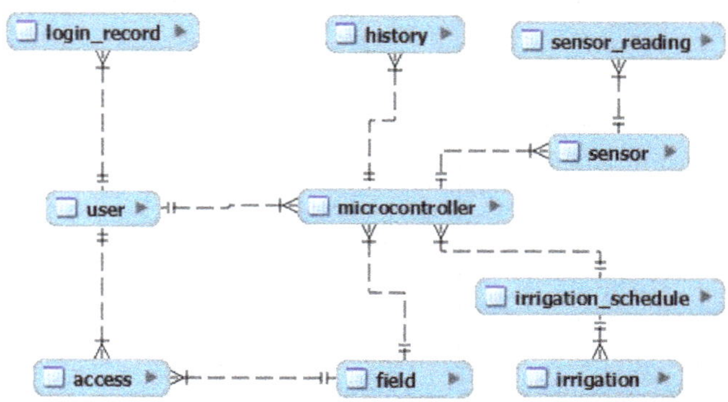

Figure 1. Entity-Relationship Diagram (ERD) of the system.

In the diagram depicted in Figure 1, the "user" table holds information about the system maintainers. This table exhibits a one-to-many (1:N) relationship with the "microcontroller" table, signifying that a single user can register multiple microcontrollers, while each microcontroller can be registered by only one user. Additionally, the "user" table maintains one-to-many (1:N) relationships with two other tables, the "login_record," which stores user log entries, and "access," which records the agricultural fields that a user is authorized to manage.

The "microcontroller" table is related to four other tables in the following manner: It exhibits a one-to-many (1:N) relationship with the "field" table, representing arable fields. This signifies that one microcontroller can belong to only one arable field, while a single arable field can be associated with multiple microcontrollers; it maintains a one-to-many relationship with the "history" table, which is used to store data every time the irrigation actuators are activated. There is a connection with the "sensor" table, facilitating the link between each microcontroller and its respective sensors; and then it is associated with the "irrigation_schedule" table, where the irrigation routines to be followed by that microcontroller are stored. Additionally, each microcontroller is uniquely identified by a Media Access Control (MAC) physical address, as denoted by the unique constraint.

The "irrigation" table specifies the soil moisture threshold at which irrigation should stay active and the time of its activation. This is connected to the "irrigation_schedule" table in a many-to-one (N:1) relationship. This means that one irrigation can be associated with only one irrigation routine, and an irrigation routine generally encompasses multiple irrigation actions.

The "sensor" table contains data regarding the latitude and type of sensor. It is related to the "microcontrolador" table in a one-to-many (1:N) relationship. This means one sensor can be linked to only one microcontroller, but a single microcontroller can be associated with numerous sensors.

The "sensor_reading" table stores the values recorded by the microcontrollers, and it is related in a one-to-many (1:N) manner with the "sensor" table. These data from the "sensor_reading" table are used to generate maps.

3. Results and Discussion

The project in question comprises a structure divided into three parts, each playing a fundamental role in the system's operation. The first part consists of the data processing and visualization server, which serves as a central hub for interactions with the users and microcontrollers. This server hosts user interface screens, providing a platform for users' interactions. Additionally, it is equipped with an Application Programming Interface (API) that acts as a receiver for the data collected by the access clients. This API is crucial for ensuring seamless communication among the various system components.

The second phase of the project entails the deployment of the microcontrollers. These components are tasked with sending HTTP requests to the primary server. Upon receiving these requests, the microcontrollers orchestrate the execution of the designated assignments, such as activating the irrigation valve when the soil moisture level falls below the threshold specified in the "irrigation" table.

In conclusion, the third part of the project focuses on creating clients that facilitate interconnection between the database server and the microcontrollers. By bridging centralized data storage and distributed processing devices, the clients play a crucial role in the integrity and smooth operation of the project as a whole. The diagram shown in Figure 2 illustrates communication between the entities within the system. The only part with direct access to the Database Management System (DBMS) is the PHP Web Server, which serves as the core of the system. This is important to ensure the integrity of the information stored in the database and prevent SQL injection vulnerabilities.

To access the information, the user first logs in on the system's homepage, and is then redirected to the interactive maps page. This page features four selection fields, allowing the user to choose the specific measurement period they wish to view. The PHP system conducts a database search for the records collected by the microcontrollers during that period, calculates the averages, and generates an interactive map using JavaScript, displaying the calculated measurement averages. Subsequently, a second layer is added to the map to showcase data interpolation, improving the user's ability to visualize the field's status.

Figure 3 illustrates a temperature map generated using the data collected by three microcontrollers on 29 August 2023. The microcontroller positioned outside the "IFPR Bloco Administrativo" recorded an average temperature of 28.4 °C during the one-hour interval when the map was generated. The one located underneath the "IFPR Bloco dos Estudantes" registered 28.8 °C, while the microcontroller inside the "IFPR Bloco de Alimentos" measured 21.8 °C. The system calculates an average of the values recorded within a specific time frame and performs data interpolation. The users can explore different positions by interacting with the map by clicking and dragging on the screen. To obtain temperature readings, they simply click on the microcontroller icons. The initial position has been set to the location of IFPR—Campus Jacarezinho.

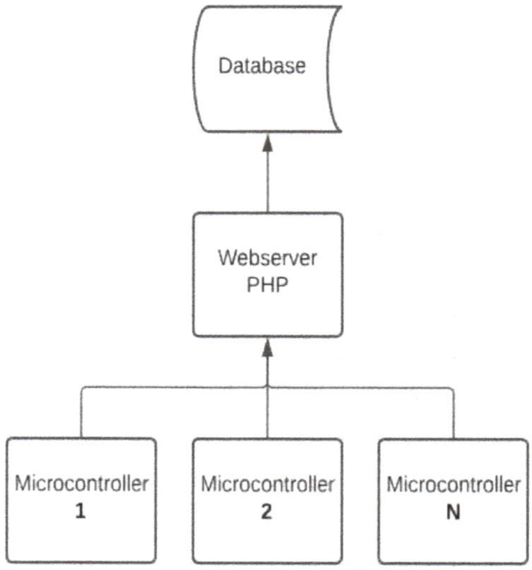

Figure 2. Diagram illustrating the network architecture of the developed system.

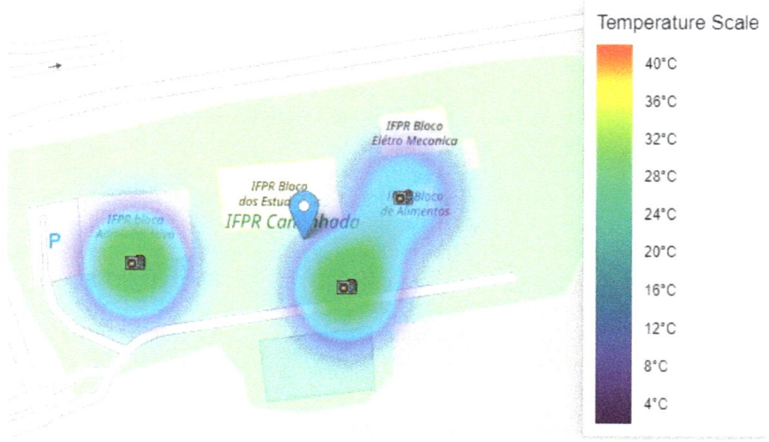

Figure 3. Map generated from the research.

4. Conclusions

Throughout this paper, strategic choices were made that molded the infrastructure and operation of this project. The careful selection of programming languages, frameworks, and devices, combined with the Waterfall development methodology, led to a prototype that encompasses data collection, processing, and visual representation. The proposed network architecture and the interaction among system entities have resulted in an agricultural monitoring system capable of performing fundamental high-level operations, such as data storage and retrieval.

The objectives have been partially achieved. The developed system is currently capable of performing tasks such as:

1. Registering, deleting, and modifying microcontrollers;
2. Listing registered arable fields and microcontrollers;

3. Sending and receiving messages from embedded systems that conduct measurements;
4. Generating interactive maps with average values within a user-defined time interval;
5. Conducting measurements of the air temperature, air humidity, and soil humidity;
6. Executing data interpolation for the information presented on the maps;
7. Performing irrigation through irrigation routines or manually.

To assess whether the users' needs have been adequately addressed, testing in agricultural environments will be essential. For future implementations, there is a plan to replace the current method of communication between the embedded systems and the PHP server. Instead of using a direct HTTP connection, each ESP module will be equipped with a radio communication module. These devices will send their requests to a central microcontroller Gateway, which will then forward them to the server. This approach will enhance the system's robustness, allowing the sensor-equipped modules to operate at greater distances from points with internet connectivity.

Author Contributions: All authors contributed to the study conception and design. Material preparation, data collection, and analysis were performed by R.J.C., L.C.d.C., R.O.A.T., A.L.S.M., L.H.R.L. and W.N.L. All authors commented on previous versions of the manuscript. All authors have read and agreed to the published version of the manuscript.

Funding: The authors would like to thank to the Pro-Rectory of Extension, Research, Post-Graduation, and Innovation of the Federal Institute of Education, Science, and Technology of Paraná–Proeppi/IFPR. Through the Directorate of Extension, Art, and Culture–Diext, as well as the Araucária Foundation Paraná, for granting financial aid scholarships to undergraduate students to promote extension activities that resulted in the development of this research.

Institutional Review Board Statement: Not applicable.

Informed Consent Statement: Not applicable.

Data Availability Statement: Data are contained within the article.

Conflicts of Interest: The authors declare no conflict of interest.

References

1. Wilmoth, J. United Nations Global Population Projections: A Critical Analysis of Key Methods and Assumptions. In Proceedings of the African Population Conference, Entebbe, Uganda, 18 November 2019.
2. Spier, H.; Rabbinge, R. Agriculture and Resource Management: Future Needs in Europe and Developing Countries. In Proceedings of the IFA Agricultural Conference on Managing Plant Nutrition, Barcelona, Spain, 29 June–2 July 1999; pp. 1–5.
3. Surian, J.; Nicochelli, L. Apostila de Banco de Dados e SQL. 2010. Available online: http://www.josevalter.com.br/download/sql/ApostiladeBancodeDadoseSQL.pdf (accessed on 29 July 2023).
4. Bootstrap. 2023. Available online: https://getbootstrap.com/ (accessed on 29 July 2023).
5. de Souza Barreto, M.V. Curso de Linguagem de PHP. Aracaju, 2000. Available online: http://www.etelg.com.br/paginaete/downloads/informatica/php.pdf (accessed on 29 July 2023).
6. LeafletJS. Leaflet—A JavaScript Library for Interactive Maps. Available online: https://leafletjs.com/ (accessed on 29 July 2023).
7. Espressif. Get Started—ESP32—ESP-IDF Programming Guide Latest Documentation. 2023. Available online: https://docs.espressif.com/projects/esp-idf/en/latest/esp32/get-started/ (accessed on 29 July 2023).

engineering proceedings

MDPI

Proceeding Paper

Gait Segmentation and Grouping in Daily Data Collected from Wearable IMU Sensors †

Zhuoli Wang [1], Chengshuo Xia [2] and Yuta Sugiura [3,*]

1 Graduate School of Science and Technology, Keio University, Yokohama 223-8522, Japan; zhuoli.wang@keio.jp
2 Advanced Manufacturing Technology Innovation Center, Guangzhou Institute of Technology, Xidian University, Guangzhou 510555, China; xiachengshuo@xidian.edu.cn
3 Department of Information and Computer Science, Faculty of Science and Technology, Keio University, Yokohama 223-8522, Japan
* Correspondence: sugiura@keio.jp

† Presented at the 10th International Electronic Conference on Sensors and Applications (ECSA-10), 15–30 November 2023; Available online: https://ecsa-10.sciforum.net/.

Abstract: Gait analysis plays a vital role in medicine as it can help diagnose illnesses, monitor recovery, and measure physical performance. Related work in gait analysis has primarily utilized laboratory data due to the inherently low noise and ease of preprocessing. Daily data, gathered through wearable sensors, can also significantly impact medical care. Nonetheless, working with such data poses numerous challenges. This paper proposes an algorithm to solve the problems associated with gait segmentation of daily data obtained by inertial measurement units (IMUs) in wearable devices. The proposed algorithm can handle time-series data collected by wearable IMU sensors, including noise and different gaits. The proposed algorithm within this paper can identify the start and end points of each gait segment within the time series, and the same type of gait will be grouped together.

Keywords: gait analysis; step segmentation; wearable sensors; inertial measurement unit (IMU)

Citation: Wang, Z.; Xia, C.; Sugiura, Y. Gait Segmentation and Grouping in Daily Data Collected from Wearable IMU Sensors. *Eng. Proc.* **2023**, *58*, 46. https://doi.org/10.3390/ecsa-10-16192

Academic Editor: Francisco Falcone

Published: 15 November 2023

1. Introduction

Gait detection technology plays a vital role in the field of medical health [1]. Gait serves as a valuable indicator of an individual's physical well-being, encompassing substantial information regarding their overall health [2]. Monitoring people's gait in daily life facilitates the timely identification of alterations in their underlying health status. The collection of human gait data can be achieved through the utilization of inertial measurement unit (IMU) sensors. With the increasing prevalence of smart wearable devices, IMU sensors are progressively gaining traction within everyday life.

Gait segmentation using daily gait data collected by IMUs poses several difficulties and challenges. Firstly, the daily gait data often contain motion information that is not solely related to gait. Secondly, individuals employ various walking patterns throughout daily life. Additionally, each person exhibits unique movement characteristics. The complexity and variability of human gait across individuals and activities further increase the difficulty of accurately segmenting the gait cycle.

An IMU is commonly employed for gait data collection. The segmentation of gait data is conducted by considering specific characteristics and regular patterns observed in human gait data obtained by the IMU [3–5]. Several studies have implemented a methodology for segmenting continuous gait data from a dataset according to the features of a set of stride patterns [6]. This technological approach has been utilized in evaluating individuals with Parkinson's disease, enabling them to conduct tests in a home environment without the need for direct medical supervision [7].

The current existing solutions for gait segmentation mostly focus on processing data that only include single gait patterns, mainly handling walking gait data. Additionally, these methods rely on data preprocessing techniques. Most of these approaches are based on laboratory data rather than real-world daily life data.

The processing of daily gait data presents some challenges, because the data will contain multiple gait types and noise. In order to overcome these challenges, a method for processing individuals' daily gait data is proposed in this paper. This proposed method effectively addresses the issue of noise within the data and achieves segmentation for different gait patterns, all without the need for pre-set template gaits or intensive model training. The output is a set of segmented gaits, categorized according to their respective gait types. It is worth noting that this proposed method finds behavioral patterns that appear frequently throughout the time series. In this study, a method is proposed for processing data that contain multiple gait types simultaneously, including walking, running and going up and down stairs. The results are presented for separate processing of data containing each of the four gaits mentioned above. Moreover, in this study, the influence of varying sensor numbers and sensor placements on the segmentation outcome is analyzed.

2. Methods

The Figure 1 below illustrates the main processing steps of the proposed method. The first step is data collection and use of the PCA method [8] to reduce the dimensions. Then, the sliding window algorithm is used to find the optimal sample gait window, which can be used to match the same type of gait segmentation from the dataset.

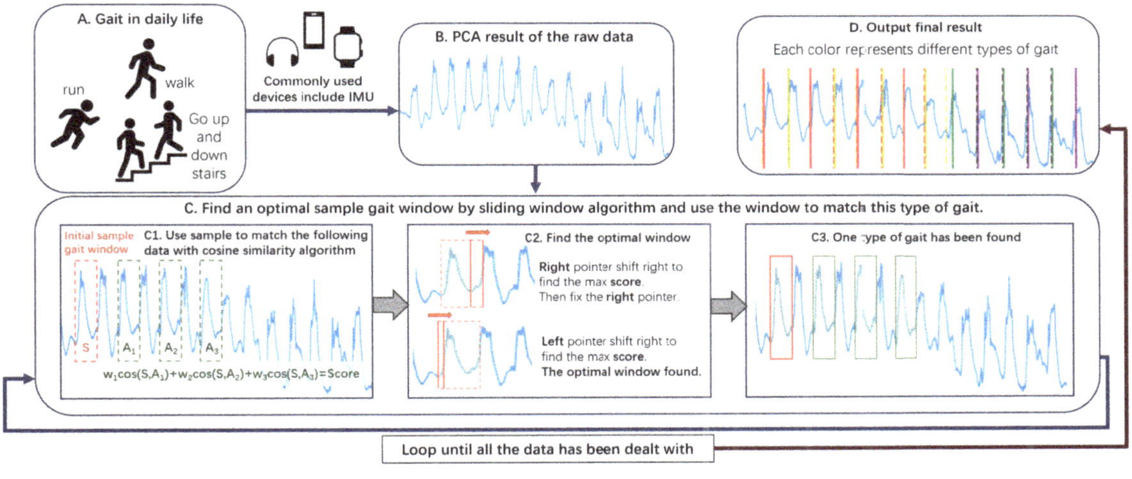

Figure 1. In this experiment, the data include walking, running, and stair climbing. Please note that the graph displayed represents a small portion of the experimental data, and thus, the plotted line does not include all types of gaits mentioned above.

2.1. Data Collection

Four young people (two males and two females; average age of 25) were recruited to collect data. All participants signed the agreement and complied with the ethical review.

Participants in the study were fitted with IMUs (MTw; Xsens Technologies Inc., Shanghai, China), aiming to gather the necessary acceleration and angular velocity data. A total of seventeen sensors were placed in specific places on the person's body, which can be seen in Figure 2a.

Participants were instructed to engage in walking, running, and going up and down stairs, each at a self-determined comfortable pace. These four distinct gaits were selected to encompass a wide range of typical human locomotion patterns in daily life. Each data point collected during each time frame included measurements for velocity, acceleration, angular velocity, and angular acceleration at a frequency of 60 Hz. Each data point encompasses measurements for each IMU in the x, y, and z directions of global coordinates. Specifically, the x-axis indicates the positive direction towards the local magnetic (north), the y-axis aligns with the right-handed coordinates (west), and the z-axis is positive when pointing upwards [9] .

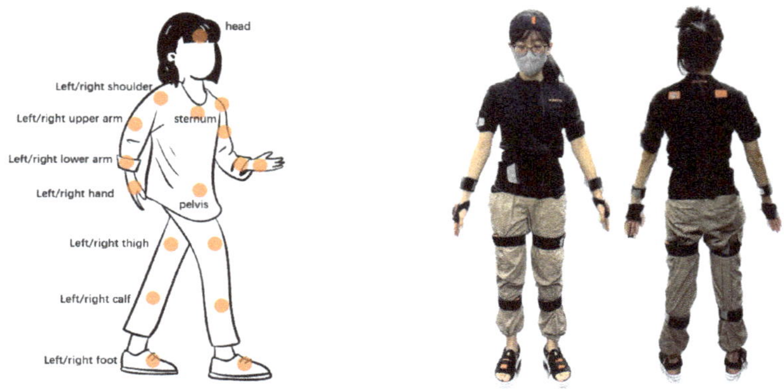

(**a**) (**b**)

Figure 2. (**a**) displays the position of the 17 IMUs [10]. The two photographs in (**b**) illustrate the participants when wearing the devices.

2.2. Data Processing

First, the moduli of velocity, acceleration, angular velocity, and angular acceleration in the horizontal plane of the obtained data are determined. This can be accomplished by employing the following formula: $|V| = \sqrt{V_x^2 + V_y^2}$. The input data for the four moduli, along with their respective vertical dimensions, will be considered in this study.

This is due to the fact that for human gait, vertical speed, acceleration, and angular velocity are important features [11]. In the field of gait analysis, the specific direction of human locomotion holds no significance. When individuals move with the same gait in different directions, these movements should not be considered as different gaits. Subsequently, the dataset was standardized to ensure that the features have comparable scales [3]. This was performed to ensure that each feature carries equal importance during the subsequent step of principal component analysis (PCA). Then, the PCA method [8] was to reduce the dataset containing all IMU data to one dimension.

The program will remove the data in a static state based on the given value. If a continuous sequence of data points in a static state reaches a certain threshold (in this experiment, it is set as 15), this segment of data will be excluded.

2.3. Initialize Sample Gait Window and Use It to Match the Following Data

The initial sample gait window is selected from the beginning of the current valid data based on a given parameter (in this experiment, it is set as 18 time frames). The sample gait window gradually slides backwards and is compared with the valid data using the cosine similarity algorithm. When the calculated similarity exceeds the given threshold (in this experiment, it is set as 0.8), the corresponding data will be marked as matching and the corresponding similarity value will be recorded. This will be used to score the current sample gait window. Previously compared data are not re-evaluated to avoid duplication.

2.4. Find the Optimal Sample Gait Window

The proposed method employs the sliding window algorithm to identify the optimal sample gait window. The continuous occurrence of matched segmented data with high similarity values indicates that the sample gait window demonstrates a higher degree of accuracy or quality, resulting in a higher score. The right pointer traverses a given range, and a score is computed in accordance with the aforementioned method. Finally, the position of the right pointer is located at the point that receives the highest score. The same process is applied to the left pointer. By determining the positions of the left and right pointers, the current optimal sample gait window can be determined. The threshold of the range is set to prevent windows with excessively large sizes. The threshold is determined as twice the average step length base on experiments. Furthermore, if the initial window contains content that is not related to any gait, such as noise, it will be unable to match similar data, resulting in a significantly low score. The program will filter out the sample windows with low scores.

2.5. One Group of Gait Was Found

The data that match with the optimal sample gait window will be considered as a group of gaits. They will be labeled as invalid during subsequent iterations in order to prevent redundant marking.

The previous two steps will be iterated to search for additional gaits. The iteration will end when all data have been traversed. Then, an output is obtained in the form of gait groups.

3. Results and Discussion

The ground truth is annotated by the 3D model movements generated by analyzing the data of human movements through the supporting software of the IMU used in this study. The visualization results of ground truth are shown in Figure 3.

A visualization of the results is shown in Figure 4. It illustrates the results obtained by the proposed method. The gait segmentation results are represented in the figure. It can be observed that each detected gait segmentation is marked with starting and ending points. The corresponding gait type is also labeled in Figure 4 according to the ground truth.

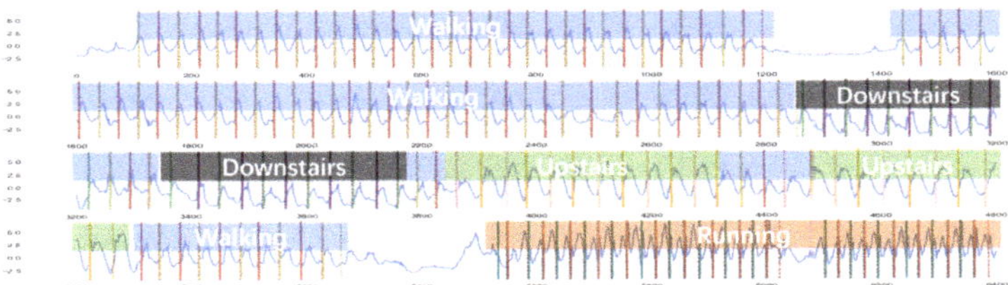

Figure 3. Ground truth. The time series gait data post-PCA are denoted by the blue line on the graph. The start of a gait is indicated by a solid line, while the termination of a gait is indicated by a dashed line. Different colored lines represent different groups of gaits.

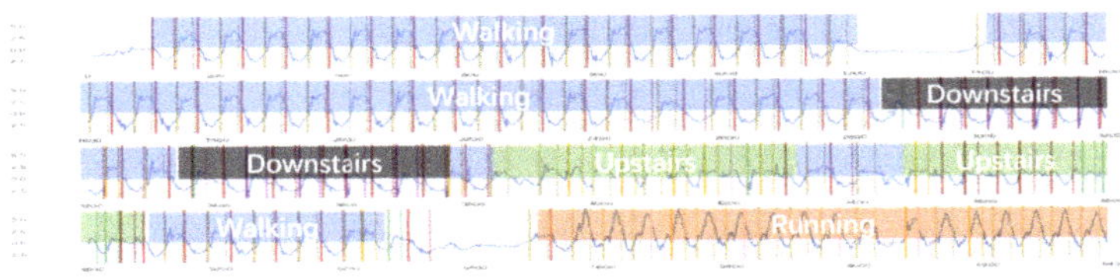

Figure 4. The results. The legend is the same as the figure above.

The subsequent figures show the validation results obtained from different combinations of sensors across various types of datasets and includes multiple distinct gaits or singular gaits exclusively. The validation results include accuracy and the Rand index. The accuracy of gait segmentation was computed using the following equation:

$$Accuracy = \frac{Accurate \quad segmentation \quad number}{Ground \quad truth \quad segmentation \quad number} \tag{1}$$

The resulting segmentations were compared with the ground truth gait segmentation and were marked as accurate. The Rand index is a method used to compare the similarity of two clusters [8]. A value close to 1 denotes a higher level of similarity between the compared clusters. In this study, the Rand index was employed to assess the resemblance between the grouped outcomes attained by the algorithm and the ground truth.

The outcomes obtained from the dataset containing four distinct gaits are presented as follows. Across various sensor combinations, the accuracy values range from 0.82 to 0.90, while the Rand index is between 0.76 and 0.83. It is notable that that among all sensor combinations, the combination of three IMUs on the feet and head leads to the best results, with an accuracy and Rand index of 0.9 and 0.83, respectively.

The "corresponding equipment" in the Figure 5 represents the IMU placed at a specific position during the experiment, which may correspond to devices used in daily life.

Body Part	Corresponding Equipment	Number of sensors	Accuracy	Rand index
Double feet	IMUs on feet	2	0.89	0.78
Double feet + head	IMUs on feet and ear phones	3	0.9	0.83
Double feet + Right hand	IMUs on feet and smart watch	3	0.83	0.76
Double feet + Left hand	IMUs on feet and smart watch	3	0.82	0.76
Double feet + Pelvis	IMUs on feet and IMUs in the phone put in the user's pocket	3	0.88	0.77

Figure 5. Testing results for all types of gait used along with the accuracy and Rand index.

The validation results for data segmentation, focusing on the dataset containing a single gait, are illustrated in the following Figure 6. Notably, the utilization of different sensor combinations leads to improved final outcomes.

Body Part	Corresponding Equipment	Number of sensors	Accuracy	Rand index
Double feet	IMUs on feet	2	0.97	1
Double feet + head	IMUs on feet and ear phones	3	0.97	1
Double feet + Right hand	IMUs on feet and smart watch	3	0.97	1
Double feet + Left hand	IMUs on feet and smart watch	3	0.97	1
Double feet + Pelvis	IMUs on feet and IMUs in the phone put in the user's pocket	3	0.97	1

Figure 6. Testing result for only walking gait used along with the accuracy and Rand index.

Table 1 below compares the proposed method in this paper with existing alternatives. It can be observed that the proposed method performs comparably well with the existing methods in terms of accuracy when using the dataset containing only walking gaits.

Table 1. Comparative analysis of our proposed methodology with existing approaches.

Reference	Methodology	Result
Jens B. et al. [12]	Subsequence Dynamic Time Warping	97.7% accuracy in walking gait segmentation
A. R. Anwary et al. [13]	Peak detection	95.47% accuracy in walking gait segmentation
Jagos H. et al. [14]	Autocorrelation	96% accuracy in walking gait segmentation
Proposed method	Matching gait by cosine similarity in a sliding window	97% accuracy in walking gait segmentation. 90% accuracy in mixed gait segmentation.

4. Conclusions

This study introduces a practical methodology capable of segmenting and grouping daily gait data obtained by IMUs. Data containing the acceleration, angular velocity, etc., of various parts of the human body were collected. Then, the method proposed in this paper was validated. The results indicated that the method performed according to expectations and demonstrated the ability to perform both gait segmentation and grouping. The findings showed that a better performance was achieved when the sensors were placed on both feet and the head. It was observed that separate gaits tended to offer better performance compared to mixed gaits in the same dataset.

In conclusion, this study proposes a method that establishes a foundation for analyzing daily gait patterns. It has potential applications in the detection of anomalies, monitoring of sports-related activities, and other areas. This includes the development of a more accurate pedometer for different gaits, saving time on manual data annotation for large datasets and aiding in the rehabilitation of Parkinson's patients through gait segmentation and recording.

Author Contributions: Conceptualization, Z.W.; formal analysis, Z.W.; methodology, Z.W.; software, Z.W.; writing—original draft preparation, Z.W.; writing—review and editing, C.X.; supervision, Y.S. All authors have read and agreed to the published version of the manuscript.

Funding: This work was funded in part by JSPS KAKENHI (grant number JP21H03485) and JST SPRING (grant number JPMJSP2123).

Institutional Review Board Statement: The study was conducted in accordance with the Declaration of Helsinki, and approved by the Institutional Review Board of Faculty of Science and Technology, Keio University (protocol code: 2023-108, date of approval: 10 October 2023).

Informed Consent Statement: Informed consent was obtained from all subjects involved in the study.

Data Availability Statement: Data are contained within the article.

Conflicts of Interest: The authors declare no conflict of interest.

References

1. Anwary, A.R.; Yu, H.; Vassallo, M. Gait quantification and visualization for digital healthcare. *Health Policy Technol.* **2020**, *9*, 204–212. [CrossRef]
2. Maki, B.E. Gait Changes in Older Adults: Predictors of Falls or Indicators of Fear? *J. Am. Geriatr. Soc.* **1997**, *45*, 313–320. [CrossRef] [PubMed]
3. Jain, R.; Semwal, V.B.; Kaushik, P. Stride segmentation of inertial sensor data using statistical methods for different walking activities. *Robotica* **2022**, *40*, 2567–2580. [CrossRef]
4. Gujarathi, T.; Bhole, K. Gait analysis using imu sensor. In Proceedings of the 2019 10th International Conference on Computing, Communication and Networking Technologies (ICCCNT), IEEE, Kanpur, India, 6–8 July 2019; pp. 1–5.
5. Ullrich, M.; Küderle, A.; Hannink, J.; Del Din, S.; Gassner, H.; Marxreiter, F.; Klucken, J.; Eskofier, B.M.; Kluge, F. Detection of gait from continuous inertial sensor data using harmonic frequencies. *IEEE J. Biomed. Health Inf.* **2020**, *24*, 1869–1878. [CrossRef] [PubMed]
6. Han, J.; Jeon, H.S.; Jeon, B.S.; Park, K.S. Gait detection from three dimensional acceleration signals of ankles for the patients with Parkinson's disease. In Proceedings of the IEEE The International Special Topic Conference on Information Technology in Biomedicine, Ioannina, Greece, 26–28 October, 2006; Volume 2628.
7. Ullrich, M.; Mücke, A.; Küderle, A.; Roth, N.; Gladow, T.; Gaßner, H.; Marxreiter, F.; Klucken, J.; Eskofier, B.M.; Kluge, F. Detection of unsupervised standardized gait tests from real-world inertial sensor data in Parkinson's disease. *IEEE Trans. Neural Syst. Rehabil. Eng.* **2021**, *29*, 2103–2111. [CrossRef]
8. Pedregosa, F.; Varoquaux, G.; Gramfort, A.; Michel, V.; Thirion, B.; Grisel, O.; Blondel, M.; Prettenhofer, P.; Weiss, R.; Dubourg, V.; et al. Scikit-learn: Machine Learning in Python. *J. Mach. Learn. Res.* **2011**, *12*, 2825–2830.
9. Roetenberg, D.; Luinge, H.; Slycke, P. Xsens MVN: Full 6DOF human motion tracking using miniature inertial sensors. *Xsens Motion Technol. BV Tech. Rep.* **2009**, *1*, 1–7.
10. Xia, C.; Maruyama, T.; Toda, H.; Tada, M.; Fujita, K.; Sugiura, Y. Knee Osteoarthritis Classification System Examination on Wearable Daily-Use IMU Layout. In Proceedings of the 2022 ACM International Symposium on Wearable Computers, Cambridge, UK, 11–15 September 2022; pp. 74–78.
11. Van Nguyen, L.; La, H.M. Real-Time Human Foot Motion Localization Algorithm With Dynamic Speed. *IEEE Trans.-Hum.-Mach. Syst.* **2016**, *46*, 822–833. [CrossRef]

12. Barth, J.; Oberndorfer, C.; Kugler, P.; Schuldhaus, D.; Winkler, J.; Klucken, J.; Eskofier, B. Subsequence dynamic time warping as a method for robust step segmentation using gyroscope signals of daily life activities. In Proceedings of the 2013 35th Annual International Conference of the IEEE Engineering in Medicine and Biology Society (EMBC), IEEE, Osaka, Japan, 3–7 July 2013; pp. 6744–6747.
13. Anwary, A.R.; Yu, H.; Vassallo, M. Optimal foot location for placing wearable IMU sensors and automatic feature extraction for gait analysis. *IEEE Sens. J.* **2018**, *18*, 2555–2567. [CrossRef]
14. Jagos, H.; Reich, S.; Rattay, F.; Mehnen, L.; Pils, K.; Wassermann, C.; Chhatwal, C.; Reichel, M. Determination of gait parameters from the wearable motion analysis system eSHOE. *Biomed. Eng.* **2013**, *58*, 000010151520134241. [CrossRef]

engineering proceedings

MDPI

Proceeding Paper

Causality Inference for Mitigating Atmospheric Pollution in Green Ports: A Castellò Port Case Study †

Rosa Martínez *, Juan Carlos Sanz-González, Ivan Felis and Eduardo Madrid

Centro Tecnológico Naval y del Mar, 30320 Fuente Álamo, Murcia, Spain; jcarlossanz@ctnaval.com (J.C.S.-G.); ivanfelis@ctnaval.com (I.F.); eduardomadrid@ctnaval.com (E.M.)

* Correspondence: rosamartinez@ctnaval.com; Tel.: +34-968-19-75-21

† Presented at the 10th International Electronic Conference on Sensors and Applications (ECSA-10), 15–30 November 2023; Available online: https://ecsa-10.sciforum.net/.

Abstract: Green Ports have emerged due to the increase in air pollution from emissions generated by maritime traffic and the dispersion of particles, as well as water pollution from spills. The primary objective of this study is to anticipate episodes of atmospheric pollution related to cargo-handling activities and assess the quantitative causality between these variables. We employ a causality inference based on time series analysis to investigate the applicability and validity of these techniques in a real-world problem setting. Specifically, methods such as the Granger Test and PCMCI are evaluated and compared with these data. The results demonstrate that cargo handling at the port under study has some causal influence on the PM (particulate matter) measurements. Finally, the PCMCI method is proposed as the most robust among the algorithms considered in this study.

Keywords: Green Ports; sustainability; atmospheric pollution; causal inference; causality; air quality monitoring; cargo handling; Castellò Port

Citation: Martínez, R.; Sanz-González, J.C.; Felis, I.; Madrid, E. Causality Inference for Mitigating Atmospheric Pollution in Green Ports: A Castellò Port Case Study. *Eng. Proc.* **2023**, *58*, 47. https://doi.org/10.3390/ecsa-10-16159

Academic Editor: Francisco Falcone

Published: 15 November 2023

1. Introduction

The growth of commercial activities and the need for competitiveness in the global market are forcing ports around the world to evaluate all possibilities systematically and continuously for optimization and the reduction of related costs and externalities. Among the main adverse effects between port activity and the environment, air pollution (due to emissions from maritime traffic or the dispersion of particles) and water pollution (spills) stand out.

Atmospheric pollutants are emitted from ports through various sources and are both directly and indirectly associated with port activities. The coexistence of multiple transportation modes, including vessels, bulk cargo, handling equipment, and rail locomotives, collectively contributes to emissions of particulate matter (PM) and greenhouse gases linked to maritime operations [1].

These problems have led to the emergence of a new port paradigm: the Green Port, in which sustainability (in its three aspects: social, economic and environmental) is the central pillar. The Green Port concept introduces these three aspects in the development and operation of ports in order to find a balance between them, resulting in ports that are competitive and integrated with both the city that hosts them and the environment [2].

This text describes the implementation of a study aimed at predicting when an episode of airborne particle pollution may occur due to port activities related to the loading and unloading of bulk cargo. To achieve this, two different methodologies are employed: firstly, a simpler method like the Granger test, and secondly, a more sophisticated method like PCMCI. These studies are applied to real data obtained from bulk cargo activities in four specific areas of the Castellò Port along with data on pollutants collected from five different stations located within the port.

2. Materials and Methods

2.1. The Study Site, Castellò Port

The present study focuses on five air quality monitoring stations in the Castellò Port (Tramontana, Gregal, Levante, Poniente and Siroco) and the four docks in color (CS06 in red, CS26 in yellow, CS05 in green and CS09 in blue). These zones are reflected in the figure below (Figure 1).

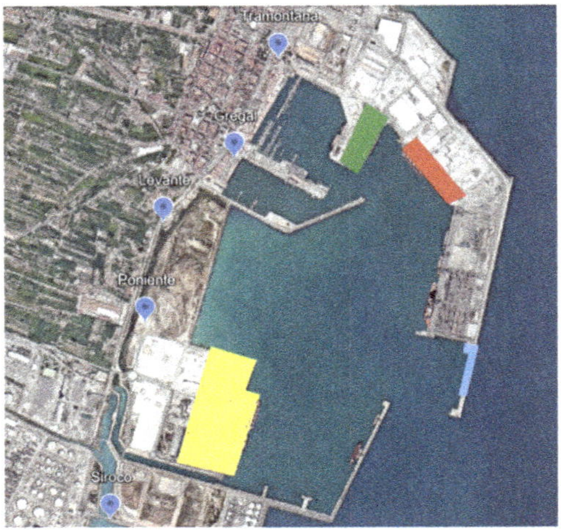

Figure 1. Study site located at Castellò Port (Castellón, Spain).

2.2. The Data Understanding

Data were acquired by Port Authority of Castellò (PAC). As the raw data are very heterogeneous, considerable work has been carried out to transform them into versions suitable for further processing. As the useful port operations data only covered the years 2020 and 2021, these two years were chosen for both datasets, comprising air quality parameters and port operations.

2.2.1. Air Quality

Since not all the stations have the same sensors and therefore do not measure the same variables, the parameters that are common to all have been selected. Regarding the pre-processing of the data, only one datum in the time series of the maximum wind speed for the year 2021 at the Poniente station was replaced by the average value of the immediate neighbors, as it was clearly an outlier. Therefore, we can say that these time series are clearly very good in terms of data quality.

Finally, the final air quality data consists of 5 multidimensional time series, each associated with a monitoring station and where the dimensions correspond to the following variables for 2020 and 2021: $PM_{2.5}$ [$\mu g/m^3$], PM_{10} [$\mu g/m^3$], wind direction [°], hourly mean wind speed [m/s] and maximum hourly wind speed [m/s].

Given the inherent cleanliness of these time series, it was not considered necessary to perform any pre-processing beyond reformatting the data or grouping them in a more convenient way for the study.

2.2.2. Port Operations

The port operations data required much more pre-processing than the air quality data. The data itself did not consist of a time series as such, but of a series of records of ship arrivals and departures at the docks, where the following parameters were monitored: tons unloaded, dock, type of goods, vessel, hands, date of arrival and departure, among others. The period covered is from 2019 to 2021, with data received with a time resolution of 6 h.

2.3. Causal Analysis Techniques

Inferring causality is a remarkably important problem because, unlike descriptive statistical analysis, it allows decisions to be made in a rational and justified way. However, there is no universal definition of causality, although several attempts have been made [3]. The first quantitative notion of causality came from N. Wiener in 1956, in the context of the study of temporal signals. Later, Clive W.J. Granger implemented and popularized this formulation (called Granger causality in his honor), which even won him the Nobel Prize in Economics in 2003 and it is the first family of methods to be studied in this paper (Section 2.2.1). However, it is a method that, as will be seen below, is limited to linear relationships between time series. It did not take long, therefore, for a large number of generalizations (and other approaches) of the same method to appear, taking into account non-linearity, both from the point of view of econometrics and from that of physics and non-linear dynamical systems [4]. On the other hand, concepts from information theory, first introduced by Schreiber in terms of transfer entropy [5], have been used to infer causality, notably in the work of Palus [6], such as conditional mutual information. In addition, other types of causality and inference methods have appeared; one can read about current causality techniques in [7,8]. The methods selected for investigation in this paper draw on several of these concepts. In particular, following an overview of different methods, this paper presents two methods that correspond to two different but related approaches, which are described in detail in the following sections.

2.3.1. Granger Causality

Granger's method assesses causality in that the addition of a variable as a component of the predictive model, the target variable, increases the predictive power of the model. Intuitively, this means that, in Granger's sense, a process X causes a process Y if predictions about future values of Y are more accurate when information from X's past is considered. More specifically, we want to test the null hypothesis of the process "X does not (Granger) cause Y". To this end, both VAR(T) (Vector Auto Regressive) model with and without the functional dependence of X are considered (1) and (2):

$$y_t = \alpha_1 + \sum_{k=1}^{T} \phi_k y_{t-k} + \sum_{k=1}^{T} \psi_k x_{t-k} + \epsilon_t \tag{1}$$

where α_1, ϕ_i and ψ_i are the coefficients of the model and ϵ is a component characterizing the signal noise, while T is the order of the model, i.e., the maximum time delay considered. In this model, the process X is assumed to influence Y if the coefficients ψ_k are not null.

$$y_t = \alpha_2 + \sum_{k=1}^{T} \beta_k y_{t-k} + u_t \tag{2}$$

where the coefficients of the model are the same as in the previous case.

Thus, the following (null) hypotheses are to be tested:

$$H_0 : \psi_1, \ldots, \psi_p = 0 \tag{3}$$

in comparison with the alternative scenarios:

$$H_1 : \psi_i \neq 0 \; i \in \{1, 2, \ldots, p\} \tag{4}$$

Then, after adjusting the coefficients (e.g., by least squares) of both models, a statistical test of significance, such as Fisher's F-test, is applied (which compares the variance of the residuals of the model including only Y with that of the model including both X and Y):

$$F = \frac{(\mathrm{RSS_r} - \mathrm{RSS_u})/q}{\mathrm{RSS_u}/(T - 3p - 1)} \tag{5}$$

where RSS_r and RSS_u are the sum of the squared residuals of the restricted and unrestricted model, respectively (Equations (1) and (2)), q is the number of null coefficients and p is the number of observations. Thus, depending on the value of F, the null hypothesis H_0 can be rejected (or not) and X can be considered to cause Y (or not) in the Granger sense.

Before applying the Granger test, it is necessary to check that the stationarity condition of the time series in question is satisfied, in order to construct time series models that assume this property. To do this, Augmented Dickey–Fuller (ADF) and Kwiatkowski–Phillips–Schmidt–Shin (KPSS) tests are evaluated.

2.3.2. PCMCI

This method of causal inference between time series, proposed by Runge in [8], is based on the concept of conditional independence to estimate the strength and directionality of causal relationships between highly interdependent multivariate time series. This comes from PC (Park and Clark) and MCI (Momentary Conditional Independence) methods. Let us consider a dynamic system $\boldsymbol{X}_t = \left(X_t^1, \ldots, X_t^N\right)$ (therefore, multivariable, where the index t indicates the time instant and the upper index N distinguishes the different variables that compose it) in which the following is true:

$$X_t^j = f_j\left(\mathcal{P}\left(X_t^j\right), \eta_t^j\right) \tag{6}$$

where f_j is a possible nonlinear functional dependence, η_t^j is a mutually independent dynamic, noise, and $\mathcal{P}\left(X_t^j\right) \subset \boldsymbol{X}_t^- = (\boldsymbol{X}_{t-1}, \boldsymbol{X}_{t-2}, \ldots)$ denotes the "causal parents" of the variable X_t^j in the entire history of the N variables; thus, a causal link $X_{t-\tau}^i \longrightarrow X_t^j$ exists if $X_{t-\tau}^i \in \mathcal{P}\left(X_t^j\right)$, where τ is a time delay. The PCMCI algorithm then attempts to find the causal parent of different time series with different time lags (τ). For this purpose, and as already indicated, this method presents two different stages [8]:

1. Identification of some relevant initial relatedness conditions $\hat{\mathcal{P}}\left(X_t^j\right)$ for all time series X_t^j by means of a PC algorithm (Markov discovery type). After this step, an approximation of the true parent distribution $\mathcal{P}$ is obtained, possibly including false positives.
2. Refinement of the identification of $\mathcal{P}$ (control of false positives) by means of a Momentary Conditional Independence MCI analysis.

3. Results

The results can be seen in Figures 2 and 3, where they have been classified according to the operating dock. Gregal, Poniente and Siroco stations were chosen among all the stations for comparisons between them, being the ones that are more relevant.

4. Conclusions

When investigating causality in relation to air quality in port areas, air quality can influence the concentration of particulate matter and thus the quality of the surrounding air. Two common approaches to address this relationship are Granger analysis and the use of the Multivariate Principal Component Method (PCMCI). Regarding the causal inference algorithms, a considerable variability of results is found among all the methods. According to the Granger method, wind has a causal strength on the PM variables (in general, stronger relationships seem to be observed for the same station where it is measured). Similarly, the bulk series of the CS06, CS05 and CS09 docks stand out as influencing the PM measurements, especially at the Siroco and Poniente stations.

According to the PCMCI algorithm, which is in principle more robust than the others, the bulk discharges of CS05 and CS09 stand out as the most important variables causing the PM dynamics at the Poniente and Gregal stations (for $PM_{2.5}$), respectively. However, these are not the closest stations to each of these terminals, which could be related to the prevailing winds.

On the other hand, data collection in ports could lead to issues related to Big Data in terms of volume data, variety data and velocity of data acquisition. The solutions proposed in the paper (Granger and PCMCI) can address Big Data issues to some extent, but their capabilities may vary, depending on the type of the analysis.

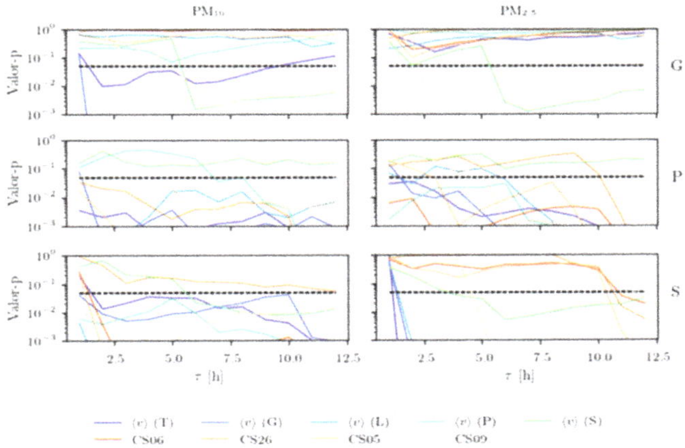

Figure 2. Granger causality between the variables of average speed and tons per hour discharged at the four docks (by color, indicated in the legend, and the variables of granulated material in suspension PM_{10} y $PM_{2.5}$).

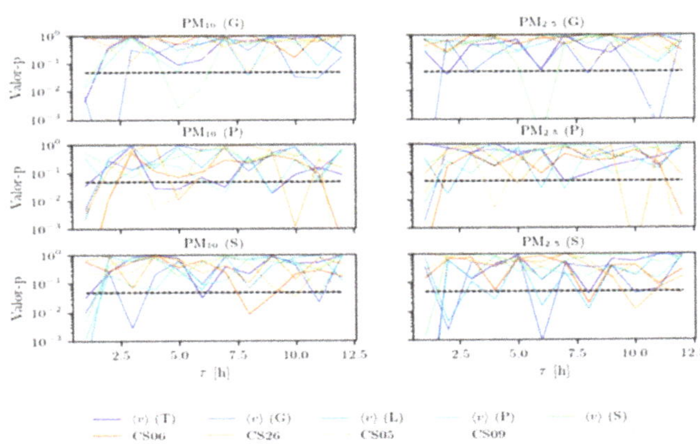

Figure 3. *p*-values according to the PCMCI algorithm for the different cause and effect variables. The statistical significance value of 0.05 is indicated by the dashed line.

4.1. Granger

In particular, the docks are identified by their initials (G: Gregal; P: Poniente; S: Siroco) in the rows, and the causal relationships from the variables indicated in the legend to the $PM_{2.5}$ and PM_{10} variables (in the columns) are shown. Recall that the Granger test indicates causality in cases where the *p*-value is less than 0.05.

For both docks, the *p*-value curves as a function of the delay time τ have a decreasing shape, with few instances of causality for delays of 1 h. The wind speed at Siroco stands out as the variable that, on average, has the least causal effect (in the Granger sense) on the others. Similarly, the variables representing tons discharged discharged per hour show mixed results, with slightly more causal relationships appearing towards the PM_{10} variable than towards the $PM_{2.5}$ variable.

In Figure 2, in short, there are many causal relationships that can be complex to identify among so many variables. In addition, the high number of causal detections is remarkable, even though there are some monitoring stations (Gregal) that show practically no causality for the tons of bulk discharged. A priori, the parameter that is most causally influenced by bulk discharge is PM_{10} (P), followed by PM_{10} (S).

4.2. PCMCI

Following the same methodology as in the previous cases, it was applied to all possible pairs of cause and target variables defined above, for a maximum time lag of 12 h. P-values below a statistical significance level of 0.05 are considered to indicate a causal relationship.

In conclusion, the choice between Granger and PCMCI for the causality analysis of air quality in ports such as Castellò Port depends on the complexity of the relationships to be investigated. PCMCI offers a significant advantage in allowing the detection of causal relationships beyond linear ones, which may lead to a complete understanding.

Author Contributions: Conceptualization and methodology, R.M., J.C.S.-G. and I.F.; data curation, E.M. and J.C.S.-G.; formal analysis, E.M., J.C.S.-G. and R.M.; validation, R.M. and I.F., writing—original draft preparation, J.C.S.-G. and R.M.; writing—review and editing I.F. All authors have read and agreed to the published version of the manuscript.

Funding: This research was funded by the *Instituto de Fomento de la Región de Murcia* (INFO) under the Program of grants aimed at Technological Centers of the Region of Murcia for the realization of non-economic R&D activities. Modality 1: Independent R&D Projects, with File No.: 2021.08.CT01.000044.

Institutional Review Board Statement: Not applicable.

Informed Consent Statement: Not applicable.

Data Availability Statement: The data presented in this study are not publicly available due to privacy reasons.

Acknowledgments: We thank Castellò Port for providing us with the necessary information and relevant data to carry out this study, especially to Inés Lopez, Maria José Rubio and Bernat Ibañez.

Conflicts of Interest: The authors declare no conflicts of interest.

References

1. Chang, C.C.; Wang, C.M. Evaluating the effects of green port policy: Case study of Kaohsiung harbor in Taiwan. *Transp. Res. Part D Transp. Environ.* **2012**, *17*, 185–189. [CrossRef]
2. Davarzani, H.; Fahimnia, B.; Bell, M.; Sarkis, J. Greening ports and maritime logistics: A review. *Transp. Res. Part D Transpor Environ.* **2016**, *48*, 473–487. [CrossRef]
3. Hlaváčková-Schindler, K.; Paluš, M.; Vejmelka, M.; Bhattacharya, J. Causality detection based on information-theoretic approaches in time series analysis. *Phys. Rep.* **2007**, *441*, 1–46. [CrossRef]
4. Hiemstra, C.; Jones, J.D. Testing for Linear and Nonlinear Granger Causality in the Stock Price-Volume Relation. *J. Financ.* **1994**, *49*, 1639–1664.
5. Schreiber, T. Measuring Information Transfer. *Phys. Rev. Lett.* **2000**, *85*, 461. [CrossRef] [PubMed]
6. Palus, M.; Hoyer, D. Detecting nonlinearity and phase synchronization with surrogate data. *IEEE Eng. Med. Biol. Mag.* **1998**, *17*, 40–45. [CrossRef] [PubMed]
7. Eichler, M. Causal inference with multiple time series: Principles and problems. *Philos. Trans. R. Soc. A Math. Phys. Eng. Sci.* **2013**, *371*, 20110613. [CrossRef] [PubMed]
8. Runge, J.; Bathiany, S.; Bollt, E.; Camps-Valls, G.; Coumou, D.; Deyle, E.; Glymour, C.; Kretschmer, M.; Mahecha, M.D.; Muñoz-Marí, J.; et al. Inferring causation from time series in Earth system sciences. *Jakob Zscheischler* **2019**, *10*, 19. [CrossRef] [PubMed]

engineering proceedings

MDPI

Proceeding Paper

Development of an Android-Based, Voice-Controlled Autonomous Robotic Vehicle †

Abubakar Umar [1,*], Mohammed Abdulkadir Giwa [2], Abduljalal Yusha'u Kassim [2], Muhammad Usman Ilyasu [3], Ibrahim Abdulwahab [4], Ezekiel Ehime Agbon [2] and Matthew T. Ogedengbe [5]

1 Computer Engineering Department, Faculty of Engineering, Ahmadu Bello University, Zaria 810107, Nigeria
2 Electronics and Telecommunications Engineering Department, Faculty of Engineering, Ahmadu Bello University, Zaria 810107, Nigeria; engineermohammedagiwa@mail.com (M.A.G.); aykassim2019@gmail.com (A.Y.K.); eagbonehime1@gmail.com (E.E.A.)
3 Department of Electrical and Electronics Engineering, School of Engineering and Engineering Technology, Federal Polytechnic, Bali 672102, Nigeria; jamilmammailmanjen@yahoo.com
4 Electrical Engineering Department, Faculty of Engineering, Ahmadu Bello University, Zaria 810107, Nigeria; iabdulwahabb@gmail.com
5 Department of Computer Science, Faculty of Science, Joseph Sarwuan Tarka University, Makurdi 970101, Nigeria; matt.oged@uam.edu.ng
* Correspondence: abubakaru061010@gmail.com
† Presented at the 10th International Electronic Conference on Sensors and Applications (ECSA-10), 15–30 November 2023; Available online: https://ecsa-10.sciforum.net/.

Abstract: This research presents the development of an android-based, voice-controlled autonomous robotic vehicle. This article was developed in a way that the robotic vehicle was controlled using voice commands. An android application combined with an android microcontroller was used to achieve this task. The connection between the android app and the autonomous vehicle was facilitated using Bluetooth technology. The vehicle was controlled with either the aid of the buttons on the app, or by spoken commands from the user. The movement of the vehicle was achieved by using four DC motors connected with the microcontroller at the receiver side. The commands from the app were converted into digital signals using the Bluetooth RF transmitter within a specific range (of around 100 m) of the autonomous vehicle. At the receiver end, the data gets decoded by the receiver and is fed to the microcontroller which moves the DC motors of the vehicle for navigation. The voice-controlled autonomous robotic vehicle performed navigational tasks by listening to the commands of the user. This was achieved by converting voice commands into text strings which are readable by the Arduino microcontroller on the android app in order to control the navigation of the robot. The vehicle was tested under different conditions and was observed to perform better using this technique and also the results were satisfactory when compared with other previous research that has been conducted in this area.

Keywords: voice control; autonomous robot; android; Arduino; Bluetooth; microcontroller

Citation: Umar, A.; Giwa, M.A.; Kassim, A.Y.; Ilyasu, M.U.; Abdulwahab, I.; Agbon, E.E.; Ogedengbe, M.T. Development of an Android-Based, Voice-Controlled Autonomous Robotic Vehicle. *Eng. Proc.* **2023**, *58*, 48. https://doi.org/10.3390/ecsa-10-16026

Academic Editor: Stefano Mariani

Published: 15 November 2023

1. Introduction

Robotics is growing fast and becoming one of the most researchable areas in engineering, even though robotics emanated from advancement of technology. This is contrary to human perception, as they are usually amazed by the concept of artificial intelligence [1,2]. The reason for using robots is that they provide cheap labor, and because of their high accuracy of output. They have different applications which all possess the problem of finding a technique to accurately and efficiently control robots [3] that can be applied in medicinal, industrial, transportation, military and surveillance contexts [4,5]. The framework of the autonomous voice-controlled vehicle consists of the chassis, the Bluetooth module, DC motors, Arduino Uno microcontroller and other important components. The control of a vehicle is realized using a Bluetooth communication module [3]. This is connected to the

DC motors and the microcontroller. The vehicle is operated by giving wireless commands from the app using the functions that were initially programmed for it. The vehicle moves in four directions: right, left, forwards or backwards. In the case of the right direction, the four motors move in the clockwise direction, while for the left direction, the motors move in the anticlockwise direction. For forward movement, the motors move in the forward direction, while for backwards movement, they move in reverse directions. Also, in order to stop the movement of autonomous vehicles, the motion of the motors is stopped, and the vehicle stops moving [6]. However, the autonomous vehicle receives command through the android app using Bluetooth medium to control the movement of the motors.

There are some research works in the field of voice-controlled autonomous robotic vehicles. Ref. [7] developed and designed a voice-controlled talking robot using a mobile phone based on an Arduino Uno microcontroller. The robot's movement was based on the voice commands that were given to it, and, in turn, responded to the command that was given. A memory card was provided in the design for the robot to save prerecorded audio files that were used for developing the talking system of the robot. Also, ref. [8] developed a robotic vehicle which was controlled by voice commands through a smart phone using an Arduino Uno microcontroller and a Bluetooth sensor. The VoiceBot application and Google Voice were used as voice commands. A low-cost autonomous robotic vehicle was presented by [9] which was controlled by voice command. The user of the robotic vehicle may be located at a particular location and as long as the user is connected to the internet, the vehicle follows the voice instructions. The concept was successful using Arduino IDE, Adafruit, NodeMCU ESP866 and Google Assistant and If-This-Then-That (IFTTT). A voice-controlled robotic car was also produced by [10], the robotic car working principle was based on the Arduino Uno microcontroller, Bluetooth and motor drivers. The hardware component of the car was developed, while the software component was simulated, and later the two were interfaced in order to achieve effective coordination and control of the vehicle. In [11], a voice-controlled robot (VCR) which was mainly designed using Arduino IDE. The user provides voice commands to the vehicle using Google thing speech through an ESP32 microcontroller which forwards the commands to the Arduino on the robotic vehicle. The VCR was able to move in either the forward, backward, left and right directions. The robotic vehicle was designed using some predefined voice commands by [12]. The commands could be either forward, backward, left or right. A camera with an LCD screen was mounted on the vehicle to view distances of obstacles between the vehicles. Also, a GPS was attached to the vehicle to locate the position of the vehicle at any time. Finally, ref. [13] designed a robotic vehicle which uses voice commands using an android application for the navigation of the vehicle. The commands spoken were translated to text using speech-to-text in the android app before they were sent to the vehicle. The android app controls the vehicle using Bluetooth medium, which was embedded in the Arduino Uni microcontroller on the vehicle.

The rest of the paper is structured as follows: Section 1 discusses the introduction; Section 2 presents the materials used for the successful achievement of this article; Section 3 provides the results and discussion; and finally the conclusion is given in Section 4.

2. Materials

This subsection outlines the materials used for the design of the android-based voice-controlled robotic vehicle.

2.1. Arduino Uno Microcontroller

This is a programmable circuit board that is used in building electronic projects. It is an open-source platform which consist of software or an integrated development environment (IDE) that runs on a computer system. The codes are written in the IDE language and uploaded to the physical circuit board. The board contains a microcontroller which is programmed for controlling and sensing of objects in the physical world. Figure 1 shows the Arduino Uno microcontroller.

Figure 1. Arduino Uno [14].

2.2. *Direct Current (DC) Motor*

This is a rotary electrical machine which converts direct current electrical energy into mechanical energy. Most DC motors rely on the forces produced by magnetic fields. They possess some internal mechanisms which is either electronics or mechanical to periodically change the direction of the current flow of the motor. DC motors can be controlled over a wide range either by using a variable voltage supply or by changing the field winding current strength. The DC motor is shown in Figure 2.

Figure 2. DC motor [14].

2.3. *Four-Wheel Robot Chassis*

This is a vehicle framework used for the control of the autonomous robot. The DIY four-wheel-drive robot chassis is the mechanical platform that is used for this robot. The kits include all the hardware and mechanical components which are required to mount the motors, wheels, bolts and nuts. Figure 3 shows the four-wheel robot chassis.

Figure 3. Four wheel robot chassis [15].

2.4. Battery

This is a collection of one or more cells whose chemical reactions create a flow of electrons in a circuit. Batteries are made up of three basic components: the anode, the cathode and an electrolyte [16]. Figure 4 shows a typical battery.

Figure 4. Battery [17].

3. Results and Discussion

This subsection discusses the results that was obtained when the voice-controlled autonomous vehicle was coupled and tested. Firstly, the hardware interface, which is shown in Figure 5, was designed as the complete setup.

Figure 5. Complete setup.

The software component was simulated and embedded into the setup using the sequence shown in Figure 6.

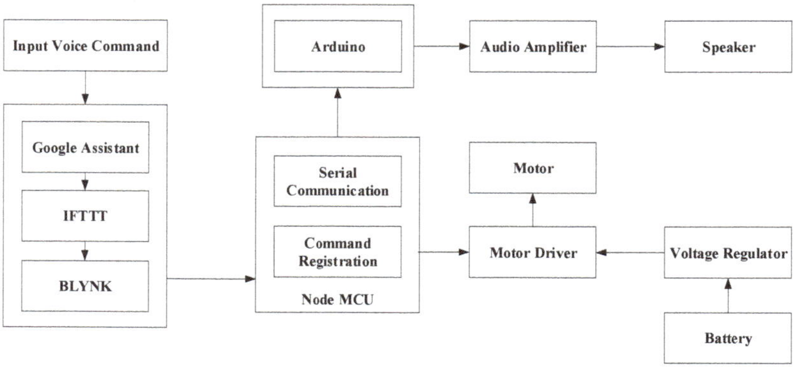

Figure 6. Robot voice control sequence.

In Figure 6, how the movement of the autonomous robot was achieved using voice control is shown whereby the command was given to the robot using a Bluetooth medium with the aid of an android application developed for the robot movement. Figure 7 shows the android app for the robot.

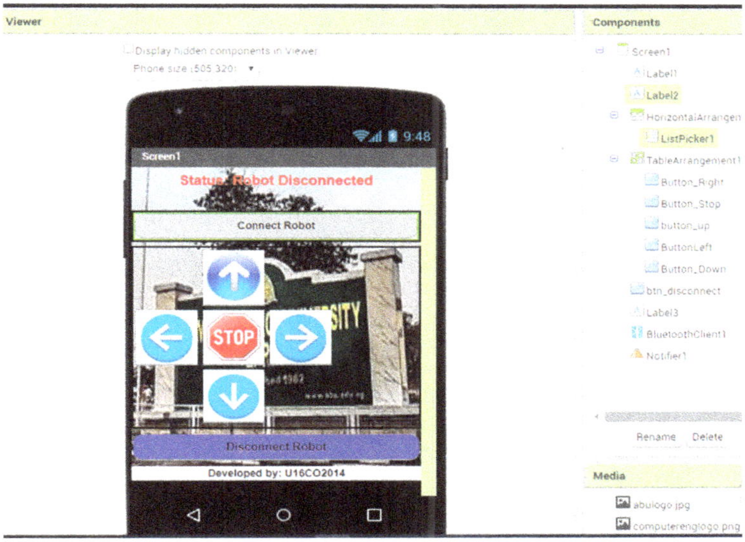

Figure 7. Android application for the autonomous robot.

In Figure 7, it is shown that the movement of the robot could be forward, backward, right and left. The robot had the capability of either going in the forward direction, backward direction, left direction and right direction. These all depended on the user's discretion.

Also, the vehicle was subjected to a voice command test, Bluetooth range test, maximum weight lift test and maximum lift angle test. It was observed that for the voice command test, the vehicle responded to all the commands accordingly. Also, for the Bluetooth test, it was observed that the maximum range obtained was 8.5–12 m depending on the location. For the maximum weight lift test, different weights were attached to the motor arm and lifted automatically and the maximum mass of the lifted objects was 50 g. For the maximum lift angle test, it was observed that the maximum lift angle the robotic vehicle could raise was 1.9 cm, this is due to the jack rod length constraint. The summary of the test carried out on the vehicle is shown in Table 1.

Table 1. Summary of the command actions obtained from the vehicle.

	Vehicle		**Appendage**	
Command	**Right Motor**	**Left Motor**	**Arm Motor**	**Gripper Motor**
	Action	**Action**	**Action**	**Action**
"Backward"	Rotate Anticlockwise	Rotate Anticlockwise	Off	Off
"Forward"	Rotate Clockwise	Rotate Clockwise	Off	Off
"Right"	Off	Rotate Clockwise	Off	Off
"Left"	Rotate Anticlockwise	Off	Off	Off
"Down"	Off	Off	Rotate Anticlockwise	Off
"Up"	Off	Off	Rotate Clockwise	Off
"Close"	Off	Off	Off	Rotate Anticlockwise
"Open"	Off	Off	Off	Rotate Clockwise
"Stop"	Off	Off	Off	Off

In Table 1, it is shown that the autonomous robotic vehicle performed satisfactorily when controlled with the android app, and also when executing the commands that was set for it to execute experimentally.

4. Conclusions

An autonomous android-based, voice-controlled robotic vehicle was developed in this article. Autonomous robots have become an integral aspect of human life, as, since the advent of these devices, they have made things easier. The robotic vehicle was controlled using an android app through the medium of Bluetooth. An Arduino Uno microcontroller was used to achieve this mission. The interface on the android app used button or spoken word commands. Theses buttons were forward, backward, left and right buttons. The movements of the vehicle were achieved using four DC motors based on the command given by the user. These commands on the android were converted into digital signals using a Bluetooth RF transmitter, where, at the receiver end, the data was decoded and fed to the Arduino microcontroller to move the DC motors for navigation of the vehicle. Testing was conducted on the vehicle using a Bluetooth range test, maximum weight lift and maximum weight angle test. They showed that the robotic vehicle performed satisfactorily. This could be made and applied to real-world applications. Further research should consider how to incorporate artificial intelligence into the robotic vehicle.

Author Contributions: Conceptualization A.U., I.A. and E.E.A.; methodology, A.U., M.A.G. and A.Y.K.; software, A.U., M.U.I. and M.T.O.; supervision, A.U., I.A. and E.E.A. All authors have read and agreed to the published version of the manuscript.

Funding: This research received no external funding.

Institutional Review Board Statement: Not applicable.

Informed Consent Statement: Not applicable.

Data Availability Statement: Data sharing is not applicable.

Acknowledgments: The authors are grateful to the entire staffs and students of control/computer research group and computer engineering department, ABU Zaria Nigeria.

Conflicts of Interest: The authors declare no conflict of interest.

References

1. Rawshan Habib, M.; Sunny, K.; Vadher, A.; Rahaman, A.; Tushar, A.N.; Mossihur Rahman, M.; Rashedul Arefin, M.; Ahmed, M.A. Design and Implementation of Voice Command-Based Robotic System. In *Inventive Systems and Control: Proceedings of ICISC 2022*; Springer: Berlin/Heidelberg, Germany, 2022; pp. 273–282.
2. Nayak, S.K.; Hota, S.P.; Behera, S. Voice Controlled Robotic Car. *UGC Care Group I J.* **2022**, *12*, 593–598.
3. Chaudhry, A.; Batra, M.; Gupta, P.; Lamba, S.; Gupta, S. Arduino Based Voice Controlled Robot. In Proceedings of the 2019 International Conference on Computing, Communication, and Intelligent Systems (ICCCIS), Greater Noida, India, 18–19 October 2019; pp. 415–417.
4. Haruna, Z.; Musa, U.; Mu'azu, M.B.; Umar, A. A Path Planning Technique for Autonomous Mobile Robot. *Int. J. Mechatron. Electr. Comput. Technol. (IJMEC)* **2020**, *10*, 4483–4492.
5. Haruna, Z.; Musa, U.; Mu'azu, M.B.; Umar, A. A Dynamic Path Planning Technique for Autonomous Mobile Robot in Unkwown Static Environment. In Proceedings of the 2019 IEEE 1st International Conference on Mechatronics and Cyber-Physical Computer Systems, Owerri, Nigeria, 7–8 March 2019; pp. 1–6.
6. Saravanan, D.; Parthiban, R.; Archanaa, G.I. Voice Controlled Robotic Car Using Arduino for Smart Agriculture. *Int. J. Pure Math* **2018**, *118*, 2097–2105.
7. Rashid, H.; Ahmed, I.U.; Osman, S.B.; Newaz, Q.; Rasheduzzaman, M.; Reza, S.M.T. Design and Implementation of a Voice Controlled Robot with Human Interaction Ability. In Proceedings of the International Conference on Computer, Communication, Chemical, Materials and Electronic Engineering, Rajshahi, Bangladesh, 26–27 January 2017; Volume 65, pp. 148–151.
8. Shalini, A.; Jayasuruthi, L.; VinothKumar, V. Voice Recognition Robot Control Using Android Device. *J. Comput. Theor. Nanosci.* **2018**, *15*, 2197–2201. [CrossRef]
9. Sachdev, S.; Macwan, J.; Patel, C.; Doshi, N. Voice-Controlled Autonomous Vehicle Using IoT. *Procedia Comput. Sci.* **2019**, *160*, 712–717. [CrossRef]
10. Srivastava, S.; Singh, R. Voice Controlled Robot Car Using Arduino. *Int. Res. J. Eng. Technol. (IRJET)* **2020**, *7*, 2356–2395.

11. Gupta, M.; Kumar, R.; Chaudhary, R.K.; Kumari, J. IoT Based Voice Controlled Autonomous Robotic Vehicle Through Google Assistant. In Proceedings of the 2021 3rd International Conference on Advances in Computing, Communication Control and Networking (ICAC3N), Greater Noida, India, 17–18 December 2021; pp. 713–717.
12. Korti, M.; Shettar, G.B.; Hadagali, G.A.; Shettar, S.; Shettar, S. Voice-Based Direction Control of a Robotic Vehicle through User Commands. *Int. Res. J. Adv. Sci. Hub* **2022**, *4*, 51–56. [CrossRef]
13. Kuriakose, S.; Harshitha, M.M.; Keerthana, D.G.; Adarsh, S.; Harshitha, K. Wireless Voice Controlled Robot. In Proceedings of the 2023 9th International Conference on Advanced Computing and Communication Systems (ICACCS), Coimbatore, India, 17–18 March 2023; Volume 1, pp. 189–194.
14. Haruna, S.H.; Umar, A.; Haruna, Z.; Ajayi, O.-O.; Zubairu, A.Y.; Rayyan, R. Development of an Autonomous Floor Mopping Robot Controller Using Android Application. In Proceedings of the 2022 5th Information Technology for Education and Development (ITED), Abuja, Nigeria, 1–3 November 2022; pp. 1–6.
15. Cornejo, J.; Palomares, R.; Hernández, M.; Magallanes, D.; Gutierrez, S. Mechatronics Design and Kinematic Simulation of a Tripteron Cartesian-Parallel Agricultural Robot Mounted on 4-Wheeled Mobile Platform to Perform Seed Sowing Activity. In Proceedings of the 2022 First International Conference on Electrical, Electronics, Information and Communication Technologies (ICEEICT), Trichy, India, 16–18 February 2022; pp. 1–7.
16. Chakraborty, S.; De, N.; Marak, D.; Borah, M.; Paul, S.; Majhi, V. Voice Controlled Robotic Car Using Mobile Application. In Proceedings of the 2021 6th International Conference on Signal Processing, Computing and Control (ISPCC), Solan, India, 7–9 October 2021; pp. 1–5.
17. Agwunedu, N.O.; Oshiga, O.; Chizea, L.O.; Oluwafemi, O.A.; Thomas, S. Arduino Based Voice Controlled Delivery System (Robot). In Proceedings of the 2021 1st International Conference on Multidisciplinary Engineering and Applied Science (ICMEAS), Abuja, Nigeria, 15–16 July 2021; pp. 1–5.

engineering proceedings

MDPI

Proceeding Paper

Developing and Validating Ensemble Classifiers for At-Home Sleep Apnea Screening †

Zilu Liang [1,2]

1 Ubiquitous and Personal Computing Lab, Faculty of Engineering, Kyoto University of Advanced Science (KUAS), Kyoto 615-8577, Japan; liang.zilu@kuas.ac.jp
2 Institute of Industrial Science, The University of Tokyo, Kyoto 153-8505, Japan
† Presented at the 10th International Electronic Conference on Sensors and Applications (ECSA-10), 15–30 November 2023; Available online: https://ecsa-10.sciforum.net/.

Abstract: In this paper, we developed ensemble classifiers with SpO2 signals for sleep apnea screening. The ensemble classifiers (eclf) were built on top of five base classifiers, including logistic regression (LR), random forest (RF), support vector machine (SVM), linear discrimination analysis (LDA), and light gradient boosting machine (LGMB). Performance evaluation showed that when heavier weights were assigned to the LR and SVM classifiers, the ECLF achieved a better balance between sensitivity (0.81 ± 0.02) and specificity (0.80 ± 0.02) while maintaining the overall performance as measured by AUC (0.81 ± 0.01).

Keywords: sleep apnea; ubiquitous computing; SpO2; ensemble learning

1. Introduction

Due to lifestyle changes, more and more people are suffering sleep disorders nowadays. Among all the different types of sleep disorders, sleep apnea is a most common one around the globe. Sleep apnea is characterized by brief stops of breathing during sleep. If that happens, on average, more than five times per hour throughout the night, then the person is clinically considered sleep apnea-positive. Sleep apnea is often accompanied by snoring during sleep and excessive daytime sleepiness. The long-term consequence of untreated sleep apnea includes increased risks of cardiovascular diseases, metabolic disorders, mental problems, and cognitive impairment [1]. Developing sleep apnea screening methods that are easy to use at home has been the objective of many studies. The main challenge of home sleep apnea screening is that most of the signal modalities in medical settings, such as EEG, ECG, and breath effort, are not readily measurable with home-use sensors. To address this challenge, several studies have attempted to develop apnea screening methods based solely on the blood oxygen levels, or SpO2, during sleep. This signal modality is ideal for home sleep apnea screening because it can be collected using consumer sleep tracking gadgets with reasonable accuracy [2]. This study aims to develop ensemble classifiers that takes the SpO2 signals as input to predict whether a user is sleep apnea-positive or -negative.

2. Related Work

Polysomnography (PSG) is the gold standard for sleep apnea diagnosis. A PSG test measures many physiological signals which allow doctors to calculate the apnea-hypopnea index (AHI), defined as the average number of obstructive respiratory events occurred per hour during sleep, for diagnosing sleep apnea. An AHI of larger than 5 events/h is considered out of the normal range and thus sleep apnea-positive. Other AHI cutoff thresholds, such as 15 and 30, are also widely used to determine the severity of sleep apnea [3]. The PSG test is expensive, invasive, and not always available. Many studies have attempted to develop alternative methods for sleep apnea screening using information stored in medical records. Instead of using physiological signals, those methods rely on

Citation: Liang, Z. Developing and Validating Ensemble Classifiers for At-Home Sleep Apnea Screening. *Eng. Proc.* **2023**, *58*, 49. https://doi.org/10.3390/ecsa-10-16184

Academic Editor: Stefano Mariani

Published: 15 November 2023

information such as demographic characteristics, blood tests, and questionnaire scores to assess the likelihood of sleep apnea [4]. In a related vein, machine learning and deep learning models that use SpO2 signals for sleep apnea screening were also developed. While early works have used only small datasets for model development [5], a very recent publication trained and tested their models on several large datasets [6]. In this study, we developed and validated our models using one of the largest sleep datasets.

3. Methodology

In this section, we present the pipeline for developing the ensemble classifier for sleep apneas screening and the dataset used. Signal processing and machine learning were performed using Python 3.10.5. Several Python modules were used, including Numpy, Pandas, Matplotlib, Scikit-learn, and Scipy.

3.1. Dataset

The Sleep Heart Health Study (SHHS) dataset used in this study consists of multi-modal physiological signals collected from 5786 subjects during sleep [7]. Permission was granted from the National Sleep Research Resource (NSRR) to export the SpO2 signals as well as some demographic information (i.e., age, gender. BMI) from the dataset. Data usage in this study was compliant with the Data Assess and Use Agreement (DAUA). The SpO2 signals were preprocessed to remove readings of zeros, below 50% or above 100%. Sudden changes between consecutive readings (>4%) were also removed. The '*nsrr_ahi_ph3r_aasm15*' variable was used as the ground truth values for AHI.

3.2. Ensemble Classifier

Ensemble is a method in machine learning that combines multiple base classifiers to improve the performance of the classifications. In this study, we used five base classifiers, including linear regression (LR), random forest (RF), support vector machine (SVM), linear discrimination analysis (LDA), and light gradient boosting machine (LGMB). The final output of an ensemble classifier is the most frequently occurring label (i.e., the mode) of the weighted outputs of the based classifiers. Let f_{LR}, f_{RF}, f_{SVM}, f_{LDA}, f_{LGMB} be the 5 base classifiers, and the final output of the ensemble classifiers f_{eclf} is calculated using Equation (1). We built two ensemble classifiers: f_{eclf1} uses a weight vector of [1, 1, 1, 1, 1], while f_{eclf2} uses a weight vector of [2, 1, 2, 1, 1]. We assigned heavier weights to the LR and SVM classifiers in the latter case because those two classifiers yielded better performance in our previous study [8].

$$f_{eclf} = \text{mode}\,(w_{LR} \cdot f_{LR},\ w_{RF} \cdot f_{RF},\ w_{SVM} \cdot f_{SVM},\ w_{LDA} \cdot f_{LDA},\ w_{LGMB} \cdot f_{LGMB}), \quad (1)$$

Following the common practice in machine learning, 30% of the dataset was used for model training and tuning, and the rest was used for model testing. Hyper-parameter tuning was achieved through 5-fold cross validation. Classifier performances were measured using AUC, sensitivity, specificity. We repeated the dataset split 50 times with different random seeds. Boxplots were generated to show the distribution of the performance measures over the 50 repetitions.

4. Result

The AUC, sensitivity, and specificity of the base classifiers and the ensemble classifiers are shown in Figures 1–3. It is shown that among the five base classifiers, LR achieved the highest AUC and specificity. RF had the highest sensitivity but the lowest specificity, which translated to the lowest AUC among all base classifiers. Compared to ECLF, ECLF2 had similar AUC to LR but a better trade-off between sensitivity and specificity.

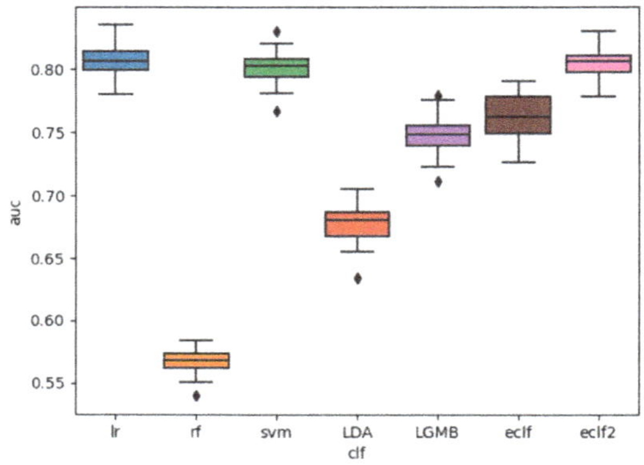

Figure 1. Performance of the classifiers for sleep apnea screening (AUC).

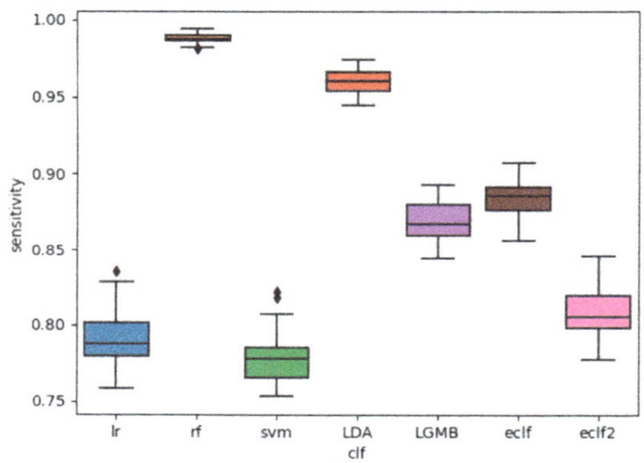

Figure 2. Performance of the classifiers for sleep apnea screening (sensitivity).

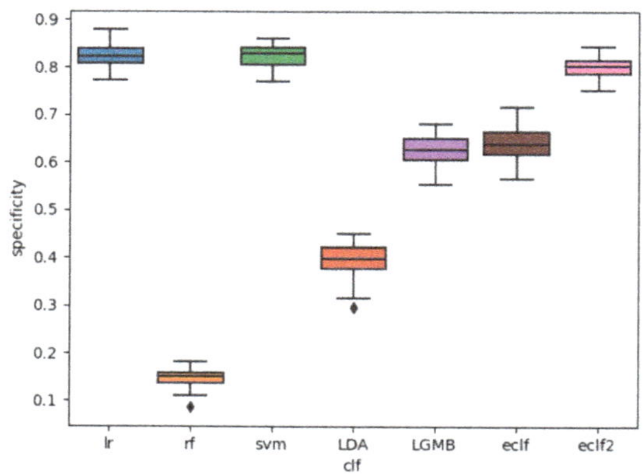

Figure 3. Performance of the classifiers for sleep apnea screening (specificity).

5. Discussion

We have presented the development and validation of ensemble classifiers with hard voting for sleep apnea screening. Performance evaluation analysis showed that when heavier weights were assigned to the LR and SVM classifiers, the ECLF achieved a better balance between sensitivity (0.81 ± 0.02) and specificity (0.80 ± 0.02) while maintaining the overall performance as measured by AUC (0.81 ± 0.01). RF and LDA achieved high sensitivity (>0.95) at the sacrifice of specificity, while LR and SVM achieved high specificity (>0.80) at the sacrifice of sensitivity. LGMB demonstrated mediocre performance on both sensitivity and specificity. In our future work, we plan to apply other ensemble techniques, such as soft voting, and to tune the weights given to the base classifiers.

Funding: This research was funded by the JSPS KAKENHI, grant number 21K17670.

Institutional Review Board Statement: The study was conducted in accordance with the Declaration of Helsinki and approved by the Ethics Review Board of Kyoto University of Advanced Science.

Informed Consent Statement: Informed consent was obtained from all subjects involved in the SHHS project.

Data Availability Statement: Restrictions apply to the availability of the data used in this study. Data were obtained from the National Sleep Research Resource (NSRR) and are available at https://sleepdata.org/datasets/shhs (accessed on 10 October 2023) with the permission of NSRR.

Acknowledgments: The author would like to thank the NSRR for sharing the SHHS dataset.

Conflicts of Interest: The authors declare no conflicts of interest.

References

1. Franklin, K.A.; Lindberg, E. Obstructive sleep apnea is a common disorder in the population-a review on the epidemiology of sleep apnea. *J. Thorac. Dis.* **2015**, *7*, 1311–1322.
2. Liang, Z.; Chapa-Martell, M.A. Accuracy of Fitbit wristbands in measuring sleep stage transitions and the effect of user-specific factors. *JMIR Mhealth Uhealth* **2019**, *7*, e13384. [CrossRef] [PubMed]
3. Berry, R.B.; Brooks, R.; Gamaldo, C.E.; Harding, S.M.; Marcus, C.; Vaughn, B.V. *The AASM Manual for the Scoring of Sleep and Associated Events: Rules, Terminology, and Technical Specifications*; Version 2.4 ed.; American Academy of Sleep Medicine: Darien, IL, USA, 2017.
4. Huang, W.C.; Lee, P.L.; Liu, Y.T.; Chiang, A.A.; Lai, F. Support vector machine prediction of obstructive sleep apnea in a large-scale Chinese clinical sample. *Sleep* **2020**, *43*, zsz295. [CrossRef] [PubMed]
5. Mostafa, S.S.; Mendonça, F.; Morgado-Dias, F.; Ravelo-García, A. SpO2 based sleep apnea detection using deep learning. In Proceedings of the 2017 IEEE 21st International Conference on Intelligent Engineering Systems (INES), Larnaca, Cyprus, 20–23 October 2017.
6. Levy, J.; Álvarez, D.; Del Campo, F.; Behar, J.A. Deep learning for obstructive sleep apnea diagnosis based on single channel oximetry. *Nat. Commun.* **2023**, *14*, 4881. [CrossRef] [PubMed]
7. Quan, S.F.; Howard, B.V.; Iber, C.; Kiley, J.P.; Nieto, F.J.; O'Connor, G.T.; Rapoport, D.M.; Redline, S.; Robbins, J.; Samet, J.M.; et al. The Sleep Heart Health Study: Design, rationale, and methods. *Sleep* **1997**, *20*, 1077–1085. [PubMed]
8. Liang, Z. Novel Method Combining Multiscale Attention Entropy of Overnight Blood Oxygen Level and Machine Learning for Easy Sleep Apnea Screening. *Digital Health* **2023**, *9*, 1–19. [CrossRef] [PubMed]

engineering proceedings

MDPI

Proceeding Paper

Novel Approach for Asthma Detection Using Carbon Monoxide Sensor †

Masoodhu Banu Noordheen Mohamed Musthafa [1], Udayakumar Anantharao [2], Dapheinkiru Dkhar [1], Ahamed Fathima Firdouse Mayiti. Jamal [1], Sabitha Prabha Murugan [1] and Pavan Sai Kiran Reddy Pittu [1,*]

[1] Department of Biomedical Engineering, Vel Tech Rangarajan Dr. Sagunthala R&D Institute of Science and Technology, Chennai 600062, India; drmasoodhubanu@veltech.edu.in (M.B.N.M.M.); dapheinkirudkhar00@gmail.com (D.D.); ahmedfathimafirdouse@gmail.com (A.F.F.M.J.); sabithaprabha49@gmail.com (S.P.M.)

[2] Medcuore Medical Solutions Private Limited, Chennai 600013, India; operations@medcuore.com

* Correspondence: pittupavansai2002@gmail.com; Tel.: +91-709-334-6948

† Presented at the 10th International Electronic Conference on Sensors and Applications (ECSA-10), 15–30 November 2023; Available online: https://ecsa-10.sciforum.net/.

Abstract: Around 339 million people suffer from asthma worldwide. An acute asthma attack causes difficulties in daily life activities and can sometimes be fatal. The unnecessary challenges faced by asthmatics signifies the need for a device that helps people monitor and control asthma to prevent possible attacks. A number of studies have reported an elevation of carbon monoxide in exhaled breath (eCO) of asthma patients and suggest that this can be used as an effective biomarker of lung inflammation. By making use of the reported results, this project aims to make use of the eCO biomarker to design a carbon monoxide (CO) asthma monitoring system. The system consists of a Raspberry Pi 3 microcontroller and a MQ 7 CO sensor for processing and detecting the carbon monoxide concentration in parts per million. For accurate results, a face mask is attached to the sensor to mitigate environmental CO. The working of the sensor circuit is validated using a carbon monoxide source. With more researchers focusing on the threshold level of CO for an imminent asthma attack, this CO sensor could eventually save lives and improve standards of living while being an affordable and user-friendly device for active lifestyles.

Keywords: asthma monitor; exhaled carbon monoxide (eCO); CO sensor; Raspberry Pi 3

Citation: Noordheen Mohamed Musthafa, M.B.; Anantharao, U.; Dkhar, D.; Mayiti. Jamal, A.F.F.; Murugan, S.P.; Pittu, P.S.K.R. Novel Approach for Asthma Detection Using Carbon Monoxide Sensor. *Eng. Proc.* **2023**, *58*, 50. https://doi.org/10.3390/ecsa-10-16002

Academic Editor: Stefano Mariani

Published: 15 November 2023

1. Introduction

Asthma, a respiratory ailment, is categorized as chronic inflammation in the pulmonary airways, along with oxidative stress in the bronchial tubes. A significant characteristic is the swelling in both proximal and distal lung airways [1]. An asthma patient can experience sudden or unexpected asthma attacks that can be triggered by any factor, and the factors vary from person to person. The intensity of an acute asthma attack is unforeseeable and has the possibility to be fatal [2].

Asthma affects human beings of all ages, and it generally begins in childhood. Asthma is a medical issue increasingly affecting people globally. As reported by the Global Asthma Report in 2018, it was believed that around 339 million people had asthma, and there were 417,918 deaths due to asthma worldwide [3]. India has around 15–20 million asthmatics, 6% of the child and 2% of the adult population. The phase 3 International Study of Asthma & Allergy in Children (ISAAC) investigated the nationwide predominance of the presence of wheeze as being 7% in Indian children aged between 6 and 7 years and between 13 and 14 years. Certainly, more than 50% of the people had terrible asthma [4].

Asthma is an incurable disease. While some asthma cases are mortal, the majority are less serious and cause hardships in day-to-day life. It is very challenging for an asthmatic to maintain a healthy lifestyle since most people with asthma encounter symptoms of exercise-induced lung constriction while performing exercises, and engaging in exercise and sports can be very burdensome and may sometimes even be dangerous. The inability to participate in daily life activities can make them inactive, affecting their social well-being, which leads to stress, sleep deprivation, depression, etc. [5].

Spirometry and Peak Expiratory Flow are employed in diagnosing and monitoring the severity of asthma exacerbation. However, physician's supervision is mandatory for both techniques, yet visiting a healthcare center is impracticable and tiresome for asthmatics. This inefficacy paved the way for advanced portable device technology [6]. These portable devices assist in diagnosing and continuously monitoring disease symptoms at earlier stages, reducing unnecessary hospitalization costs, allowing remote monitoring and helping maintain better patient-to-doctor ratios.

In recent times, chemical biomarkers like exhaled nitric oxide (eNO) and exhaled carbon monoxide (eCO) have been employed by healthcare professionals as another capable means of analyzing, evaluating and framing asthma treatment plans [2]. Carbon monoxide (CO) is a biomarker of oxidative stress induced by the stress protein heme-oxygenase [(HO)-1] and also due to inflammation [1]. Exhaled CO (eCO) is like exhaled NO, which has been assessed as a candidate breath biomarker of pathophysiological states elevated in asthmatics and decreases with steroid therapies. As mentioned in [1], NO is of airway origin, whereas CO is of alveoli origin. The main advantage of utilizing eCO as a breath biomarker is that it is observed in greater concentrations than nitric oxide (NO). This allows for less sophisticated and more reliable monitoring devices [7]. Many of the currently available asthma monitoring devices are either unaffordable, unreliable or not suitable for an active lifestyle [5].

When a person is experiencing an asthma attack, the body might be triggered by a stimuli, which results in bronchial inflammation. One method to detect these phenomena is by monitoring eCO concentrations before, during and after an acute attack [5]. Prior studies [8–15] have proven that CO levels increase during an asthma attack. This project utilizes the results found in these studies to develop an asthma monitor using a CO sensor.

Thus, the detection of eCO could be achieved via an effortless, non-invasive device for monitoring the acute exacerbation of asthma. This study aims to develop a programmed device that monitors eCO levels by using a CO sensor, which could possibly reduce acute asthma attacks and help improve the quality of life for asthmatics with varying lifestyles by reducing costs and extending portability, ultimately saving lives.

2. Literature Survey

The study conducted by Kiyoshi et al. (1997) shows an elevation of eCO in asthmatics that reduces with corticosteroid therapy, and it further indicates that the changes in the concentration of exhaled CO were remarkably associated with the number of eosinophil cells present in the sputum [8]. P. Paredi et al. (1999) further inspected the possibility that allergen challenge can raise exhaled carbon monoxide (eCO) levels as an indication of HO activation in 15 asthmatics and found out that the eCO increases in asthmatic reactions occur independently of the change in lung calibre [9]. The research conducted by M. Yamaya et al. (2001) has shown that eCO concentrations in people with unstable severe asthma were notably higher than in those with stable conditions. eCO concentrations in slight and non-extreme asthma patients did not differ appreciably from those patients who are controlled and do not have a smoking habit ($p > 0.20$). The research further states a remarkable proportionality between eCO concentration and forced expiratory volume [10].

Susumu Sato et al. (2003) estimated the optimal cutoff level of exhaled CO concentration to differentiate actual smokers from nonsmokers among 161 asthmatics and 170 COPD patients. The resulting analysis demonstrated that in asthmatics and COPD patients, the eCO levels were potentially affected by underlying lung inflammation, leading to the miscategorization of smoking status due to exhaled CO. This shows that CO levels increase in asthma [11]. The research of Ildiko Horvath et al. (2015) indicates that in asthmatic lungs, the induction of HO-1 may cause increasing eCO concentrations, suggesting that it is medically applicable as a diagnostic tool in terms of evaluating asthma [12]. The prevalence of the CO biomarker was studied by Yoichiro et al. (2016), and the result signifies that exhaled CO levels in children were considerably increased while encountering asthma exacerbation and were reduced after inhalation therapy with β2-agonist and SCG in those with intermittent asthma [13].

Amanda A. Pereira et al. (2018) speculated an increase in the levels of exhaled CO in asthmatic children, specifically in children with partially stable or unstable asthma, and the mean-adjusted eCO level was 0.56 ppm greater in asthmatic children, suggesting that exhaled CO may serve as an inexpensive biomarker for asthma control [14]. Yoichiro et al. (2020) further stated that in asthmatic infants and toddlers, asymptomatic asthma had exhaled CO levels of 2.0 (1.0–2.0) ppm, and those experiencing an asthma attack had exhaled CO levels of 2.0 (2.0–3.25) ppm ($p < 0.0001$) [15].

3. Materials and Methods

3.1. Overall Architecture of the Implemented System

As shown in Figure 1, the proposed system consists of two working units. One is the hardware architecture block and another is the software architecture block.

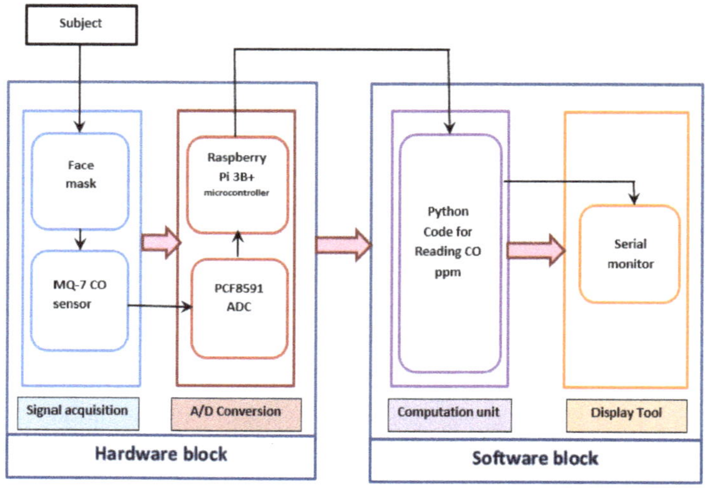

Figure 1. Block diagram of the proposed CO monitoring model.

3.2. Hardware Architecture

(1) Signal Acquisition Unit: Figure 2 shows a pictorial view of the signal acquisition unit. For accuracy and in order to avoid environmental CO, the sensor needs to be held close to the mouth. For this purpose, a face mask is used. The MQ-7 sensor contains 4 pins, namely A0, D0, Vcc and GND. The A0 pin is connected to AIN0 of PCF8591, Vcc and GND are connected to pin 2 and pin 6 of the Raspberry Pi. The CO is acquired through MQ-7 and transmitted to PCF8591. Figure 3 shows the connection of the MQ-7 sensor with the Raspberry Pi 3b+ and PCF8591.

(2) Analog-to-Digital Conversion: The MQ-7 sensor gives output in the form of analog values. These analog data are given to PCF8591 ADC. PCF8591 works on I2C communication.

The obtained Analog value is converted to digital volts by

$$V_{OUT} = A_{OUT} \times 5/256 \quad (1)$$

A_{OUT}—analog value

V_{OUT}—digital value in (0–5 V).

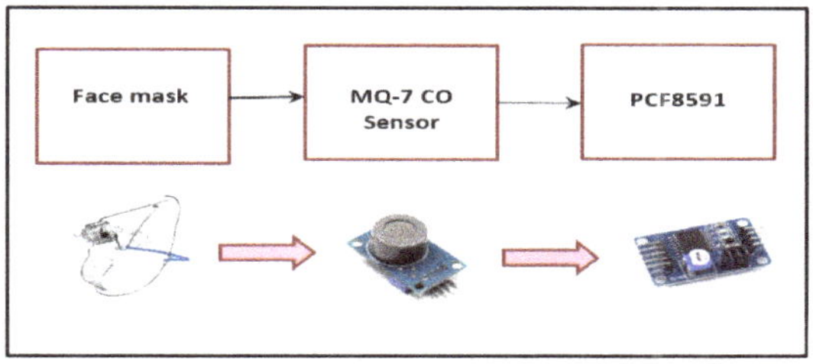

Figure 2. Signal acquisition unit.

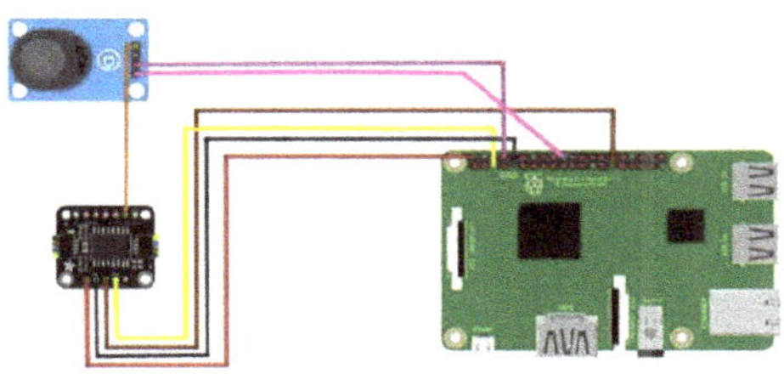

Figure 3. Schematic diagram of the CO-sensing system.

3.3. Software Architecture

The concentration of CO in parts per million (ppm) was determined in accordance with the resistance ratio ($RS/R0$). RS is the estimated resistance altered when the sensing channel encounters gas, and $R0$ is the stable sensor resistance in clean air or in the absence of gas. Employing Ohm's law and the sensor circuit diagram, the following can be obtained

$$RS = VC - RL/V_{OUT} - RL \tag{2}$$

VC—voltage current (in this condition 5Volts from pi)
V_{OUT}—output voltage (calculated analog/digital value)
RL—load resistance (here the value is 10 K).
$R0$ can then be computed with the below equation,
$R0 = RS/$fresh air ratio value from sensitivity graph of MQ-7.

For the purpose of transforming the digital signal values to CO concentration in ppm, the MQ-7 datasheet is used. The correlation coefficients (A and B) are found by performing a power regression. The sensitivity characteristic graph of the MQ-7 shown in Figure 4 is loaded on Web Plot Digitizer software v4.6. The obtained data points are then loaded to R script code to perform a power regression.

Calculation of CO ppm concentration: To calculate the CO ppm concentration, two codes are written. The first section of the code is for calibrating the MQ-7 sensor in fresh air to acquire the RS and R0 values in atmospheric air, and the second section of code is for sensing CO in the testing environment. For the purpose of calibrating in clean air, the mean of 500 values obtained by running the first part of code was taken into account. The initial value begins with 0 and allows the addition of every read to each other till the given condition fails. Thus, from the mean value, the RS value in air and the R0 value in fresh air are calculated. This code runs a single time.

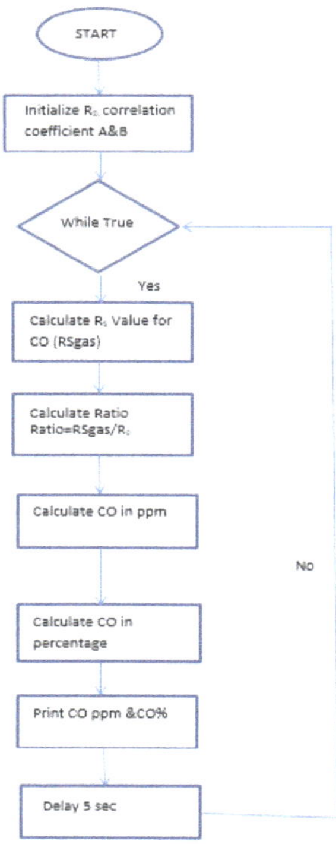

Figure 4. Flowchart for detecting the presence of CO in real time.

The next section of the code senses the presence of carbon monoxide, calculates the ratio and reads out the concentration of CO in ppm with respect to the change in voltage value. Figure 4 shows the flowchart for calculating CO concentration in real time.

Figure 5 shows the entire setup of the developed system with a face mask attached to the sensor.

Figure 5. Proposed CO-sensing system with face mask.

4. Results and Discussion

The estimated values of correlation coefficients A and B using R-Script are found to be 83.69 and −1.63 approximately. The sensor has been calibrated based on the datasheet. The sensor was calibrated to detect 40 ppm CO in air. The response from the sensor in terms of ppm, as shown in Figure 6, is obtained for healthy (six numbers) and chronic asthma (three numbers) patients, and the level of CO was higher than the threshold for asthma patients, whereas for normal patient is below the threshold. This shows the sensitivity of the designed system against CO sources. The Table 1 below shows the converted analog reading, as shown above in terms of ppm.

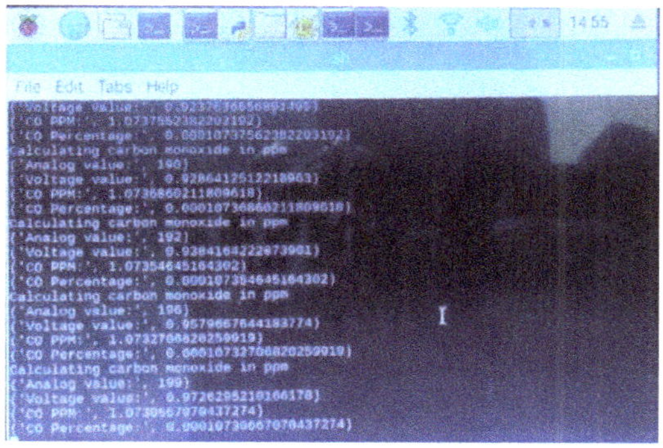

Figure 6. Result of CO concentration in ppm.

Table 1. Exhaled CO concentration in healthy and individual with asthma.

S.No	Type of Subject	Exhaled CO Concentration Range in Ppm
1	Healthy	1.5 to 2.0
2	Asthma Patient	6.0 to 6.5

5. Conclusions and Future Scope

The designed CO sensor circuit shows high sensitivity to the CO sources. The target use of this system is to alert the patient to take precautions to avoid an asthma attack in the first place by monitoring their CO levels. Once the threshold level of CO that is dangerous to an asthma patient is found via clinical trials, the system can alert patients when their CO levels cross the threshold range. In addition, the design of the system can be enhanced further by reducing the size of the system as small as possible for handy use, travel ability and wearability. This project can be improved further by incorporating an IoT structure, giving the physician access to the data anywhere at any time to have an insight into the patient's health condition. This project can be further enhanced to develop an accurate and fully functional device that justifies the necessity of every asthmatic who would make use of it.

Author Contributions: M.B.N.M.M. and U.A. conceptualized the idea of this manuscript. D.D. and designed the model. A.F.F.M.J., S.P.M. and P.S.K.R.P. contributed to the software and coding. D.D. and S.P.M. conducted the formal analysis. A.F.F.M.J. conducted the investigation. M.B.N.M.M. and U.A. supervised the work. All authors have read and agreed to the published version of the manuscript.

Funding: This research received no external funding.

Institutional Review Board Statement: Not applicable.

Informed Consent Statement: Not applicable.

Data Availability Statement: Data sharing is not applicable.

Acknowledgments: We thank our biomedical department and Meccuore Medical Solutions Private Limited for supporting this work and we are also grateful for our supervisors Masoodhu Banu and Udayakumar Anantharao for guiding us.

Conflicts of Interest: The authors declare no conflicts of interest.

References

1. Shorter, J.H.; Nelson, D.D.; McManus, J.B.; Zahniser, M.S.; Sama, S.R.; Milton, D.K. Clinical study of multiple breath biomakers of asthma and COPD (NO, CO_2, CO and N_2O) by infrared laser spectroscopy. *J. Breath Res.* **2011**, *15*, 037108. [CrossRef] [PubMed]
2. Kwan, A.M.; Fung, A.G.; Jansen, P.A.; Schivo, M.; Kenyon, N.J.; Delplanque, J.P.; Davis, C.E. Personal Lung Function Monitoring Devices for Asthma Patients. *IEEE Sens. J.* **2015**, *15*, 2238–2247. [CrossRef]
3. Global Initiative for Asthma.Global Strategy for Asthma Management and Prevention. Available online: www.ginasthma.org (accessed on 1 July 2022).
4. Krishna, M.T.; Mahesh, P.A.; Vedanthan, P.K.; Mehta, V.; Moitra, S.; Christopher, D.J. The burden of allergic diseases in the Indian subcontinent: Barriers and challenges. *Lancent Glob. Health* **2020**, *4*, e478–e479. [CrossRef] [PubMed]
5. Liu, R.; Hannah, L.; Detrick, J. Reducing Asthma Attacks with a Portable CO Sensor. In Proceedings of the 3rd International Conference on Frontiers of Biological Sciences and Engineering, Diwaniyah, Iraq, 21–22 April 2021.
6. Abinayaa, B.; Raja, A. Smart Portable Monitoring Device for Asthma Patients. *Middle-East J. Sci. Res.* **2016**, *24*, 136–142.
7. Zetterquist, W.; Marteus, H.; Johannesson, M.; Nordvall, S.L.; Ihre, E.; Lundberg, J.O.N.; Alving, K. Exhaled carbon monoxide is not Elevated in Patients with Asthma and Cystic Fibrosis. *Eur. Respir. J.* **2002**, *20*, 92–99. [CrossRef] [PubMed]
8. Zayasu, K.; Sekizawa, K.; Okinaga, S.; Yamaya, M.; Ohrui, T.; Sasaki, H. Increased Carbon Monoxide in Exhaled Air of Asthmatic patients. *Am. J. Respir. Crit. Care Med.* **1997**, *156*, 1140–1143. [CrossRef]
9. Paredi, P.; Leckie, M.J.; Horvath, I.; Allegra, L.; Kharitonov, S.A.; Barnes, P.J. Changes in Exhaled Carbon Monoxide and Nitric Oxide Levels Following Allergen Challenge in Patients with Asthma. *Eur. Respir. J.* **1999**, *13*, 48–52. [CrossRef]
10. Yamaya, M.; Hosoda, M.; Ishizuka, S.; Monma, M.; Matsui, T.; Suzuki, T.; Sekizawa, K.; Sasaki, H. Relation between exhaled carbon monoxide levels and clinical severity of asthma. *Clin. Exp. Allergy* **2001**, *31*, 417–422. [CrossRef] [PubMed]
11. Sato, S.; Nishimura, K.; Koyama, H.; Tsukino, M.; Oga, T.; Hajiro, T.; Mishima, M. Optimal Cutoff Level of Breath Carbon Monoxide for Assessing Smoking Status in Patient with Asthma and COPD", Department of Respiratory Medicine. *Chest J.* **2003**, *124*, 1749–1754. [CrossRef] [PubMed]
12. Horváth, I.; Donnelly, L.E.; Kiss, A.; Paredi, P.; Kharitonov, S.A.; Barnes, P.J. Raised levels of exhaled carbon monoxide are associated with an increased expression of hema oxygenase-1 in airway macrophages in asthma: A new maker of oxidative stress. *Thorax* **2015**, *53*, 668–672. [CrossRef] [PubMed]
13. Ohara, Y.; Ohrui, T.; Morikawa, T.; He, M.; Yasuda, H.; Yamaya, M.; Sasaki, H. Exhaled Carbon Monoxide Levels in School-Age Children With Episodic Asthma. *Pediatr. Pulmonol.* **2016**, *41*, 470–474. [CrossRef] [PubMed]
14. Pereira, A.A.; Pollard, S.L.; Locke, R.; Romero, K.; Lima, J.J.; Hansel, N.N.; Checkley, W. Association between exhaled carbon monoxide and asthma outcomes in Peruvian Children. *Respir. Med.* **2018**, *145*, 212–216. [CrossRef] [PubMed]
15. Ohara, Y.; Ohara, T.; Hashimoto, K.; Hosoya, M. Exhaled carbon monoxide levels in infants and toddlers with episodic asthma. *Fukushima J. Med. Sci.* **2020**, *66*, 78–87. [CrossRef] [PubMed]

engineering proceedings

MDPI

Proceeding Paper

Biosensor Time Response and Noise Models That Take into Account Spatial Rearrangement of Adsorbed Biomolecules †

Ivana Jokić *, Miloš Frantlović, Olga Jakšić, Katarina Radulović and Stevan Andrić

Institute of Chemistry, Technology and Metallurgy, National Institute of the Republic of Serbia, University of Belgrade, 11000 Belgrade, Serbia; frant@nanosys.ihtm.bg.ac.rs (M.F.); olga@nanosys.ihtm.bg.ac.rs (O.J.); kacar@nanosys.ihtm.bg.ac.rs (K.R.); stevan@nanosys.ihtm.bg.ac.rs (S.A.)

* Correspondence: ijokic@nanosys.ihtm.bg.ac.rs

† Presented at the 10th International Electronic Conference on Sensors and Applications (ECSA-10), 15–30 November 2023; Available online: https://ecsa-10.sciforum.net/.

Abstract: In order to improve biosensor performance, it is important to develop mathematical models of sensors' temporal response and noise, which include the effects of processes and phenomena relevant to the real applications of these devices. Here, we present a novel, more comprehensive response and noise models that consider the rearrangement process of biomolecules upon their adsorption on the sensing surface. We evaluate the extent of the influence of this process for various rates of rearrangement and adsorption–desorption processes. The development of such models is indispensable for the correct interpretation of the measurement results and also for the estimation and improvement of sensor performance limits, yielding the more reliable detection of the target agent in the analyzed samples.

Keywords: sensor noise; sensor time response; adsorption; biomolecular rearrangement; protein conformation; protein sensor; mathematical model

Citation: Jokić, I.; Frantlović, M.; Jakšić, O.; Radulović, K.; Andrić, S. Biosensor Time Response and Noise Models That Take into Account Spatial Rearrangement of Adsorbed Biomolecules. *Eng. Proc.* **2023**, *58*, 51. https://doi.org/10.3390/ecsa-10-16070

Academic Editor: Stefano Mariani

Published: 15 November 2023

1. Introduction

The growing need for high-performance in situ biosensing is driving the development of micro/nanobiosensors, which have already shown a significant potential for the highly sensitive detection of biological specimens or biologically relevant chemical substances in applications such as the real-time monitoring of biological pollution in the air, water, and food, or the health conditions of living organisms [1–3]. Research efforts are being made to push their performance further beyond the current limits. In this sense, it is important to investigate physical processes and phenomena that inevitably affect the generation of the sensor's response and its fluctuations, thus setting a fundamental performance limit. The basis of these investigations is the development and application of mathematical models of sensor time response and noise, which take into account all relevant processes.

Adsorption-based biosensing relies on the reversible adsorption process of biomolecules on a sensing surface. In addition to producing the response of the sensor, this process, stochastic in nature, is also a source of noise that affects the performance of micro/nanosensors. There are several sensor response and noise models that take into account different additional processes coupled with adsorption, depending on a specific practical case [4–7]. The spatial rearrangement of adsorbed biomolecules is an additional process that changes the binding/unbinding kinetics to a two-step process behavior and, therefore, affects both the sensor's time response and its fluctuations. We present the improved models of the sensor time response and noise, which consider biomolecular rearrangement and evaluate the extent of its influence for various rates of rearrangement and adsorption/desorption processes. The development of improved mathematical models of sensor temporal response and noise, which include the effects pronounced in the real applications of these devices, is indispensable for both a correct interpretation of the measurement results and estimation of sensor performance limits and, thus, for the achievement of the reliable detection of the target agent in analyzed samples.

2. The Models of Biosensor Response and Noise

The adsorption of biomolecules (especially proteins) on a sensing surface, or their affinity-based binding to other biomolecules that are used for sensor functionalization is often followed by a change in their structure and/or orientation from one configuration to another [8,9]. There are many different variants of such changes, which can be encompassed by the term "molecular rearrangement". Apart from the spatial configuration of adsorbed molecules, their rearrangement also alters their properties in terms of functionality and even an affinity towards surface adsorption sites: the adsorbed molecules can be bonded to the adsorption sites more or less strongly than when they are in their original configuration. Let us assume that protein molecules can be reversibly adsorbed only in one configuration (configuration A), that the rearrangement of adsorbed molecules from configuration A to another configuration (B) is reversible, that the change in the configuration of adsorbed molecules does not influence the occupancy of the sensing surface, and that the adsorbed particles in configuration B can also be desorbed (Figure 1). In that case, the instantaneous numbers of adsorbed particles in the conformations A and B, N_A and N_B, are determined by the equations of the mathematical model, which is, at the same time, the sensor's temporal response model (assuming that the response is proportional to the number of adsorbed particles):

$$\frac{dN_A}{dt} = k_{aA}C(N_m - N_A - N_B) - k_{dA}N_A + k_{BA}N_B - k_{AB}N_A \quad (1)$$

$$\frac{dN_B}{dt} = k_{AB}N_A - k_{BA}N_B - k_{dB}N_B \quad (2)$$

Here, k_{aA} and k_{dA} are the adsorption and desorption rate constants of the protein with configuration A, k_{dB} is the desorption rate constant of the protein with configuration B, k_{AB} and k_{BA} are the rate constants of the transformation of adsorbed molecules from configuration A to B and back, C is the concentration of the protein in the sample, and N_m is the total number of adsorption sites on the sensing surface. The numbers of adsorbed particles, N_{As} and N_{Bs}, obtained by solving Equations (1) and (2) assuming $dN_A/dt = 0$ and $dN_B/dt = 0$, determine the steady-state sensor response:

$$N_{As} = \frac{(k_{dB} + k_{BA})k_{aA}C}{(k_{AB} + k_{dB} + k_{BA})k_{aA}C + k_{dA}(k_{dB} + k_{BA}) + k_{dB}k_{AB}}N_m \quad (3)$$

$$N_{Bs} = \frac{k_{AB}}{k_{dB} + k_{BA}}N_{As} \quad (4)$$

Both the adsorption and rearrangement processes are stochastic in nature, so the numbers of adsorbed particles fluctuate around the values determined by Equations (3) and (4). These fluctuations, ΔN_A and ΔN_B, constitute the inevitable intrinsic sensor noise. The analysis of this noise is based on the Langevin approach for the two-variable gain-loss stochastic processes [10], and such a process is one consisting of the mutually dependent processes ΔN_A and ΔN_B. The analysis starts from the Langevin form of Equations (1) and (2), which are obtained after substituting $N_A = N_{As} + \Delta N_A$ and $N_B = N_{Bs} + \Delta N_B$ and adding the suitable Langevin stochastic source functions (η_A and η_B) on the right side of Equations (1) and (2). The resulting equations obtained for fluctuations are solved in the frequency domain for $\Delta N_A(\mathrm{j}\omega)$ and $\Delta N_B(\mathrm{j}\omega)$ (knowing that the spectra of the Langevin stochastic source functions for that class of stochastic processes are white [10]; $\omega = 2\pi f$, where f is the Fourier frequency). Since the fluctuations of the total number of adsorbed molecules are $\Delta N = \Delta N_A + \Delta N_B$, their power spectral density (PSD) is $S(\omega) = \Delta N(\mathrm{j}\omega)\Delta N(-\mathrm{j}\omega)$, and, by using the obtained frequency domain solutions for $\Delta N_A(j\omega)$ and $\Delta N_B(j\omega)$, the final form of the PSD of noise is obtained:

$$S(f) = S_{LFNM}\frac{1+(2\pi f)^2\tau_{III}^2}{[1+(2\pi f)^2\tau_I^2][1+(2\pi f)^2\tau_{II}^2]} \tag{5}$$

where

$$\tau_{I,II} = \frac{2}{K_{11}+K_{22}\pm\sqrt{(K_{11}-K_{22})^2+4K_{12}K_{21}}} \tag{6}$$

$$\tau_{III} = \frac{\sqrt{d_{As}+d_{Bs}}}{\sqrt{(K_{22}-K_{21})^2 d_{As}+(K_{11}-K_{12})^2 d_{Bs}}} \tag{7}$$

$$S_{LFNM} = 4(d_{As}+d_{Bs})\frac{\tau_I^2\tau_{II}^2}{\tau_{III}^2} \tag{8}$$

$$K_{11} = k_{aA}C + k_{dA} + k_{AB},\ K_{12} = k_{aA}C - k_{BA},\ K_{21} = -k_{AB},\ K_{22} = k_{dB} + k_{BA} \tag{9}$$

$$d_{As} = (k_{dA}+k_{AB})N_{As},\ d_{Bs} = (k_{dB}+k_{BA})N_{Bs}. \tag{10}$$

This completes the models of sensor response and noise.

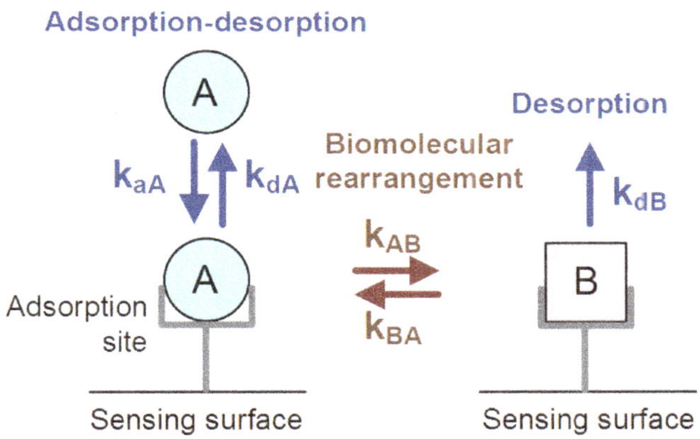

Figure 1. Schematic representation of interaction processes of biomolecules and adsorption sites. The rearrangement process of adsorbed molecules changes their spatial configuration from A to B and back. The adsorption and desorption rate constants, as well as the forward and reverse rearrangement rate constants, are shown near the corresponding arrows.

3. Results of the Analysis and Discussion

We present the results of a case study with the kinetic parameter values of the exemplary protein adsorption: $k_{aA} = 2\cdot10^7$ 1/(Ms), $k_{dA} = 10$ 1/s, $k_{dB} = 0.1$ 1/s, $k_{BA} = 0.01$ 1/s, $N_m = 10^{12}$ adsorption sites, $C = 5\cdot10^{-7}$ M, and three different values of the forward rearrangement rate constant k_{AB}: 0.25 1/s, 0.5 1/s, and 2 1/s (unit 1 M = mol/dm^3). These parameter values are in the range characteristic for the detection of proteins in biological samples [9,11–13], and they were chosen to illustrate the possible effects of biomolecular rearrangement processes on both sensor response and noise. The results were obtained using the response and noise models presented in Section 2.

Figure 2 shows the time evolution of the number of adsorbed molecules for three cases when a two-way rearrangement process occurs and for the case when the rearrangement does not occur (solid green line). The three cases differ in the ratio of forward and reverse rearrangement rate constants, k_{AB}/k_{BA}: 25 (dash-dotted blue line), 50 (dashed red line), and 200 (dotted line). A significant change caused by the rearrangement is visible in the response kinetics: the response is slowed down, and the steady-state value increases compared to the case when the rearrangement does not occur. Also, the change in the response kinetics can be seen: from single-exponential (characteristic of AD processes without rearrangement) to two-exponential (two-step kinetics), with a fast starting rise, followed by a slower approach to the steady state. Such two-step response kinetics is experimentally observed [14]. Clearly, such experimental results could not be interpreted by a model that neglects biomolecular rearrangement.

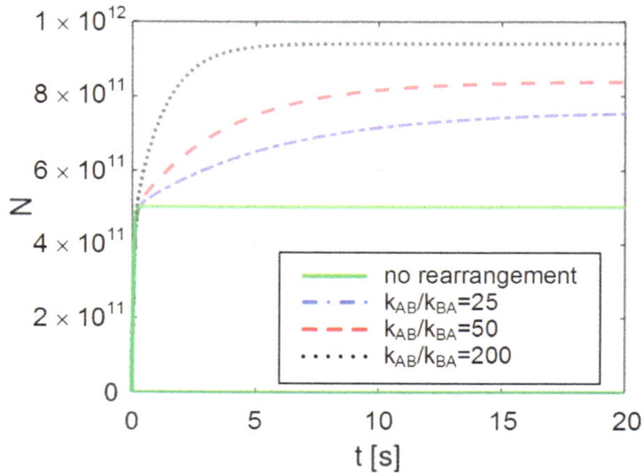

Figure 2. The time evolution of the number of adsorbed molecules, which determines the sensor response kinetics. The three cases when the reversible rearrangement process occurs (with different ratios of forward and reverse rearrangement rate constants, k_{AB}/k_{BA}) are shown, as well as the case when the configuration of biomolecules does not change after adsorption.

Figure 3 shows the noise power spectral density for the same four cases of protein adsorption as in Figure 2. As can be seen, the rearrangement of adsorbed molecules changes the noise spectrum. Instead of one characteristic frequency in the spectrum, there are three characteristic frequencies that determine the positions of two knees and one valley. LFNM also changes (for the given set of parameter values, LFNM increases). The change in the k_{AB}/k_{BA} ratio significantly alters the value of the lower characteristic frequency (it increases with the increase in k_{AB}/k_{BA}), while the influence on the highest frequency is negligible. It is clear that the three characteristic frequencies and the LFNM magnitude contain information about the values of the rearrangement process rate constants. Data about the values of these constants are scarce in the literature in spite of the fact that they can describe the interactions of biomolecules with solid surfaces or other biomolecules, whose characterization is important for the development of medications, implants, and biosensors, as well as for the research of various biochemical processes. The presented noise model can, therefore, be used for the development of methods for the characterization of biomolecular rearrangement processes based on the analysis of experimentally obtained noise spectra.

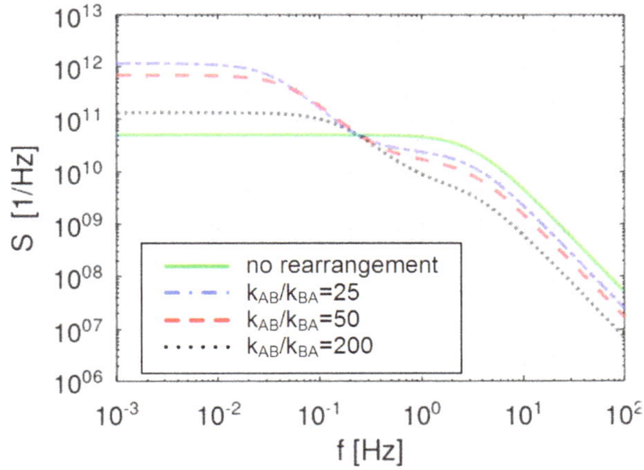

Figure 3. The power spectral density of fluctuations of the number of adsorbed biomolecules, representing the noise spectrum. The same cases are shown in Figure 2.

4. Conclusions

In this work, we developed the response and noise models of adsorption-based biosensors, taking into account the rearrangement processes of adsorbed biomolecules. We obtained the analytical expressions for their characteristic parameters. These models are used for the analysis of the influence of rearrangement processes on the sensor response kinetics and on the sensor's noise described by the spectral density of fluctuations of the number of adsorbed molecules.

Under the influence of the rearrangement process, the response kinetics becomes two-exponential (in contrast to the single-exponential kinetics of biomolecular adsorption, which is not followed by a rearrangement), with a fast starting rise and slow approach to the steady state. The sensor response slows down, and its steady-state value increases.

Due to the random change in the spatial configuration of adsorbed molecules, the noise spectrum changes from the single-knee to the two-knee shape. Three characteristic frequencies of the noise power spectral density and the low-frequency noise magnitude contain information about the rates of the reversible rearrangement process.

The presented sensor response model enables a more accurate interpretation of measurement results, which leads to more reliable biosensing. The noise model provides a better estimation of the ultimate sensor performance. Both the response and noise models are useful for the development of methods for characterization of biomolecular rearrangement processes. This would make it possible to compensate for the lack of data on the values of rearrangement rate constants, which is important for many applications in medicine, pharmacy, biosensing, and fundamental biochemical research.

Author Contributions: Conceptualization, I.J.; methodology, I.J. and M.F.; software, O.J., I.J. and K.R.; validation, M.F., O.J. and K.R.; formal analysis, I.J. and M.F.; investigation, I.J. and S.A.; writing—original draft preparation, I.J.; writing—review and editing, M.F.; visualization, I.J., O.J. and S.A.; supervision, I.J. All authors have read and agreed to the published version of the manuscript.

Funding: This research was funded by the Ministry of Science, Technological Development and Innovations of the Republic of Serbia, grant number 451-03-47/2023-01/200026.

Institutional Review Board Statement: Not applicable.

Informed Consent Statement: Not applicable.

Data Availability Statement: Data are contained within the article.

Conflicts of Interest: The authors declare no conflicts of interest. The funders had no role in the design of the study; in the collection, analyses, or interpretation of data; in the writing of the manuscript; or in the decision to publish the results.

References

1. Kulkarni, M.B.; Ayachit, N.H.; Aminabhavi, T.M. A Short Review on Miniaturized Biosensors for the Detection of Nucleic Acid Biomarkers. *Biosensors* **2023**, *13*, 412. [CrossRef] [PubMed]
2. Maduraiveeran, G.; Sasidharan, M.; Ganesan, V. Electrochemical sensor and biosensor platforms based on advanced nanomaterials for biological and biomedical applications. *Biosens. Bioelectron.* **2018**, *103*, 113–129. [CrossRef] [PubMed]
3. Kulkarni, M.B.; Ayachit, N.H.; Aminabhavi, M. Biosensors and microfluidic biosensors: From fabrication to application. *Biosensors* **2022**, *12*, 543. [CrossRef] [PubMed]
4. Djurić, Z.; Jokić, I.; Frantlović, M.; Jakšić, O. Fluctuations of the number of particles and mass adsorbed on the sensor surface surrounded by a mixture of an arbitrary number of gases. *Sens. Actuators B Chem.* **2007**, *127*, 625–631. [CrossRef]
5. Jokić, I.; Djurić, Z.; Frantlović, M.; Radulović, K.; Krstajić, P.; Jokić, Z. Fluctuations of the number of adsorbed molecules in biosensors due to stochastic adsorption-desorption processes coupled with mass transfer. *Sens. Actuators B Chem.* **2012**, *166–167*, 535–543. [CrossRef]
6. Frantlović, M.; Jokić, I.; Djurić, Z.; Radulović, K. Analysis of the Competitive Adsorption and Mass Transfer Influence on Equilibrium Mass Fluctuations in Affinity-Based Biosensors. *Sens. Actuators B Chem.* **2013**, *189*, 71–79. [CrossRef]
7. Jokić, I. Microfluidic Adsorption-Based Biosensors: Mathematical Models of Time Response and Noise, Considering Mass Transfer and Surface Heterogeneity. In *Biosensors—Current and Novel Strategies for Biosensing*; Villarreal-Gómez, L., Ed.; IntechOpen: London, UK, 2021; pp. 1–25.
8. Plikusiene, I.; Balevicius, Z.; Ramanaviciene, A.; Talbot, J.; Mickiene, G.; Balevicius, S.; Stirke, A.; Tereshchenko, A.; Tamosaitis, L.; Zvirblis, G.; et al. Evaluation of affinity sensor response kinetics towards dimeric ligands linked with spacers of different rigidity: Immobilized recombinant granulocyte colony-stimulating factor based synthetic receptor binding with genetically engineered dimeric analyte derivatives. *Biosens. Bioelectron.* **2020**, *156*, 112112.
9. Daniels, K.G.; Suo, Y.; Oas, T.G. Conformational kinetics reveals affinities of protein conformational states. *Biophys. Comput. Biol.* **2015**, *112*, 9352–9357. [CrossRef] [PubMed]
10. Van Vliet, K.M.; Fasset, J.R. Fluctuations due to electronic transitions and transport in solids. In *Fluctuation Phenomena in Solids*; Burgess, R.E., Ed.; Academic Press: New York, NY, USA, 1965; pp. 267–352.
11. Li, Z.; Beeram, S.R.; Bi, C.; Suresh, D.; Zheng, X.; Hage, D.S. High-Performance Affinity Chromatography: Applications in Drug-Protein Binding Studies and Personalized Medicine. In *Advances in Protein Chemistry and Structural Biology*; Doven, R., Ed.; Academic Press: Cambridge, MA, USA, 2016; Volume 102, pp. 1–39.
12. Bi, C.; Beeram, S.; Li, Z.; Zheng, X.; Hage, D.S. Kinetic analysis of drug-protein interactions by affinity chromatography. *Drug Discov. Today Technol.* **2015**, *17*, 16–21. [CrossRef] [PubMed]
13. Canziani, G.; Zhang, W.; Cines, D.; Rux, A.; Willis, S.; Cohen, G.; Eisenberg, R.; Chaiken, I. Exploring Biomolecular Recognition Using Optical Biosensors. *Methods* **1999**, *19*, 253–269. [CrossRef]
14. Kim, J. Mathematical modeling approaches to describe the dynamics of protein adsorption at solid interfaces. *Colloids Surf. B Biointerfaces* **2018**, *162*, 370–379. [CrossRef]

engineering proceedings

MDPI

Proceeding Paper

Development of a MEMS Multisensor Chip for Aerodynamic Pressure Measurements †

Žarko Lazić, Milče M. Smiljanić, Dragan Tanasković, Milena Rašljić-Rafajilović, Katarina Cvetanović, Evgenija Milinković, Marko V. Bošković, Stevan Andrić, Predrag Poljak and Miloš Frantlović *

Institute of Chemistry, Technology and Metallurgy (ICTM), National Institute of the Republic of Serbia, University of Belgrade, Njegoševa 12, 11000 Belgrade, Serbia; zlazic@nanosys.ihtm.bg.ac.rs (Ž.L.); smilce@nanosys.ihtm.bg.ac.rs (M.M.S.); dragant@nanosys.ihtm.bg.ac.rs (D.T.); milena@nanosys.ihtm.bg.ac.rs (M.R.-R.); katarina@nanosys.ihtm.bg.ac.rs (K.C.); evgenija@nanosys.ihtm.bg.ac.rs (E.M.); boskovic@nanosys.ihtm.bg.ac.rs (M.V.B.); stevan@nanosys.ihtm.bg.ac.rs (S.A.); predrag.poljak@nanosys.ihtm.bg.ac.rs (P.P.)

* Correspondence: frant@nanosys.ihtm.bg.ac.rs

† Presented at the 10th International Electronic Conference on Sensors and Applications (ECSA-10), 15–30 November 2023; Available online: https://ecsa-10.sciforum.net/.

Abstract: The existing instruments for aerodynamic pressure measurements are usually built around an array of discrete pressure sensors, placed in the same housing together with a few discrete temperature sensors. However, this approach is limiting, especially regarding miniaturization, sensor matching, and thermal coupling. In this work, we intend to overcome these limitations by proposing a novel MEMS multisensor chip, which has a monolithically integrated matrix of four piezoresistive MEMS pressure-sensing elements and two resistive temperature-sensing elements. After finishing the preliminary chip design, we performed computer simulations in order to assess its mechanical behavior when measured pressure is applied. Subsequently, the final chip design was completed, and the first batch was fabricated. The used technological processes included photolithography, thermal oxidation, diffusion, sputtering, micromachining (wet chemical etching), anodic bonding, and wafer dicing.

Keywords: MEMS multisensor; pressure sensing; chip fabrication

Citation: Lazić, Ž.; Smiljanić, M.M.; Tanasković, D.; Rašljić-Rafajilović, M.; Cvetanović, K.; Milinković, E.; Bošković, M.V.; Andrić, S.; Poljak, P.; Frantlović, M. Development of a MEMS Multisensor Chip for Aerodynamic Pressure Measurements. *Eng. Proc.* **2023**, *58*, 52. https://doi.org/10.3390/ecsa-10-16071

Academic Editor: Jean-marc Laheurte

Published: 15 November 2023

1. Introduction

Sensors that belong to micro-electromechanical systems (MEMS) constitute one of the most popular and promising classes of sensors. MEMS sensors are miniature sensing devices fabricated using microelectronic and micromachining technologies. First appearing shortly after the development of monolithic integrated circuits in the 1960s, they have been adopted for a wide range of applications during the decades since. Today, they are indispensable in vehicles, medical devices, mobile phones, and many other modern industrial and commercial applications. Piezoresistive MEMS pressure sensors constitute a large portion of the MEMS sensor market today, with the demand for them still growing. They have significant advantages over other types of pressure sensors, including miniature dimensions, good measurement performance, high reliability, and low cost due to efficient mass production. Although piezoresistive MEMS pressure sensors may nowadays be considered a mature technology, new applications continue to arise, and there are new scientific and technological challenges to be addressed [1,2].

The aerodynamic testing of various objects is often performed by measuring pressure at a multitude of points on aerodynamic surfaces or structural elements. Typically, the air pressure from the measurement points is transferred via flexible tubing to multichannel pressure-sensing instruments. Such instruments are usually built around an array of discrete pressure sensors and a few discrete temperature sensors placed in the same housing. That approach has some limitations regarding miniaturization, sensor matching, and thermal coupling. The goal of this work was to overcome these limitations by developing a novel MEMS multisensor chip with several monolithically integrated pressure- and temperature-sensing elements. Although reports on chips with multiple pressure-sensing elements exist in the literature [3], their concept and design are not suitable for this application.

After finishing the preliminary multisensor chip design, we performed computer simulations of the mechanical influences that the pressure applied to one sensing element can have on the whole chip structure. Subsequently, the final chip design was created, and the first batch of chips was manufactured and tested.

2. Materials and Methods

The concept of the proposed monolithically integrated multisensor structure is illustrated in Figure 1a. It is envisioned as a rectangular silicon chip with four piezoresistive MEMS pressure-sensing elements and two resistive temperature-sensing elements, as well as the necessary interconnects and wire bonding pads. The temperature-sensing elements are intended for direct temperature sensing on the chip. The thermal coupling between the temperature- and pressure-sensing elements on the same chip is much better than is possible in the case of pressure and temperature sensors on separate substrates. This is important for the temperature compensation of the pressure-sensing elements. Another important advantage of the concept is that it enables better matching of characteristics of the pressure-sensing elements on the same chip, because they are fabricated simultaneously during the same technological processes.

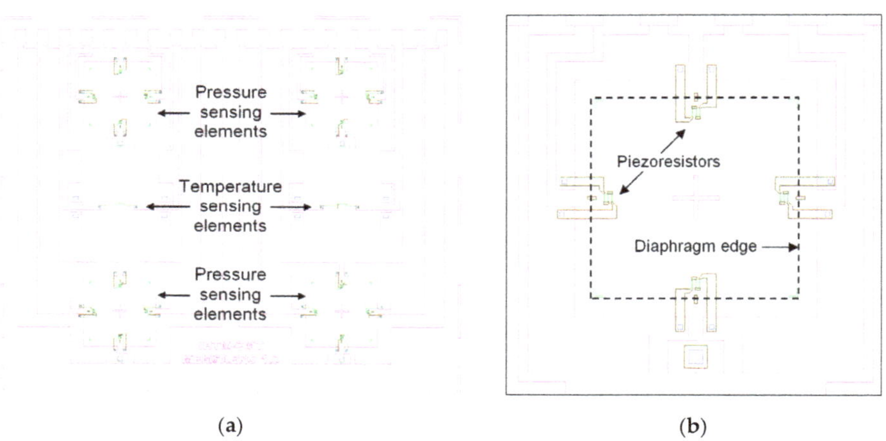

Figure 1. Topographic image of photolithographic masks used for the multisensor chip fabrication: (**a**) whole chip; (**b**) details of a pressure-sensing element.

Silicon piezoresistive MEMS pressure sensor chips are typically fabricated on single-crystal silicon wafers with the crystallographic orientation (100). In general, a pressure-sensing element consists of a precisely defined mechanical structure (i.e., a diaphragm), and piezoresistors formed near the diaphragm's edges, in the crystallographic direction <110>. The piezoresistors are placed at the locations where the highest mechanical stress occurs during the diaphragm deflection caused by the measured pressure. Usually, four such piezoresistors are used, of which two are in the radial direction and the remaining two in the tangential direction relative to the diaphragm which is typically quadratic in shape. The four piezoresistors form a Wheatstone bridge, so that a differential voltage signal proportional to the measured pressure value is generated when sensor excitation is applied. The structure of a pressure-sensing element is shown in more detail in Figure 1b.

2.1. Computer Simulations

Computer simulations based on the finite element method (FEM) were performed with the main goal of investigating the parasitic influence that the mechanical stress caused by the pressure applied to one of the pressure-sensing elements may have on the other sensing elements on the same multisensor chip. In this case, the displacement profile and von Mises stress of the chip's surface were observed, and the applied pressure was 100 kPa. The results are shown visually in Figure 2.

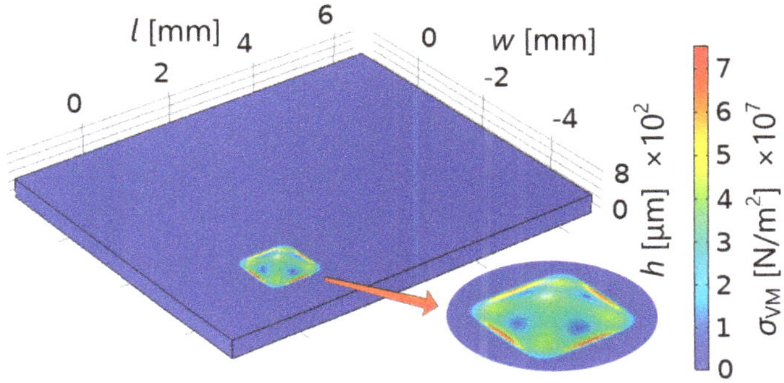

Figure 2. Visual representation of the computer simulation results: the displacement of the multisensor chip's top surface caused by the pressure of 100 kPa applied to one pressure-sensing element (in the bottom corner); color grading represents von Mises stress, according to the color scale on the right.

The simulations confirmed that the above-mentioned influence is negligible. Hence, the preliminary design of the multisensor chip was accepted. Based on it, the final design was created and subsequently used for chip fabrication.

2.2. Multisensor Chip Fabrication

The MEMS multisensor chip is developed based on ICTM's extensive experience in the field [4–6], including the ICTM SP-12 pressure sensor chip intended for the measurement of medium pressures. The multisensor chip is produced using 1 μm technology. The photolithographic masks were created using the following equipment: EVG 620 double sided mask aligner (EV Group Europe & Asia/Pacific GmbH, St. Florian am Inn, Austria) and Microtech LaserWriter LW405 (MICROTECH srl, Palermo, Italy). The fabrication started with silicon wafers of the following characteristics: n-type (resistivity from 3 Ωcm to 5 Ωcm), polished on both sides, 76.2 mm (3″) in diameter, (100) orientation. The piezoresistors were formed by the diffusion of p-type impurities (boron), resulting in electrical resistance in the range from 2 kΩ to 4 kΩ. Metallic interconnects and bonding pads were formed by the deposition of aluminum (high vacuum sputtering at $\approx 2 \cdot 10^{-4}$ Pa). The diaphragms of the pressure-sensing elements were produced by anisotropic wet chemical etching in 30% KOH water solution.

3. Results and Discussion

Photographs of the produced multisensor chip are shown in Figure 3. For the purposes of testing and characterization, the chip was anodically bonded to a flat Pyrex glass plate using the AML AWB-04 aligner wafer bonder (Applied Microengineering Ltd., Didcot, Oxfordshire, UK). As the bonding is performed in a vacuum, the pressure-sensing elements' diaphragms are deflected by the atmospheric pressure, which is noticeable in Figure 3a for the lower two pressure-sensing elements. A photograph of the bottom side of the chip, showing the etched MEMS diaphragms, is shown in Figure 3b.

(**a**)

(**b**)

Figure 3. Photographs of the realized MEMS multisensor chip: (**a**) top side; (**b**) bottom side (four etched MEMS diaphragms).

Three silicon wafers were processed, each with 30 chips. Electrical tests were carried out using a probing station and a semiconductor parameter analyzer.

A preliminary characterization of the chip was performed by connecting the pressure-sensing elements to a constant voltage excitation of 2.5 V, and measuring their output voltages at several pressure and temperature values. The applied relative pressure was in the range from −700 hPa to 700 hPa. It was set using the Mensor APC 600 automated pressure calibrator (Mensor LP, San Marcos, Texas, USA). The temperature from 2 °C to 50 °C was set using the Heraeus Vötsch VMT 04/140 temperature chamber (Weiss Technik GmbH, Reiskirchen-Lindenstruth, Germany). The results are presented in a graph in Figure 4. They indicate that the chip performs as expected.

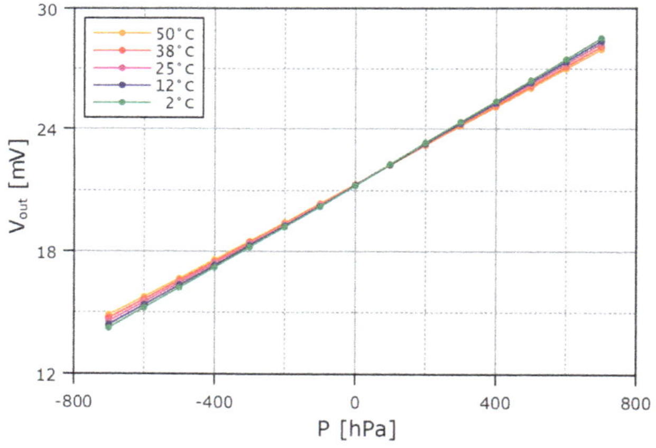

Figure 4. Results of the preliminary characterization of the obtained multisensor chip: the dependence of the output voltage of a tested pressure-sensing element on the applied pressure, with temperature as a parameter.

4. Conclusions

In this paper, we presented the development of the MEMS multisensor chip intended for aerodynamic pressure measurements. We briefly explained the concept and the methodology, which included the chip design, computer simulations and, finally, the fabrication processes that resulted in the finished chip. We also performed the preliminary characterization of the chip.

The development of the multisensor chip is an important step toward the higher integration, higher performance, and further miniaturization of aerodynamic pressure measurement devices. Our future work in this field will include a detailed characterization of the developed multisensor chip, and the development of signal-processing methods in order to optimize the pressure measurement performance. Also, the possibility of increasing the number of sensing elements on the chip will be considered.

Author Contributions: Conceptualization, M.F. and Ž.L.; methodology, Ž.L., M.M.S., D.T., M.R.-R., K.C., E.M., M.V.B. and S.A.; validation, Ž.L., D.T. and P.P.; formal analysis, Ž.L. and M.F.; investigation, Ž.L., M.M.S. and D.T.; resources, M.F.; writing—original draft preparation, M.F., Ž.L. and D.T.; writing—review and editing, P.P.; visualization, Ž.L. and D.T.; supervision, M.F.; project administration, M.F.; funding acquisition, M.F. All authors have read and agreed to the published version of the manuscript.

Funding: This research was supported by the Science Fund of the Republic of Serbia, Grant Number 7754287, MEMS Multisensor Instrument for Aerodynamic Pressure Measurements—MEMSAERO. It was also supported by the Ministry of Science, Technological Development and Innovation of the Republic of Serbia, Grant Number 451-03-47/2023-01/200026.

Data Availability Statement: Data are contained within the article.

Conflicts of Interest: The authors declare no conflicts of interest. The funders had no role in the writing of the manuscript or in the decision to publish the results.

References

1. Kumar, S.S.; Pant, B.D. Design principles and considerations for the 'ideal' silicon piezoresistive pressure sensor: A focused review. *Microsyst. Technol.* **2014**, *20*, 1213–1247. [CrossRef]
2. Fiorillo, A.S.; Critello, C.D.; Pullano, S.A. Theory, technology and applications of piezoresistive sensors: A review. *Sens. Actuators A* **2018**, *281*, 156–175. [CrossRef]
3. Zhang, J.; Chen, J.; Li, M.; Ge, Y.; Wang, T.; Shan, P.; Mao, X. Design, Fabrication, and Implementation of an Array-Type MEMS Piezoresistive Intelligent Pressure Sensor System. *Micromachines* **2018**, *9*, 104. [CrossRef] [PubMed]
4. Djurić, Z.; Matić, M.; Matović, J.; Petrović, R.; Simičić, N. Experimental determination of silicon pressure sensor diaphragm deflection. *Sens. Actuators* **1990**, *24*, 175–179. [CrossRef]
5. Simičić, N.; Tanasković, D.; Djurić, Z.; Lazić, Ž.; Petrović, R.; Matović, J.; Popović, M.; Matić, M.; Totovski, D. Investigation of impurity distribution for pressure sensor piezoresistors. *J. Serb. Chem. Soc.* **1993**, *58*, 951–957.
6. Smiljanić, M.M.; Jović, V.; Lazić, Ž. Maskless convex corner compensation technique on a (1 0 0) silicon substrate in a 25 wt% TMAH water solution. *J. Micromech. Microeng.* **2012**, *22*, 115011. [CrossRef]

engineering proceedings

Proceeding Paper

A Comparative Design Analysis of Internal and External Frame Structures for MEMS Vibrating Ring Gyroscopes †

Waqas Amin Gill *, Ian Howard, Ilyas Mazhar and Kristoffer McKee *

Department of Mechanical Engineering, Curtin University, Perth, WA 6845, Australia; i.howard@curtin.edu.au (I.H.); i.mazhar@curtin.edu.au (I.M.)
* Correspondence: waqasamin.gill@postgrad.curtin.edu.au (W.A.G.); k.mckee@curtin.edu.au (K.M.)
† Presented at the 10th International Electronic Conference on Sensors and Applications (ECSA-10), 15–30 November 2023; Available online: https://ecsa-10.sciforum.net/.

Abstract: This research presents a comparative analysis of the two important design methodologies involved in developing microelectromechanical system (MEMS) vibrating ring gyroscopes, namely, internal and external ring gyroscopes. Internal ring gyroscopes are constructed with the outside placement of support pillars connected with the semicircular beams that are attached to the vibrating ring structure. The design importance of this particular setting effectively isolates the vibrating ring structure from any external mechanical vibrations, significantly improving the gyroscope's performance. The internal ring structure provides exceptional precession and reliability, making this design an ideal candidate for harsh conditions, as they can sustain substantial amounts of unwanted and external vibrations without degrading the performance of the gyroscope. On the other hand, external ring gyroscopes include the placement of the support pillars within the vibrating ring structure. This particular design setting is quite convenient in terms of fabrication and provides higher gyroscopic sensitivity. However, this design may lead to coupling of the vibrational modes and potentially compromise the performance of the gyroscope. This research discusses and compares the findings of a modal analysis of the two distinguished design approaches for the MEMS vibrating ring gyroscopes.

Keywords: MEMS; MEMS design; vibrating ring gyroscope; internal ring; external ring; ring resonator

Citation: Gill, W.A.; Howard, I.; Mazhar, I.; McKee, K. A Comparative Design Analysis of Internal and External Frame Structures for MEMS Vibrating Ring Gyroscopes. *Eng. Proc.* **2023**, *58*, 53. https://doi.org/10.3390/ecsa-10-16182

Academic Editor: Stefano Mariani

Published: 15 November 2023

1. Introduction

Microelectromechanical system (MEMS) inertial sensors have emerged as essential components of inertial measurement units (IMUs) and have experienced substantial proliferation in the field of space applications. The utilisation of MEMS vibrating ring gyroscopes [1–6] has been the subject of substantial discourse in the realm of research and development, particularly in the context of space applications [7–9]. There are various other uses for MEMS vibrating ring gyroscopes in contemporary electrical systems. In the realm of smartphone technology, the utilisation of inertial sensors that rely on MEMS vibrating gyroscopes has become increasingly prevalent. The utilisation of this technology enhances the detection of angular movement in digital cameras, while also finding applications in other industries such as biomedical, missile technology, and rollover detections in automotive applications.

MEMS vibrating ring gyroscopes have favourable characteristics for their utilisation in IMUs within space applications due to their symmetrical configuration, exceptional precision, and resilience in challenging environmental conditions [10,11]. The gyroscope's ring-shaped form, using semicircular springs, exhibits a symmetrical construction that enables it to withstand challenging environmental conditions and high-vibration industries [12,13]. This work aims to analyse a comparative design study on two different designs for MEMS vibrating ring gyroscope. This paper presents the comparative design methodology and early design modelling of the vibrating ring gyroscope. The present study focuses on the examination of modal analysis pertaining to the proposed vibrating ring gyroscope design.

2. Design Methodology

In order to improve performance and stability and decouple vibrational modes, the frame structures of MEMS vibrating ring gyroscopes are important design considerations. There are, in particular, two types of frame structures, the inner ring gyroscope design and the outer ring gyroscope design. Both of the mentioned design frames will be discussed in detail below.

2.1. Internal Ring Gyroscope

In this type of design frame, the anchor support is designed outside of the vibrating ring structure. The vibrating ring structure is attached with the semicircular beams and the whole vibrating structure is attached to the pillars placed outside. This design configuration of an interior ring has the ability to improve the performance of the gyroscope by protecting the vibrating structure from any disruptive external vibrations. Because of their symmetric structure, inertial sensors with internal vibrating ring gyroscopes are indispensable in extreme conditions. They can withstand extreme conditions and vibrations, making them ideal inertial sensors for use in spacecraft applications. A schematic depiction of the internal ring gyroscope structure is shown in Figure 1.

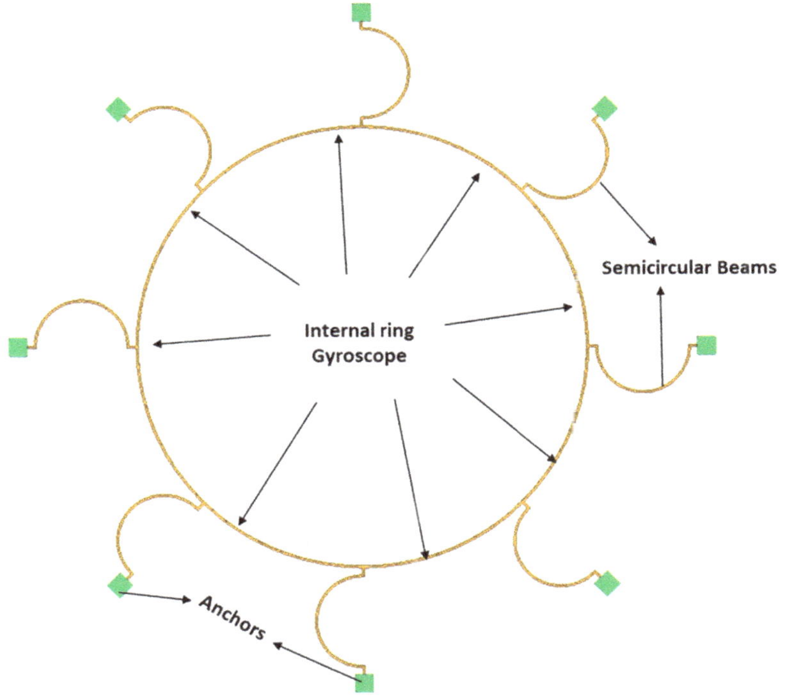

Figure 1. A schematic representation of the internal ring gyroscope.

2.2. External Ring Gyroscope

When designing and developing MEMS vibrating ring gyroscopes, it is usual practice to incorporate an external ring design structure. The external ring gyroscope is designed with the support anchor and semicircular beam construction integrated into its interior. This design strategy has the advantage of being able to handle bigger vibrating ring gyroscope structures, which significantly improves the gyroscope's overall sensitivity. However, unlike the internal ring gyroscope design, this method of design has various difficulties, most notably in terms of its resilience under extreme conditions and its heightened sensitivity to unwanted external vibrations. The external ring gyroscope is shown in Figure 2.

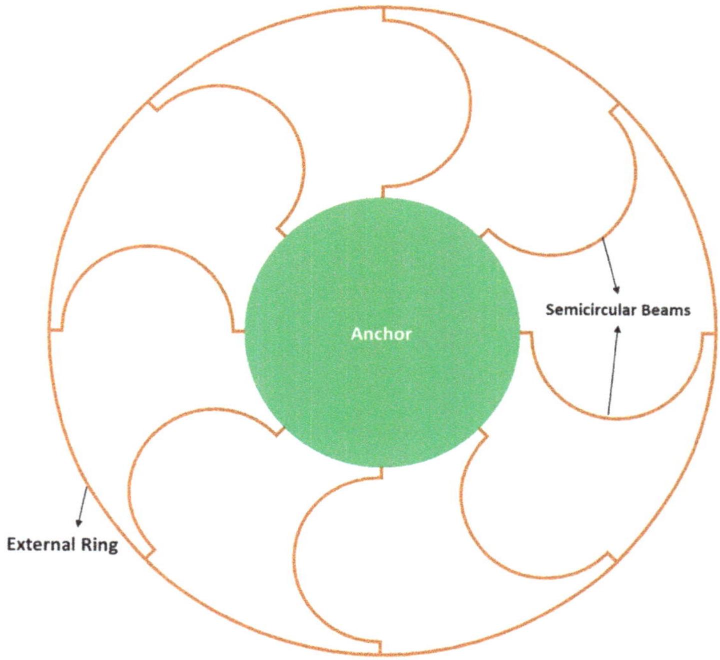

Figure 2. A schematic representation of the external ring gyroscope.

Internal and external ring gyroscopes can be used in a wide variety of vibrating ring gyroscope configurations. A few examples of these design configurations are gyroscopes with one or more rings, resonator discs, cylinders, shells, and stars. The external ring gyroscope can fit any of these gyroscope designs due to its flexible construction. Single ring gyroscopes, on the other hand, provide many benefits due to their internal design structure. Internal space in the internal ring gyroscope could be utilised for the installation of extra tuning electrodes or the planning and implementation of a microscale inertial sensor.

3. Simulation Analyses

The utilisation of finite element method (FEM) modal analysis is of utmost importance in computational analyses, particularly in the determination of vibration characteristics pertaining to mechanical structures such as the MEMS vibrating ring gyroscope. This methodology facilitates a comprehensive analysis of the dynamics and modal characteristics shown by the vibrating ring gyroscope. The modal analysis offers a comprehensive understanding of the vibrational modes and frequencies of a vibrating ring gyroscope structure. This is achieved by discretising the structure into finite elements and solving the resulting equations. The modal analysis using the FEM based software Ansys^{TM} has been conducted for both internal and external vibrating ring gyroscopes.

3.1. Modal Analysis of Internal Ring Gyroscope

The modal analysis of the internal ring gyroscope is conducted based on its distinctive structural alteration. The design comprises a total of eight semicircular beams, all of which are connected to the internal ring resonator. The entirety of the vibrating ring structure is upheld by externally positioned support pillars around the vibrating structure. The presented design technique effectively implements measures to isolate the vibrating ring structure from potential external disturbances that may have adverse effects on the overall performance of the gyroscope. The design features comprise several key parameters. These include a ring radius of 1000 μm, a semicircular beam radius of 200 μm, a ring and beam thickness of 10 μm, and a structural height of 100 μm. The frequencies depicted in Figure 3 for the internal ring design are 48.78 kHz and 48.80 kHz for mode 1 and mode 2, respectively. The mode mismatch recorded only 17 Hz, which shows a good prospect of high sensitivity. The modal results are shown in Figure 3.

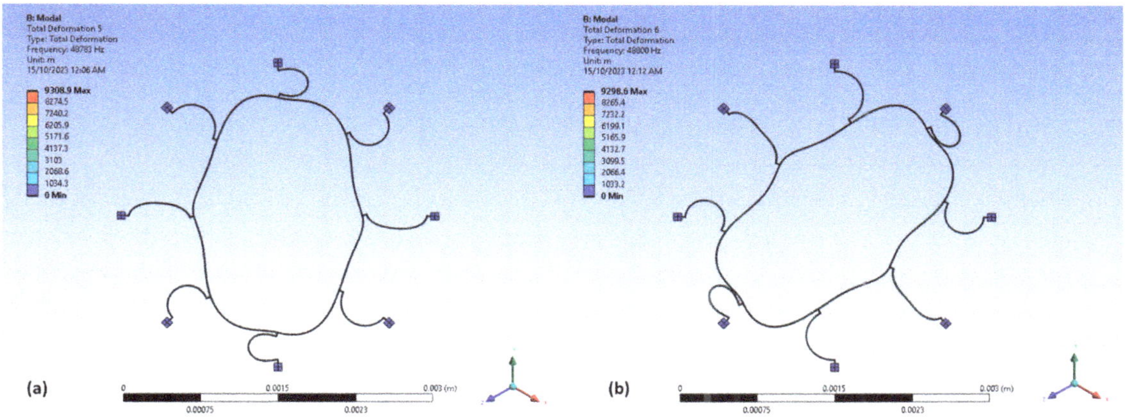

Figure 3. The modal analysis for the internal ring gyroscope: (**a**) driving elliptical shape; (**b**) sensing elliptical shape.

3.2. Modal Analysis of External Ring Gyroscope

In contrast, the external ring gyroscope exemplifies a conventional methodology for the vibrating ring gyroscope design. The external ring gyroscope exhibits the presence of eight semicircular beams that are connected to the external ring resonator, and the entire vibrating structure is interconnected to the centrally positioned circular pillar. The design features include a ring radius measuring 1000 μm, a semicircular beam radius of 200 μm, thicknesses of 10 μm for both the ring and the beams, and the structure has a height of 100 μm. The modal resonance frequency results are illustrated in Figure 4, with mode 1 exhibiting a resonance frequency of 40.59 kHz and mode 2 displaying a resonance frequency of 40.63 kHz. The mode mismatch recorded 34 Hz, which is slightly higher than the internal ring gyroscope design.

Both designs possess identical design characteristics, with the exception being the design geometry. The radii of the vibrating ring and the semicircular beams are same in both gyroscopes. However, it is observed that the internal ring gyroscope exhibited elliptical vibrations at a frequency of around 49 kHz, whereas the external ring gyroscope displayed elliptical vibrations at a frequency of 40 kHz. The mode mismatches for the internal and external ring gyroscopes are observed at frequencies of 17 Hz and 34 Hz, as seen in Table 1. The results indicate that the internal ring gyroscope exhibits higher sensitivity and low mode mismatch between elliptical shapes of vibrations, rendering it a preferable candidate over the external ring design for vibrating ring gyroscopes.

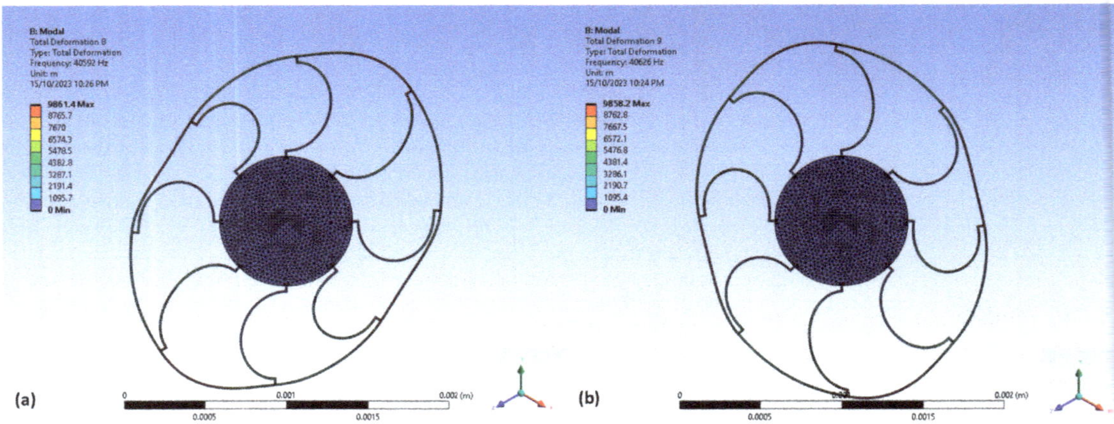

Figure 4. The modal analysis for the external ring gyroscope: (**a**) driving elliptical shape; (**b**) sensing elliptical shape.

Table 1. Modal analysis results for internal and external ring gyroscope.

Vibrational Modes	Internal Ring Design	External Ring Design	Ring Radius (μm)	Ring Width (μm)	Semicircular Radius (μm)	Device Height (μm)
Driving (kHz)	48.78	40.59 kHz	1000	10	200	100
Sensing (kHz)	48.80	40.63 kHz	1000	10	200	100
Mode Mismatch (Hz)	17	34				

4. Conclusions

We have successfully demonstrated two design types of MEMS vibrating ring gyroscopes, the internal ring and the external ring gyroscope. Comparatively, the internal ring vibrating gyroscope with its outer pillar support shows a good stability against harsh environments. This makes it notably suitable for spacecraft applications. On the other hand, the external ring gyroscope with its centrally positioned pillar offers higher flexibility in design structure but with reduced robustness against high-vibration environments. The FEM modal analysis shows differences in the vibrational mode frequencies of both designs. Despite having identical design features, the internal ring gyroscope demonstrated higher sensitivity with higher resonance frequency responses. The mode mismatch was observed to be just 17 Hz, in contrast to the 34 Hz for the external ring gyroscope. Therefore, the internal ring gyroscope emerges as the more favourable candidate for our further research on gyroscopes.

Author Contributions: The research idea was proposed by W.A.G. and subsequently examined in collaboration with K.M. and W.A.G. conducted all aspects of the research project, including data collection, analysis, and the creation of graphic representations. The manuscript was written by W.A.G. The final manuscript was edited and reviewed by I.H., I.M. and K.M. All authors have read and agreed to the published version of the manuscript.

Funding: This research received no external funding.

Institutional Review Board Statement: Not applicable.

Informed Consent Statement: Not applicable.

Data Availability Statement: Data are contained within the article.

Conflicts of Interest: The authors declare no conflicts of interest.

References

1. Yoon, S.W.; Lee, S.; Najafi, K. Vibration sensitivity analysis of MEMS vibratory ring gyroscopes. *Sens. Actuators A Phys.* **2011**, *171*, 163–177. [CrossRef]
2. Cao, H.; Liu, Y.; Kou, Z.; Zhang, Y.; Shao, X.; Gao, J.; Huang, K.; Shi, Y.; Tang, J.; Shen, C.; et al. Design, fabrication and experiment of double U-beam MEMS vibration ring gyroscope. *Micromachines* **2019**, *10*, 186. [CrossRef] [PubMed]
3. Ayazi, F.; Najafi, K. A HARPSS polysilicon vibrating ring gyroscope. *J. Microelectromech Syst.* **2001**, *10*, 169–179. [CrossRef]
4. Gill, W.A.; Howard, I.; Mazhar, I.; McKee, K. A Review of MEMS Vibrating Gyroscopes and Their Reliability Issues in Harsh Environments. *Sensors* **2022**, *22*, 7405. [CrossRef] [PubMed]
5. Jia, J.; Ding, X.; Qin, Z.; Ruan, Z.; Li, W.; Liu, X.; Li, H. Overview and analysis of MEMS Coriolis vibratory ring gyroscope. *Measurement* **2021**, *182*, 109704. [CrossRef]
6. Hyun An, B.; Gill, W.A.; Lee, J.S.; Han, S.; Chang, H.K.; Chatterjee, A.N.; Choi, D.S. Micro-electromechanical vibrating ring gyroscope with structural mode-matching in (100) silicon. *Sens. Lett.* **2018**, *16*, 548–551. [CrossRef]
7. Kou, Z.; Liu, J.; Cao, H.; Han, Z.; Sun, Y.; Shi, Y.; Ren, S.; Zhang, Y. Investigation, modeling, and experiment of an MEMS S-springs vibrating ring gyroscope. *J. Micro/Nanolithography MEMS MOEMS* **2018**, *17*, 015001. [CrossRef]
8. Gill, W.A.; Ali, D.; An, B.H.; Syed, W.U.; Saeed, N.; Al-shaibah, M.; Elfadel, I.M.; Al Dahmani, S.; Choi, D.S. MEMS multi-vibrating ring gyroscope for space applications. *Microsyst. Technol.* **2020**, *26*, 2527–2533. [CrossRef]
9. Dussy, S.; Durrant, D.; Moy, A.; Perriault, N.; Celerier, B. (Eds.) MEMS gyros for space applications-overview of european activities. In Proceedings of the AIAA Guidance, Navigation, and Control Conference and Exhibit, San Francisco, CA, USA, 15–18 August 2015.
10. Chikovani, V.V.; Tsiruk, H.V. Digital rate MEMS vibratory gyroscope modeling, tuning and simulation results. *Comput. Sist.* **2017**, *21*, 147–159. [CrossRef]
11. Khan, I.; Ting, D.S.; Ahamed, M.J. Design and development of a MEMS vibrating ring resonator with inner rose petal spring supports. *Microsyst. Technol.* **2021**, *27*, 985–995. [CrossRef]
12. Kou, Z.; Liu, J.; Cao, H.; Shi, Y.; Ren, J.; Zhang, Y. A novel MEMS S-springs vibrating ring gyroscope with atmosphere package. *AIP Adv.* **2017**, *7*, 125301. [CrossRef]
13. Gill, W.A.; Howard, I.; Mazhar, I.; McKee, K. Design and Considerations: Microelectromechanical System (MEMS) Vibrating Ring Resonator Gyroscopes. *Designs* **2023**, *7*, 106.

engineering proceedings

MDPI

Proceeding Paper

A Parsimonious Yet Robust Regression Model for Predicting Limited Structural Responses of Remote Sensing †

Alireza Entezami *, Bahareh Behkamal, Carlo De Michele and Stefano Mariani

Department of Civil and Environmental Engineering, Politecnico di Milano, 20133 Milano, Italy; bahareh.behkamal@polimi.it (B.B.); carlo.demichele@polimi.it (C.D.M.); stefano.mariani@polimi.it (S.M.)
* Correspondence: alireza.entezami@polimi.it
† Presented at the 10th International Electronic Conference on Sensors and Applications (ECSA-10), 15–30 November 2023; Available online: https://ecsa-10.sciforum.net/.

Abstract: Small data analytics, at the opposite extreme of big data analytics, represent a critical limitation in structural health monitoring based on spaceborne remote sensing technology. Besides the engineering challenge, small data is typically a demanding issue in machine learning applications related to the prediction of system evolutions. To address this challenge, this article proposes a parsimonious yet robust predictive model obtained as a combination of a regression artificial neural network and of a Bayesian hyperparameter optimization. The final aim of the offered strategy consists of the prediction of structural responses extracted from synthetic aperture radar images in remote sensing. Results regarding a long-span steel arch bridge confirm that, although simple, the proposed method can effectively predict the structural response in terms of displacement data with a noteworthy overall performance.

Keywords: bridge health monitoring; machine learning; artificial neural network; Bayesian hyperparameter optimization

Citation: Entezami, A.; Behkamal, B.; De Michele, C.; Mariani, S. A Parsimonious Yet Robust Regression Model for Predicting Limited Structural Responses of Remote Sensing. *Eng. Proc.* **2023**, *58*, 54. https://doi.org/10.3390/ecsa-10-16028

Academic Editor: Francisco Falcone

Published: 15 November 2023

1. Introduction

Structural health monitoring (SHM) of bridge structures [1–4] must fully account for various environmental actions such as ambient temperature, wind, moisture, and possible chemical attacks [5]. In most cases, bridges, especially long-span ones, are slender and are therefore susceptible to vibrations [6–8]. The SHM of such structures is indeed of paramount importance for our interconnected communities [9].

Health monitoring has to exploit sensor equipment and data measurements through data analysis and decision-making strategies. The choice of an appropriate sensing technology and of the measurement of the structural response to different natural or man-made excitation sources is critical to provide data sensitive to the structural state. The process of data analytics is often conducted through data cleaning, compression, fusion [10], data augmentation [11], data prediction [12], data normalization [13], and feature extraction [14]. Different machine learning algorithms within the realms of unsupervised learning [15–18] and supervised learning [19] can be adopted for decision-making about whether the bridge has suffered damage or can still operate normally.

Recently, the technology of remote sensing has opened a golden window to the SHM of bridge structures [20–23]. With this technology, it is possible to access synthetic aperture radar (SAR) images and extract structural displacements at different spots of the structures without any sensor installation or labor-intensive field measurements. Despite such important benefits, there are some limitations related to this technology. First, the products of spaceborne remote sensing can be claimed to be Big Data, requiring ad hoc analysis. In most cases, speckle noise in a SAR image is a major challenge for displacement extraction. Second, unlike the contact-based sensing systems, it is not trivial to collect structural responses hourly. Hence, small data is the other important challenge related to the SAR-based SHM. Third, it may be unnecessary to use feature extraction tools like interferometric techniques to obtain information in terms of local displacements with remote sensing, particularly in view of recent progress of machine learning algorithms. All in all, the most appropriate solution to such challenges can be to leverage machine learning-aided prediction capabilities.

A small dataset is one of the challenges of SHM via remote sensing, as a machine learner with insufficient data cannot operate properly. For the problem of data prediction, the same issue can affect the applicability. Since most of the predictive models are developed from regression techniques, the use of small datasets increases the probability of underfitting or overfitting. The best solution in this case is to take advantage of parsimonious yet robust predictive models, featuring simple configurations but providing reliable and robust predictions without any concern related to the underfitting and overfitting problems.

From the aforementioned discussion, it stems that the main goal of this research is to propose a parsimonious and robust regression model to predict partial structural displacements retrieved from a few SAR images. The proposed method is a coupling of a regression artificial neural network (RANN) featuring a fully connected architecture and Bayesian hyperparameter optimization (BHO). The RANN undertakes the prediction of the structural response to temperature variability, while BHO tunes the hyperparameters of the RANN. To assess the effectiveness of the proposed model, partial displacement responses of a long-span bridge are adopted. Results show that the proposed RANN–BHO method is quite successful in predicting the bridge response, even in the presence of small training datasets.

2. Supervised Artificial Neural Network for Regression

2.1. Network Configuration

A RANN is an ANN specifically tailored for regression problems. It is a feedforward, fully connected neural network showing the standard input layer, a number of hidden fully connected layers, and the output layer. The network input is defined as the predictor data. Each fully connected layer handles the input data by means of a weight matrix and a bias vector; an activation function (e.g., the rectified linear unit, hyperbolic tangent, sigmoid function, and linear function) can provide nonlinear transformations of the information/data, see [24–26]. A backpropagation algorithm is adopted to tune the weights of the RANN, managing a loss function (as a prediction error between the input and the output) to be minimized with the stochastic gradient descent algorithm. Finally, the predicted response is given as the network output. Figure 1 shows a graphical representation of the RANN, wherein N denotes the number of fully connected (hidden) layers.

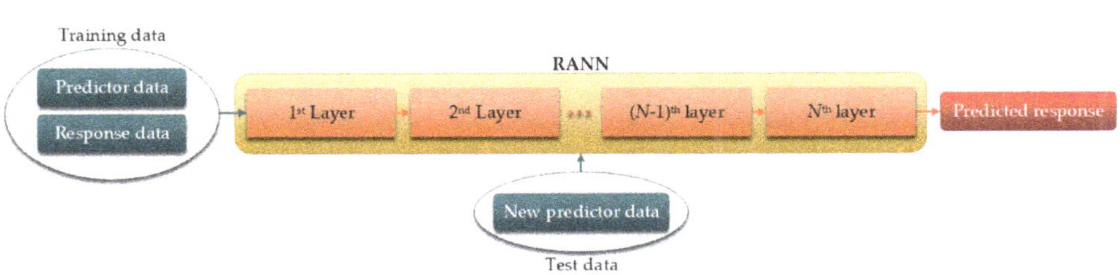

Figure 1. Graphical sketch of the RANN.

The strength of the RANN is the ability to learn both linear and nonlinear relationships between the predictor and the response data due to employing different activation functions, see Hagan et al. [27]. The main hyperparameters of the RANN are the number N of hidden layers, and the number of neurons of each layer. These hyperparameters must be determined before the learning stage.

2.2. Model Tuning via Bayesian Hyperparameter Optimization

Hyperparameter optimization in machine learning deals with the choice of the best values of key parameters of a given machine learning model so that the highest overall performance on a validation set can be attained [28]. Although there are different approaches to hyperparameters tuning, Bayesian hyperparameter optimization (BHO) is the most useful in a case when reaching a reliable overall performance is challenging, like in the presence of small datasets to learn from. Bayesian optimization is an approach that uses Bayes theorem to find the minimum of an objective function. The BHO keeps track of past evaluation outputs and uses them to develop a probabilistic model through the mapping of hyperparameters to a probability of a score on the objective function. For the problem of data prediction, the BHO is designed to minimize the following objective function:

$$F = \log(1 + E_{MSE}) \tag{1}$$

where E_{MSE} denotes the cross-validation mean-squared-error (MSE) between observation and estimation; this is achieved iteratively. At each iteration, the objective function F in Equation (1) yields a logarithmic transformed validation loss-value computed for the regression model, along with the relevant optimal set of hyperparameters. As mentioned, the BHO not only handle this function, but also incorporates a probability distribution model to be updated at each iteration. BHO thus defines an acquisition function and the next set of hyperparameters. Hence, it can be considered to deliver a posterior probability distribution model for each hyperparameter. The best hyperparameter values can be selected after reaching a good match between real and predicted data.

In relation to the proposed predictive model, it has been already reported that BHO makes attempts to tune two key hyperparameters of the RANN: the number N of the fully connected layers and the number of neurons in each layer. Apart from hyperparameters, a machine learning model may rest upon other unknown elements, which are termed model parameters. The main difference between hyperparameters and model parameters is that the former should be determined before the learning stage, while the latter can be adjusted during the same [29]. For instance, the weight and bias of each neuron of the RANN represents its model parameters.

3. Method Performance Evaluation for a Steel Arch Bridge

The Lupu Bridge is a steel arch bridge crossing the Huangpu River in Shanghai as shown in Figure 2. It has a total length of 750 m, comprising a main span of 550 m and two side spans of 100 m. Figure 2b provides the elevation view of the Lupu bridge as well as its main sizes. The girder in the side span is a closed steel box, with a width of 41 m and a height of 2.7 m [30].

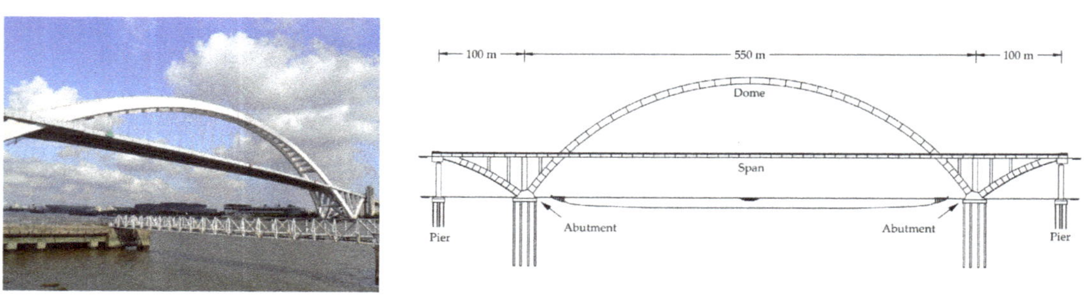

(a) (b)

Figure 2. (**a**) A picture of the Lupu Bridge, and (**b**) elevation view of the bridge with main dimensions.

A long-term SHM program of the bridge was carried out by Qin et al. [30] with the aid of spaceborne remote sensing to inspect the variability in the displacement response of the bridge dome and main span. Fifty-five SAR images from TerraSar-X were collected to extract the displacements of the mentioned bridge components. During the monitoring period, the air temperature was also recorded to incorporate seasonal and thermal effects in the SHM program. Figure 3 illustrates the collected displacement and temperature data. From the regression viewpoint, the displacement and temperature samples are the main dependent (response or output) and independent (predictor or input) data, respectively.

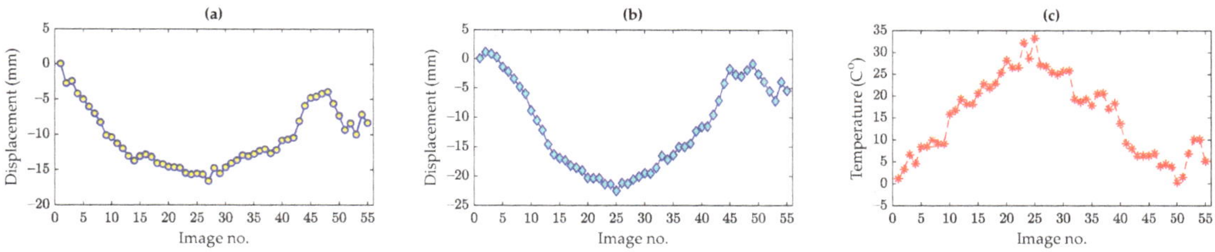

Figure 3. Dependent and independent data for prediction: (**a**) displacement of the dome, (**b**) displacement of the main span, and (**c**) temperature.

By dividing the dataset with a ratio 80:20, respectively referring to the training and test subsets, the total numbers of training and test samples turn out to be 44 and 11. For the learning process, the training samples are also subdivided on their own into the training and validation sets, leading to 35 and 9 samples to handle respectively. Using these datasets, the BHO is adopted to tune the number of layers and the corresponding neuron sizes. The outcome of the optimization procedure is collected in Table 1. Figure 4 shows the good convergence rate of the trained RANNs adopted for the dome and main span of the bridge, as obtained with the minimization of the objective function *F* after 30 iterations. Furthermore, Figure 5 depicts the results of displacement predictions obtained with these trained models at the same dome and main span locations. As it can be observed, target and predicted data points match well with each other. To also provide a quantitative evaluation of the agreement, in the figure it is reported that the proposed predictive model yields R-squared values are equal to 0.8509 and 0.9223 at the two locations.

Table 1. Tuned hyperparameters of the RANNs via BHO.

Element	Number of Layers	Neuron Sizes		
		1st Layer	2nd Layer	3rd Layer
Dome	2	3	2	–
Span	3	2	2	3

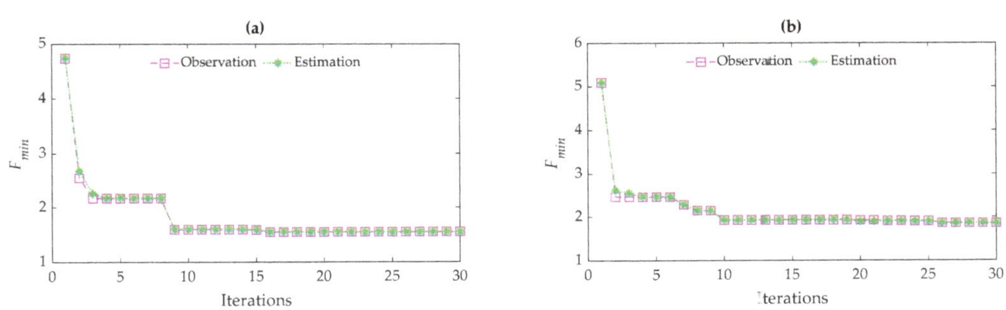

Figure 4. Convergence rate of the RANNs via BHO: (**a**) dome, and (**b**) main span.

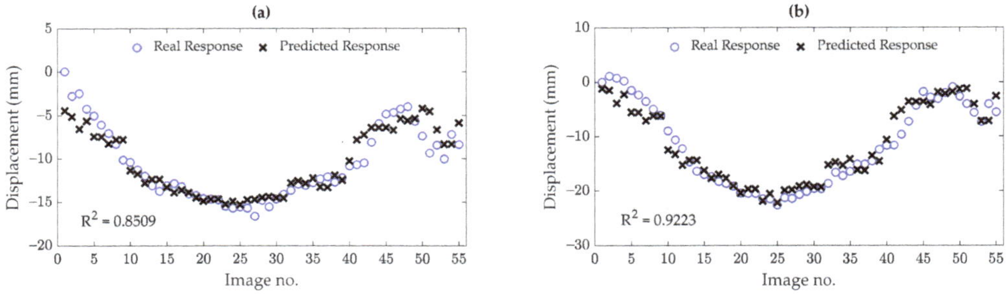

Figure 5. Prediction of the displacement response by the proposed RANN-BHO method: (**a**) dome, and (**b**) main span.

4. Conclusions

This paper has discussed the issues linked to the health monitoring of bridges in cases of limited/small datasets, like those typically collected with remote sensing systems. To address the limitations of small data for prediction, a parsimonious yet robust predictive model has been proposed by combining RANN and BHO. BHO has been exploited to tune the main hyperparameters of the RANN—the number of hidden layers and their neuron sizes.

Displacements along with air temperature related to a long-span steel arch bridge, have been used to verify the capability and performance of the proposed method. The results have demonstrated that the RANN–BHO-based method is an effective and simple predictive tool, featuring reliable estimations in the presence of small datasets to be exploited for the prediction of the structural health.

Author Contributions: Conceptualization, A.E. and B.B.; methodology, A.E., B.B., C.D.M. and S.M.; software, A.E. and B.B.; validation, A.E. and B.B.; formal analysis, A.E., B.B., C.D.M. and S.M.; investigation, A.E., B.B., C.D.M. and S.M.; resources, A.E. and B.B.; data curation, A.E. and B.B.; writing—original draft preparation, A.E. and B.B.; writing—review and editing, A.E., B.B., C.D.M. and S.M.; visualization, A.E., B.B., C.D.M. and S.M.; supervision, C.D.M. and S.M.; project administration, A.E., B.B., C.D.M. and S.M. All authors have read and agreed to the published version of the manuscript.

Funding: This research has been partially funded by the European Space Agency (ESA) under contract no. 4000132658/20/NL/MH/ac.

Institutional Review Board Statement: Not applicable.

Informed Consent Statement: Not applicable.

Data Availability Statement: Data is unavailable due to privacy or ethical restrictions.

Conflicts of Interest: The authors declare no conflicts of interest. The funders had no roles in the design of the study; in the collection, analyses, or interpretation of data; in the writing of the manuscript; or in the decision to publish the results.

References

1. Limongelli, M.P.; Gentile, C.; Biondini, F.; di Prisco, M.; Ballio, F.; Zonno, G.; Borlenghi, P.; Bianchi, S.; Capacci, L.; Anghileri, M.; et al. Bridge structural monitoring: The Lombardia regional guidelines. *Struct. Infrastruct. Eng.* **2022**, 1–24. [CrossRef]
2. Entezami, A.; Sarmadi, H.; Behkamal, B. Long-term health monitoring of concrete and steel bridges under large and missing data by unsupervised meta learning. *Eng. Struct.* **2023**, *279*, 115616. [CrossRef]
3. Entezami, A.; Sarmadi, H.; Behkamal, B.; De Michele, C. On continuous health monitoring of bridges under serious environmental variability by an innovative multi-task unsupervised learning method. *Struct. Infrastruct. Eng.* **2023**, 1–19. [CrossRef]
4. Lan, Y.; Li, Z.; Lin, W. A Time-Domain Signal Processing Algorithm for Data-Driven Drive-by Inspection Methods: An Experimental Study. *Materials* **2023**, *16*, 2624. [CrossRef] [PubMed]
5. Torzoni, M.; Rosafalco, L.; Manzoni, A. A combined model-order reduction and deep learning approach for structural health monitoring under varying operational and environmental conditions. *Eng. Proc.* **2020**, *2*, 94. [CrossRef]

6. Sarmadi, H.; Entezami, A.; Salar, M.; De Michele, C. Bridge health monitoring in environmental variability by new clustering and threshold estimation methods. *J. Civ. Struct. Health Monit.* **2021**, *11*, 629–644. [CrossRef]
7. Soleymani, A.; Jahangir, H.; Nehdi, M.L. Damage detection and monitoring in heritage masonry structures: Systematic review. *Constr. Build. Mater.* **2023**, *397*, 132402. [CrossRef]
8. da Silva, S.; Figueiredo, E.; Moldovan, I. Damage Detection Approach for Bridges under Temperature Effects using Gaussian Process Regression Trained with Hybrid Data. *J. Bridge Eng.* **2022**, *27*, 04022107. [CrossRef]
9. Ardani, S.; Eftekhar Azam, S.; Linzell, D.G. Bridge Health Monitoring Using Proper Orthogonal Decomposition and Transfer Learning. *Appl. Sci.* **2023**, *13*, 1935. [CrossRef]
10. Farrar, C.R.; Worden, K. *Structural Health Monitoring: A Machine Learning Perspective*; John Wiley & Sons Ltd.: Chichester, UK, 2013.
11. Entezami, A.; Arslan, A.N.; De Michele, C.; Behkamal, B. Online hybrid learning methods for real-time structural health monitoring using remote sensing and small displacement data. *Remote Sens.* **2022**, *14*, 3357. [CrossRef]
12. Behkamal, B.; Entezami, A.; De Michele, C.; Arslan, A.N. Investigation of temperature effects into long-span bridges via hybrid sensing and supervised regression models. *Remote Sens.* **2023**, *15*, 3503. [CrossRef]
13. Sarmadi, H.; Entezami, A.; Magalhães, F. Unsupervised data normalization for continuous dynamic monitoring by an innovative hybrid feature weighting-selection algorithm and natural nearest neighbor searching. *Struct. Health Monit.* **2023**, *22*, 4005–4026. [CrossRef]
14. Zhang, C.; Mousavi, A.A.; Masri, S.F.; Gholipour, G.; Yan, K.; Li, X. Vibration feature extraction using signal processing techniques for structural health monitoring: A review. *Mech. Syst. Sig. Process.* **2022**, *177*, 109175. [CrossRef]
15. Daneshvar, M.H.; Sarmadi, H.; Yuen, K.-V. A locally unsupervised hybrid learning method for removing environmental effects under different measurement periods. *Meas.* **2023**, *208*, 112465. [CrossRef]
16. Entezami, A.; Sarmadi, H.; Behkamal, B. A novel double-hybrid learning method for modal frequency-based damage assessment of bridge structures under different environmental variation patterns. *Mech. Syst. Sig. Process.* **2023**, *201*, 110676. [CrossRef]
17. Figueiredo, E.; Brownjohn, J. Three decades of statistical pattern recognition paradigm for SHM of bridges. *Struct. Health Monit.* **2022**, *21*, 3018–3054. [CrossRef]
18. Akintunde, E.; Azam, S.E.; Linzell, D.G. Singular value decomposition and unsupervised machine learning for virtual strain sensing: Application to an operational railway bridge. *Structures* **2023**, *58*, 105417. [CrossRef]
19. Lan, Y.; Zhang, Y.; Lin, W. Diagnosis algorithms for indirect bridge health monitoring via an optimized AdaBoost-linear SVM. *Eng. Struct.* **2023**, *275*, 115239. [CrossRef]
20. Biondi, F.; Addabbo, P.; Ullo, S.L.; Clemente, C.; Orlando, D. Perspectives on the Structural Health Monitoring of Bridges by Synthetic Aperture Radar. *Remote Sens.* **2020**, *12*, 3852. [CrossRef]
21. Farneti, E.; Cavalagli, N.; Costantini, M.; Trillo, F.; Minati, F.; Venanzi, I.; Ubertini, F. A method for structural monitoring of multispan bridges using satellite InSAR data with uncertainty quantification and its pre-collapse application to the Albiano-Magra Bridge in Italy. *Struct. Health Monit.* **2023**, *22*, 353–371. [CrossRef]
22. Giordano, P.F.; Turksezer, Z.; Previtali, M.; Limongelli, M.P. Damage detection on a historic iron bridge using satellite DInSAR data. *Struct. Health Monit.* **2022**, *21*, 2291–2311. [CrossRef]
23. Giordano, P.F.; Previtali, M.; Limongelli, M.P. Monitoring of a Metal Bridge Using DInSAR Data. In *European Workshop on Structural Health Monitoring*; Springer: Cham, Switzerland, 2022. [CrossRef]
24. Gatti, F.; Rosafalco, L.; Colombera, G.; Mariani, S.; Corigliano, A. Multi-storey shear type buildings under earthquake loading: Adversarial learning-based prediction of the transient dynamics and damage classification. *Soil Dyn. Earthquake Eng.* **2023**, *173*, 108141. [CrossRef]
25. Torzoni, M.; Manzoni, A.; Mariani, S. Structural health monitoring of civil structures: A diagnostic framework powered by deep metric learning. *Comput. Struct.* **2022**, *271*, 106858. [CrossRef]
26. Torzoni, M.; Rosafalco, L.; Manzoni, A.; Mariani, S.; Corigliano, A. SHM under varying environmental conditions: An approach based on model order reduction and deep learning. *Comput. Struct.* **2022**, *266*, 106790. [CrossRef]
27. Hagan, M.T.; Demuth, H.B.; Beale, M.H.; De Jesús, O. *Neural Network Design*; Martin Hagan: Lexington, KY, USA, 2014.
28. Yang, L.; Shami, A. On hyperparameter optimization of machine learning algorithms: Theory and practice. *Neurocomputing* **2020**, *415*, 295–316. [CrossRef]
29. Sarmadi, H. Investigation of machine learning methods for structural safety assessment under variability in data: Comparative studies and new approaches. *J. Perform. Constr. Facil.* **2021**, *35*, 04021090. [CrossRef]
30. Qin, X.; Zhang, L.; Yang, M.; Luo, H.; Liao, M.; Ding, X. Mapping surface deformation and thermal dilation of arch bridges by structure-driven multi-temporal DInSAR analysis. *Remote Sens. Environ.* **2018**, *216*, 71–90. [CrossRef]

engineering proceedings

MDPI

Proceeding Paper

A Distributed Sensor Network (DSN) Employing a Raspberry Pi 3 Model B Microprocessor and a Custom-Designed and Factory-Manufactured Multi-Purpose Printed Circuit Board for Future Sensing Projects †

Alan Ibbett and Yeslam Al-Saggaf *

School of Computing, Mathematics and Engineering, Charles Sturt University, Wagga Wagga, NSW 2678, Australia; aibbett@csu.edu.au
* Correspondence: yalsaggaf@csu.edu.au

† Presented at the 10th International Electronic Conference on Sensors and Applications (ECSA-10), 15–30 November 2023; Available online: https://ecsa-10.sciforum.net/.

Abstract: This paper presents a detailed design of an inexpensive, simple, and scalable Distributed Sensor Network (DSN). Each sensor's hardware consists of a Raspberry Pi 3 Model B microprocessor, a specifically designed and factory-made Printed Circuit Board (PCB), an Uninterruptible Power Supply (UPS) Hat based on a High-Capacity Lithium Polymer battery (LiPo), a Power over Ethernet Splitter, a GPS receiver, and a LoRaWAN module. Each sensor is built to capture GPS, Wi-Fi, and Bluetooth signals and sends this information to a network controller implementing a LoRaWAN gateway. Each sensor's software is developed so that all applications run on top of a Linux operating system. The layer above it includes system daemon applications, such as Air-mon, HCI tools, GPSd, and networking support. An SQLite database sits on top of the daemon applications and records the captured information. After the DSN was successfully tested, it was deployed in a research study. The novelty of this study is that this was the first time that a DSN was used in high schools to detect leakage from IoT devices to educate students about cyber safety.

Keywords: distributed sensor network; Raspberry Pi; printed circuit board; LoRaWAN gateway; inter-integrated circuit

Citation: Ibbett, A.; Al-Saggaf, Y. A Distributed Sensor Network (DSN) Employing a Raspberry Pi 3 Model B Microprocessor and a Custom-Designed and Factory-Manufactured Multi-Purpose Printed Circuit Board for Future Sensing Projects. *Eng. Proc.* **2023**, *58*, 55. https://doi.org/10.3390/ecsa-10-16187

Academic Editor: Stefano Mariani

Published: 15 November 2023

1. Introduction

This paper presents the detailed design of the various components of a Distributed Sensor Network (DSN). The DSN design is inspired by the WiFiScanMap [1]. The paper begins by first describing the design of the sensor nodes and then the network controller. The paper then presents the key features of the software design of the nodes and the software for the dashboards. The paper then reports the results of a novel study in which the DSN was used in a school setting to educate students about cyber safety.

The DSN comprises a number of nodes, designed to monitor Wi-Fi and Bluetooth traffic within the range of the node [2]. Each node is an independent module, and each node controls and manages information flows between the node and the DSN network controller [2]. Control and management information shared between nodes and the DSN controller includes node status and other telemetry information [2]. Figure 1 below depicts the overall architecture of the DSN.

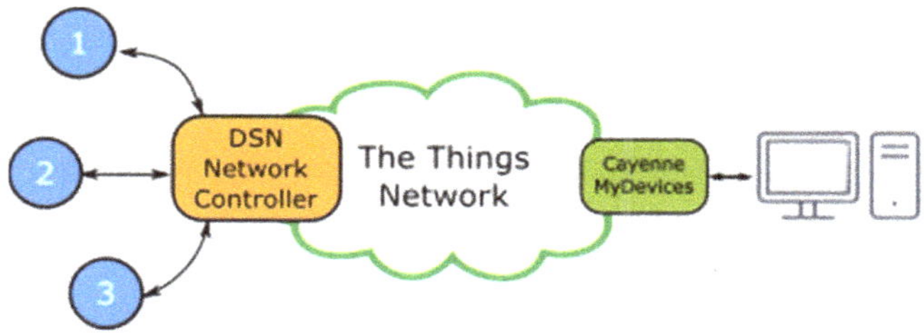

Figure 1. Key design blocks of the DSN [2].

The sensor network monitors Wi-Fi and Bluetooth connection requests, beacons and other broadcast-type packets, including the Basic Service Set Identifier (BSSID), Service Set Identifier (SSID), Media Access Control address of client (MAC), Channel Number, Received Signal Strength Information (RSSI), Encryption Capabilities, UTC time and date of capture, and location information of capture [2].

To capture and manage these packets, a number of open-source tools were used, including iwlist (an application to display detailed wireless information), airmon-ng (a packet capturing applications for Linux), gpsd gps (GPS support for Linux), SQLite (an open-source database), python 3.0 (an open-source programming language), and DB Browser for SQLite (an open-source database browser). Other supporting software such as the Re4son Kernel and the Raspbian Linux distribution were also used to support the DSN [2].

2. Hardware Design

Each sensor's hardware consists of a Raspberry Pi 3 Model B microprocessor, a specifically designed and factory-made Printed Circuit Board (PCB), an Uninterruptible Power Supply (UPS) Hat based on a High-Capacity Lithium Polymer battery (LiPo), a Power over Ethernet Splitter allowing for 5 V power via micro-USB (from a PoE injector), a GPS receiver, a LoRaWAN module, control push buttons, an OLED screen, a Four-wire I2C connector, a 40-pin Raspberry Pi connector, a status LED, and five GPIO connectors configured for Signal, Voltage, and Ground (SVG) [2]. Each sensor is built to capture GPS, Wi-Fi, and Bluetooth signals and sends this information to a network controller implementing a LoRaWAN gateway [2].

The Raspberry Pi Model 3B microprocessor had Wi-Fi and Bluetooth built onto the main board. The Wi-Fi standard is based on the frequency of operation being in the so-called Industrial Scientific and Medical (ISM) band, which is unlicensed and free to use. The ISM band covers many frequencies, and Wi-Fi has been defined to work on the 2.4 GHz and 5 GHz part of the ISM bands. The Wi-Fi on the Raspberry Pi 3B was limited to the 2.4 Ghz band [2].

A prototype Printed Circuit Board (PCB) was designed and sent for printing to OSH-PARK, a PCB fabrication house based in the United States that specializes in producing high-quality low-cost small-batch runs of PCBs. All node components were designed to be on a single PCB that can be stacked onto other boards as required [2]. By using this "stackable" architecture, it was possible to quickly assemble and, if necessary, disassemble a node. The sensor board [2] contains the following items:

- Organic light-emitting diode (OLED) display to indicate system status: the OLED provides a way to display a simple menu-based user interface;
- RFM9x LoRa radio: this provides a network connection to LoRaWAN or other LoRa nodes;
- Global Position System GPS receiver: this allows the node to accurately collect coordinated universal time (UTC) time and date as well as location information of other data captured from the GPS;
- DC power supply input connector to supply regulated 5VDC to the system;

- Status LED to show system status at a glance;
- Five GPIO ports of the Raspberry Pi have been broken up into an SVG layout for connecting to external devices. These can include relays, sensors, switches, and so forth;
- A three-button keypad for interacting with the OLED menu user interface.

The addition of I2C and GPIO breakout connections meant the node could be used with a wide variety of sensors in a number of remote locations [2].

The DSN node has been designed to be self-contained and to be used in a fixed location [2]. Uninterruptable power supplies (UPSs) based on the high-capacity 3.7 Lithium Polymer (LiPo) battery technology provided power supply to the central processing unit (CPU) and sensors. LiPo batteries are high-capacity, easy to use, and relatively inexpensive. The LiPo battery is paired with a purpose-built UPS card (or "hat", as expansion cards for the Raspberry Pi are known) that has the appropriate circuitry to detect a low battery level and initiate a controlled shutdown of the node. The UPS card also handles the charging of the unit. Five volts from the PoE splitter are plugged into the GPS unit which then feeds the entire node [2]. Figure 2 shows the key elements of the DSN node before assembly. Figure 3 shows the elements of the DSN node after assembly.

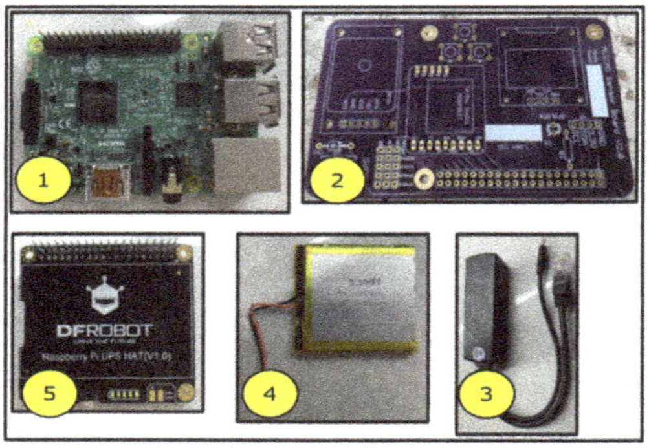

Figure 2. The key elements of the DSN node. Taken from top row moving clockwise: (1) CPU (Raspberry Pi Model 3B); (2) Sensor PCB, shown here unpopulated; (3) Power over Ethernet Splitter; (4) High-Capacity LiPo battery; (5) UPS Hat [2].

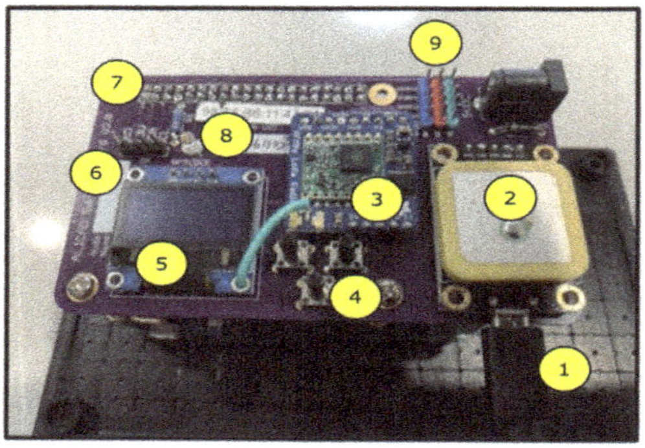

Figure 3. A complete DSN node. (1) 5V power in via micro-USB from the PoE injector; (2) GPS unit (the chip antenna is visible in the picture); (3) LoRaWAN module; (4) Control push buttons; (5) OLED screen; (6) Four-wire I2C connector; (7) 40-pin Raspberry Pi connector; (8) Status LED; (9) Five GPIO connectors configured for Signal, Voltage, and Ground (SVG) [2].

As the nodes in this design were placed in fixed locations, an "out of band" monitoring service was needed to check the status of each node. In this case, LoRa radios and the LoRaWAN network were chosen to provide this remote monitoring [2]. The LoRa network is completely separate in frequency and operation to both the Wi-Fi and Bluetooth networks that the nodes are monitoring, thereby separating node management functions from the networks being monitored [2].

The DSN network controller (DSNNC) is designed to provide real-time feedback on the operation of each of the distributed sensor nodes. The role of the DSNNC is to act as a gateway from each of the DSN nodes to a central dashboard. Here, the status of each node can be monitored and, if necessary, the node can be restarted [2]. LoRa is a simple technology and permits easy integration of telemetry and control data. LoRa is also completely out of band, which means that LoRa transmissions do not share the same IP network space as the Wi-Fi that the DSN is monitoring [3].

The RAK7249 LoRaWAN gateway provided a wide range of connectivity and power supply options including Gigabit Ethernet, Wi-Fi, 4G, and PoE power supply [2]. The gateway's geographical and time values are set by an integrated GPS receiver. The gateway has a (LoRa) range of over 15 km. The free Cayenne MyDevices (CMD) dashboard service was selected to provide the application server services. The CMD was selected because it is secure, easy to set up, allows for shared viewing of dashboards, and is free [2].

3. Software Design

The original Lauters code [1] published on GitHub was used as the template for the software design after translating the Python 2.0 code to Python 3.0, but it was necessary to add in additional features to support the DSN hardware, including (1) LoRa support, (2) Display Support, (3) Menu Support, and (4) Keypad Support [2]. As the nodes were placed in fixed locations, in order to allow each node to start up autonomously, it was necessary to create a start-up script that was unique to each node [2].

The software architecture of a DSN node is simple (see Figure 4 below). All applications run on top of the Linux operating system [2]. For additional information, see the Supplementary Materials section below. The next layer includes system daemon applications such as Air-mon, HCI tools, GPSd, and networking support. The SQLite database sits on top of the daemon applications, and information is delivered to the user by the web server. Networking tools are made available to all layers of the system. The web server is implemented by a custom Python script that also controls the way that the daemon processes access the database [2].

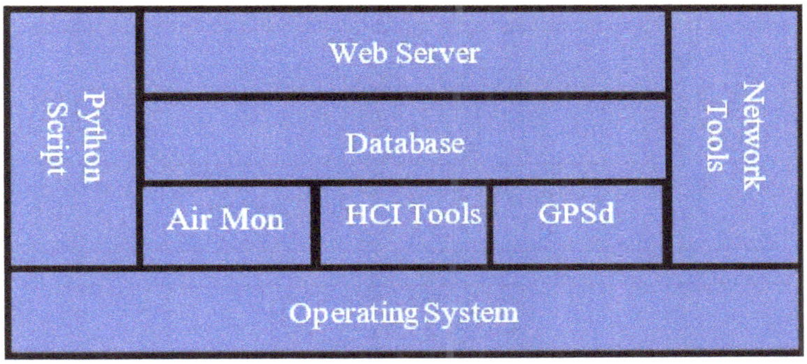

Figure 4. DSN node software architecture [2].

The nodes are designed to capture and log GPS, Wi-Fi, and Bluetooth transmission, but provision has been made to connect other digital instruments for monitoring and logging via an Inter-Integrated Circuit (I2C) and one-wire data busses [2].

The DSN node is a general-purpose device and can support a number of different software options [2]. The DSN node is based on a Linux Distribution with a number of local tasks running such as Virtual Network Computing (VNC), as well as the main data acquisition task. The main data acquisition task is written as a single class in Python 3.0. The instantiation of this class provides all of the data structure initialization required, as well as the database creation and setup if needed [2].

There are five tasks associated with the DSN implementation, namely, the GPS Poller, Wi-Fi Poller, Bluetooth Poller, LCD Poller, and LoRa Poller [2], as depicted in Figure 5. The GPS Poller task runs to regularly collect, parse, and validate GPS data. The validated data are passed back to the main thread where they are written to the database. The GPS Poller thread looks for latitude and longitudinal data, as well as precision and time information. The Wi-Fi Poller thread uses data collected from the airmon-ng process to collect Wi-Fi packets from the surrounding area. These packets are validated by the thread and passed back to the main thread. The Bluetooth Poller thread is similar to the Wi-Fi Poller thread but uses the hcitool to capture Bluetooth packets [2].

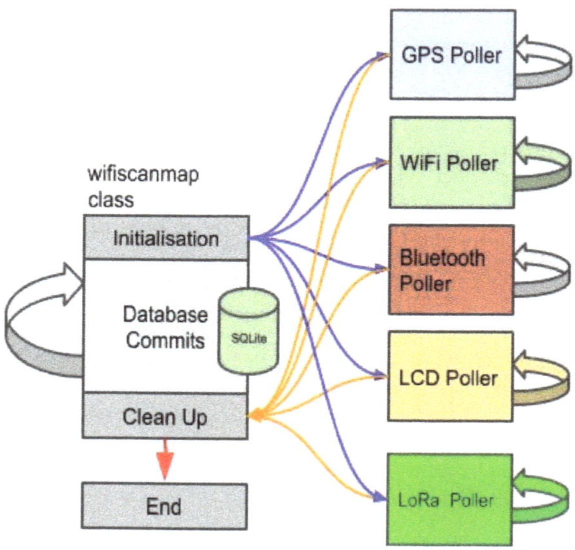

Figure 5. DSN node software architecture [2].

The LCD Poller task handles the LCD and the user interface push buttons [2]. The LCD Poller thread uses the rpi-128 × 64-oled -menu Python library [4]. This library, in turn, uses the RPi.GPIO library. A number of simple menus were set up to allow the user to check the state of the DSN node. The reliability of the data gathered by this experiment depends on the ability of the DSN to capture and record packets. The status of this packet capture was monitored by the DSNNC. The DSNNC regularly sends LoRa messages to the user via the RAK gateway. If these signals are lost, or indicate that a node has failed, then two actions can occur: the node will attempt to reconnect with the controller and, if necessary, will reboot [2].

The Wi-Fi and Bluetooth Poller threads rely on the Airmon-ng script [2]. Airmon-ng enables monitor mode on suitable wireless interfaces. The script also allows an interface to be switched from monitor mode (the mode used in this application) to managed mode (traditional Wi-Fi mode). Airmon-ng allows for the capture and classification of packets. Airmon-ng requires that the Wi-Fi chip on the Raspberry Pi be put into monitoring mode, which was achieved with the help of the drivers from the nexmon project [5]. More details on the re4son project can be found here [6]. This workaround allowed the Airmon-ng code to collect Wi-Fi packets as expected. These data were then recorded into an SQLite Database, which sat on top of the daemon applications. Networking tools are made available to all layers of the system. Information from each sensor forming the DSN that is received by the LoRaWAN gateway is communicated to the user via a web server. The web server is implemented by a custom Python script which also controlled the way that the daemon processes accessed the database [2].

4. Study Results and Conclusions

A series of tests were carried out to ensure that the DSN was functioning as expected [2]. Based on the successful outcome of these tests, the DSN was used in a first research study of its kind, in which the DSN was deployed on a high school ground in Australia to educate students about cyber safety [2]. After informing the students about the presence of a DSN on campus detecting leakage from IoT devices, the DSN was turned on. The traffic that was captured by the DSN at that time served as the standard for a normal traffic at that school. Then, during a cyber safety lesson, the students were shown the results of the DSN monitoring and were trained on how to modify their device settings to reduce data leakage. At that time, the DSN continued to detect leakage from IoT devices to see if children changed their device settings to protect themselves online. The results of the study revealed that the amount of data leaked from smartphones after the cyber safety lesson was significantly smaller than the traffic captured before the cyber safety lesson, suggesting that the cyber safety lesson was effective in reducing data leakage [2].

The DSN was simple, scalable, and inexpensive. The total cost to build a single node (a sensor), including the cost of the Raspberry Pi, was less than AUD 160 [2]. The inclusion of a LoRaWAN module, control push buttons, OLED screen, and status LED made accessing and controlling the DSN remotely a simple process [2]. The scalability of the sensor's PCB, which has a Four-wire I2C connector, makes the device capable of performing numerous types of sensing and control functions.

Supplementary Materials: The following supporting information can be downloaded at https://github.com/alanibbett (accessed on 8 October 2023). Files S1: Distributed Sensor Network—PCB Files; Files S2: Distributed Sensor Network Python Script.

Author Contributions: Conceptualization, methodology, software, validation, formal analysis, investigation, resources, data curation, and visualization A.I.; writing—original draft preparation, writing—review and editing, and supervision Y.A.-S. All authors have read and agreed to the published version of the manuscript.

Funding: This research received no external funding.

Institutional Review Board Statement: This study was conducted in accordance with the Declaration of Helsinki and approved by the Human Research Ethics Committee) of Charles Sturt University (protocol code H19072 and date of approval 6 June 2020).

Informed Consent Statement: Not applicable.

Data Availability Statement: Data are contained within the article and supplementary materials.

Conflicts of Interest: The authors declare no conflicts of interest.

References

1. Bordeaux: A Digital Urban Exploration. Available online: https://github.com/mehdilauters/wifiScanMap/blob/master/Results.md (accessed on 8 October 2023).
2. Ibbett, A. An Examination of Real-World Data Leakage from IoT Devices. Ph.D. Thesis, Charles Sturt University, Wagga Wagga, Australia, 2022.
3. Vangelista, L.; Zanella, A.; Zorzi, M. Long-Range IoT Technologies: The Dawn of LoRa™. In *Future Access Enablers for Ubiquitous and Intelligent Infrastructures: Proceedings of the First International Conference, FABULOUS 2015, Ohrid, Republic of Macedonia, 23–25 September 2015*; Revised Selected Papers 2015; Springer International Publishing: Berlin/Heidelberg, Germany, 2015; pp. 51–58. [CrossRef]
4. rpi-128x64-Oled-Menusystem. Available online: https://github.com/jpuk/rpi-128x64-oled-menusystem (accessed on 8 October 2023).

5. Schulz, M.; Wegemer, D.; Hollick, M. Nexmon: Build your own wi-fi testbeds with low-level mac and phy-access using firmware patches on off-the-shelf mobile devices. In Proceedings of the 11th Workshop on Wireless Network Testbeds, Experimental Evaluation and Characterization, Snowbird, UT, USA, 20 October 2017.
6. Boeving, C.; Klimaszewski, S.; Rebillout, A.; Wilson, B.; Ruiz de Alegría, D.; O'Gorman, J.; O'Gorman, J.; Hertzog, R.; Re4son-Kernel for Raspberry Pi. Kali Linux Development Team. 12 May 2019. Available online: https://re4son-kernel.com/re4son-pi-kernel/ (accessed on 8 October 2023).

engineering proceedings

MDPI

Proceeding Paper

Regression Tree Ensemble to Forecast Thermally Induced Responses of Long-Span Bridges †

Alireza Entezami *, Bahareh Behkamal, Carlo De Michele and Stefano Mariani

Department of Civil and Environmental Engineering, Politecnico di Milano, 20133 Milano, Italy; bahareh.behkamal@polimi.it (B.B.); carlo.demichele@polimi.it (C.D.M.); stefano.mariani@polimi.it (S.M.)
* Correspondence: alireza.entezami@polimi.it

† Presented at the 10th International Electronic Conference on Sensors and Applications (ECSA-10), 15–30 November 2023; Available online: https://ecsa-10.sciforum.net/.

Abstract: The ambient temperature is a critical factor affecting the deformation of long-span bridges, due to its seasonal fluctuations. Although there exist various sensor technologies and measurement techniques to extract the actual structural response in terms of the displacement field, this is a demanding task in long-term monitoring. To address this challenge, data prediction looks to be the best solution. In this paper, the thermally induced response of a long-span bridge is forecasted with a regression tree ensemble method in conjunction with Bayesian hyperparameter optimization, adopted to tune the proposed regressor. Results testify that the offered method is reliable when there is a linear correlation between the temperature and the induced structural deformation, hence in terms of the thermally induced displacement field.

Keywords: long-span bridges; supervised learning; regression tree ensemble; temperature effects; remote sensing

Citation: Entezami, A.; Behkamal, B.; De Michele, C.; Mariani, S. Regression Tree Ensemble to Forecast Thermally Induced Responses of Long-Span Bridges. *Eng. Proc.* **2023**, *58*, 56. https://doi.org/10.3390/ecsa-10-16030

Academic Editor: Stefan Bosse

Published: 15 November 2023

1. Introduction

Ambient temperature affects the response of civil structures to external excitations [1,2], especially long-span bridges [3]. The critical impact of this environmental factor is related to its seasonal variability [4], regardless of the type, size, and material the structures are made of [5,6]. Notably, temperature variability is an important challenge for long-term structural health monitoring (SHM) [7]. Seasonal temperature fluctuations in summer and winter profoundly affect the structural response and the physical properties, leading to changes to the structural characteristics. Several experimental and field monitoring studies demonstrated that varying temperature conditions may have a more significant effect on the structural behavior than operational loadings [8–10]. This is because spatial and temporal temperature variations cause thermal loads, unpredictable internal stresses and forces in elements, and also changes in the boundary conditions, particularly for large-scale bridges, all of which lead to damages such as cracks in concrete and yielding of steel elements [11]. Therefore, it is essential to regularly monitor structures like long-span bridges under varying ambient temperature effects to remove such effects [12–15].

The routine and simple monitoring process usually consists of conducting field measurements. Accordingly, structures are equipped with sensors to measure structural and environmental data with accelerometers, strain gauges, thermocouples, anemometers, etc. Recently, the technology of spaceborne remote sensing has helped civil engineers to leverage remote sensors mounted on satellites (e.g., Sentinel-1, TerraSar-X, COSMO-SkyMed) to capture optical or synthetic aperture radar (SAR) images [16–18] and extract structural displacements [19–23]. The great advantage of this technology is the lack of contact sensors, which represent a considerable cost to the SHM system and may be broken in cases of extreme environmental events (e.g., typhoons). Despite the development of new sensing technologies for data measurement, a field measurement process may show several limitations. First, it is not simple to perform a long-term field measurement program. Second, although spaceborne remote sensing facilitates the extraction of structural displacement responses, the outcome of this technology (i.e., optical and SAR images) often has a huge size (in the unit of GB), leading to limitations linked to memory storage. Third, spaceborne remote sensing cannot supply rich data/features as obtained with SHM via contact-based sensing. In most long-term SHM projects, the monitoring problem is based on small data [24]. However, it is difficult to provide such a large data stream from remote sensors. One of the most appropriate solutions to this challenge is to implement data forecasting or prediction.

According to the concept of data forecasting, it suffices to define the main predictors (input or independent data) and relevant responses (output or dependent data). In health monitoring of civil structures, the environmental conditions (i.e., temperature, humidity, wind, etc.) are related to the predictors, while changes in the structures themselves pertain to the responses. Using such information, supervised regression modeling is the best approach to data forecasting [3,25,26].

The main objective of this paper is to propose a regression ensemble tree (RTE) in conjunction with the Bayesian hyperparameter optimization (BHO), here called RTE–BHO, to forecast the thermally induced responses of long-span bridges. In this method, BHO supports RTE to select the best ensemble learning algorithm between bagging and boosting strategies. To evaluate the performance of the supervised regressor, small sets of displacement responses and ambient temperature records related to a long-span bridge were considered. The displacement histories were extracted from 55 SAR images of TerraSar-X between 2009 and 2010. Results demonstrate that the proposed RTE–BHO provides an automated data forecasting approach, able to predict the displacement responses retrieved from spaceborne remote sensing technology.

2. Regression Tree Ensemble

Ensemble learning is an advanced machine learning algorithm that can significantly improve the overall performance of any model. In particular, it has been an effective strategy for SHM [27,28]. For the problem of regression, the RTE method develops a supervised predictive model as a weighted combination of multiple regression trees. Indeed, the RTE benefits the concept of ensemble learning to enhance the decision tree regression that may suffer from high variance, bias, and overfitting. Therefore, multiple regression trees based on the concept of ensemble learning can be combined to make an ensemble of trees and improve the prediction performance.

Bagging and boosting are two popular and tried-and-tested ensemble learning algorithms adopted in RTE. Bagging or bootstrap aggregating utilizes multiple separate sets from the original, randomly generated training data with replacements to train the different regression trees. Let B denote the number of trees to generate B different bootstrapped training datasets. The predictions of B trees, defined as $\{\hat{y}_1, \ldots, \hat{y}_B\}$, can be computed and then averaged to reduce the variance with respect to a single regression tree. The averaging process in the bagging method is given by

$$\hat{y}_{Bag} = \frac{1}{B}\sum_{i=1}^{B} \hat{y}_i \qquad (1)$$

where $\hat{y}_{Bag}$ denotes the average prediction of all the regression trees. The boosting technique increases the number of trees B sequentially so that each tree uses a modified version of the whole dataset accounting for information from the previously developed trees.

The fundamental principle of the boosting technique is to increase the number of trees sequentially, wherein each tree uses a modified version of the whole dataset through information from the previously developed tree. This technique somehow resembles the bagging method except that the trees are grown sequentially, which means that each tree is grown using information from previously developed trees. The boosting technique does not involve bootstrap sampling; instead, each tree is fit to a modified version of the original dataset. The boosting learning process improves the prediction performance of each tree from $\hat{y}_1$ to $\hat{y}_B$ by updating the weights of the training samples, without the bootstrap sampling. In this regard, the boosted model output $\hat{y}_{Boost}$ is defined as follows:

$$\hat{y}_{Boost} = \sum_{i=1}^{B} \lambda \hat{y}_i \quad (2)$$

where λ is the shrinkage parameter (i.e., a constant value) that controls the rate of the boosting process for learning.

3. Bayesian Hyperparameter Optimization

In machine learning, a hyperparameter is an unknown element that affects the overall performance of a model [29]. Among the various hyperparameter optimization techniques, BHO is one of the most effective ones. This is because it can tune any type of unknown parameter based on Bayes' theorem. In the forecasting problem by regression modeling, the fundamental principle of BHO is to minimize the following objective function, along with its hyperparameter(s), in a bounded domain:

$$F = \log(1 + E_{MSE}) \quad (3)$$

where E_{MSE} stands for the cross-validation mean squared error (MSE) between the observation (i.e., real data) and estimation (i.e., predicted data) values. The function F denotes a logarithmic transformed validation loss, computed for the regressor and hyperparameter values at each iteration. Using this function, BHO considers a probability distribution model for the objective function of interest, and the model is then updated at each new evaluation/iteration. In the following, an acquisition function is defined to maximize and also determine the next values(s) of the hyperparameter(s) of the objective function. One can also incorporate a posterior probability distribution for each hyperparameter. If there is any error in the iterative process, one should take more iterations until convergence is reached or consider a stopping condition. By minimizing the objective function F, the best hyperparameter set is the one leading to good convergence (i.e., a minimum difference) between the observed and estimated data.

Regarding RTE, BHO undertakes the tuning of some important elements, such as the type of ensemble learning algorithm, the minimum number of lead nodes, the maximum number of decision splits (branch nodes), and the shrinkage parameter in Equation (2).

4. Case Study: The Lupu Bridge

The Lupu Bridge is a steel arch bridge that crosses the Huangpu River in Shanghai; see Figure 1a. This structure features a total length of 750 m, due to a main span of 550 m and two side spans of 100 m. Figure 1b provides the side view of the bridge, along with its main dimensions and structural components. The girder in the side span is characterized by a closed steel box with a width of 41 m and a height of 2.7 m. The box girder was fixed with the arch ribs, columns, and end cross-beam of the side spans. More structural details can be found in Qin et al. [30].

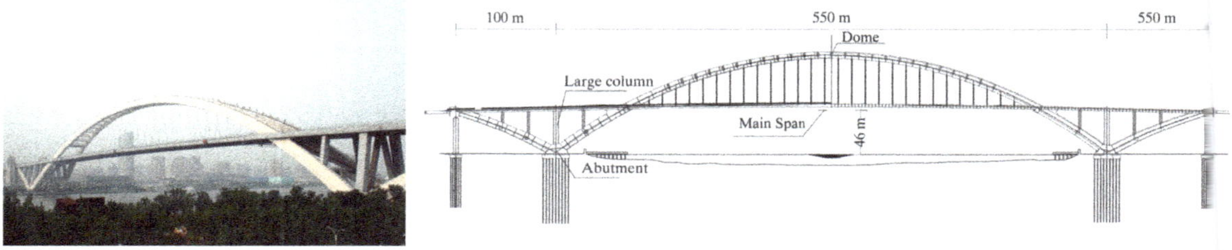

(a) (b)

Figure 1. (**a**) The Lupu Bridge; (**b**) side view and main dimensions.

4.1. Thermally Induced Structural Responses

Due to the sensitivity of the Lupu Bridge to geological and environmental conditions, a field measurement based on spaceborne remote sensing and limited SAR images was conducted to obtain the displacement responses of two critical components of the bridge in a long-term monitoring program [30]: the main span and the dome. Accordingly, 55 SAR images from TerraSar-X between 2009 and 2010 were analyzed to extract the displacement points of the mentioned components. During this field measurement, air temperature data were also recorded to provide rich information for the prediction problem. Figure 2 shows the time histories of the displacements associated with the dome and the main span, along with the temperature data.

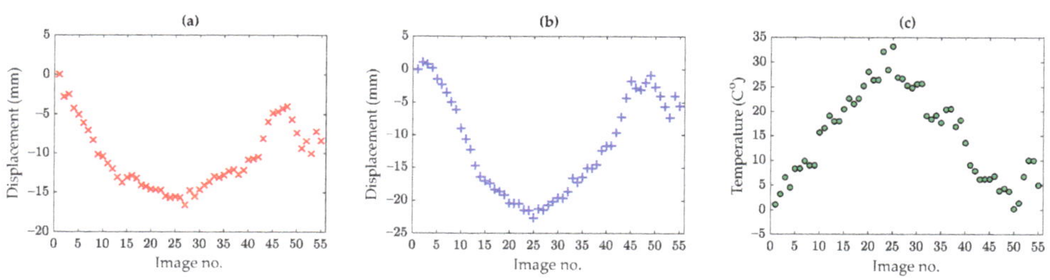

Figure 2. Displacement responses of the Lupu Bridge (**a**) at the dome, and (**b**) the main span; (**c**) relevant temperature data.

4.2. Forecasted Results

The measured temperature and extracted displacement evolutions were collected to be arranged as the training and test datasets. Eighty percent of the temperature and displacement samples were used to generate the training set, while the remaining twenty percent of the samples were considered the test set. The number of the training and test points were therefore equal to 44 and 11, respectively. Using BHO with 50 sample iterations, Table 1 reports the main hyperparameters of the RTE models to forecast the displacement responses.

Table 1. Bayesian hyperparameter optimization of the RTE model.

Hyperparameter	Bridge Element	
	Dome	**Span**
Ensemble learning algorithm	Bagging	Bagging
Number of leaves	5	4
Number of decision nodes	49	34

Using the trained RTE models, Figure 3 compares the predicted and real displacement samples, including those considered in the training process and those related to the test set. As it can be observed, the predicted and real values are in good agreement with each other. Moreover, Figure 4 displays the regression plot by comparing the actual and predicted data: it can be seen that there are small deviations of all the data points from the straight line, providing a perfect fit. To get additional insights, Table 2 gathers the values of the R-squared metric relevant to the RTE–BHO method, in cases of different training ratios. The regressors developed for the dome and span have large R-squared values, close to 1, when the training ratio is 80%. However, if the training rate is reduced, the R-squared values only slightly reduce.

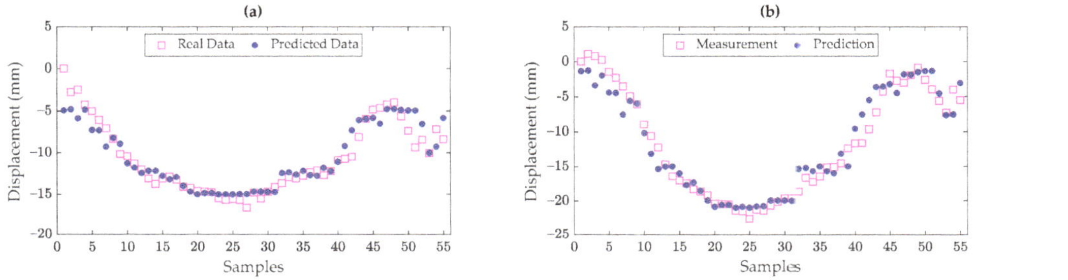

Figure 3. Measured and predicted displacements by the RTE model: (**a**) dome, and (**b**) main span.

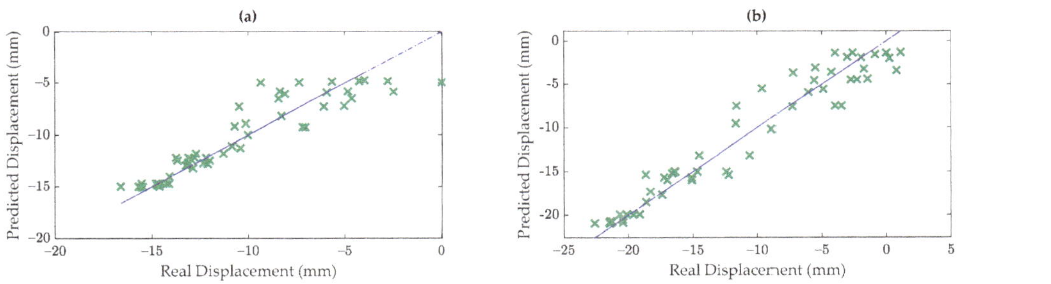

Figure 4. Predicted versus actual displacements, based on the RTE method: (**a**) dome, and (**b**) main span.

Table 2. Performance evaluation of the RTE–BHO method, in terms of the R-squared measure under a varying training ratio.

Bridge Element	**Training Ratios**		
	80%	**50%**	**30%**
Dome	0.8625	0.8453	0.7513
Span	0.9365	0.9121	0.8547

5. Conclusions

This paper has addressed the problem of forecasting structural displacements with remote sensing technology. The forecasting problem has been intended to evaluate the effect of ambient temperature on long-span bridges in long-term monitoring programs. A supervised regressor has been developed by combining a regression ensemble tree and Bayesian hyperparameter optimization.

Using the displacement responses at the dome and main span of the Lupu Bridge, the results have shown that the supervised regressor succeeds in forecasting the response data with a high overall ability, even in the case of a small dataset, to learn the main features of said structural response.

Author Contributions: Conceptualization, A.E. and B.B.; methodology, A.E., B.B., C.D.M. and S.M.; software, A.E. and B.B.; validation, A.E. and B.B.; formal analysis, A.E., B.B., C.D.M. and S.M.; investigation, A.E., B.B., C.D.M. and S.M.; resources, A.E. and B.B.; data curation, A.E. and B.B.; writing—original draft preparation, A.E. and B.B.; writing—review and editing, A.E., B.B., C.D.M. and S.M.; visualization, A.E., B.B., C.D.M. and S.M.; supervision, C.D.M. and S.M.; project administration, A.E., B.B., C.D.M. and S.M. All authors have read and agreed to the published version of the manuscript.

Funding: This research is partially funded by the European Space Agency (ESA) under contract no. 4000132658/20/NL/MH/ac.

Institutional Review Board Statement: Not applicable.

Informed Consent Statement: Not applicable.

Data Availability Statement: Data are unavailable due to privacy or ethical restrictions.

Conflicts of Interest: The authors declare no conflicts of interest. The funders had no role in the design of the study; in the collection, analyses, or interpretation of data; in the writing of the manuscript; or in the decision to publish the results.

References

1. Han, Q.; Ma, Q.; Xu, J.; Liu, M. Structural health monitoring research under varying temperature condition: A review. *J. Civ. Struct. Health Monit.* **2021**, *11*, 149–173. [CrossRef]
2. Soleymani, A.; Jahangir, H.; Nehdi, M.L. Damage detection and monitoring in heritage masonry structures: Systematic review. *Constr. Build. Mater.* **2023**, *397*, 132402. [CrossRef]
3. Behkamal, B.; Entezami, A.; De Michele, C.; Arslan, A.N. Investigation of temperature effects into long-span bridges via hybrid sensing and supervised regression models. *Remote Sens.* **2023**, *15*, 3503. [CrossRef]
4. Xia, Q.; Zhang, J.; Tian, Y.; Zhang, Y. Experimental study of thermal effects on a long-span suspension bridge. *J. Bridge Eng.* **2017**, *22*, 04017034. [CrossRef]
5. Entezami, A.; Sarmadi, H.; Behkamal, B. Long-term health monitoring of concrete and steel bridges under large and missing data by unsupervised meta learning. *Eng. Struct.* **2023**, *279*, 115616. [CrossRef]
6. Entezami, A.; Sarmadi, H.; Behkamal, B. A novel double-hybrid learning method for modal frequency-based damage assessment of bridge structures under different environmental variation patterns. *Mech. Syst. Sig. Process.* **2023**, *201*, 110676. [CrossRef]
7. Torzoni, M.; Rosafalco, L.; Manzoni, A. A combined model-order reduction and deep learning approach for structural health monitoring under varying operational and environmental conditions. *Eng. Proc.* **2020**, *2*, 94. [CrossRef]
8. Wang, D.; Liu, Y.; Liu, Y. 3D temperature gradient effect on a steel–concrete composite deck in a suspension bridge with field monitoring data. *Struct. Contr. Health Monit.* **2018**, *25*, e2179. [CrossRef]
9. da Silva, S.; Figueiredo, E.; Moldovan, I. Damage Detection Approach for Bridges under Temperature Effects using Gaussian Process Regression Trained with Hybrid Data. *J. Bridge Eng.* **2022**, *27*, 04022107. [CrossRef]
10. Akintunde, E.; Azam, S.E.; Linzell, D.G. Singular value decomposition and unsupervised machine learning for virtual strain sensing: Application to an operational railway bridge. *Structures* **2023**, *58*, 105417. [CrossRef]
11. Figueiredo, E.; Park, G.; Farrar, C.R.; Worden, K.; Figueiras, J. Machine learning algorithms for damage detection under operational and environmental variability. *Struct. Health Monit.* **2011**, *10*, 559–572. [CrossRef]
12. Daneshvar, M.H.; Sarmadi, H.; Yuen, K.-V. A locally unsupervised hybrid learning method for removing environmental effects under different measurement periods. *Meas.* **2023**, *208*, 112465. [CrossRef]
13. Sarmadi, H.; Entezami, A.; Magalhães, F. Unsupervised data normalization for continuous dynamic monitoring by an innovative hybrid feature weighting-selection algorithm and natural nearest neighbor searching. *Struct. Health Monit.* **2023**, *22*, 4005–4023. [CrossRef]
14. Entezami, A.; Mariani, S.; Shariatmadar, H. Damage detection in largely unobserved structures under varying environmental conditions: An autoregressive spectrum and multi-level machine learning methodology. *Sensors* **2022**, *22*, 1400. [CrossRef] [PubMed]
15. Figueiredo, E.; Brownjohn, J. Three decades of statistical pattern recognition paradigm for SHM of bridges. *Struct. Health Monit.* **2022**, *21*, 3018–3054. [CrossRef]
16. Cavalagli, N.; Kita, A.; Farneti, E.; Falco, S.; Trillo, F.; Costantini, M.; Fornaro, G.; Reale, D.; Verde, S.; Ubertini, F. Remote sensing and in-situ measurements for the structural monitoring of historical monuments: The Consoli Palace of Gubbio, Italy. In *European Workshop on Structural Health Monitoring*; Springer: Cham, Switzerland, 2021. [CrossRef]

17. Macchiarulo, V.; Milillo, P.; Blenkinsopp, C.; Giardina, G. Monitoring deformations of infrastructure networks: A fully automated GIS integration and analysis of InSAR time-series. *Struct. Health Monit.* **2022**, *21*, 1849–1878. [CrossRef]
18. Milillo, P.; Giardina, G.; Perissin, D.; Milillo, G.; Coletta, A.; Terranova, C. Reply to Lanari, R.; et al. comment on "pre-collapse space geodetic observations of critical infrastructure: The morandi bridge, Genoa, Italy" by Milillo et al. (2019). *Remote Sens.* **2020**, *12*, 4016. [CrossRef]
19. Giordano, P.F.; Turksezer, Z.; Previtali, M.; Limongelli, M.P. Damage detection on a historic iron bridge using satellite DInSAR data. *Struct. Health Monit.* **2022**, *21*, 2291–2311. [CrossRef]
20. Giordano, P.F.; Previtali, M.; Limongelli, M.P. Monitoring of a Metal Bridge Using DInSAR Data. In *European Workshop on Structural Health Monitoring*; Springer: Cham, Switzerland, 2022. [CrossRef]
21. Farneti, E.; Cavalagli, N.; Venanzi, I.; Salvatore, W.; Ubertini, F. Residual service life prediction for bridges undergoing slow landslide-induced movements combining satellite radar interferometry and numerical collapse simulation. *Eng. Struct.* **2023**, *293*, 116628. [CrossRef]
22. Farneti, E.; Cavalagli, N.; Costantini, M.; Trillo, F.; Minati, F.; Venanzi, I.; Ubertini, F. A method for structural monitoring of multispan bridges using satellite InSAR data with uncertainty quantification and its pre-collapse application to the Albiano-Magra Bridge in Italy. *Struct. Health Monit.* **2023**, *22*, 353–371. [CrossRef]
23. Farneti, E.; Cavalagli, N.; Costantini, M.; Trillo, F.; Minati, F.; Venanzi, I.; Salvatore, W.; Ubertini, F. Remote Sensing Satellite Data and Progressive Collapse Analysis for Structural Monitoring of Multi-span Bridges. In *European Workshop on Structural Health Monitoring*; Springer: Cham, Switzerland, 2022. [CrossRef]
24. Entezami, A.; De Michele, C.; Arslan, A.N.; Behkamal, B. Detection of partially structural collapse using long-term small displacement data from satellite images. *Sensors* **2022**, *22*, 4964. [CrossRef] [PubMed]
25. Kromanis, R.; Kripakaran, P. Predicting thermal response of bridges using regression models derived from measurement histories. *Comput. Struct.* **2014**, *136*, 64–77. [CrossRef]
26. Laory, I.; Trinh, T.N.; Smith, I.F.C.; Brownjohn, J.M.W. Methodologies for predicting natural frequency variation of a suspension bridge. *Eng. Struct.* **2014**, *80*, 211–221. [CrossRef]
27. Sarmadi, H.; Entezami, A.; Saeedi Razavi, B.; Yuen, K.-V. Ensemble learning-based structural health monitoring by Mahalanobis distance metrics. *Struct. Contr. Health Monit.* **2021**, *28*, e2663. [CrossRef]
28. Ardani, S.; Eftekhar Azam, S.; Linzell, D.G. Bridge Health Monitoring Using Proper Orthogonal Decomposition and Transfer Learning. *Appl. Sci.* **2023**, *13*, 1935. [CrossRef]
29. Yang, L.; Shami, A. On hyperparameter optimization of machine learning algorithms: Theory and practice. *Neurocomputing* **2020**, *415*, 295–316. [CrossRef]
30. Qin, X.; Zhang, L.; Yang, M.; Luo, H.; Liao, M.; Ding, X. Mapping surface deformation and thermal dilation of arch bridges by structure-driven multi-temporal DInSAR analysis. *Remote Sens. Environ.* **2018**, *216*, 71–90. [CrossRef]

engineering proceedings

Proceeding Paper

A Comparative Study on Structural Displacement Prediction by Kernelized Regressors under Limited Training Data †

Alireza Entezami *, Bahareh Behkamal, Carlo De Michele and Stefano Mariani

Department of Civil and Environmental Engineering, Politecnico di Milano, 20133 Milano, Italy; bahareh.behkamal@polimi.it (B.B.); carlo.demichele@polimi.it (C.D.M.); stefano.mariani@polimi.it (S.M.)
* Correspondence: alireza.entezami@polimi.it
† Presented at the 10th International Electronic Conference on Sensors and Applications (ECSA-10), 15–30 November 2023; Available online: https://ecsa-10.sciforum.net/.

Abstract: An accurate prediction of the structural response in the presence of limited training data still represents a big challenge if machine learning-based approaches are adopted. This paper investigates and compares two state-of-the-art kernelized supervised regressors to predict the structural response of a long-span bridge retrieved from spaceborne remote sensing technology. The kernelized supervised procedure is either based on a support vector regression or on a Gaussian process regression. A small set of displacement time histories and corresponding air temperature data are fed into the regressors to predict the actual structural response. Results demonstrate that the proposed regression techniques are reliable, even when only 30% of the training data are used at the learning stage.

Keywords: remote sensing; structural displacements; machine learning; supervised regression; long-span bridges

Citation: Entezami, A.; Behkamal, B.; De Michele, C.; Mariani, S. A Comparative Study on Structural Displacement Prediction by Kernelized Regressors under Limited Training Data. *Eng. Proc.* **2023**, *58*, 57. https://doi.org/10.3390/ecsa-10-16031

Academic Editor: Francisco Falcone

Published: 15 November 2023

1. Introduction

Structural health monitoring (SHM) has brought a practical methodology for ensuring the safety and integrity of civil structures [1–5]. This methodology is based on sensor deployment over the structure to be monitored, data acquisition, modeling, feature extraction, and feature analysis [6–8]. The modeling stage can be either physics- or data-based [9–12]. Sensors are obviously important to any SHM process because the acquired data from the structures provide information on their behavior and current state. Recently, spaceborne remote sensing has become an emerging and practical technology for monitoring large-scale civil structures [13,14] by using synthetic aperture radar (SAR) images [15–19]. Despite some limitations such as speckle noise and low spectral and resolution information, SAR images have become important data for the SHM process to rely on [20,21]. The main product of the remote sensing for SHM is the extraction of structural displacements from said SAR images [22–24].

Even if recent progress in SAR-based SHM using the aforementioned displacement responses can be exploited, especially for huge civil structures, some limitations cause obstacles to fully take advantage of this methodology. First, as for any SHM program, the in situ/field measurements are not always trivial. In most cases, field testing and measurements entail high costs, low efficiency, impact on traffic, and damage to the structures. Although the use of non-contact-based sensors, particularly spaceborne remote sensing, significantly copes with the limitations of contact-based sensing and its difficulties regarding in situ measurement, SAR images produce Big Data of huge sizes (in the unit of GB), leading to issues related to memory storage. Second, the displacement is a feature extracted from SAR images. This means that such information is not provided directly from sensor recordings, and feature extraction techniques (like interferometric approaches) appear therefore necessary. In some cases, this results in poor and unreliable displacement data. Third, spaceborne remote sensing cannot provide as rich data/features as other contact-based sensing methods, which may be installed permanently. More precisely, in most practical and long-term SHM projects based on installed contact sensors, it is possible to supply large datasets measured hourly; however, it is difficult to provide so much data from remote sensing. Fourth, it is probable that any in situ/field measurement may contain missing data, which leads to incomplete information for SHM purposes. To address these limitations, the most appropriate solution is to predict the structural response based on measured excitation or input parameters.

In machine learning, data prediction is a well-known problem. To this purpose, it is possible to exploit various regression models based on predictor (independent or input) and response (dependent or output) data [25]. In relation to the SAR-based SHM strategy, the major challenge to face is that structural displacements extracted from the SAR images are limited. In other words, due to the size of such images, few observations are often considered to extract displacement responses. In this case, the use of any regression model for small data may be problematic. The best solution is thus to leverage data expansion techniques. From the viewpoint of regression modeling, support vector regression (SVR) and Gaussian process regression (GPR) are two supervised regressors developed from the concept of kernel trick that expand a low-dimensional feature space to a high-dimensional one with a different kernel function [26,27]. However, the performance of these techniques in the presence of small datasets and the consideration of a limited training ratio have not been explored properly for SAR-based SHM.

This paper mainly intends to compare the SVR and GPR methods for predicting structural responses obtained from a few SAR images, under a tiny and unusual ratio of training data. For this goal, data related to the structural response of a long-span bridge have been considered along with ambient temperature recordings with contact-based sensors. The structural responses have been considered for some areas of the bridge, exploiting 29 SAR images only of Sentinel-A1 in a long-term monitoring scheme. Accordingly, the temperature and displacement samples are divided into training and test sets with a ratio of 30:70. The recorded ambient temperature stems as the major predictor datum, while the structural displacements are considered as the main response for prediction. Results demonstrate that SVR outperforms GPR.

2. Supervised Regression Techniques

2.1. Support Vector Regression

The fundamental principle of SVR is to map the original training data to a higher-dimensional space, and then apply an optimization approach to find a hyperplane that can separate the training data in the transformed space. This hyperplane resembles a function that can predict a target value within a tolerance margin or a decision boundary based on the training data points [28]. Given the predictor data $\mathbf{x} = \{x_1,\ldots,x_n\}$ and response data $\mathbf{y} = \{y_1,\ldots,y_n\}$, which, in the present case, gather the temperature and displacement points, respectively, the general form of the SVR model can be expressed as $y = \mathbf{w}^T\mathbf{x} + b$, where $\mathbf{w}$ denotes the weight vector and b is the bias. Based on this general form, SVR intends to exploit the training data to predict the response data, moving through the following steps: (i) separating the training data into support vectors, (ii) mapping the support vectors into high-dimensional space via a kernel function, and (iii) developing a regression model containing estimated parameters through an optimization process. Based on Mercer's theorem, the mapping procedure is performed by using different kernel functions. The procedure aims to minimize a convex function subject to constraints; for more details, readers are referred to [28].

To deal with the nonlinear regression problem, the low-dimensional parameter space needs to be mapped into the high-dimensional one by a kernel function $\boldsymbol{\phi}(\mathbf{x})$, which computes inner product values of mapped points in the feature space stored in a matrix. Therefore, the final SVR model based on any kernel function can be expressed as $y = \mathbf{w}^T\boldsymbol{\phi}(\mathbf{x}) + b$.

2.2. Gaussian Process Regression

GPR is a supervised regression model that predicts data based on the development of a kernel-based probabilistic algorithm and the theory of Gaussian processes [29]. Given the predictor and response data **x** and **y**, GPR predicts a response point using new predictor data by introducing latent variables, $L(x_1),\ldots,L(x_n)$ from a GP, and an explicit basis function. For each $i = 1,\ldots,n$, if $L(x_i)$ and x_i conform to this process, the joint distribution of the random variables $L(x_1),\ldots, L(x_n)$ is Gaussian. If these variables are from a zero mean Gaussian process, one can derive the GPR model as $\mathbf{h}(\mathbf{x})^T\mathbf{a} + \mathbf{L}(\mathbf{x})$, where $\mathbf{h}(\mathbf{x})$ denotes a basis function that transforms the predictor data $\{x_1,\ldots,x_n\}$ into a new vector, and **a** is the set of the coefficients of this function. As the GPR model is based on the probability theory, $\mathbf{h}(\mathbf{x})^T\mathbf{a} + \mathbf{L}(\mathbf{x})$ can be re-written as

$$\Pr(\mathbf{y}|\,\mathbf{L}(\mathbf{x}),\mathbf{x}) \sim N\left(\mathbf{x}|\,\mathbf{h}(\mathbf{x})^T\mathbf{a} + \mathbf{g}(\mathbf{x}), \sigma^2\right) \quad (1)$$

where $\mathbf{g}(\mathbf{x}) \sim GP(0, \boldsymbol{\phi}(\mathbf{x}))$ is equivalent to a zero-mean GP and $\boldsymbol{\phi}(\mathbf{x})$ denotes the kernel function (matrix) of the predictor data. Once the GPR model has been developed via the training data, it can predict any new response by means of the conditional distribution in Equation (1).

3. Application

The Dashengguan Bridge is a long-span high-speed railway steel bridge that crosses the Yangtze River in Nanjing, China [20]. The bridge features a large-span continuous steel arch truss with a length of 1615 m. This work focuses on the six main parts of the bridge, with a total length of 1272 m, as shown in Figure 1. The arches consist of three truss planes above the deck. The main truss has a welded, monolithic joint. The members and gusset plates were welded together in the fabrication yard and then transported to the site and spliced outside the joint with high-strength bolts.

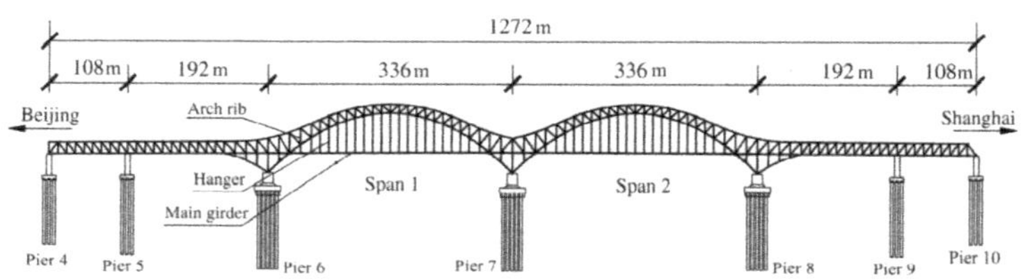

Figure 1. Side view and main dimensions of the Dashengguan Bridge.

3.1. Predictor and Response Data

In a long-term monitoring program between 25 April 2015 and 5 August 2016, 29 SAR images from Sentinel-A1 were used to extract displacement responses in the unit of mm at some critical areas of the bridge, including piers 4–6 and 8–10; see Figure 1. Figure 2 shows the mentioned structural responses at the six piers of the bridge, which were extracted with the persistent scatterer interferometry technique [20]. The ambient temperature was also recorded by contact-based sensors and is shown in Figure 3.

3.2. Prediction Results

To predict the displacement responses, the predictor and response data are divided into the training and test sets. For the training process, a small ratio of 30% is considered so that the training data consist of nine samples, including both the recorded temperature and the extracted displacement points. The remaining 20 samples are incorporated into the test dataset. The Bayesian hyperparameter optimization is adopted to tune the unknown elements of the SVR and GPR, especially the kernel function. On this basis, the linear and squared exponential kernel functions are selected for SVR and GPR, respectively, to map the limited training points into the high-dimensional feature space.

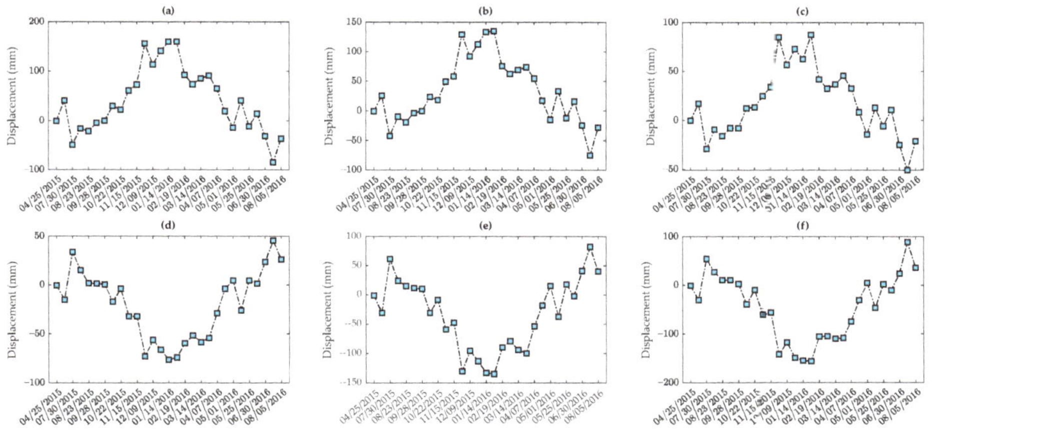

Figure 2. Structural displacements of the Dashengguan Bridge from 29 SAR images of Sentinel-A1: (**a**–**c**) Piers 4–6, (**d**–**f**) Piers 8–10.

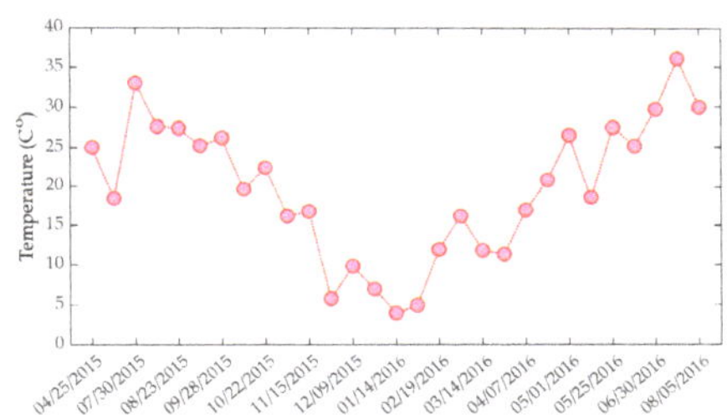

Figure 3. Air temperature records.

The results of displacement response prediction via SVR and GPR are shown in Figures 4 and 5, respectively. These charts display the scatter plots of the predicted displacements versus their real values, as extracted from the SAR images. When the scatter points are close to the reported straight line, representing a perfect match between the two datasets, one can infer that the prediction process operates well. The comparison between Figures 4 and 5 testifies that SVR outperforms GPR in predicting the displacement data in the case of a small training ratio. For further investigation, Table 1 compares the numerical outputs of the regression modeling based upon the R-squared (R^2) and root-mean-square-error (RMSE) measures; an R-squared value close to 1 is indicative of a good prediction. In Table 1, one can observe that R^2 and RMSE values relevant to the SVR model are closer to 1 and smaller than the corresponding values concerning the GPR model. Therefore, both the graphical and numerical assessments confirm a better performance of SVR compared to GPR.

Table 1. Performance evaluation of the kernelized supervised regressor.

Pier No.	Metrics			
	R^2	RMSE	R^2	RMSE
	SVR	GPR	SVR	GPR
4	0.9732	0.8413	10.6044	25.8231
5	0.9475	0.8231	12.3346	22.6547
6	0.9210	0.7603	9.5984	16.7244
8	0.9680	0.9632	6.1634	6.6125
9	0.9875	0.8155	6.7403	25.9005
10	0.9548	0.9701	14.1607	11.5007

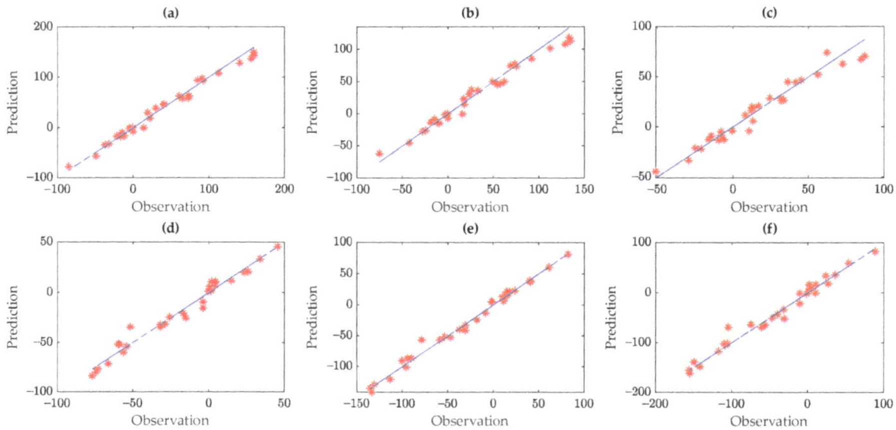

Figure 4. Predicted versus real displacements based on SVR: (**a**) Pier 4, (**b**) Pier 5, (**c**) Pier 6, (**d**) Pier 8, (**e**) Pier 9, (**f**) Pier 10.

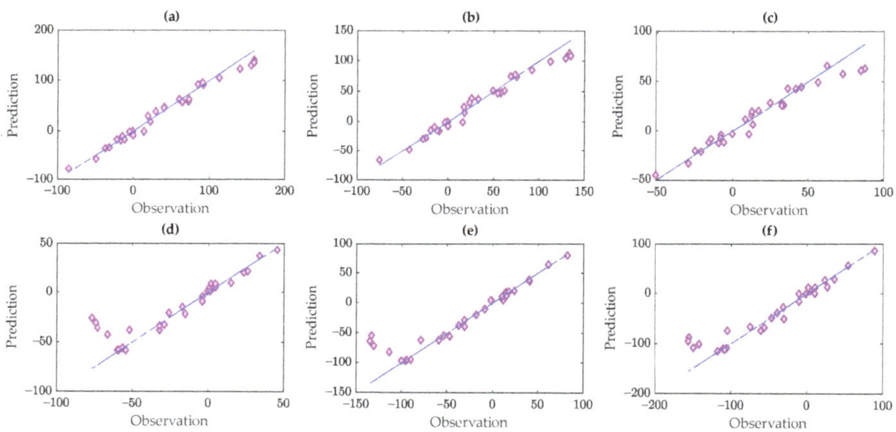

Figure 5. Predicted versus real displacements based on GPR: (**a**) Pier 4, (**b**) Pier 5, (**c**) Pier 6, (**d**) Pier 8, (**e**) Pier 9, (**f**) Pier 10.

4. Conclusions

This paper has compared two kernelized supervised regressors, namely SVR and GPR, to predict the structural displacements extracted from a few SAR images obtained with remote sensing technology in the case of a small ratio of training data. Bayesian hyperparameter optimization has also been applied to tune the unknown elements of the SVR and GPR models, especially their kernel functions.

A limited number of displacement responses of a long-span steel bridge coupled with the relevant ambient temperature values have been considered, to evaluate the capability of the regressors. The results have demonstrated that SVR provides better prediction results than GPR, in the case of only a small training dataset used in the data analysis stage of the SHM strategy.

Author Contributions: Conceptualization, A.E. and B.B.; methodology, A.E., B.B., C.D.M. and S.M.; software, A.E. and B.B.; validation, A.E. and B.B.; formal analysis, A.E., B.B., C.D.M. and S.M.; investigation, A.E., B.B., C.D.M. and S.M.; resources, A.E. and B.B.; data curation, A.E. and B.B.; writing—original draft preparation, A.E. and B.B.; writing—review and editing, A.E., B.B., C.D.M. and S.M.; visualization, A.E., B.B., C.D.M. and S.M.; supervision, C.D.M. and S.M.; project administration, A.E., B.B., C.D.M. and S.M. All authors have read and agreed to the published version of the manuscript.

Funding: This research was partially funded by the European Space Agency (ESA) under contract no. 4000132658/20/NL/MH/ac.

Institutional Review Board Statement: Not applicable.

Informed Consent Statement: Not applicable.

Data Availability Statement: Data are unavailable due to privacy or ethical restrictions.

Conflicts of Interest: The authors declare no conflicts of interest. The funders had no role in the design of the study; in the collection, analyses, or interpretation of data; in the writing of the manuscript; or in the decision to publish the results.

References

1. Entezami, A.; Sarmadi, H.; Behkamal, B.; Mariani, S. Health Monitoring of Large-Scale Civil Structures: An Approach Based on Data Partitioning and Classical Multidimensional Scaling. *Sensors* **2021**, *21*, 1646. [CrossRef] [PubMed]
2. Daneshvar, M.H.; Sarmadi, H. Unsupervised learning-based damage assessment of full-scale civil structures under long-term and short-term monitoring. *Eng. Struct.* **2022**, *256*, 114059. [CrossRef]
3. Entezami, A.; Sarmadi, H.; Behkamal, B. A novel double-hybrid learning method for modal frequency-based damage assessment of bridge structures under different environmental variation patterns. *Mech. Syst. Sig. Process.* **2023**, *201*, 110676. [CrossRef]
4. Entezami, A.; Sarmadi, H.; Behkamal, B. Long-term health monitoring of concrete and steel bridges under large and missing data by unsupervised meta learning. *Eng. Struct.* **2023**, *279*, 115616. [CrossRef]
5. Soleymani, A.; Jahangir, H.; Nehdi, M.L. Damage detection and monitoring in heritage masonry structures: Systematic review. *Constr. Build. Mater.* **2023**, *397*, 132402. [CrossRef]
6. Ardani, S.; Eftekhar Azam, S.; Linzell, D.G. Bridge Health Monitoring Using Proper Orthogonal Decomposition and Transfer Learning. *Appl. Sci.* **2023**, *13*, 1935. [CrossRef]
7. Akintunde, E.; Azam, S.E.; Linzell, D.G. Singular value decomposition and unsupervised machine learning for virtual strain sensing: Application to an operational railway bridge. *Structures* **2023**, *58*, 105417. [CrossRef]
8. Leyder, C.; Dertimanis, V.; Frangi, A.; Chatzi, E.; Lombaert, G. Optimal sensor placement methods and metrics–comparison and implementation on a timber frame structure. *Struct. Infrastruct. Eng.* **2018**, *14*, 997–1010. [CrossRef]
9. Sarmadi, H.; Entezami, A.; Ghalehnovi, M. On model-based damage detection by an enhanced sensitivity function of modal flexibility and LSMR-Tikhonov method under incomplete noisy modal data. *Eng. Comput.* **2022**, *38*, 111–127. [CrossRef]
10. Entezami, A.; Sarmadi, H.; Behkamal, B.; De Michele, C. On continuous health monitoring of bridges under serious environmental variability by an innovative multi-task unsupervised learning method. *Struct. Infrastruct. Eng.* 2023; 1–19, *in press*. [CrossRef]
11. Torzoni, M.; Rosafalco, L.; Manzoni, A. A combined model-order reduction and deep learning approach for structural health monitoring under varying operational and environmental conditions. *Eng. Proc.* **2020**, *2*, 94. [CrossRef]
12. Figueiredo, E.; Brownjohn, J. Three decades of statistical pattern recognition paradigm for SHM of bridges. *Struct. Health Monit.* **2022**, *21*, 3018–3054. [CrossRef]
13. Farneti, E.; Cavalagli, N.; Costantini, M.; Trillo, F.; Minati, F.; Venanzi, I.; Ubertini, F. A method for structural monitoring of multispan bridges using satellite InSAR data with uncertainty quantification and its pre-collapse application to the Albiano-Magra Bridge in Italy. *Struct. Health Monit.* **2023**, *22*, 353–371. [CrossRef]
14. Laflamme, S.; Ubertini, F.; Di Matteo, A.; Pirrotta, A.; Perry, M.; Fu, Y.; Li, J.; Wang, H.; Hoang, T.; Glisic, B. Roadmap on measurement technologies for next generation structural health monitoring systems. *Meas. Sci. Technol.* **2023**, *34*, 093001. [CrossRef]
15. Plank, S. Rapid Damage Assessment by Means of Multi-Temporal SAR—A Comprehensive Review and Outlook to Sentinel-1. *Remote Sens.* **2014**, *6*, 4870–4906. [CrossRef]
16. Giordano, P.F.; Turksezer, Z.; Previtali, M.; Limongelli, M.P. Damage detection on a historic iron bridge using satellite DInSAR data. *Struct. Health Monit.* **2022**, *21*, 2291–2311. [CrossRef]
17. Giordano, P.F.; Previtali, M.; Limongelli, M.P. Monitoring of a Metal Bridge Using DInSAR Data. In *European Workshop on Structural Health Monitoring*; Springer: Cham, Switzerland, 2021. [CrossRef]
18. Farneti, E.; Cavalagli, N.; Venanzi, I.; Salvatore, W.; Ubertini, F. Residual service life prediction for bridges undergoing slow landslide-induced movements combining satellite radar interferometry and numerical collapse simulation. *Eng. Struct.* **2023**, *293*, 116628. [CrossRef]
19. Cavalagli, N.; Kita, A.; Farneti, E.; Falco, S.; Trillo, F.; Costantini, M.; Fornaro, G.; Reale, D.; Verde, S.; Ubertini, F. Remote sensing and in-situ measurements for the structural monitoring of historical monuments: The Consoli Palace of Gubbio, Italy. In *European Workshop on Structural Health Monitoring*; Springer: Cham, Switzerland, 2021. [CrossRef]
20. Huang, Q.; Crosetto, M.; Monserrat, O.; Crippa, B. Displacement monitoring and modelling of a high-speed railway bridge using C-band Sentinel-1 data. *ISPRS J. Photogramm. Remote Sens.* **2017**, *128*, 204–211. [CrossRef]

21. Entezami, A.; Arslan, A.N.; De Michele, C.; Behkamal, B. Online hybrid learning methods for real-time structural health monitoring using remote sensing and small displacement data. *Remote Sens.* **2022**, *14*, 3357. [CrossRef]
22. Farneti, E.; Cavalagli, N.; Costantini, M.; Trillo, F.; Minati, F.; Venanzi, I.; Salvatore, W.; Ubertini, F. Remote Sensing Satellite Data and Progressive Collapse Analysis for Structural Monitoring of Multi-span Bridges. In *European Workshop on Structural Health Monitoring*; Springer: Cham, Switzerland, 2022. [CrossRef]
23. Macchiarulo, V.; Milillo, P.; Blenkinsopp, C.; Giardina, G. Monitoring deformations of infrastructure networks: A fully automated GIS integration and analysis of InSAR time-series. *Struct. Health Monit.* **2022**, *21*, 1849–1878. [CrossRef]
24. Milillo, P.; Giardina, G.; Perissin, D.; Milillo, G.; Coletta, A.; Terranova, C. Reply to Lanari, R., et al. comment on "pre-collapse space geodetic observations of critical infrastructure: The morandi bridge, Genoa, Italy" by Milillo et al. (2019). *Remote Sens.* **2020**, *12*, 4016. [CrossRef]
25. Figueiredo, E.; Park, G.; Farrar, C.R.; Worden, K.; Figueiras, J. Machine learning algorithms for damage detection under operational and environmental variability. *Struct. Health Monit.* **2011**, *10*, 559–572. [CrossRef]
26. Behkamal, B.; Entezami, A.; De Michele, C.; Arslan, A.N. Investigation of temperature effects into long-span bridges via hybrid sensing and supervised regression models. *Remote Sens.* **2023**, *15*, 3503. [CrossRef]
27. da Silva, S.; Figueiredo, E.; Moldovan, I. Damage Detection Approach for Bridges under Temperature Effects using Gaussian Process Regression Trained with Hybrid Data. *J. Bridge Eng.* **2022**, *27*, 04022107. [CrossRef]
28. Smola, A.J.; Schölkopf, B. A tutorial on support vector regression. *Stat. Comput.* **2004**, *14*, 199–222. [CrossRef]
29. Schulz, E.; Speekenbrink, M.; Krause, A. A tutorial on Gaussian process regression: Modelling, exploring, and exploiting functions. *J. Math. Psychol.* **2018**, *85*, 1–16. [CrossRef]

engineering proceedings

MDPI

Proceeding Paper

Simulation of ZnO Nanofils as Applications for Acetone Gas Sensors †

Asmaa Zeboudj *, Mokhtar Zardali *, Asmaa Tadji * and Saad Hamzaoui *

LMESM Laboratory, Physics Department, University of Science and Technology Mohamed Boudiaf (USTO-MB), Oran 31000, Algeria

* Correspondence: asmaa.zeboudj@univ-usto.dz (A.Z.); mokhtarzerdali@gmail.com (M.Z.); tadjiasma@yahoo.fr (A.T.); hamzaoui.saad@gmail.com (S.H.)

† Presented at the 10th International Electronic Conference on Sensors and Applications (ECSA-10), 15–30 November 2023; Available online: https://ecsa-10.sciforum.net/.

Abstract: Our objective is to present a valuable contribution towards designing more efficient sensors using undoped ZnO nanofils. The utilization of nanostructures based on ZnO has led to significant enhancements in sensor performance, due to the excellent chemical and thermal stability exhibited at its high melting temperature, our work, we focused on modeling the behavior of ZnO semiconductors by employing the Schottky defect model as a source of free carriers. Specifically, we examined the theoretical model of oxygen molecule adsorption and desorption. We explored two types of molecules responsible for adsorbing reducing gases, taking acetone gas as an example. Through the use of the COMSOL software, we found that the interaction between the solid and gas occurs at a considerably lower temperature of 295 °C, compared to ZnO thin films, which typically require temperatures as high as 500 °C. This outcome can be attributed to the behavior of ZnO nanostructures, where the influence of side surfaces (1010) is predominant, along with their lower activation energy compared to (0002) surfaces. These ZnO nanofils exhibit numerous active and thermodynamically favorable surfaces, which facilitate the adsorption of reducing gases. Employing simulation methods, such as COMSOL, offers an effective approach for achieving an optimal device design, thereby ensuring superior device performance. This research demonstrates the potential of using undoped ZnO nanofils for the development of highly efficient sensors with enhanced operational characteristics.

Keywords: semiconductor; ZnO; Nanostructures; the Schottky defect; adsorption; desorption; acetone gas

Citation: Zeboudj, A.; Zardali, M.; Tadji, A.; Hamzaoui, S. Simulation of ZnO Nanofils as Applications for Acetone Gas Sensors. *Eng. Proc.* **2023**, *58*, 58. https://doi.org/10.3390/ecsa-10-16027

Academic Editor: Stefano Mariani

Published: 15 November 2023

1. Introduction

This research focuses on the development of environmentally conscious materials with strong capabilities in gas detection. This poses a significant challenge for scientists, who are aiming to create durable solutions in the realm of material science. Among the various possibilities offered by metal oxides, those that exhibit semi-conductive properties have emerged as particularly promising candidates for gas-sensing applications.

One standout example within this category is zinc oxide (ZnO), which has garnered substantial attention due to its wide-ranging potential, spanning applications in both renewable energy and environmental preservation. ZnO, found abundantly in vivid ruby-red ores, possesses remarkable attributes that make it highly suitable for gas-sensing purposes. Moreover, ZnO's versatility extends to pioneering novel materials for the realm of renewable energy.

While naturally occurring ZnO displays a distinctive ruby-red color, artificially synthesized versions appear colorless or white. This compound finds practical utility across various domains, encompassing solar cell technology, light-emitting diodes (LEDs), and the field of gas sensors the main objective of this study is to develop sensors based on these nanowires for their potential use in medical devices, enabling early diagnosis of certain conditions, including the measurement of acetone in the urine of diabetic individuals.

2. Materials and Methods

- **Nanowire-Based Sensors: Configurations and Operations:**

Sensors constructed using solid metal oxide materials typically comprise the following [1]: A sensitive layer of oxide/semiconductor that directly interacts with the targeted gas. electrodes that facilitate electrical measurements and monitor the interaction process.And an optional heating system, to manage and regulate the temperature of the sensitive layer.

Nanowires have garnered substantial interest in the realm of gas detection applications due to their elevated surface-to-volume ratio, which holds the promise of achieving ulta-sensitivity. This emphasis on nanowires has underscored the fact that the operational efficacy of these sensors is intricately tied to both the selection of material for the nanowires and the specific configuration chosen [2,3].

In our specific context, we are focusing on the resistance-based configuration for fabricating gas sensors utilizing ZnO nanowires (Figure 1).

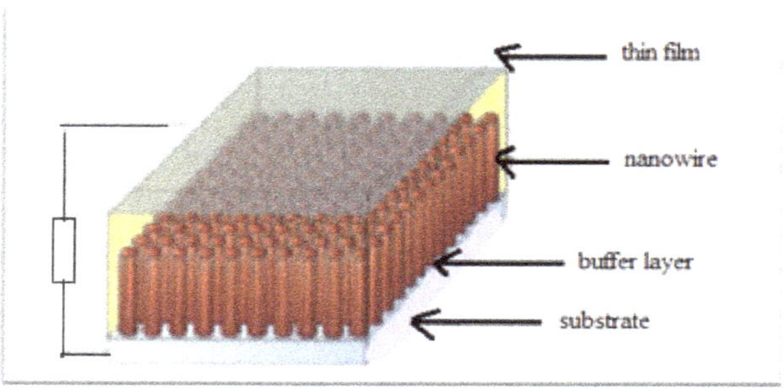

Figure 1. Schematic representation of the ZnO nanowire-based sensor [4,5].

In this arrangement, the nanowire functions as a channel, with its ends connected to the source and drain. Upon exposure of the ZnO nanowires' surface (0002) to the gas, a distinct reaction ensues, inducing alterations in their electrical behavior and generating an electrical signal that correlates with the gas detection phenomenon.

- **Gas Detection Principle**

The fundamental principle of gas detection relies on the alteration of semiconductor surface resistance, typically involving modifications in the conductivity of the sensor's constituent materials. This change originates from fluctuations in the density of electron species caused by the adsorption of gas molecules onto the surface of the host material [4,5]. In our context, we are particularly concerned with detecting acetone gas.

Acetone, a commonly used solvent in research laboratories and the paint and adhesive industry, falls into the category of highly volatile organic compounds (VOCs) and is recognized as a toxic compound. Its high volatility is concerning, due to its significant absorption rate (approximately 75%) through human respiratory pathways [6]. Exposure to concentrated acetone poses risks to living organisms.

The detection of acetone using a semi-conductive surface has been extensively explored in existing literature when ZnO is exposed to the reducing gas acetone.

(CH_3COCH_3), the acetone molecules react with adsorbed oxygen (O) species on the surface of ZnO (0002), resulting in the formation of CO_2 and H_2O. This chemical reaction leads to an increase in carrier concentration and a subsequent reduction in electrical resistance [7].

$$CH_3\ COCH_3 + 8O^- \rightarrow 3CO_2 + 3H_2O + 8e^-$$

3. Results and Discussion

We utilized the COMSOL software (5.4.a.) to develop our research. Our primary objective was to design an undoped ZnO nanowire gas sensor, where the nanowire size is defined as 100 nm. This sensor was integrated with an embedded mini furnace crafted from platinum and deposited onto a glass wafer. The integration aimed to efficiently control and regulate the temperature. Moreover, we positioned a sensitive ZnO layer between two titanium electrodes. The ZnO layer's thickness was established at 0.5 µm. The outcome obtained from the experiment, as indicated in (Table 1).

Table 1. The various concentrations employed to ascertain their response concerning gas concentrations.

V_{in}	T (°C)	R_a (ppm = 0)	R_a/R_G (ppm = 10)	R_a/R_G (ppm = 100)	R_a/R_G (ppm = 500)
1	56	1.24×10^9	1.00	1.00	1.01
2.5	187	4.69×10^6	3.22	1.02	1.08
2.6	198	5.14×10^6	1.52	1.11	1.71
3.5	295	2.46×10^6	1.69	7.93	3.57×10^1
4.8	444	1.87×10^3	1.19	2.95	1.07×10^1
5	468	8.99×10^2	1.19	2.93	1.07×10^1

The R_a/R_G: ratio illustrates the sensor's sensitivity or response to the detected gas, where

V_{in}: input voltage;

R_a: air resistance;

R_G: gas resistance.

The data are translated into a profile illustrating the sensor's temperature against the R_a/R_G response (Figure 2).

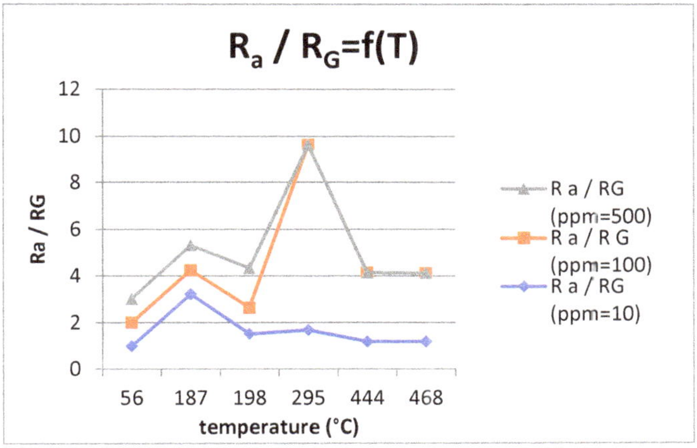

Figure 2. The sensitivity of gas sensors based on ZnO nanowires in relation to temperature.

Figure 2 depicts how temperature influences the sensitivity of undoped ZnO nanowires to different levels of acetone gas. The profile clearly demonstrates that the sensitivity of the sensor, as indicated by the ratio of resistances, varies with rising temperature. The sensor's sensitivity is quantified by the specific acetone-driven ratio, Ra/Rs (air resistance/gas resistance). The sensor maintains a consistent response, relative to air across temperatures, but its sensitivity becomes more pronounced around Ts = 295 °C for the various acetone concentrations simulated.

4. Conclusions

ZnO, a non-toxic semiconductor, is gaining increasing significance within the scientific community. This is primarily due to its remarkable physicochemical properties, making it a promising material for metal oxide gas sensors. In this context, it is utilized to optimize sensor components through electrical measurements (I–V), conducted using Multiphysics simulation software (COMSOL). We studied the effect of operating temperature on the sensitivity of ZnO nanowires to acetone. The results clearly demonstrate that this temperature plays a major role in determining the film's sensitivity to different concentrations of acetone.

Author Contributions: A.Z. wrote the main manuscript text, prepared the figures, and performed all experiments. M.Z., A.T. and S.H. oversaw the project and assisted with the writing of the overall manuscript. All authors have read and agreed to the published version of the manuscript.

Funding: This research received no external funding.

Institutional Review Board Statement: Not applicable.

Informed Consent Statement: Not applicable.

Data Availability Statement: Data are provided in the figures of the article.

Acknowledgments: We would like to thank the LMESM Laboratory, Physics Department, University of Science and Technology Mohamed Boudiaf in "Instrumentation aux limites".

Conflicts of Interest: The authors declare no conflict of interest.

References

1. Moustaghfir, A. Elaboration et Caractérisation de Couches Minces D'oxyde de Zinc: Application à la Photoprotection du Polycarbonate. Ph.D. Thesis, Université Blaise Pascal, Clermont-Ferrand, France, 2006.
2. Look, D.C.; Joues, R.L.; Sizelove, J.R.; Graces, N.Y.; Giles, N.C.; Halliburton, L.E. The parth to ZnO devices; donor and acceptors dynamic. *Phys. Status Solidi* **2003**, *195*, 171. [CrossRef]
3. Erhart, P.; Albe, K.; Klein, A. First-principles study of intrinsic point defects in ZnO: Role of band structure, Volume relaxation, and finite-size effects. *Phys. Rev. B* **2006**, *73*, 205203. [CrossRef]
4. Li, S.S. *Semiconductor Physical Electronics*; Springer: Berlin/Heidelberg, Germany, 2006.
5. Kaszynsky, A. Etude des Phénomènes de Transport dans les Matériaux Semi-Conducteurs par la Méthode Monte Carlo: Application a L'arséniure de Gallium de Type N. Ph.D. Thesis, Université des Sciences et Techniques de Lille, Villeneuve d'Ascq, France, 1979.
6. Gomri, S.; Seguin, J.L.; Guerin, J.; Aguir, K. Adsorption-desorption noise in gas sensors: Modelling using Langmuir and Wolkenstein models for adsorption. *Sens. Actuators B* **2006**, *114*, 451–459. [CrossRef]
7. Oba, F.; Choi, M.; Togo, A.; Tanaka, I. Point defects in ZnO: An approach from first principles. *Sci. Technol. Adv. Mater.* **2011**, *12*, 034302. [CrossRef] [PubMed]

engineering proceedings

Proceeding Paper

Changes in Trunk Kinematics in People with Chronic Non-Specific Low Back Pain Using Wearable Inertial Sensors †

Batlkham Dambadarjaa [1], Batbayar Khuyagbaatar [2,*], Damdindorj Boldbaatar [3], Baljinnyam Avirmed [4] and Munkh-erdene Bayartai [1,*]

[1] Department of Physical Therapy, School of Nursing, Mongolian National University of Medical Sciences, Ulaanbaatar 14191, Mongolia; batlkham@mnums.edu.mn
[2] Biomechanical Research Laboratory, Mongolian University of Science and Technology, Ulaanbaatar 14191, Mongolia
[3] Research and International Affairs, Mongolian National University of Medical Sciences, Ulaanbaatar 14191, Mongolia; damdindorj@mnums.edu.mn
[4] Department of Rehabilitation Medicine, School of Medicine, Mongolian National University of Medical Sciences, Ulaanbaatar 14191, Mongolia; baljinnyam.a@mnums.edu.mn
* Correspondence: batbayarkh@must.edu.mn (B.K.); munkh-erdene@mnums.edu.mn (M.-e.B.); Tel.: +976-86075099 (B.K.); +976-99059026 (M.-e.B.)
† Presented at the 10th International Electronic Conference on Sensors and Applications (ECSA-10), 15–30 November 2023; Available online: https://ecsa-10.sciforum.net/.

Abstract: Low back pain (LBP) is one of the most common musculoskeletal conditions and the leading cause of disability. It is estimated that at least 8 out of 10 people experience low back pain during their lifetime. The purpose of this study was to determine trunk kinematics in individuals with and without non-specific chronic LBP during flexion–extension and hurdle step tests. A total of 90 participants (45 participants with LBP and 45 without LBP), aged between 18 and 50, participated in this study. The wearable inertial sensors were used to capture three-dimensional movements during both trunk flexion–extension and the hurdle step test. Altered trunk kinematics during the flexion–extension and the hurdle step test were observed in individuals with non-specific chronic low back pain.

Keywords: low back pain; kinematics; wearable sensors

Citation: Dambadarjaa, B.; Khuyagbaatar, B.; Boldbaatar, D.; Avirmed, B.; Bayartai, M.-e. Changes in Trunk Kinematics in People with Chronic Non-Specific Low Back Pain Using Wearable Inertial Sensors. *Eng. Proc.* **2023**, *58*, 59. https://doi.org/10.3390/ecsa-10-16204

Academic Editor: Stefan Bosse

Published: 15 November 2023

1. Introduction

Low back pain (LBP) is one of the most common musculoskeletal conditions and the leading cause of disability [1]. It is estimated that 8 out of 10 people experience LBP during their lifetime [1]. Non-specific LBP, where no pathological or anatomical changes [2] are found, accounts for 85 percent of all LBP cases [1]. Muscle stiffness and movement impairment or limitation are the common symptoms of non-specific LBP [1]. Researchers have reported changes in the lumbar lordosis and spinal range of motion (ROM) in persons with LBP when compared with controls [3–5]. Zubierer et al. [3] studied the convergence and discriminant validity for lumbar range of motion tests and LBP. Alaa Haj et al. [4] reported on the ROM, average speed, maximum speed, and maximum acceleration of lumbar rotation in the neutral position and full flexion. Ng et al. [5] compared the lumbar kinematics of a flexion–extension and lateral flexion ROM test between persons with LBP and controls. Other researchers have measured lumbar kinematics using a hurdle step test [2,6]. However, few studies have focused on trunk kinematics during both ROM and the hurdle step test using wearable motion capture systems, although these systems have been extensively used in human motion analyses [7,8]. The purpose of this study was to determine trunk kinematics in individuals with and without non-specific chronic LBP during flexion–extension and hurdle step tests.

2. Material and Methods

2.1. Participant Information

A cross-sectional study design was conducted with a total of 90 participants (45 participants with LBP and 45 without LBP), aged between 18 and 50. The study was approved by the Ethics Committee of the Mongolian National University of Medical Sciences (N° 2022\3-7).

2.2. Experimental Procedure

The full-body wearable, Xsens motion capture system (MVN, Xsens Technologies BV, the Netherlands) was used to capture the three-dimensional (3D) movements of the trunk during flexion–extension while standing and during the hurdle step test at a sampling rate of 120 Hz. Each participant performed both movement tasks three times. Trunk flexion–extension was performed in the standing position by bending forward and backward with the knees locked in extension (Figure 1a). In the hurdle step test, participants started in a standing position and stepped over the hurdle (Figure 1b). The height of the hurdle was adjusted to equal the height of the person's tibial tuberosity [2].

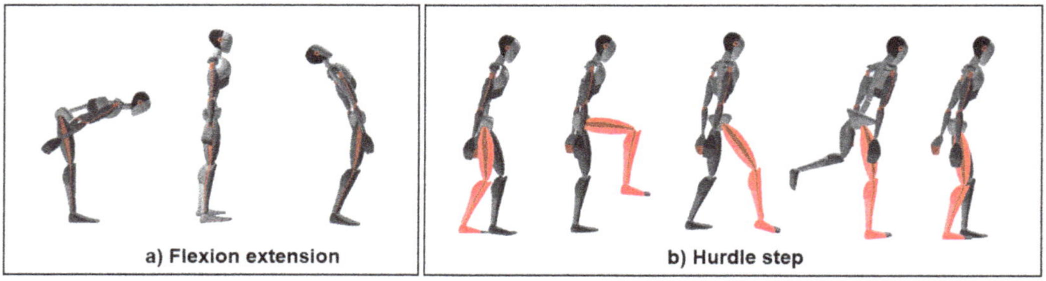

Figure 1. Trunk flexion–extension and hurdle step tests were performed by the participants: (**a**) from a standing position, bending forward and backward with locked knees. Repeat 3 times; (**b**) step over the hurdle from a standing position (start on right leg). Repeat 3 times. The height of the hurdle is equal to the height of the person's tibial tuberosity (5).

The wearable captain system includes 15 inertial measurement unit (IMU) sensors, which were attached to the head, sternum, pelvis, left/right shoulder, upper- and forearm, upper and lower leg, and foot. With sensors attached, each participant performed trunk flexion–extension and the hurdle step test three times, according to the protocol of previous studies [2]. The trunk joint angles and velocity in the sagittal (flexion–extension), frontal (lateral bending), and transversal (axial rotation) planes were calculated using a relative orientation between pelvis and thorax segments, and averaged for the LBP and control groups [9].

2.3. Statistical Analysis

The statistical analysis of participant characteristics and trunk kinematics was performed using IBM SPSS Statistics Version 25. Mean values and standard deviations for age, weight, height, and gender were calculated using descriptive statistics. Differences in parameters between people with and without non-specific LBP were determined using an age- and weight-adjusted analysis of variance (ANOVA).

Flexion–extension range of movement and hurdle step tests for spine motion according to the protocol in [2] were provided by the Xsens software (Figure 1).

3. Results

Table 1 shows the characteristics of the participants. We compared the measurement of trunk range of motion (ROM) and velocity in three planes during the flexion–extension ROM test (Table 2) and the hurdle step test (Table 2) between the LBP and control groups.

We found that some trunk kinematics were different between people with and without LBP (Table 2). For instance, during the flexion–extension test, trunk lateral bending and rotation range of motion angles had statistically significant differences between the group with LBP and the control group, as highlighted in red in Table 2. Additionally, during the hurdle step test, there were significant differences in trunk rotation velocity between the group with LBP and the control group.

Table 1. Participants' characteristics.

Subjects	LBP n = 45	Control n = 45
Age (years)	33 ± 10.1	24.5 ± 8.0 *
Height (cm)	162.9 ± 8.1	163 ± 8.1
Weight (kg)	67.4 ± 16.5	59.4 ± 10.2 *
Gender (female)	36 (80%)	31 (69%)

* $p \leq 0.05$—a significant difference. Mean ± SD.

Table 2. Comparison of trunk ROM and velocity during Flexion–Extension and Hurdle Step Tests in LBP and control groups.

	Variables	Control (n = 45)	LBP (n = 45)	Differences in ROM
Flexion–Extension Test	Flexion (degree)	11.3 (1.2)	7.7 (1.2)	3.6 (−0.03 to 7.2)
	Extension (degree)	−14.2 (1.0)	−17.2 (1.0)	−2.9 (−0.2 to 6.1)
	Lateral bending (degree)	***12.6 (0.7)***	***8.8 (0.7) ****	***3.7 (1.5 to 5.9)***
	Rotation (degree)	***8.2 (0.5)***	***6.0 (0.5) ****	***2.1 (0.6 to 3.6)***
	Flexion–extension velocity (m/s)	0.61 (0.09)	0.69 (0.09)	−0.07 (−0.3 to 0.2)
	Lateral bending velocity (m/s)	0.2 (0.03)	0.2 (0.03)	0.03 (−0.08 to 0.14)
	Rotation velocity (m/s)	0.29 (0.04)	0.25 (0.04)	0.03 (−0.08 to 0.16)
Hurdle Step Test	Flexion–extension (degree)	4.4 (1.0)	3.5 (1.0)	0.9 (−2.2 to 4.0)
	Lateral bending (degree)	12.0 (1.1)	12.3 (1.1)	−0.3 (−3.8 to 3.2)
	Rotation (degree)	66.8 (3.2)	63.7 (3.2)	3.0 (−6.6 to 12.6)
	Flexion–extension velocity (m/s)	37.2 (3.2)	41.3 (3.2)	−4.1 (−13.6 to 5.3)
	Lateral bending velocity (m/s)	47.6 (3.4)	46.3 (3.4)	1.2 (−5.8 to 11.3)
	Rotation velocity (m/s)	***56.6 (3.9)***	***40.8 (3.9 *)***	***15.7 (4.1 to 27.3)***

* $p \leq 0.05$—significantly different. Mean (standard error). Mean—covariates of age and weight.

4. Discussion and Conclusions

In this study, we used a wearable IMU sensor to compare trunk kinematics during flexion–extension and hurdle step tests between LBP and control groups (Table 2). We used flexion–extension ROM to determine the range of motion and velocity of the trunk. During trunk flexion–extension, lateral bending, and rotation, the range of motion angles showed statistically significant differences between the LBP and control groups. Some studies of persons with LBP, using IMU sensor measures, have shown a decreased range of flexion, extension, lateral bending, and axial rotation [9–11], and these results are consistent with the results of our current study. The results indicate that trunk kinematic changes in the spine can be evaluated using an IMU sensor in persons with non-specific LBP. In addition, the trunk joint angle and velocity measured during the hurdle step test showed significantly less trunk rotation in the LBP group than in the control group. The hurdle stepping task requires stability and coordination between the hips and torso during the stepping motion. Ko, et al. suggested that patients with chronic LBP lack this stability and coordination [6]. The low score of patients with chronic LBP on the hurdle step task confirms that spine and hip mobility is limited in chronic LBP [12]. In LBP, the movement of the trunk and hips may limit the range of motion velocity of the trunk rotation during functional tasks such as the hurdle step test. The main limitation of the present study is that the subjects in the control group were younger and leaner than those in the LBP group. However, the statistical analysis was adjusted for age and weight. In conclusion, altered trunk kinematics during the flexion–extension and hurdle step test were observed in individuals with non-specific chronic LBP. This result may be useful in further investigations into movement analyses of persons with low back pain and potentially support development of kinematic outcome measures.

Author Contributions: Conceptualization, B.K. and B.D.; methodology, B.K. and B.D.; software, B.K. and B.D.; validation, B.K., B.D. and M.-e.B.; formal analysis, B.K. and B.D.; investigation, B.D.; resources, B.D. and M.-e.B.; data curation, B.D. and M.-e.B.; writing—original draft preparation, B.D.; writing—review and editing, B.K., D.B., M.-e.B. and B.A.; visualization, B.K. and B.D.; supervision, D.B. and B.A.; project administration, B.K. and M.-e.B.; funding acquisition, B.K. All authors have read and agreed to the published version of the manuscript.

Funding: This work was supported by the Mongolian University of Science and Technology (mfund-052022) and the "Mongolia-Japan Engineering Education Development" project (J24C16), Mongolia.

Institutional Review Board Statement: The study was conducted according to the guidelines of the Declaration of Helsinki and approved by the Ethics Committee of the Mongolian National University of Medical Sciences (N° 2022\3-7).

Informed Consent Statement: Informed content was obtained from all subjects involved in the study.

Data Availability Statement: Data are contained within the article.

Acknowledgments: The authors would like to thank the team of Biomechanical Research Laboratory, Mongolian University of Science and Technology.

Conflicts of Interest: The authors declare no conflicts of interest.

References

1. Disease, G.B.D.; Injury, I.; Prevalence, C. Global, regional, and national incidence, prevalence, and years lived with disability for 310 diseases and injuries, 1990–2015: A systematic analysis for the Global Burden of Disease Study 2015. *Lancet* **2016**, *388*, 1545–1602. [CrossRef]
2. Bayartai, M.E.; Ferreira, P.H.; Pappas, E.; Pinheiro, M.B.; Dambadarjaa, B.; Khuyagbaatar, E.; Sullivan, J. Genetic and environmental effects on lumbar posture, flexibility and motion control in healthy adults. *Musculoskelet Sci. Pract.* **2020**, *50*, 102253. [CrossRef] [PubMed]
3. Zuberbier, O.A.; Kozlowski, A.J.; Hunt, D.G.; Berkowitz, J.; Schultz, I.Z.; Crook, J.M.; Milner, R.A. Analysis of the convergent and discriminant validity of published lumbar flexion, extension, and lateral flexion scores. *Spine* **2001**, *26*, E472–E478. [CrossRef] [PubMed]
4. Haj, A.; Weisman, A.; Masharawi, Y. Lumbar axial rotation kinematics in men with non-specific chronic low back pain. *Clin. Biomech.* **2019**, *61*, 192–198. [CrossRef] [PubMed]
5. Ng, J.K.; Richardson, C.A.; Kippers, V.; Parnianpour, M. Comparison of lumbar range of movement and lumbar lordosis in back pain patients and matched controls. *J. Rehabil. Med.* **2002**, *34*, 109–113. [CrossRef] [PubMed]
6. Ko, M.J.; Noh, K.H.; Kang, M.H.; Oh, J.S. Differences in performance on the functional movement screen between chronic low back pain patients and healthy control subjects. *J. Phys. Ther. Sci.* **2016**, *28*, 2094–2096. [CrossRef] [PubMed]
7. Khuyagbaatar, B.; Purevsuren, T.; Ganbat, D. Normal Range of Motion of Lower Extremity Joints in Mongolian Subjects. *Eng. Proc.* **2021**, *11*, 29.
8. Tumurbaatar, M.; Khuyagbaatar, B.; Kim, Y.H.; Danaa, G. Kinematic Characteristics of National and College Level Weightlifters during the Snatch Technique Using Wearable Inertial Sensors. *Eng. Proc.* **2021**, *10*, 22.
9. Shojaei, I.; Vazirian, M.; Salt, E.G.; Van Dillen, L.R.; Bazrgari, B. Timing and magnitude of lumbar spine contribution to trunk forward bending and backward return in patients with acute low back pain. *J. Biomech.* **2017**, *53*, 71–77. [CrossRef] [PubMed]
10. Roncarati, A.; McMullen, W. Correlates of low back pain in a general population sample: A multidisciplinary perspective. *J. Manip. Physiol. Ther.* **1988**, *11*, 158–164.
11. Troup, J.D.; Foreman, T.K.; Baxter, C.E.; Brown, D. 1987 Volvo award in clinical sciences. The perception of back pain and the role of psychophysical tests of lifting capacity. *Spine* **1987**, *12*, 645–657. [CrossRef] [PubMed]
12. McGregor, A.H.; McCarthy, I.D.; Doré, C.J.; Hughes, S.P. Quantitative assessment of the motion of the lumbar spine in the low back pain population and the effect of different spinal pathologies of this motion. *Eur. Spine J.* **1997**, *6*, 308–315. [CrossRef] [PubMed]

engineering proceedings

MDPI

Proceeding Paper

Automated Damage Detection on Concrete Structures Using Computer Vision and Drone Imagery †

Timothy Malche [1], Sumegh Tharewal [2] and Rajesh Kumar Dhanaraj [2,*]

1 Department of Computer Applications, Manipal University Jaipur, Jaipur 303007, India; timothy.malche@gmail.com

2 Symbiosis Institute of Computer Studies and Research (SICSR), Symbiosis International (Deemed) University (SIU), Pune 411016, India; sumeghtharewal@gmail.com

* Correspondence: sangeraje@gmail.com

† Presented at the 10th International Electronic Conference on Sensors and Applications (ECSA-10), 15–30 November 2023; Available online: https://ecsa-10.sciforum.net/.

Abstract: The manual inspection of concrete structures, such as tall buildings, bridges, and huge infrastructures can be time-consuming and costly, and damage assessment is a crucial task that requires the close-range inspection of all surfaces. The proposed system uses computer vision model to identify various types of damages on these structures. The computer vision model and was trained on a large dataset of drone footage, which was annotated manually to ensure accuracy. The model was then tested on new data, and the results showed that it could accurately detect and identify structural damage on concrete structures with a 94% accuracy. The system is much faster and more efficient than manual inspection, reducing the time and cost required for damage assessment. The proposed system has the potential to revolutionize the way we perform damage assessment on concrete structures. It can help to preserve and protect these valuable assets by enabling the early detection of damage and facilitating timely repairs.

Keywords: drone-based automated system; computer vision; structural damage detection; deep learning algorithms; Internet of Things (IoT); close-range footage analysis

Citation: Malche, T.; Tharewal, S.; Dhanaraj, R.K. Automated Damage Detection on Concrete Structures Using Computer Vision and Drone Imagery. *Eng. Proc.* **2023**, *58*, 60. https://doi.org/10.3390/ecsa-10-16059

Academic Editor: Stefano Mariani

Published: 15 November 2023

1. Introduction

The manual inspection of concrete structures, such as tall buildings, bridges, and huge infrastructures, is a time-consuming and risky process for human employees. Drones with sensor camera nodes have shown potential in gathering close-range footage, but the problem lies in rapidly analyzing enormous volumes of data to detect and diagnose structural deterioration. This study deals with these challenges by presenting an Internet of Things (IoT), computer-vision- and deep-learning-based automated solution. The primary issue addressed in this research is the requirement for a more efficient and reliable way of identifying structural damage on concrete structures.

The traditional manual inspection technique is time-consuming and expensive, making timely repairs and maintenance impossible. As a result, an automated solution is necessary to speed up the damage assessment process while reducing dangers to human personnel. The proposed system focuses on detecting various types of damage, such as cracks, Alkali-Silica Reactions (ASRs), concrete degradation, and others, on concrete structures using drone-captured video footage [1,2]. The system's scope includes developing a Convolutional Neural Network (CNN) architecture tailored to this specific task and implementing a seamless process for automatically obtaining video data from drones.

The primary objective of this work is to create and implement an automated damage detection system capable of identifying structural damage on concrete structures in an efficient and accurate manner. The technology intends to expedite the inspection process by utilizing IoT, computer vision, and deep learning techniques, enabling proactive maintenance and preservation activities. The novelty of the proposed system is a custom-designed CNN architecture that is optimized for detecting damage on concrete structures and a system architecture based on IoT to automatically capture data and perform analysis and reporting. The performance of the proposed automated damage detection system was evaluated using a diverse dataset of drone-captured video footage containing various types of damage on concrete structures. The CNN architecture demonstrated impressive results, achieving an accuracy of 94% in correctly identifying different types of structural damage.

The approach involves capturing close-range footage of the infrastructure with drones and processing the footage using a computer vision model to identify and classify damage. The proposed method can provide an efficient and cost-effective way to detect damage to concrete structures and help preserve these important historical structures for future generations. The working of the system is shown in Figure 1.

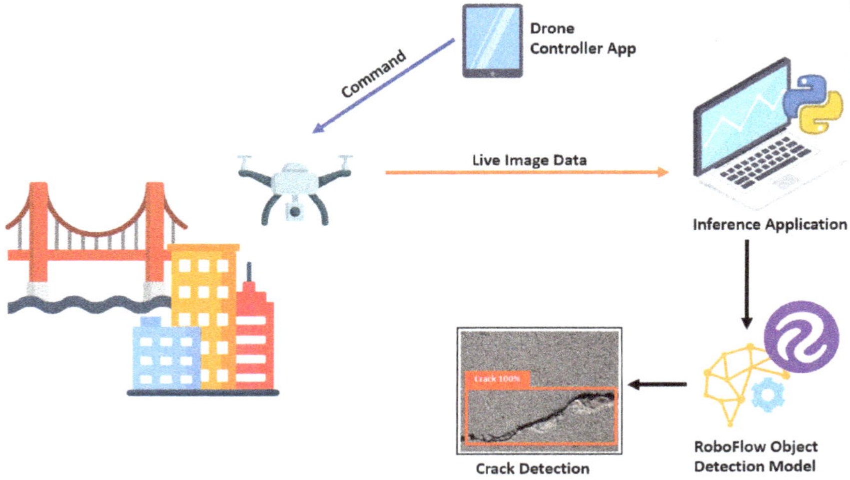

Figure 1. System Architecture.

2. Related Work

This comprehensive analysis [3] explores into the critical issue of damage detection in civil engineering, emphasizing the transformational potential of Computer Vision and Deep Learning algorithms. The research [4] presents a Deep Convolutional Neural Network-based Damage Locating (DCNN-DL) approach for steel frame inspection that outperforms existing techniques with a 99.3% accuracy. The approach uses the DenseNet architecture to properly identify and detect damaged regions, providing a fast real-time solution for visual damage evaluation in civil structures. The research in [5] investigates the use of computer vision algorithms in conjunction with remote cameras and unmanned aerial vehicles (UAVs) for non-contact civil infrastructure evaluation. The study in [6] presents a unique structural identification framework for bridge health monitoring that makes use of computer-vision-based measures. It employs a novel damage indicator, a displacement unit influence surface, and successfully identifies and localizes simulated damage on a large-scale bridge model, demonstrating its use in structural health evaluation. The thorough study in [7] addresses gaps in the current literature on computer-vision-based crack diagnosis for civil infrastructures by providing a complete evaluation of qualitative and quantitative methodologies, including deep-learning-based approaches.

Despite extensive research into image-based damage identification and quantification, this technology is still in its early phases, with limitations and gaps for further investigation. Despite efforts in improving the reliability of image-based approaches, attaining total automation in damage assessment and categorization remains a substantial task.

3. System Design

The proposed system involves the use of a drone to capture image data, which are then fed to an object detection model to identify different types of structural damage in concrete structures. In this section, we provide a detailed methodology for building and deploying the system, which involves collecting images of cracks in buildings, labeling and training the model used, and deploying the model using a Python 3.x script.

3.1. Dataset Collection

The first step in building the system is to collect images of cracks in buildings. These images are taken using a drone, which captures high-resolution images of the building's surface. The images are then annotated with labels indicating the type of damage using the Roboflow platform. The labeling process involves drawing bounding boxes around the damaged area and assigning them a label indicating the type of damage, such as cracks, ASR, or concrete degradation.

Different images of cracks were collected. The images are sized to 800 × 800 pixels which is a standard transformation applied before training. The image dataset is then split up into training, testing, and validating groups:

- 2527 images used for training.
- 2149 images used for validation.
- 279 images used for testing.

With the use of the Roboflow platform's rectangular "bounding box" tool, each picture in the dataset is labelled for object detection. The labelling for the various photos is shown in the Figure 2:

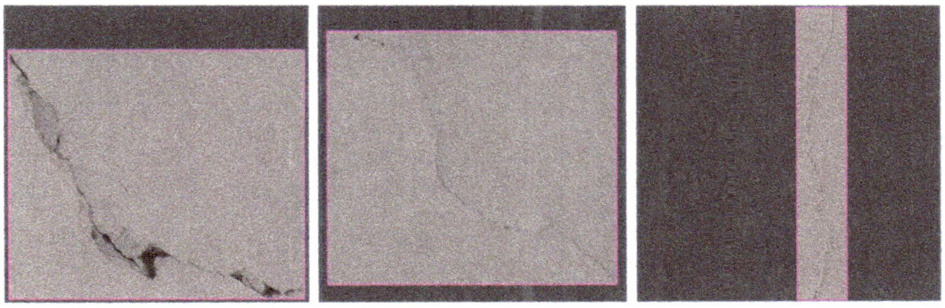

Figure 2. Annotated images in the dataset. Bounding box for cracks detection.

3.2. Model Training

The annotated data are then used to train the object detection model. The model is trained using a deep learning algorithm, YOLO.

The trained model is deployed using a Python script, which receives a live camera stream and runs the object detection model to detect the cracks. The Python script uses the Model API to query the hosted version of the model and return the results. The API provides a simple interface for sending images or video to the model and receiving the output in a standardized format. The methodology of system is presented in the Figure 3.

The selection of hardware components plays a crucial role in this research as they enable the deployment and testing of the computer vision model. Specifically, a drone and a camera module are utilized to facilitate the implementation of the model. The Figure 4 below illustrates the setup of the drone used in this research. The setup of the drone, as depicted in the figure, showcases the integration of the camera module and other necessary components to ensure seamless data capture and transmission.

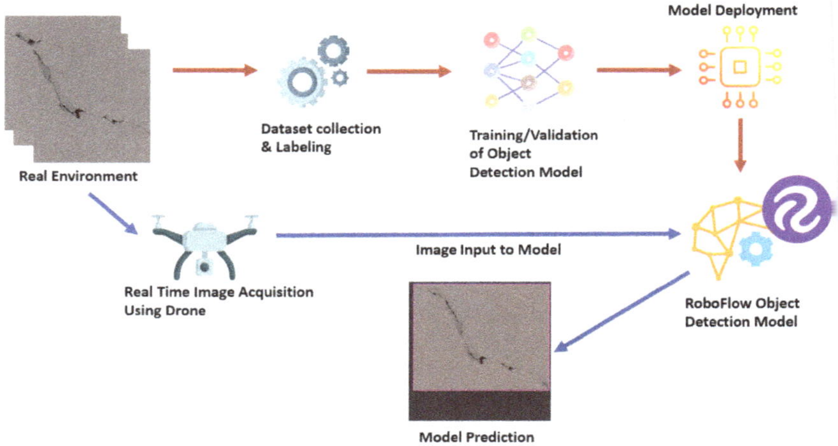

Figure 3. Methodology of the system.

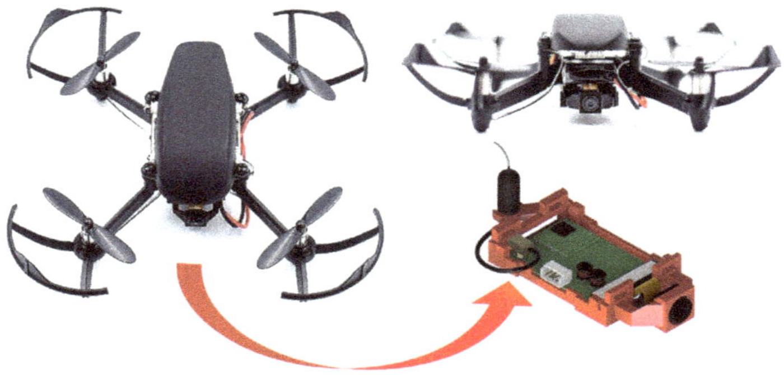

Figure 4. Drone with WiFi Camera Module.

The system incorporates several essential hardware components for its operation. It utilizes a Quadcopter Drone Kit, WiFi Camera Module, a Roboflow account, and a Python development environment.

4. Results and Discussion

In this section, the results of experiments using different object detection models for detecting physical damage in concrete structures has been discussed. The three different models, YOLOv8, YOLOv7, and YOLOv5, are used to evaluate the performance of the system on a test dataset of annotated images for the field of concrete structures monitoring and protection.

The performance of the YOLOv8, YOLOv7, and YOLOv5 models has been evaluated on the test dataset of annotated images. The models were trained using the dataset from the Roboflow platform. As transfer learning is used in this research, the following tables, Tables 1 and 2, show the pre-trained YOLO models used and the training settings.

Table 1. YOLO pre-trained models.

Model	Weights	Layers	Parameters	Gradients	GFLOPs
YOLOv8	yolov8m.pt	295	25,856,899	25,856,883	79.1
YOLOv7	Yolov7.pt	407	37,194,710	37,194,710	105.1
YOLOv5	Yolov5m.pt	291	20,871,318	20,871,318	48.2

Table 2. Training settings.

Model	Image Size	Learning Rate	Batch Size	Epochs
YOLOv8	800	0.01	16	200
YOLOv7	640	0.01	16	200
YOLOv5	800	0.01	16	200

The evaluation metrics used were precision, recall, and F1 score. Equations (1)–(4) [8] were used to evaluate the performance of machine learning models and measure their accuracy, precision, recall, F1 score, and latency:

$$\text{Accuracy} = (\text{TP} + \text{TN})/(\text{TP} + \text{TN} + \text{FF} + \text{FN}) \quad (1)$$

$$\text{Precision} = \text{TP}/(\text{TP} + \text{FP}) \quad (2)$$

$$\text{Recall} = \text{TP}/(\text{TP} + \text{TN}) \quad (3)$$

$$\text{F1Score} = (2 \times \text{Precision} \times \text{Recall})/(\text{Precision} + \text{Recall}) \quad (4)$$

Table 3 shows the results of the evaluation for each model.

Table 3. Performance matrix of different YOLO models.

Model	Precision	Recall	F1 Score
YOLOv8	0.91	0.85	0.89
YOLOv7	0.85	0.81	0.86
YOLOv5	0.71	0.75	0.82

As shown in Table 1, the YOLOv8 model achieved the highest performance in terms of precision, recall, and F1 score. The model achieved a precision of 0.93, recall of 0.85, and F1 score of 0.89. The YOLOv7 model also performed well with a precision of 0.92, recall of 0.81, and F1 score of 0.86. The YOLOv5 model achieved a precision of 0.90, recall of 0.75, and F1 score of 0.82.

The following Table 4 compares the mAP scores for each class across the YOLOv8, YOLOv7, and YOLOv5 models:

Table 4. Class-wise accuracy of different YOLO models.

Class	YOLO8 mAP Score	YOLO7 mAP Score	YOLO5 mAP Score
Plastic shrinkage cracks	0.89	0.8	0.66
Crazing and Crusting Cracks	0.93	0.85	0.73
Settling cracks	0.9	0.84	0.7
Expansion cracks	0.91	0.83	0.72
Heaving cracks	0.78	0.85	0.65
Overloading cracks	0.92	0.83	0.68
Corrosion of Reinforcement	0.93	0.86	0.71

As shown in the table, the YOLOv8 model has the highest mAP score for all classes except for "Heaving cracks", where the YOLOv7 model performs slightly better. The YOLOv7 model also shows good performance for "Crazing and Crusting Cracks", "Settling cracks", and "Corrosion of Reinforcement". The YOLOv5 model generally shows lower mAP scores for all classes compared to the other two models, but still performs relatively well for "Crazing and Crusting Cracks" and "Corrosion of Reinforcement". The following Table 5 compares the inferencing time for all three models.

Table 5. Accuracy and speed comparison of YOLO models.

Model	mAP Score	Inference Time (ms)
YOLOv8	0.91	25
YOLOv7	0.85	32
YOLOv5	0.71	18

As shown in the table, the YOLOv8 model has the highest mAP score and a relatively fast inference time, making it the best overall option for detecting structural damages. The YOLOv7 model has a moderate mAP score but a slightly slower inference time compared to the YOLOv8 model. The YOLOv5 model has the fastest inference time but the lowest mAP score, making it a suitable option for applications where speed is a priority over accuracy. The following Figure 5 depict the result from YOLOv8 model.

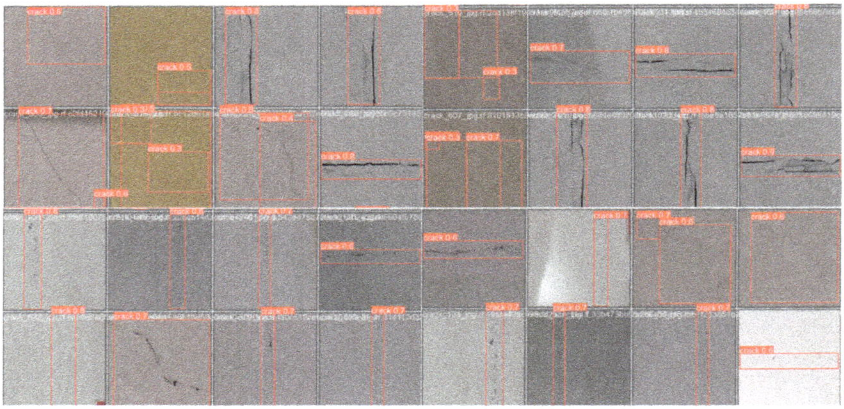

Figure 5. Inferencing result showing detection of cracks on wall.

5. Conclusions

This research demonstrates the effectiveness of using drone-captured imagery and computer vision techniques for the inspection of physical damage in in concrete structures such as buildings, bridges etc. By employing object detection models such as YOLOv8, YOLOv7, and YOLOv5, various types of damage, including plastic shrinkage cracks, crazing and crusting cracks, settling cracks, expansion cracks, heaving cracks, overloading cracks, and corrosion of reinforcement, have been successfully identified. The evaluation of the models has shown promising results, with high mAP scores across different classes.

This research has highlighted the potential of leveraging computer vision and drone technology in damage assessment, providing a safer, cost-effective, and efficient alternative to traditional manual inspections. By automating the detection process, we reduce the need for manual evaluation, which can be time-consuming and prone to human error. The integration of these technologies allows for comprehensive and detailed inspections, facilitating early detection and timely intervention to mitigate further damage to concrete structures.

Author Contributions: T.M., S.T. and R.K.D., conceptualized the idea for this manuscript. T.M. provided the resources. S.T. designed and developed the hardware for data acquisition. R.K.D. carried out the investigation and data curation of the acquired data. T.M. validated the data and results of the acquired data. S.T. prepared the original draft. R.K.D. reviewed and edited the original draft. T.M. and S.T. supervised the work. R.K.D. administered the work. All authors have read and agreed to the published version of the manuscript.

Funding: This research received no external funding.

Institutional Review Board Statement: Not applicable.

Informed Consent Statement: Not applicable.

Data Availability Statement: The data presented in this study are available on request from the first author.

Conflicts of Interest: The authors declare no conflicts of interest.

References

1. Lu, C.; Bu, S.; Zheng, Y.; Kosa, K. Deterioration of concrete mechanical properties and fracture of steel bars caused by alkali-silica reaction: A review. *Structures* **2022**, *35*, 893–902. [CrossRef]
2. Kovler, K.; Chernov, V. Types of damage in concrete structures. In *Failure, Distress and Repair of Concrete Structures*; Woodhead Publishing: Sawston, UK, 2009; pp. 32–56.
3. Lingxin, Z.; Junkai, S.; Baijie, Z. A review of the research and application of deep learning-based computer vision in structural damage detection. *Earthq. Eng. Eng. Vib.* **2022**, *21*, 1–21. [CrossRef]
4. Kim, B.; Yuvaraj, N.; Park, H.W.; Preethaa, K.S.; Pandian, R.A.; Lee, D.E. Investigation of steel frame damage based on computer vision and deep learning. *Autom. Constr.* **2021**, *132*, 103941. [CrossRef]
5. Spencer, B.F., Jr.; Hoskere, V.; Narazaki, Y. Advances in computer vision-based civil infrastructure inspection and monitoring. *Engineering* **2019**, *5*, 199–222. [CrossRef]
6. Khuc, T.; Catbas, F.N. Structural identification using computer vision–based bridge health monitoring. *J. Struct. Eng.* **2018**, *144*, 04017202. [CrossRef]
7. Deng, J.; Singh, A.; Zhou, Y.; Lu, Y.; Lee, V.C.S. Review on computer vision-based crack detection and quantification methodologies for civil structures. *Constr. Build. Mater.* **2022**, *356*, 129238. [CrossRef]
8. Cao, C.; Wang, B.; Zhang, W.; Zeng, X.; Yan, X.; Feng, Z.; Liu, Y.; Wu, Z. An improved faster R-CNN for small object detection. *IEEE Access* **2019**, *7*, 106838–106846. [CrossRef]

engineering proceedings

MDPI

Proceeding Paper

A Comparison between Different Acquisition Modes for FT-IR Spectra Collection from Human Cell Lipid Extracts †

Valeria Cardamuro [1], Bahar Faramarzi [2], Martina Moggio [3], Nadia Diano [3], Lorenzo Manti [1,4], Marianna Portaccio [3,*] and Maria Lepore [3]

1 Dipartimento di Fisica "E. Pancini", Università degli Studi di Napoli "Federico II", 80126 Napoli, Italy; v.cardamuro@studenti.unina.it (V.C.); manti@na.infn.it (L.M.)
2 Dipartimento di Matematica e Fisica, Università della Campania "Luigi Vanvitelli", 81100 Caserta, Italy; bahar.faramarzi@unicampania.it
3 Dipartimento di Medicina Sperimentale, Università della Campania "Luigi Vanvitelli", 80138 Napoli, Italy; martina.moggio@unicampania.it (M.M.); nadia.diano@unicampania.it (N.D.); maria.lepore@unicampania.it (M.L.)
4 Istituto Nazionale di Fisica Nucleare—Sezione di Napoli, 80100 Napoli, Italy
* Correspondence: marianna.portaccio@unicampania.it
† Presented at the 10th International Electronic Conference on Sensors and Applications (ECSA-10), 15–30 November 2023; Available online: https://ecsa-10.sciforum.net/.

Abstract: Lipids are organic compounds that contribute to numerous cellular functions. Fourier Transform Infrared spectroscopy can be particularly useful in investigating the biochemical features of the lipid content of cells and their changes induced by interaction with physicochemical external agents. In the present work, we aim to investigate the extract of lipids from human cells to compare the results obtained by using two different geometries: transmission and attenuated total reflectance. Multiple acquisitions of spectra were carried out and statistical criteria were applied for monitoring and comparing them. The positive and negative aspects of the two examined acquisition modes are presented and discussed.

Keywords: FT-IR spectroscopy; transmission; ATR; lipids; hepatocarcinoma cells

Citation: Cardamuro, V.; Faramarzi, B.; Moggio, M.; Diano, N.; Manti, L.; Portaccio, M.; Lepore, M. A Comparison between Different Acquisition Modes for FT-IR Spectra Collection from Human Cell Lipid Extracts. *Eng. Proc.* **2023**, *58*, 61. https://doi.org/10.3390/ecsa-10-16211

Academic Editor: Jean-marc Laheurte

Published: 15 November 2023

1. Introduction

Lipids are organic compounds widely distributed in nature and represent one of the four main classes of organic compounds of biological interest, along with carbohydrates, proteins, and nucleic acids. In eukaryotes, lipids contribute to numerous cellular functions, ranging from energy storage to cell signaling [1]. Thanks to its ability to analyze cellular components at a molecular level, Fourier Transform Infrared Spectroscopy (FT-IR) can be particularly useful in investigating the biochemical features of the lipid content of cells and their changes induced by interaction with physicochemical external agents. In the present work, we aim to investigate lipids extracted from cells to compare the results obtained by using two different geometries that are usually available for the acquisition of FT-IR spectra for liquid samples: transmission and Attenuated Total Reflectance (ATR) geometry [2]. Multiple acquisitions of spectra were carried out, and statistical criteria were applied for monitoring and comparing them. The positive and negative aspects of the two examined acquisition modes are presented and discussed.

2. Materials and Methods

Human hepatocarcinoma cells (HepG2) were cultured in Dulbecco's Modified Eagle Medium. The medium was supplemented with 10% heat-inactivated FBS, 100 U/mL penicillin, 100 μg/mL streptomycin, and 1% L-glutamine. The cells were grown in a humidified atmosphere of 95% air/5% CO_2 at 37 °C in T25 flasks.

Lipids were extracted from cells by using the Bligh and Dyer method [3]. In this method, for each 1 mL of sample, 3.75 mL 1:2 (*v*/*v*) CHCl3:MeOH were added and vortexed thoroughly. Then, 1.25 mL $CHCl_3$ was added and vortexed again. Finally, 1.25 mL of deionized water (dH2O) was added, vortexed, and the sample was then centrifuged at 1000 rpm for 5 min at room temperature to yield a two-phase system (aqueous top and organic bottom). The organic bottom layer was carefully collected using Pasteur pipettes into a clean glass vial, dried with nitrogen, and stored at −20 °C until measurements.

To obtain measurements in transmission geometry, a few microliters of lipids extracted from hepatocarcinoma cells (HepG2) were dissolved in methanol and positioned onto CaF_2 windows. Spectra were then acquired using the microscope stage of a Perkin Elmer Spectrum One spectrometer, which was equipped with a mercury cadmium telluride (MCT) detector. This approach allowed for the collection of spectra using 32 scans in the range from 4000 to 1000 cm^{-1} with a 4 cm^{-1} spectral resolution in a 100 × 100 μm^2 region. For measurements in Attenuated Total Reflectance (ATR) geometry, drops of the extracted lipids were placed on the top of the diamond crystal of the Universal ATR accessory of the abovementioned FT-IR spectrometer provided via an MIR TGS detector. In this case, spectra were collected using 32 scans in the range from 4000 to 650 cm^{-1} with a 4 cm^{-1} spectral resolution.

Preliminary subtraction of the background spectrum acquired in a free-cell zone of the slide was performed for all the spectra. For the purpose of comparing spectra, Standard Normal Variate (SNV) normalization was carried out [4]. To compare the acquisition modes, a study of the correlation among the various spectra acquired in transmittance and ATR mode was carried out using a MATLAB code.

3. Results and Discussion

In Figures 1 and 2, the average FT-IR spectra for lipids extracted from HepG2 cells are reported. For the spectrum collected in transmission mode, the investigated range is from 4000 to 1000 cm^{-1}, while for the spectrum acquired in ATR mode, the spectrum is related to the 4000–650 cm^{-1} wavenumber region. As can be observed, the two spectra exhibit very similar behavior in the high-wavenumber region, while some differences occur in the fingerprint region.

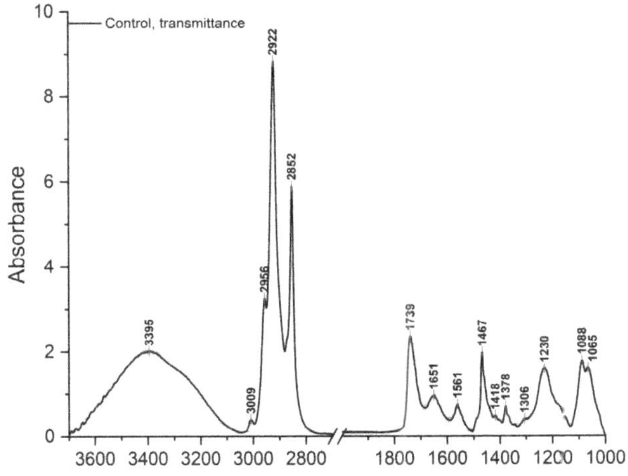

Figure 1. Average FT-IR spectrum obtained in transmittance mode from the average spectra relative to repeated measurements on lipids extracted from HepG2 cells. Data are presented as mean ± SEM. The peaks are shown, and the assignments are made by referring to texts in the literature (Table 1) [5–7].

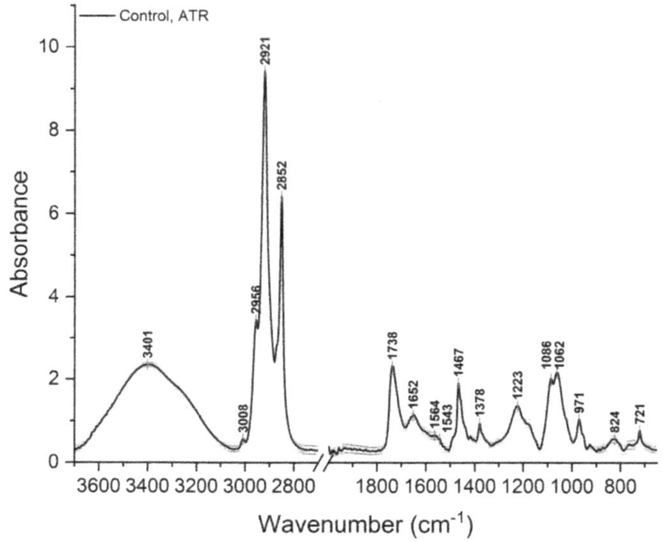

Figure 2. Average FT-IR spectrum obtained in ATR mode from the average spectra relative to repeated measurements of lipids extracted from HepG2 cells. Data are presented as mean ± SEM. The peaks are shown, and the assignments are made by referring to texts in the literature (Table 1) [5–7].

Table 1. Spectral position of main peaks of FT-IR spectra of Figures 1 and 2. Assignments are reported according to refs. [5,8,9]. Bold character indicates shifts larger than experimental spectra resolution (4 cm^{-1}). Nomenclature: ν stands for stretching, s = symmetric, as = asymmetric, δ stands for bending.

Assignments	Transmittance	ATR
	Peak (cm^{-1})	Peak (cm^{-1})
ν(N-H), ν(O-H), ν(C-H)	3395	**3401 (+6)**
ν(H-C=)	3009	3008 (−1)
$\nu as(CH_3)$	2956	2956
$\nu as(CH_2)$	2922	2921 (−1)
$\nu s(CH_2)$	2852	2852
ν(C=O)	1739	1738 (−1)
ν(C=O)	1651	1652 (+1)
δ(N-H)	1561	**1543 (−18)**
$\delta(CH_2)$	1467	1467
$\delta(CH_3)$	1378	1378
–	1306	–
$\nu as(PO_2^-)$	1230	**1223 (−7)**
$\nu s(PO_2^-)$	1088	1086 (−2)
$\nu(C-O-H)$	1065	1062 (−3)
$\nu(C-O-P)$	–	–
$N^+(-CH_3)_3$	–	971
νas(P-O)	–	824
$\rho(CH_2)$	–	721

In Table 1, the positions of the main contributions are reported together with their assignments. Upon examining the peak positions of the bands of interest, the spectra acquired in the two modes appear consistent. The positions are almost similar, and in general comparable, since the differences are lower than the spectral resolution of the measuring apparatus, i.e., 4 cm^{-1}, except for the bands around 3401 and 3395 cm^{-1} (related to the asymmetric stretching of the O-H, N-H and C-H groups) that show a shift of +6 cm^{-1}, the features at 1561 and 1541 cm^{-1} (due to the N-H bending of amide II group) that show a shift of −18 cm^{-1} and the peaks at 1230 and 1223 cm^{-1} (ascribed to the asymmetric stretching of the phosphate group), which evidence a shift of −7 cm^{-1} in the spectra associated with the two acquisition methods.

The major differences between the two acquisition modes lie in the sample preparation and the extension of the infrared range characteristic of the spectrum. As described in the Material and Methods section, samples for transmittance measurements were first diluted in methanol and then left to dry on CaF_2 windows; in contrast, samples for ATR, also dissolved in methanol, were placed directly on the diamond core and, also in this case, were left to dry. In the ATR mode, it is possible to obtain spectra of up to 650 cm^{-1}, instead of up to 1000 cm^{-1} as in transmittance. In general, the high wavenumber region is less dependent on the acquisition geometry compared with the fingerprint region.

We also investigated the spectra reproducibility given by the two different acquisition modes. From the correlation study between the various spectra acquired repeatedly in transmittance, the analysis returned a correlation coefficient R = 0.95 in the case of the high wave number range (3700–2700 cm^{-1}), and R = 0.79 in the case of the fingerprint zone (1800–1000 cm^{-1}). In the case of attenuated total reflectance (ATR) acquisition, the correlation coefficient obtained in both high and low wave numbers is R = 0.99. This analysis indicated that ATR geometry can be the most valuable approach for acquiring infrared spectra from lipid samples according to the literature [10].

4. Conclusions

The present study indicates that both transmittance and ATR acquisition geometries can be adopted for the investigated lipid sample, but ATR approach offers some advantages in terms of wavenumber range and spectra reproducibility. This can probably be due to the different sampled region. In transmission mode, the spectra were acquired on small regions ($100 \times 100\ \mu m^2$), while in ATR geometry, the spectra were acquired by exploiting the total area of the diamond crystal surface (nearly equal to 3 mm^2). In addition, this measurement method speeds up the acquisition of spectra and reduces the costs.

Author Contributions: Conceptualization, M.L., M.P., N.D. and L.M.; methodology, M.P., B.F. and V.C.; software, V.C.; investigation, B.F., L.M., V.C., M.M. and M.P.; data curation, V.C. and B.F.; writing—original draft preparation, M.L., V.C. and B.F.; writing—review and editing. M.L., V.C., B.F., M.P. and N.D. All authors have read and agreed to the published version of the manuscript.

Funding: This research received no external funding.

Data Availability Statement: Data are available on request.

Conflicts of Interest: The authors declare no conflicts of interest.

References

1. Muro, E.; Atilla-Gokcumen, G.E.; Eggert, U.S. Lipids in cell biology: How can we understand them better? *Mol. Biol. Cell* **2014**, *25*, 1819–1823. [CrossRef] [PubMed]
2. Errico, S.; Moggio, M.; Diano, N.; Portaccio, M.; Lepore, M. Different experimental approaches for Fourier-transform infrared spectroscopy applications in biology and biotechnology: A selected choice of representative results. *Biotechnol. Appl. Biochem.* **2023**, *70*, 937–961. [CrossRef] [PubMed]
3. Bligh, E.G.; Dyer, W.J. A rapid method of total lipid extraction and purification. *Can. J. Biochem. Physiol.* **1959**, *37*, 911–917. [CrossRef] [PubMed]
4. Lasch, P. Spectral pre-processing for biomedical vibrational spectroscopy and microspectroscopic imaging. *Chemom. Intell. Lab. Syst.* **2012**, *117*, 100–114. [CrossRef]

5. Dreissig, I.; Machill, S.; Salzer, R.; Krafft, C. Quantification of brain lipids by FTIR spectroscopy and partial least squares regression. *Spectrochim. Acta Part A Mol. Biomol. Spectrosc.* **2009**, *71*, 2069–2075. [CrossRef] [PubMed]
6. Abdelrazzak, A.B.; El-Bahy, G.S. FT-IR spectroscopic investigation of ionizing radiation-induced damage in the small intestine of whole-body irradiated rats. *Vib. Spectrosc.* **2018**, *99*, 146–150. [CrossRef]
7. Cakmak, G.; Miller, L.M.; Zorlu, F.; Severcan, F. Amifostine, a radioprotectant agent, protects rat brain tissue lipids against ionizing radiation induced damage: An FTIR microspectroscopic imaging study. *Arch. Biochem. Biophys.* **2012**, *520*, 67–73. [CrossRef] [PubMed]
8. Talari, A.C.S.; Martinez, M.A.G.; Movasaghi, Z.; Rehman, S.; Rehman, I.U. Advances in Fourier transform infrared (FTIR) spectroscopy of biological tissues. *Appl. Spectrosc. Rev.* **2017**, *52*, 456–506. [CrossRef]
9. Movasaghi, Z.; Rehman, S.; ur Rehman, D.I. Fourier transform infrared (FTIR) spectroscopy of biological tissues. *Appl. Spectrosc. Rev.* **2008**, *43*, 134–179. [CrossRef]
10. Fringeli, U.P.; Günthard, H.H. Infrared membrane spectroscopy. *Membr. Spectrosc.* **1981**, *31*, 270–332.

engineering proceedings

MDPI

Proceeding Paper

Vision-Based Structural Identification Using an Enhanced Phase-Based Method †

Samira Azizi [1,2,*], Kaveh Karami [2] and Stefano Mariani [1]

[1] Department of Civil and Environmental Engineering, Politecnico di Milano, 20133 Milano, Italy; stefano.mariani@polimi.it
[2] Department of Civil Engineering, University of Kurdistan, Sanandaj P.O. Box 416, Iran; ka.karami@uok.ac.ir
* Correspondence: samira.azizi@polimi.it
† Presented at the 10th International Electronic Conference on Sensors and Applications (ECSA-10), 15–30 November 2023; Available online: https://ecsa-10.sciforum.net/.

Abstract: Operational modal analysis is based on data collected using a network of sensors installed on a monitored structure to measure its response to the external stimuli. As the instrumentation can be costly, sensors are placed at a limited number of locations where damage-sensitive features can hopefully be sensed. Hence, the actual ability to detect a shift from the undamaged structural state in real time might be detrimentally affected. Non-contact measurement methods relying on, e.g., digital video cameras, which have gained interest in recent years, can instead provide high-resolution and diffused measurements/information. In this study, moving from videos of a vibrating structure, a shift in its dynamic response was assessed. By means of a phase-based optical flow methodology, a linear correlation between the phase and the structural motion was customarily assumed using, e.g., the Gabor filter. Since such a correlation does not result in being always linear, linearization is necessary for all the frames. By using the blind source separation method, mode shapes and vibration frequencies were finally obtained. The performance of the proposed method is investigated to verify the accuracy in extracting the dynamic features of the considered structure using the proposed method.

Keywords: structural health monitoring; modal analysis; blind source separation method; Gabor filter; optical flow

Citation: Azizi, S.; Karami, K.; Mariani, S. Vision-Based Structural Identification Using an Enhanced Phase-Based Method. *Eng. Proc.* **2023**, *58*, 62. https://doi.org/10.3390/ecsa-10-16036

Academic Editor: Francisco Falcone

Published: 15 November 2023

1. Introduction

Modal analysis is a vital tool used to identify the dynamic behavior of structures in terms of natural frequencies, mode shapes, and damping ratios [1–3]. By understanding how the structures respond to the external forces, it can be ensured that they withstand such loads as well as the relevant environmental conditions [4]. Modal analysis serves as a crucial tool used in model order reduction techniques [5,6] and structural health monitoring (SHM) [7]. In recent times, the use of vision-based measurements has emerged as a highly effective method for full-field identification [8,9], damage detection [10], model updating [11], and response measurement [12,13]. This innovative approach exploits the data from images to gain insights into structural behavior. Besides high-resolution response measurements, it avoids the additional weight linked to the sensing system and reduces the cost of purchasing, installing, and maintaining sensors.

Optical flow estimation is a method used in computer vision that tracks the movement of pixels between consecutive frames of a video. Fleet and Jepson [14] demonstrated that the local phase of an image, obtained through quadratic filters [15], represents motion more robustly than the intensity. Phase-based motion magnification, introduced by Wadhwa et al. [16], involves the amplification of subtle motions in a video sequence by focusing on phase information. By means of this technique, Yang et al. [17] obtained mode shapes from videos of vibrating structures. These mode shapes were subsequently employed in [18] to identify and locate damage (also refer to [19]). Southwick et al. [20] expanded this approach to extract 3D volumetric motions. Luo et al. [21] introduced a novel image-processing technique that addresses the challenge vision sensors face in outdoor structural displacement monitoring, as the conventional approach is susceptible to noise and limited in its measurement range. To enhance accuracy, Cai et al. [22] addressed the limitations of phase-based estimation and proposed a novel multi-view measurement method. Miao et al. [23], by optimizing the Gabor filter parameters, introduced a robust phase-based displacement measurement technique to capture vibration responses.

In this paper, we explore the impact of a specific filter parameter on the filtering response. As shown in the results, it is crucial to recognize that not all the pixels in an image are suitable for motion detection or identification processes; the selection of the appropriate one(s) is a critical concern. Via a defined criterion, regions are identified where the linear correlation between phase and motion is not established. This detection process enhances the accuracy of motion estimation. The phase-based displacement measurement was applied to a real case test, and phase-based identification, incorporating Independent Component Analysis (ICA) blind-source separation, was performed on videos of a vibrating structure.

2. Motion and Phase Relation

In one-dimensional signal analysis, phase delay is related to how signals evolve over time; the same concept can be applied to two-dimensional signals like images. When a feature in an image undergoes spatial movement, like a translation or a rotation, this process leads to changes in the local phase of the pixels. This shift in phase is directly related to the extent and direction of the movement.

Fleet and Jepson [14] explored the connection between local phase difference and motion. By tracking the constant phase contours in successive frames, a motion field can be obtained. Assuming that the intensity of the first frame at time t_0 in a pixel with coordinates $X(x, y)$ is $I(X)$ if it becomes displaced by $\Delta(X, t)$, the intensity profile of the next frame at time t_1 is $I(X + \Delta(X, t))$. To extract the local phase, it is necessary to use filters like the Gabor filter.

The response of the Gabor filter is a complex valued function that can be expressed as

$$R(X, t) = \rho(X, t) \times e^{i\phi(X,t)} \tag{1}$$

where $\rho(X, t)$ and $\phi(X, t)$ represent its amplitude and phase component, given by

$$\begin{aligned} \rho(X, t) &= |R(X, t)| = \sqrt{\mathbb{Re}[R(X, t)]^2 + \mathbb{Im}[R(X, t)]^2} \\ \phi(X, t) &= \arg\left[R(X, t)\right] \end{aligned} \tag{2}$$

The tracking of continuous changes in the phase contours over time, providing a reliable approximation of the motion field. In simple terms, points represented by $X(x, y)$ on these contours maintain a constant value of $\phi(X, t) = c$. The displacement in a direction θ is then derived from the movement of local phase contours.

3. Unreliable Phase Detection

Motion estimation from videos is characterized by a significant challenge: it is not feasible to extract genuine motion data from every pixel. This issue arises due to various factors, including fluctuations in lighting conditions, Gabor filter parameters, varying object scales within an image, and the presence of noise.

To provide an illustration of this problem, Figure 1a displays an image of a chimney located at Politecnico di Milano, as captured using a standard mobile phone camera. Images were then processed to create a simulated displacement scenario (ten pixels).

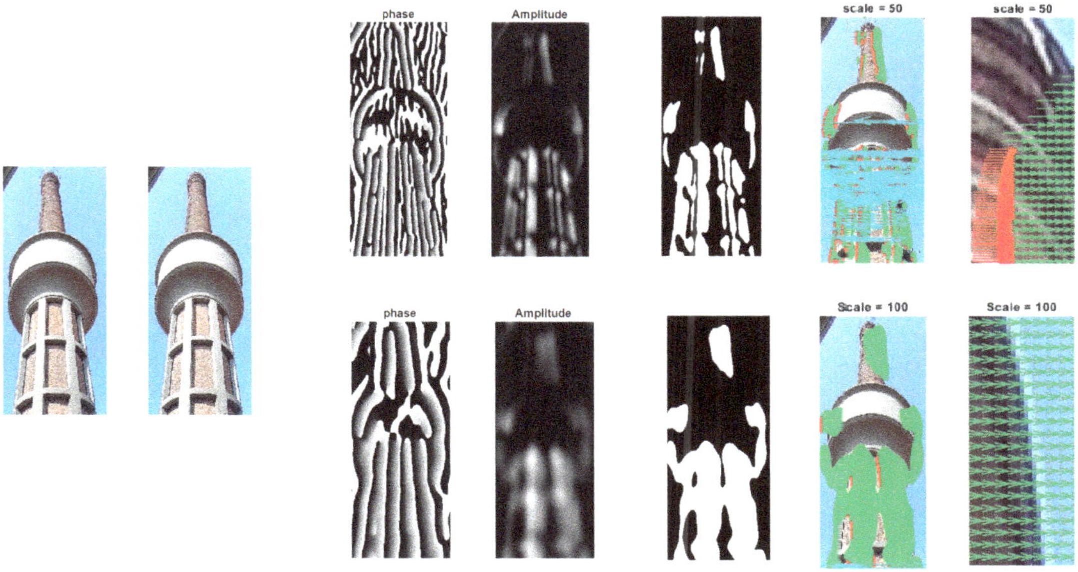

(a) (b) (c) (d) (e)

Figure 1. (**a**) Pair of images with 10-pixel displacement between them; (**b**) contours of amplitude and phase; (**c**) selected pixel locations; (**d**) displacement measurement. Here, green regions are characterized by an error smaller than 5%, red regions by an error larger than 5%, and cyan regions by an unstable phase; (**e**) close-up area. Upper row with scale = 50; lower row with scale = 100.

By applying the Gabor filter with two different scales, the phase and amplitude obtained from suitable pixels were selected for displacement measurement. When the phase varies between values of π and $-\pi$, jumps (phase wrapping) occur, but they should not be considered as unstable regions. A simple way of unwrapping the phase was adopted in this study: when the phase difference between two frames at a point is greater than π, 2π is added or subtracted.

In Figure 1d, displacement measurements with three colors are shown: the green region stands for measurements characterized by an error smaller than 5%, while the red region features measurement errors larger than 5%; the cyan region is instead related to an unstable phase. As shown in the figure, the area representing accurate measurements increases in size at a higher scale, but finer details are lost; for instance, the upper-left part of the tower is not captured by the selected pixels at the higher scale.

The phase-based motion estimation methodology is based on the linear correlation between the phase and the structural motion. However, phase nonlinearity stands out as the primary cause of inaccurate motion estimation. By detecting the region in which the phase contours are not likely to provide reliable information about the motion, the measurement error can be reduced. According to [24], two constraints may be needed to detect the points with such a behavior. This approach was adopted here, with different bounds on the output of the applied Gabor filter tuned to one scale. The first one is the frequency constraint: the instantaneous frequency of the filter output, which is the spatial gradient of the phase $\phi_x(X,t)$, was constrained such that it would be close to the value at which the filter was tuned:

$$|\phi_x(X,t) - k| < \tau \tag{3}$$

The second constraint was instead based on the amplitude: its derivative $\rho_x(X,t)$ was constrained such that it would be small, according to:

$$|\rho_x(X,t)|/\rho(X,t) < \tau \tag{4}$$

To determine the impact of these constraints, the displacement field between two frames in Figure 1a were computed at four scales. Pixels were selected using a simple threshold on the amplitude using Equations (3) and (4). Figure 2 shows that by employing these thresholds with $\tau = 0.05$, the cyan regions are completely avoided, and the red region is reduced in comparison with that linked to the phase selection solely based on the amplitude.

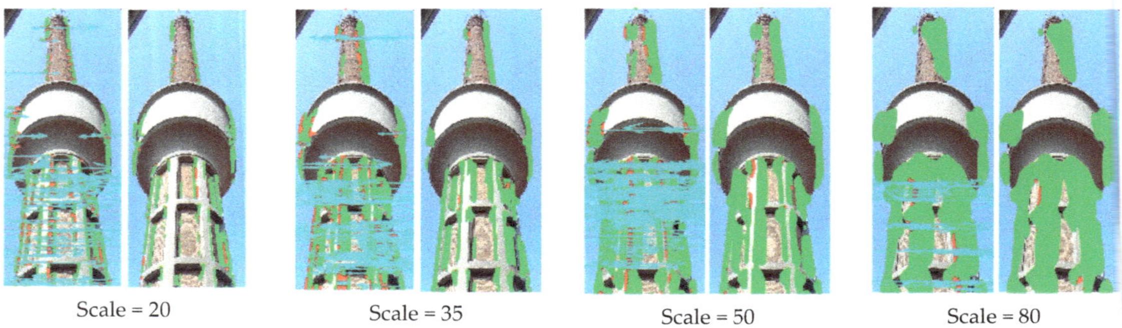

Figure 2. Effect of different thresholds on the displacement measurement between two frames at different scales (colors are defined in the caption of Figure 1).

4. Full-Field Identification

An investigation of the performance of pixel selection based on the proposed criteria is provided in this section, with the aim of identifying full-field modal shapes and vibration frequencies. A model of a ruler subject to free vibrations was captured in a video using MATLAB. Non-linearity in the phase was introduced to account for the lack of alignment between the filter orientation and the edge of the body. The process of identification is depicted in Figure 3.

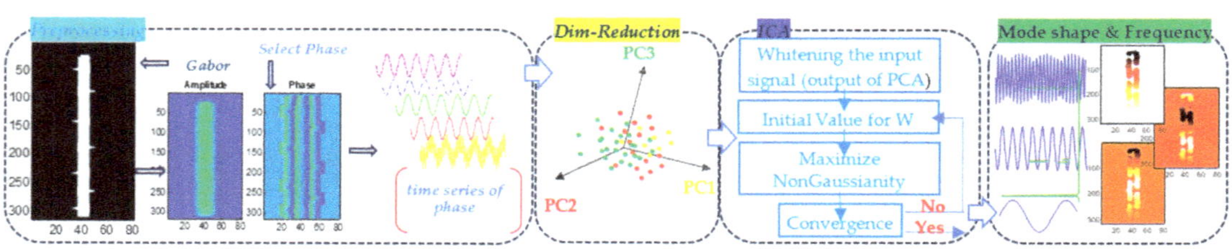

Figure 3. Full-field identification using PCA–ICA.

By defining the appropriate pixels on the basis of the discussion in Section 3, a matrix was created to represent the time series of the selected phases across all frames. By employing PCA, the dimensionality of this matrix was reduced to the number of excited modes, which turned out to be three in the present case. In general cases, this number of modes is set on the basis of the eigenvalues of the original matrix: to avoid issues related to noise, only principal components linked to eigenvalues larger than 1% of the maximum value were retained in the model. Afterward, by way of ICA [25], the frequencies of the vibrations and the corresponding mode shapes were obtained.

In Figure 4a, the identified frequencies (2.99, 18.45, and 51.87) are compared with their corresponding real values (2.93, 18.63, and 51.77) to demonstrate the remarkable accuracy obtained, amounting to 98%, 99.04%, and 99.8%, respectively. Figure 4b shows the identified mode shapes, which are compared to the real ones using the modal assurance criterion, revealing similarities of 99.1%, 96.7%, and 98.2%, respectively. In Figure 5, it is further demonstrated that the selected regions all have consistent representations of the mode shapes.

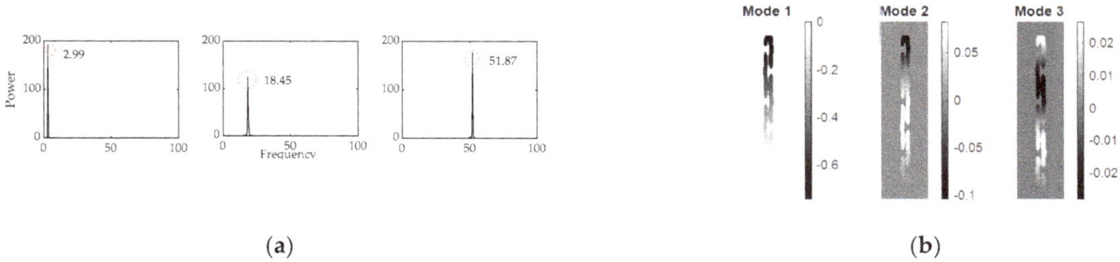

Figure 4. Identified (**a**) vibration frequencies and (**b**) mode shapes using PCA−ICA.

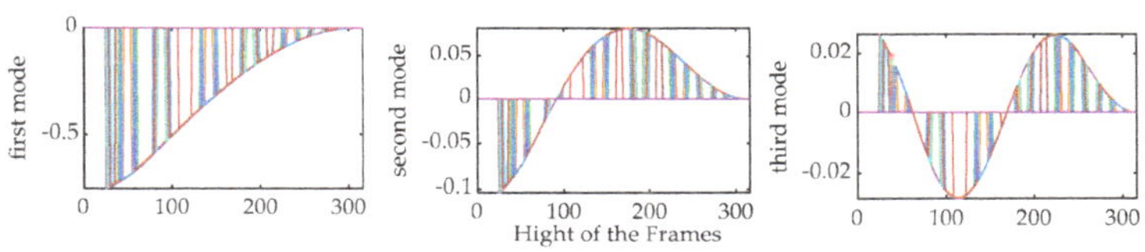

Figure 5. First, second, and third full field mode shapes with pixel intensity represented with different colors.

5. Conclusions

In this study, the impact of the bandwidth of the Gabor filter on the phase extracted from images of a video of structural vibrations was investigated. By enlarging the value of this parameter, the level of detail under examination is decreased; alternatively, by decreasing it, the linear relationship assumption between phase and motion is violated. Choosing appropriate pixels for the analysis therefore poses challenges in image processing. By imposing specific conditions on the phase and amplitude of the filtered images, artificial displacement in a real structure was examined. It was observed that these conditions allowed for the identification of unsuitable phase areas in order to attain more-accurate solutions. Moreover, the selected phases were exploited for blind modal identification after dimensionality reduction through PCA and the application of ICA. The identified frequencies and modal shapes were shown to accurately match the real ones.

In this investigation, pixel selection was performed exclusively using the first frame of the video. However, considering the potential effects of non-linearity, this choice might differ across different frames. The objective in future research will be to identify these locations across all frames, remove them, and subsequently provide accurate estimations using learning-based methods.

Author Contributions: Conceptualization, S.A., S.M. and K.K.; methodology, S.A. and K.K.; validation, S.A., S.M. and K.K.; formal analysis, S.A.; resources, K.K.; data curation, S.A.; writing—original draft preparation, S.A.; writing—review and editing, S.M.; visualization, S.M.; supervision, S.M. All authors have read and agreed to the published version of the manuscript.

Funding: This research received no external funding.

Institutional Review Board Statement: Not applicable.

Informed Consent Statement: Not applicable.

Data Availability Statement: The data presented in this study are available on request from the corresponding author.

Conflicts of Interest: The authors declare no conflict of interest.

References

1. Rosafalco, L.; Eftekhar Azam, S.; Manzoni, A.; Corigliano, A.; Mariani, S. Unscented Kalman Filter Empowered by Bayesian Model Evidence for System Identification in Structural Dynamics. *Comput. Sci. Math. Forum.* **2022**, *2*, 3.
2. Azizi, S.; Karami, K.; Nagarajaiah, S. Developing a semi-active adjustable stiffness device using integrated damage tracking and adaptive stiffness mechanism. *Eng. Struct.* **2021**, *238*, 112036. [CrossRef]
3. Amini, F.; Karami, K. Damage detection algorithm based on identified system Markov parameters (DDA/ISMP) in building structures with limited sensors. *Smart Mater. Struct.* **2012**, *21*, 055010. [CrossRef]
4. Gatti, F.; Rosafalco, L.; Colombera, G.; Mariani, S.; Corigliano, A. Multi-storey shear type buildings under earthquake loading: Adversarial learning-based prediction of the transient dynamics and damage classification. *Soil Dyn. Earthq. Eng.* **2023**, *173*, 108141. [CrossRef]
5. Torzoni, M.; Rosafalco, L.; Manzoni, A.; Mariani, S.; Corigliano, A. SHM under varying environmental conditions: An approach based on model order reduction and deep learning. *Comput. Struct.* **2022**, *266*, 106790. [CrossRef]
6. Rosafalco, L.; Manzoni, A.; Mariani, S.; Corigliano, A. Combined model order reduction techniques and artificial neural network for data assimilation and damage detection in structures. In *Computational Sciences and Artificial Intelligence in Industry: New Digital Technologies for Solving Future Societal and Economical Challenges*; Springer: Cham, Switzerland, 2022; pp. 247–259.
7. Torzoni, M.; Manzoni, A.; Mariani, S. Structural health monitoring of civil structures: A diagnostic framework powered by deep metric learning. *Comput. Struct.* **2022**, *271*, 106858. [CrossRef]
8. Feng, D.; Feng, M.Q. Experimental validation of cost-effective vision-based structural health monitoring. *Mech. Syst. Signal Process.* **2017**, *88*, 199–211. [CrossRef]
9. Yang, Y.; Dorn, C. Affinity propagation clustering of full-field, high-spatial-dimensional measurements for robust output-only modal identification: A proof-of-concept study. *J. Sound Vib.* **2020**, *483*, 115473. [CrossRef]
10. Dworakowski, Z.; Kohut, P.; Gallina, A.; Holak, K.; Uhl, T. Vision-based algorithms for damage detection and localization in structural health monitoring. *Struct. Control Health Monit.* **2016**, *23*, 35–50. [CrossRef]
11. Martini, A.; Tronci, E.M.; Feng, M.Q.; Leung, R.Y. A computer vision-based method for bridge model updating using displacement influence lines. *Eng. Struct.* **2022**, *259*, 114129. [CrossRef]
12. Yang, Y.; Jung, H.; Dorn, C.; Park, G.; Farrar, C.; Mascareñas, D. Estimation of full-field, full-order experimental modal model of cable vibration from digital video measurements with physics-guided unsupervised machine learning and computer vision. *Struct. Control Health Monit.* **2019**, *26*, e2358. [CrossRef]
13. Bhowmick, S.; Nagarajaiah, S.; Lai, Z. Measurement of full-field displacement time history of a vibrating continuous edge from video. *Mech. Syst. Signal Process.* **2020**, *144*, 106847. [CrossRef]
14. Fleet, D.J.; Jepson, A.D. Computation of component image velocity from local phase information. *Int. J. Comput. Vis.* **1990**, *5*, 77–104. [CrossRef]
15. Weldon, T.P.; Higgins, W.E.; Dunn, D.F. Efficient Gabor filter design for texture segmentation. *Pattern Recognit.* **1996**, *29*, 2005–2015. [CrossRef]
16. Wadhwa, N.; Rubinstein, M.; Durand, F.; Freeman, W.T. Phase-based video motion processing. *ACM Trans. Graph.* **2013**, *32*, 1–10. [CrossRef]
17. Yang, Y.; Dorn, C.; Mancini, T.; Talken, Z.; Kenyon, G.; Farrar, C.; Mascareñas, D. Blind identification of full-field vibration modes from video measurements with phase-based video motion magnification. *Mech. Syst. Signal Process.* **2017**, *85*, 567–590. [CrossRef]
18. Yang, Y.; Dorn, C.; Mancini, T.; Talken, Z.; Theiler, J.; Kenyon, G.; Farrar, C.; Mascareñas, D. Reference-free detection of minute, non-visible, damage using full-field, high-resolution mode shapes output-only identified from digital videos of structures. *Struct. Health Monit.* **2018**, *17*, 514–531. [CrossRef]
19. Yang, Y.; Jung, H.; Dorn, C.; Park, G.; Farrar, C.; Mascareñas, D. Estimation of full-field dynamic strains from digital video measurements of output-only beam structures by video motion processing and modal superposition. *Struct. Control Health Monit.* **2019**, *26*, e2408. [CrossRef]
20. Southwick, M.; Mao, Z.; Niezrecki, C. Volumetric Motion Magnification: Subtle Motion Extraction from 4D Data. *Measurement* **2021**, *176*, 109211. [CrossRef]
21. Luo, L.; Feng, M.Q.; Wu, Z.Y. Robust vision sensor for multi-point displacement monitoring of bridges in the field. *Eng. Struct.* **2018**, *163*, 255–266. [CrossRef]
22. Cai, E.; Zhang, Y.; Ji, X.; Lu, X.; Xie, L.; Zhuang, Y.; Zhao, T.; Lin, G. Estimating small structural motions from multi-view video measurement. *Eng. Struct.* **2023**, *275*, 115259. [CrossRef]
23. Miao, Y.; Jeon, J.Y.; Kong, Y.; Park, G. Phase-based displacement measurement on a straight edge using an optimal complex Gabor filter. *Mech. Syst. Signal Process.* **2021**, *164*, 108224. [CrossRef]

24. Jepson, A.D.; Fleet, D.J. Phase singularities in scale-space. *Image Vis. Comput.* **1991**, *9*, 338–343. [CrossRef]
25. Hyvärinen, A.; Oja, E. Independent component analysis: Algorithms and applications. *Neural Netw.* **2000**, *13*, 411–430. [CrossRef]

engineering proceedings

MDPI

Proceeding Paper

Deep Learning-Empowered Robot Vision for Efficient Robotic Grasp Detection and Defect Elimination in Industry 4.0 †

Yassine Yazid [1,2,*], Antonio Guerrero-González [2], Ahmed El Oualkadi [2] and Mounir Arioua [2]

1 Department of Automation, Electrical Engineering and Electronic Technology, Universidad Politécnica de Cartagena, Plaza del Hospital 1, 30202 Cartagena, Spain

2 Innovative Systems Engineering Laboratory (ISI), National School of Applied Sciences of Tetuan (ENSATe), Abdelmalek Essaadi University, Tétouan 93000, Morocco; antonio.guerrero@upct.es (A.G.-G.); aeloualkadi@uae.ac.ma (A.E.O.); m.arioua@uae.ac.ma (M.A.)

* Correspondence: yassine.yazid@edu.upct.es

† Presented at the 10th International Electronic Conference on Sensors and Applications (ECSA-10), 15–30 November 2023; Available online: https://ecsa-10.sciforum.net/.

Abstract: Robot vision, enabled by deep learning breakthroughs, is gaining momentum in the industry 4.0 digitization process. The present investigation describes a robotic grasp detection application that makes use of a two-finger gripper and an RGB-D camera linked to a collaborative robot. The visual recognition system, which is integrated with edge computing units, conducts image recognition for faulty items and calculates the position of the robot arm. Identifying deformities in object photos, training and testing the images with a modified version of the You Only Look Once (YOLO) method, and establishing defect borders are all part of the process. Signals are subsequently sent to the robotic manipulator to remove the faulty components. The adopted technique used in this system is trained on custom data and has demonstrated a high accuracy and low latency performance as it reached a detection accuracy of 96% with 96.6% correct grasp accuracy.

Keywords: robot vision; deep learning; Industry 4.0; robot grasp; defect detection; YOLO

Citation: Yazid, Y.; Guerrero-González, A.; El Oualkadi, A.; Arioua, M. Deep Learning-Empowered Robot Vision for Efficient Robotic Grasp Detection and Defect Elimination in Industry 4.0. *Eng. Proc.* **2023**, *58*, 63. https://doi.org/10.3390/ecsa-10-16079

Academic Editor: Stefano Mariani

Published: 15 November 2023

1. Introduction

Industry 4.0, also known as the Fourth Industrial Revolution, is a conceptual framework that is redefining the way industries operate, manufacture, and interact with the global economy. It emerges as a response to the growing need for greater efficiency, flexibility, and sustainability in manufacturing. In essence, Industry 4.0 means the integration of cyber-physical systems, the Internet of Things (IoT), Cloud Computing (CC), and Artificial Intelligence (AI) into the fabric of industrial operations [1]. The impact of this transformation is spreading throughout the manufacturing sector, impacting many aspects of production. It fosters a dynamic, connected ecosystem where machines, devices, and processes communicate seamlessly, enabling real-time data analysis and decision-making [2]. The tangible outcomes of Industry 4.0 include increased productivity, reduced production costs, improved product quality, and reduced time to market. Additionally, manufacturers will be able to respond more quickly to market fluctuations and customer demands, redefining the competitive landscape. Furthermore, in industrial production, the quality of manufactured things is critical to achieving client satisfaction. As a result, the procedure of quality control over created components should be followed before they reach their final destinations on the market. The procedure should begin on the production lines to allow for early detection of defects and deficiencies, utilizing emerging vision assessment technology to further improve the rhythm of production with quality satisfaction. Vision sensors are widely utilized to address such concerns nowadays; however, inspection quality still must be improved [3]. With that aim, we are combining deep learning with sensing cameras to create a monitoring application that enables the early detection of faulty manufactured objects on the manufacturing line using newly developed methodologies. Many works have been interested in adopting AI in industrial vision applications. The work carried out in [4] provides a methodology to recognize the class of an object while estimating its 6D pose with RGB-D data. Specifically, the proposed model adopts a global approach, first recognizing an object and the region of interest (ROI) from RGB images. The pose of the feature is then estimated using relative depth information. Several works have focused on adopting robotic arms to induce many benefits in different applications. For instance, the authors of [5] have proposed a robotic grasping system for automatically sorting garbage based on machine vision in complex backgrounds. To achieve the accurate grabbing of the target object, they adopted a deep learning scheme interested in the Region Proposal Generation and the VGG-16 model for object recognition and pose estimation. The machine vision system sends the information about the geometric center coordinates and the angle of the long side of the target object to the manipulator which completes the classification and grabbing of the target object.

The vision methods were also adopted for industrial welding applications. Vision-aided robotic welding has been applied in the industrial field for decades for the procedures of vision-aided robotic welding. Various methods in vision sensor calibration and hand–eye calibration have been illustrated. For example, the authors of [6] have presented a simple and robust hand–eye calibration strategy that requires minimal user interaction and makes use of a single planar calibration artifact. A simulation approach is used to further investigate and optimize the impact of robot position on the calibration process, and the resulting optimal robot positions are then experimentally validated for a real robot-mounted laser line sensor. Using the proposed optimal method, a semi-automatic calibration process, which requires only four manually scanned lines, is defined, and experimentally demonstrated.

On the other hand, robots have experienced an important evolution to deal with applications within fruit and vegetable harvesting. Using different end-effectors, acceptable results have been reached for apples, tomatoes, sweet peppers, and cucumbers [7]. However, the performances of the used final effectors (harvesting speed, success rate, costs, etc.) compared with the performances of human operators have shown less efficiency than human operators in the harvesting processes [8]. The authors of [9] have combined an advanced RGB-D camera and DL policies with a robot arm for the assembly of mobile phone items. This system exploits a modified version of the YOLO (You Only Look Once) scheme to arbitrarily detect the pieces in industrial working space. To make the UR5 robot plan and execute its movement intelligently without any human operator training, the authors of [10] used a stereo vision camera and DL Faster-RCNN for object classification and recognition. A series of investigations have been devoted to enhancing industrial manufacturing quality and efficiency, relying on vision inspection and emerging DL methods [11]. T.P. Nguyen et al. have proposed a smart industrial vision inspecting system based on CNN to detect product defects like blow holes, chipping, and cracks [12].

An approach based on learning eye–hand coordination for robotic grasping from monocular images was proposed in [13], to learn hand–eye coordination for grasping. A large CNN model was trained to predict the probability of the gripper movement which resulted in successful grasps. Another autonomous rock-stacking method dedicated to construction in an indoor environment using a robot arm was investigated in [14]. This method employs gradient descent with random initial orientation to detect randomly placed objects in a stacking scene. Therefore, in every industrial manufacturing scenario, ensuring excellent quality of manufactured commodities is essential. Early detection of product flaws is critical in quality control in manufacturing to better meet client needs. As a result, in this work, we examined the performance of YOLOv5 for defect inspection in our custom data. Then, the system is linked to a UR5 robot arm related to an RGB-D camera application.

The remainder of this paper is structured as follows. In Section 2, we summarize the proposed system application. Then, the YOLOv5 algorithm is described in Section 3. A description and annotation of the dataset used are depicted in Section 4 and the performance assessment metrics are discussed in this section. We comment on and discuss the obtained results in Section 5. Finally, we provide the conclusion in Section 6.

2. The Proposed System Architecture

This section provides an overview of the overall application system, as depicted in Figure 1. The intelligent grasping system comprises three principal components: a robotic arm equipped with an integrated two-finger gripper for item manipulation, an RGB-D camera affixed to the arm for data acquisition, and an industrial conveyor belt for the transportation of manufactured items. The central element of this developed application is the collaborative UR5 robot. Due to its inherent capabilities, this robot serves as an asset for assisting human operators in assembly or inspection tasks. It possesses a 6-degree-of-freedom (6-DOF) configuration, accompanied by a control box and a user-friendly programming interface tailored for light assembly duties and automated workbench scenarios. Additionally, it is equipped with a two-finger gripper capable of handling objects weighing up to 5 kg, featuring a total pivotal motion range of 360 degrees. The RGB-D camera in this context fulfills a dual purpose. RGB images are utilized as data inputs, subject to processing within the processing units, aimed at feature extraction and the classification of regions of interest (ROIs) for the detection of any defective areas. Simultaneously, the depth component, utilizing grayscale annotations, serves to estimate pose points for coordinating the 6D robot arm's position. On the depth image, the X, Y, and Z coordinates are delineated on the corresponding side of the ROI. Depth information is derived from RGB images, enhancing pose estimation capabilities. To optimize system efficiency and practicality, we place a significant emphasis on decentralizing processing units, relocating the processing tasks to Edge units to enhance the localization and recognition of monitored data. Specifically, the proposed YOLOv5 model is designed to facilitate real-time data processing in the Edge unit, thereby improving the system's overall performance.

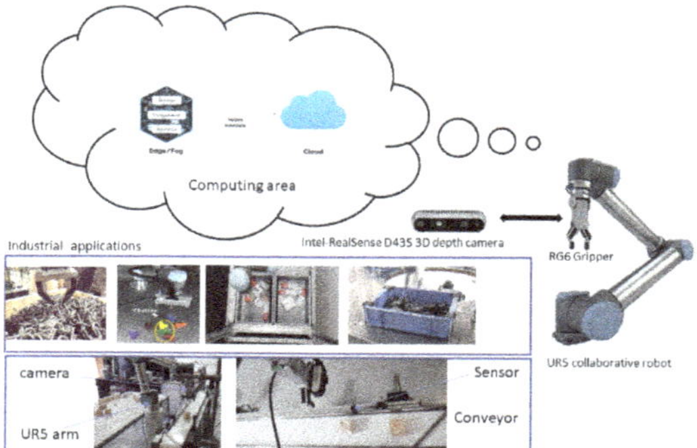

Figure 1. The proposed system architecture.

3. You Only Look Once Algorithm

YOLOv5 [15] is an object detection model built on CNN. In a single forward pass, the YOLOv5 model takes an image as an input and outputs the bounding boxes and class probabilities for all objects detected in the image. YOLOv5 has several improvements over previous versions, including a smaller model size, faster inference speed, and improved accuracy. It employs an innovative architecture that combines a CSP (Cross Stage Partial) backbone with an SPP (Spatial Pyramid Pooling) head, as well as other optimization techniques like multi-scale training and hyper-parameter optimization. YOLOv5 has achieved cutting-edge performance on several object detection benchmarks, including COCO and Open Images, and is widely used in industry and academia for a variety of computer vision applications. Figure 2 illustrates the YOLOv5 architecture's four basic components: the input, the backbone, the neck, and the output. The input block primarily contains data pre-processing, such as mosaic data augmentation and adaptive image filling. YOLOv5 integrates adaptive anchor structure calculations into the input to adapt to different datasets, making it possible to instantaneously set the initial anchor frame size whenever the dataset changes.

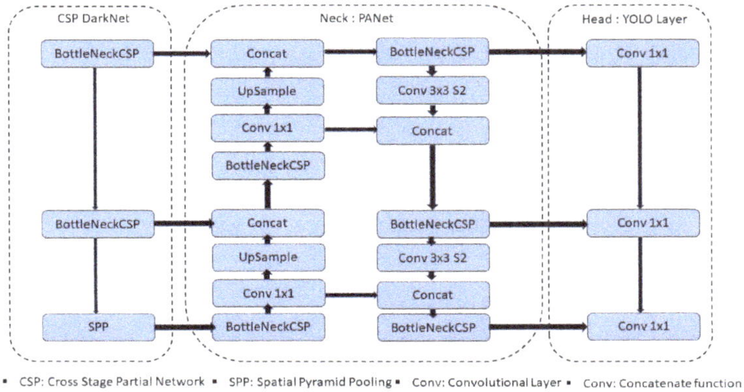

Figure 2. Yolov5 basic architecture.

The backbone network primarily employs a CSP and an SPP to retrieve feature maps of varying sizes from the input image via multiple convolutions and pooling. The bottleneck CSP structure serves to lower computation and increase inference speed, whereas the SPP structure realizes feature extraction from multiple scales for the same feature map and can generate three-scale feature maps, which improves detection accuracy.

The neck network employs the FPN and PAN pyramid topologies. The FPN structure communicates contextual information characteristics from top to bottom feature maps. At the same time, the PAN structure transmits strong localization features from lower to higher feature maps. The above two structures collaborate to boost detection capabilities by fortifying features gathered from distinct network levels in backbone fusion. Finally, the head output is primarily used as a final detection step to predict targets of various sizes on feature maps.

4. Data-Set Collection and Annotation

4.1. Dataset

To investigate the performance of the YOLOv5 model for our application, we collected a custom dataset comprising both defective and non-defective coils. This dataset was categorized into two distinct classes. Seventy percent of the data was allocated for training, while the remaining 30% was designated for testing. To be precise, the dataset comprises a total of approximately 1294 images, encompassing both "OK" and "Non-OK" pieces. These images were acquired using an RGB camera, capturing the object under study, primarily a metal part, from various angles. Additionally, the influence of lighting conditions on image quality and object characteristics was considered. We conducted image capturing under five different lighting setups: one from the top and four from different positions, each focusing on centralizing the object. To accommodate the input requirements of the deep learning model, we adjusted the image size during the training process. It is important to note that, to mimic real-world conveyor applications, the distance between the camera and the pieces varied along the vertical axis in the captured images. For the generation of the dataset, we classified the data into two classes: non-defective coils and defective ones. The definition of an ideal piece for production involves a green spool with 12 yellow coils. Consequently, any deviation from these specifications is categorized as a defective piece. Defective pieces are identified if they contain either more or fewer than 12 coils, or if the coils and spools exhibit different colors. Our simulation and testing procedures commenced with the training of the network, initially using a learning rate of 0.001 for 100 epochs. Subsequently, we continued training for an additional 40 epochs with a reduced learning rate of 0.0001. In Figure 3, we present sample images of faulty coils (Non-Ok) and non-defective coils (Ok) from our training and test dataset used in our simulation. These images serve to highlight the uniqueness and distinct characteristics of each piece within our dataset.

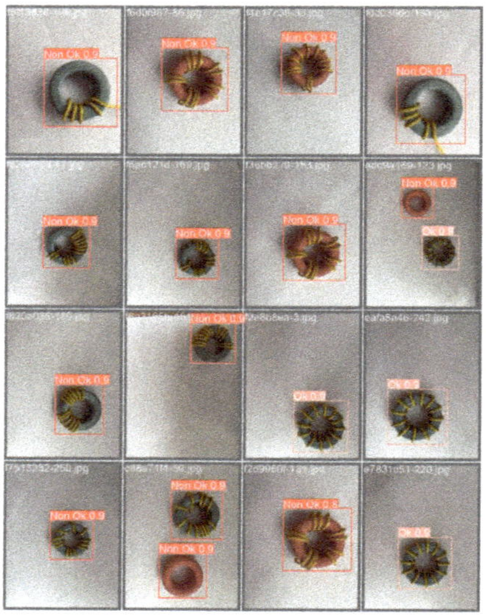

Figure 3. Training batches for defective and non-defective pieces.

4.2. Machine Learning Metric Measurements

During the YOLOv5 algorithm evaluation process, we used accuracy as the mean measurement; the detection model's evaluation indicators are precision (P), recall (R), and F1 score. True Positives (TP), False Positives (FP), True Negatives (TN), and False Negatives (FN) are also discrimination values (FN). The predicted results are compared to the actual labels for evaluation purposes. If the model's output is a correct prediction of a positive class, it is referred to as TP, and if it is a correct prediction of a negative class, it is referred to as TN. While False Positive refers to the incorrect prediction of a positive class, False Negative refers to the incorrect prediction of a negative class. On the other hand, accuracy is a performance criterion that shows the ratio of correct predictions to the total number of occurrences. Measures are also evaluated using precision and recall rates. Precision is the ratio of TP predictions to total positive predictions, and recall rate is the ratio of TP predictions to the sum of True Positive and False Negative predictions. F1 score is another measure used to report results that show the harmonic mean of recall rate and precision.

4.3. Robot Kinematics and Successful Grasp Accuracy

Time to detection is the amount of time needed to identify an object in an input image and create a plan for where to place or capture it. Given the temporal and substantive issues in the system, a real-time response must be given within the allotted time limits. In order to extract features and classify the images captured by the RGB-D camera, the employed YOLOv5 algorithm is performed on a computer that functions as an Edge Computing center. Rapid processing speed and high accuracy are two critical features of our system, so the model that meets these needs is selected as the final scheme.

The processing and robot grasping time are very important for our application as they accelerate the overall task. To measure the total time required for each cycle we calculate the time needed for the RGB-D camera to take the picture, the processing time on the edge unit, and the robot movement when required. Therefore, the total time needed for each detection cycle is:

$$T_{total} = T_{pros} + T_{camera} + T_{UR5} + T_{pose}$$

where T_{total}, T_{pros}, T_{camera}, T_{UR5}, and T_{pose} are the duration of processing, the time needed for the RGB-D camera to take the picture of the piece, the required time for the UR5 to move and take the piece from the conveyor, and the time needed from the camera to estimate the pose of the piece. If, after processing, the piece is detected to be non-defective, the time required for both T_{UR5} and T_{pose} is null. It should be noted that the time for the robot's movement is adjustable and can be set as desired.

We have selected YOLOv5 for our concrete application using the UR5 robot after assessing its accuracy and simulation time. We employ a computer that acts as an interface for Edge Computing. The UR5 controlling box is linked to a platform that incorporates the YOLOv5 concept. We utilize an intermediary tool that is Python-programmed to control the movement of the UR5 robot. The system operates as shown in Figure 4. After the conveyor is turned on, the components under study move until a sensor mounted on the conveyor detects them. The sensor then signals the conveyor to stop, and the camera snaps an image, which is then sent to processing center to apply the YOLOv5. There are two ways in which the inspected piece could be defective or not when the processing runs out. When a piece is defective, the system signals the UR5 robot to move, pick it up, and place it in a designated spot. If not, the conveyor receives a signal to move and prepares to receive a new piece, at which point the cycle is repeated.

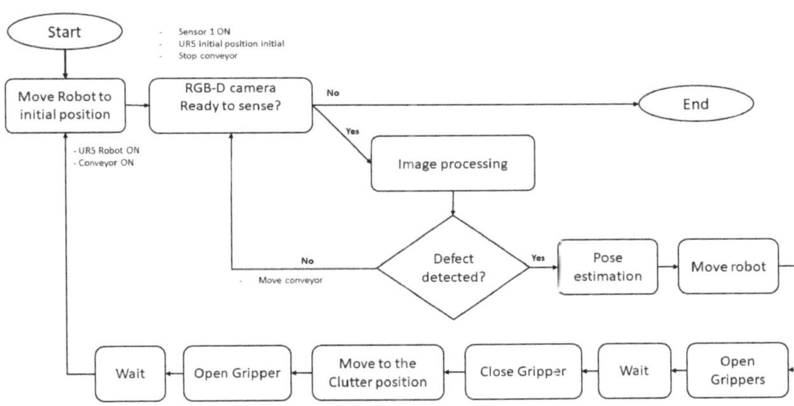

Figure 4. Robot inspection system flowchart.

5. Results and Discussions

The experimental tests and training were carried out on the Google Colab platform. Different comparable datasets were utilized in this work for different power inductor parts with varied faults. A suitable sample size of varied faulty and non-defect components ensured the flexibility and dependability of the training results, as well as enhanced the defect detection accuracy.

5.1. YOLOv5 Defect Detection Accuracy

Figure 5 illustrates how to evaluate the YOLOv5 data by utilizing bounding boxes to identify identified items, their class labels, confidence scores, and locations inside the image. It also evaluates the algorithm's performance using performance measures, which is essential for assessing the accuracy and reliability of the algorithm.

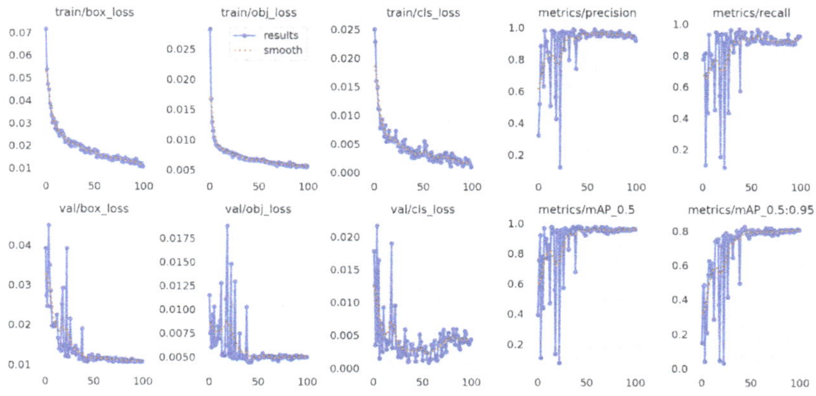

Figure 5. Yolov5 performance over epochs.

The term train box loss shown in Figure 5 typically refers to the loss function associated with the bounding box predictions during the training of an object detection model using YOLOv5. This loss function is crucial in guiding the model to learn how to accurately predict the locations of objects within an image. As shown during training, the value decreases with an increasing number of epochs, which means that the model is learning to predict bounding box coordinates accurately and assign appropriate confidence scores. Generally, when the model's predictions align closely with the ground truth annotations in the training data, the box loss decreases. Typically, the phrase Train obj loss refers to the object loss that occurs during the training of an object detection model. This loss function is crucial in helping the model identify the proper items in the image. Classification loss and object loss are two of this function's main components. As is apparent, this function decrements with epochs, which is in line with the training goal. By reducing this loss, the model is better able to categorize objects inside its bounding boxes. The object loss minimizes as the model's predictions move closer to the ground truth class labels and the object ratings in the training data.

In addition, the mAP score also shown in Figure 5 provides a summary of the model's overall performance in terms of object detection accuracy, considering all the classes or categories of objects in our custom dataset. As shown, the metric/mAP reaches high scores near 1 at an IoU of 0.5 which indicates a better object detection performance. This also means that the model is accurate in locating and identifying objects with at least a 50% overlap with the ground truth.

Figure 6 presents the confusion matrix which is especially useful for understanding how well the model performs in terms of True Positives, False Positives, True Negatives, and False Negatives. Figure 7 depicts F1 versus confidence performance for the defective and non-defective dataset. Typically, it shows how the F1 score, a standard assessment metric for object identification, evolves as confidence levels are changed. This sort of graph provides useful information on the trade-off between accuracy and recall in an object detection model at various confidence levels. It enables us to make informed decisions regarding the trade-off between accuracy and recall, maximizing the performance of your object detection model for your unique job or application. As illustrated, the optimal F1 score is obtained with a confidence level of 0.415.

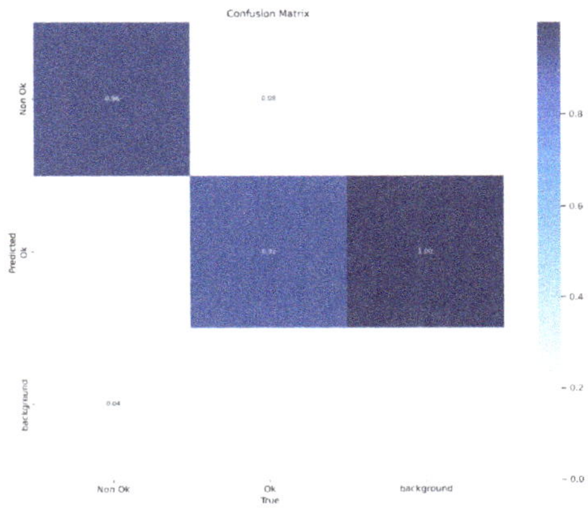

Figure 6. Confusion matrix performance.

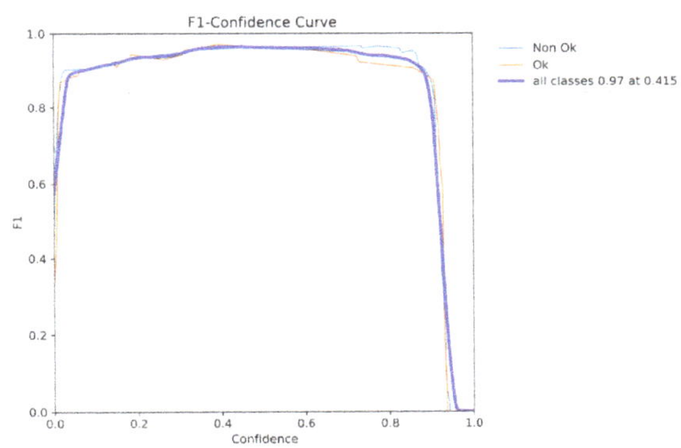

Figure 7. F1 performance versus confidence.

Figure 8 shows how the performance curve changes for precision versus confidence. As shown, the curve typically starts high and increases until it reaches 1, at the confidence of 0.883. In addition, Figure 9 provides an insight into how well YOLOv5 balances the trade-off between precision and recall across different confidence thresholds. In YOLOv5, as in many object detection models, precision represents the accuracy of positive predictions, and recall measures the model's ability to detect all the true positive objects.

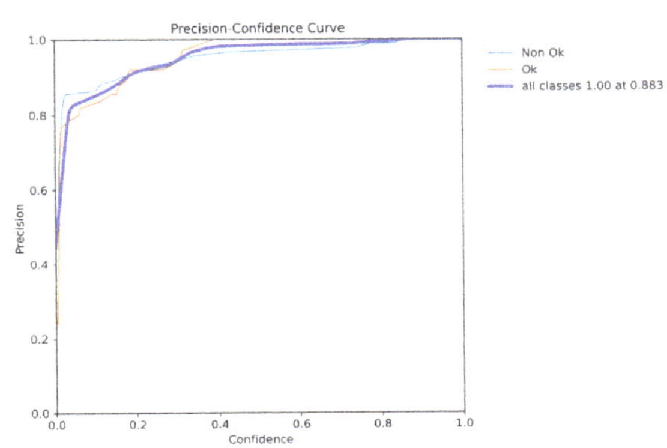

Figure 8. Performance curve for precision versus confidence.

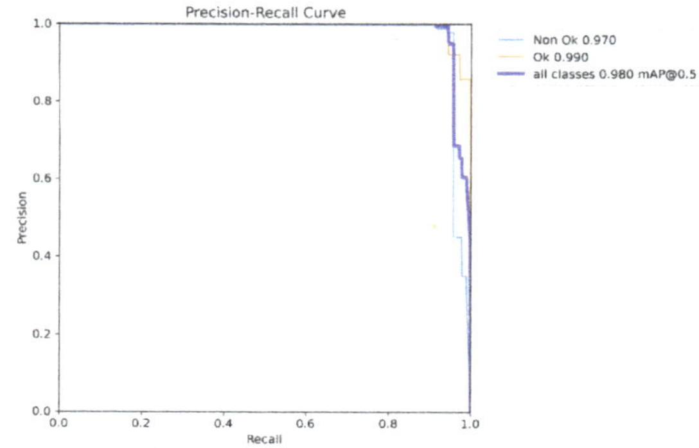

Figure 9. Performance curve for precision recall.

Figure 10 depicts how the model's recall rate increases at various confidence levels. It illustrates the trade-off between recall and accuracy. As illustrated, the model becomes more selective in accepting predictions as the confidence threshold increases (going from left to right on the curve). This increases accuracy (fewer false positives) but decreases recall (more false negatives).

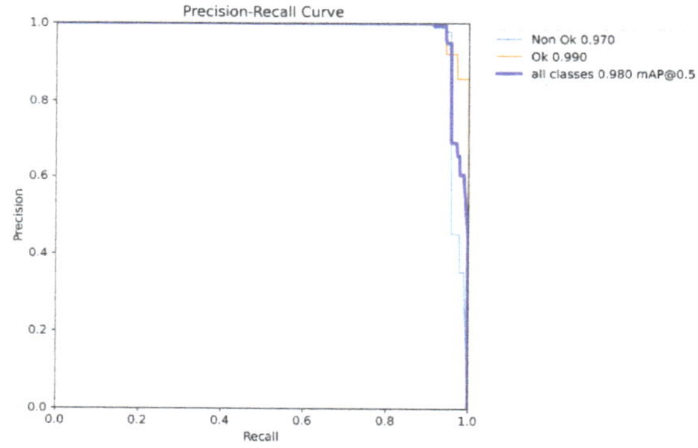

Figure 10. Confidence curve for recall performance.

5.2. Robot Grasp and Delay Accuracy

The statistical findings of the practical assessment of the suggested fault identification application based on the YOLOv5 algorithm are shown in Table 1. The assessment was conducted using five distinct scenario situations, with the category of the item being supervised on the conveyor being changed each time, and the procedure being repeated 100 times for each scenario. In the first scenario, we were utilizing a non-defective component to conduct the investigation. In addition to this, we provided several defective items with names ranging from defect type 1 to defect type 4 in the same condition. We have selected a distinct defect type for every single case. Meanwhile, it should be mentioned that each time we conducted the experiment, we tried to modify the piece's position, the robot arm's position, and the brightness of the lighting. It should be mentioned that, every time we run the experience, we attempt to adjust the item's location, the robot arm's position, and the illumination from various corners that surround the piece. The table illustrates that the robot is 100% accurate in identifying the non-defective item out of 100 distinct experiences. The defect type 2 produced the same results, as it achieved an accuracy of 100%. In the case of defect type 1, the robot misidentified two pieces because it did not recognize them well. In those tries, the defective parts were detected as non-defective, yielding a 98% proper-grip accuracy for that experience. In the case of defect type 3, a

95% proper detection accuracy was achieved despite five incorrect tries. We attempted to increase the complexity of the last scenario, defect type 4, by adding the hardest scenario possible to the system. Therefore, out of 100 attempts, 10 incorrect recognitions were made, yielding a 90% accuracy in the grasp. In summary, the system's total accuracy for each of the five scenarios that were tested for over 100 repetitions was 96.6% grasp success accuracy. This can be explained by the significant accuracy of the YOLOv5 algorithm that was utilized to identify and categorize the pieces, as well as the accurate performance of the RGBD camera that is used to determine the exact location of the piece on the conveyor.

Table 1. Evaluation of correct grasp accuracy on UR robot.

Scenario	Average Time T_{total} [s]	Correct Grasp	Incorrect Grasp	Correct Grasp Accuracy
Non-defect	35.12	100	0	100%
Defect Type 1	72.25	98	2	98%
Defect Type 2	71.07	100	0	100%
Defect Type 3	58.60	95	5	95%
Defect Type 4	80.32	90	10	90%

6. Conclusions

In industrial areas, AI algorithms are being adopted for various reasons, like enhancing and speeding up productivity and quality. In this paper, we are interested in investigating the performance and accuracy of the YOLOv5 algorithm in detecting defective features on a specific piece. The process relies on different conditions, like if the number of coiled circles fits with that required by the production charge. In addition, it can detect faults even if the spills and the color are as defined in the desired piece. Furthermore, the distance between the coiled circles may cause a problem for the dedicated application. Therefore, this is also considered by our system. In general, the power inductance pieces are of different shapes and forms, each one is specifically dedicated to a given application. Thus, considering this issue, is mandatory to make the process of production more rapid and intelligent relying on the effectiveness of the emerging AI algorithms. The use of YOLOv5 in our proposed industrial system has shown a high performance in terms of both accuracy and latency, which, nowadays, is required to improve the rate of production and quality. The YOLOv5 has reached an accuracy of 96% for our custom data and the correct grasp accuracy for studied scenarios reached an average of 96.6%.

Author Contributions: Y.Y., A.G.-G., A.E.O. and M.A. contributed to the design and implementation of the research work, upon the analysis, simulations, and writing of the manuscript. All authors have read and agreed to the published version of the manuscript.

Funding: This research received no external funding.

Institutional Review Board Statement: Not applicable.

Informed Consent Statement: Not applicable.

Data Availability Statement: The data used may be requested directly from the respective author.

Conflicts of Interest: The authors declare that there are no conflicts of interest regarding the publication of this paper.

References

1. Chi, H.R.; Wu, C.K.; Huang, N.F.; Tsang, K.F.; Radwan, A. A Survey of Network Automation for Industrial Internet-of-Things Towards Industry 5.0. *IEEE Trans. Ind. Inform.* **2022**, *19*, 2065–2077. [CrossRef]
2. Nizam, H.; Zafar, S.; Lv, Z.; Wang, F.; Hu, X. Real-time deep anomaly detection framework for multivariate time-series data in industrial IoT. *IEEE Sens. J.* **2022**, *22*, 22836–22849. [CrossRef]

3. Futai, M.M.; Bittencourt, T.N.; Carvalho, H.; Ribeiro, D.M. Challenges in the application of digital transformation to inspection and maintenance of bridges. *Struct. Infrastruct. Eng.* **2022**, *18*, 1581–1600. [CrossRef]
4. Hajari, N.; Lugo Bustillo, G.; Sharma, H.; Cheng, I. Marker-Less 3d Object Recognition and 6d Pose Estimation for Homogeneous Textureless Objects: An RGB-D Approach. *Sensors* **2020**, *20*, 5098. [CrossRef]
5. Zhihong, C.; Hebin, Z.; Yanbo, W.; Binyan, L.; Yu, L. A vision-based robotic grasping system using deep learning for garbage sorting. In Proceedings of the 2017 36th Chinese Control Conference (CCC), Dalian, China, 26–28 July 2017; IEEE: Piscataway, NJ, USA, 2017; pp. 11223–11226.
6. Sharifzadeh, S.; Biro, I.; Kinnell, P. Robust hand-eye calibration of 2D laser sensors using a single-plane calibration artefact. *Robot. Comput.-Integr. Manuf.* **2020**, *61*, 101823. [CrossRef]
7. Wang, C.; Tang, Y.; Zou, X.; SiTu, W.; Feng, W. A robust fruit image segmentation algorithm against varying illumination for vision system of fruit harvesting robot. *Optik* **2017**, *131*, 626–631. [CrossRef]
8. Morar, C.; Doroftei, I.; Doroftei, I.; Hagan, M. Robotic applications on agricultural industry. A review. In Proceedings of the 9th International Conference on Advanced Concepts in Mechanical Engineering—ACME 2020, Iași, Romania, 4–5 June 2020.
9. Olesen, A.S.; Gergaly, B.B.; Ryberg, E.A.; Thomsen, M.R.; Chrysostomou, D. A collaborative robot cell for random bin-picking based on deep learning policies and a multi-gripper switching strategy. *Procedia Manuf.* **2020**, *51*, 3–10. [CrossRef]
10. Chen, X.; Guhl, J. Industrial robot control with object recognition based on deep learning. *Procedia CIRP* **2018**, *76*, 149–154. [CrossRef]
11. Ren, Z.; Fang, F.; Yan, N.; Wu, Y. State of the art in defect detection based on machine vision. *Int. J. Precis. Eng. Manuf.-Green Technol.* **2021**, *9*, 661–691. [CrossRef]
12. Eklund, K.; Aros-O'Malley, M.; Murrieta, I. Multicultural supervision: What difference does difference make? *Contemp. Sch. Psychol.* **2014**, *18*, 195–204. [CrossRef]
13. Levine, S.; Pastor, P.; Krizhevsky, A.; Ibarz, J.; Quillen, D. Learning hand-eye coordination for robotic grasping with deep learning and large-scale data collection. *Int. J. Robot. Res.* **2018**, *37*, 421–436. [CrossRef]
14. Furrer, F.; Wermelinger, M.; Yoshida, H.; Gramazio, F.; Kohler, M.; Siegwart, R.; Hutter, M. Autonomous robotic stone stacking with online next best object target pose planning. In Proceedings of the 2017 IEEE International Conference on Robotics and Automation (ICRA), Singapore, 29 May–3 June 2017; IEEE: Piscataway, NJ, USA, 2017; pp. 2350–2356.
15. Redmon, J.; Divvala, S.; Girshick, R.; Farhadi, A. You only look once: Unified, real-time object detection. In Proceedings of the IEEE Conference on Computer Vision and Pattern Recognition, Las Vegas, NV, USA, 27–30 June 2016; pp. 779–788.

engineering proceedings

MDPI

Proceeding Paper

Study of the Temperature Influence on an Electret Microphone in the Monitoring of Fused Deposition Modeling †

Thiago Glissoi Lopes *, Paulo Roberto Aguiar *, Cristiano Soares Junior, Reinaldo Götz de Oliveira Junior, Paulo Monteiro Carvalho Monson and Gabriel Augusto David

Faculty of Engineering, Electrical Engineering Department, Sao Paulo State University—UNESP, Bauru 17033-360, Sao Paulo, Brazil; cristiano.soares@unesp.br (C.S.J.); reinaldo.gotz@unesp.br (R.G.d.O.J.); paulo.monson@unesp.br (P.M.C.M.); gabriel.david@unesp.br (G.A.D.)

* Correspondence: thiago.glissoi@unesp.br (T.G.L.); paulo.aguiar@unesp.br (P.R.A.)

† Presented at the 10th International Electronic Conference on Sensors and Applications (ECSA-10), 15–30 November 2023; Available online: https://ecsa-10.sciforum.net/.

Abstract: The evaluation of the response of sensors fixed to the print bed in the fused filament fabrication (FFF) process has been the subject of recent studies due to the increasing use of the FFF process. Many of these studies focus on topics related to monitoring the FFF process through the signals collected by sensors. Recently, some works employing piezoelectric diaphragm and electret microphones can be found in the monitoring of the FFF process, but the influence of the transducer response due to the variation of temperature has not been addressed. Thus, this work presents a study of the response of a low-cost electret microphone attached to the print bed under different temperature values. A 3D printer with polylactic acid (PLA) filament was used in the tests, which consisted of applying the pencil lead break method (PLB on the heated print bed at temperature values ranging from 25 °C to 65 °C. The acoustic waves generated by the tests were captured by the electret microphone attached near the breakage point, and the signals were sampled using an oscilloscope at a frequency of 2 MHz. The signals were processed in the time and frequency domains, followed by comparative analyses between the signals obtained for different temperature values. The results showed that it was not possible to determine a single temperature value at which the response of the electret microphone starts to undergo significant changes, but rather there is inconsistent change in the transducer's response across all frequency bands, indicating that the influence of temperature takes place in a complex way as frequency varies. This complexity is further evidenced by the non-linear behavior of RMSD values for the evaluated temperatures. Thus, the results can be helpful to those who use this type of transducer attached to the printing bed for monitoring purposes.

Keywords: electret microphone; temperature influence; fused filament fabrication; monitoring; pencil lead breakage method

Citation: Lopes, T.G.; Aguiar, P.R.; Junior, C.S.; de Oliveira Junior, R.G.; Monson, P.M.C.; David, G.A. Study of the Temperature Influence on an Electret Microphone in the Monitoring of Fused Deposition Modeling. *Eng. Proc.* **2023**, *58*, 64. https://doi.org/10.3390/ecsa-10-16041

Academic Editor: Stefan Bosse

Published: 15 November 2023

1. Introduction

The fused filament fabrication (FFF) manufacturing process, also known as 3D printing, consists of adding successive layers of melted filament on a heated bed [1]. The filaments used in the FFF process can be made of different types of materials such as acrylonitrile butadiene styrene (ABS), polyethylene terephthalate glycol (PETG), polylactic acid (PLA), copolyester TRITAN, among others [2]. However, certain printer models are not capable of manufacturing parts with all types of materials, as these printers do not have the necessary heated print chambers for certain material types.

The temperature at which the print bed should be heated varies depending on the material used in the FFF process and is usually defined in a range of values. One of the reasons for defining a range of temperature values is the relationship with the characteristics of the material itself, which allows, in some cases, manufacturing under different print bed temperature values. The temperature range based on the material can be quite broad, as is the case with PLA, which can be manufactured at print bed temperatures ranging from 0 °C to 60 °C [3]. This allows even printers without heated print chambers to manufacture parts with PLA filament, since the print bed temperature range for this material includes ambient temperature values.

Up until now, there have been several studies aimed at monitoring the FFF process using sensors located on the print bed. Notably, the use of acoustic emission sensors can be observed in the works developed by [4,5]. Additionally, a literature review reveals works proposing the use of low-cost electret microphones [6]. However, it is known that elevated temperature values, such as those that can be reached on the print bed, have various effects on a sensor's responses [7]. In the study referenced in [8], the impact of temperature on the mechanical/acoustic properties of condenser microphones was analyzed. The results revealed that temperature significantly influences the resonance of the microphone's membrane. It was discovered that certain microphones and preamplifiers function at temperatures below 100 K, a previously unobserved phenomenon.

One of the methods used to evaluate sensor responses under controlled conditions is the Hsu–Nielsen method, also known as the pencil lead break (PLB) method [9]. In the PLB method, an artificial source of acoustic emission is generated by breaking a graphite pencil lead with specific hardness, dimensions, and angle defined by the ASTM E976-10 standard [10]. The obtained signal can then be processed and analyzed for its frequency components, as conducted in the study by [7].

A literature review revealed studies aimed at evaluating the influence of print bed temperature on sensor responses used for monitoring the FFF process through the PLB method [11]. However, no study was found that covered multiple temperature values within a specific range to determine the temperature value at which the sensor response would undergo significant changes.

Thus, the primary objective of this study is to investigate the influence of print bed temperature on the response of a low-cost electret microphone affixed to the print bed for FFF process monitoring purposes. The PLB method was employed, conducting pencil lead break tests at 5 °C intervals within the temperature range suitable for PLA filament printing, ranging from 25 °C (ambient temperature) to 65 °C. The aim is to determine at which temperature value the response of the low-cost sensor starts to undergo significant changes, following the frequency band analysis method used by [7] to determine if such influence on the response is limited to a specific frequency band of the sensor's response.

2. Materials and Methods

The following section will present the test bench and the digital signal processing methodology employed to achieve the objectives of this study.

2.1. Test Bench

To accomplish the proposed objectives in this study, tests were conducted using a Cartesian 3D printer model Graber i3, from GTMax3D, Americana, Brazil, as shown in Figure 1a. The print bed of the Graber i3 has dimensions of 200 × 200 mm and a heating system that can reach up to 70 °C, but it does not have a heated chamber for printing. To accommodate this restriction, the present study evaluated temperatures ranging from 25 °C to 65 °C, the recommended range for PLA filament printing, using the PLB method. It is worth mentioning that the acoustic emission sensor with a white color base shown in Figure 1 was not used in this work.

The pencil lead break tests were conducted at the print bed's center, adjusting the temperature in 5 °C steps and allowing a minute for stabilization. A mechanical pencil with a rubber hose at the tip ensured a 45-degree angle with the print bed, and therefore avoided contact of the metal cone cap and lead sleeve of the pencil with the surface under test. The 2H hardness, 0.5 mm diameter graphite is shown in Figure 1b. The print bed's temperature was computer-controlled via USB using Repetier-host® software, version 1.4. To ensure accuracy, three test repetitions were carried out for each temperature.

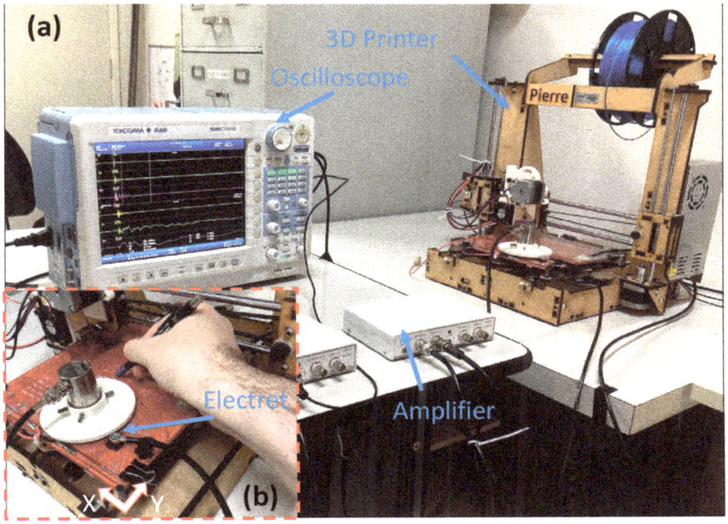

Figure 1. Test bench. (**a**) Complete test bench, (**b**) PLB method representation.

An electret microphone with dimensions of 9.7 × 4.5 mm was affixed at 37 × 50 mm coordinates on the print bed using silicone adhesive. It was directly attached without heat isolation and connected to a transimpedance amplifier, as shown in Figure 1a. The microphone's signals were captured using a DL 850 oscilloscope, from Yokogawa®, Tokyo, Japan, at 2 MHz sampling frequency, chosen for comprehensive signal analysis. Data were stored in the oscilloscope's hard drive right after data acquisition, and then downloaded for processing in MATLAB®, version R2022a.

2.2. *Signal Processing*

Initially, the data downloaded from the oscilloscope were converted from binary format into MATLAB® format to be digitally processed in the software. In order to evaluate the temperature's influence on the electret microphone's response, analyses focused on the frequency spectra were conducted. Firstly, Welch's power spectral density estimates for each PLB signal were computed through the MATLAB's pwelch function, using a Hamming window of 4096 data points. Subsequently, the spectral averages for each temperature condition were calculated from the three repetitions.

From the spectra results, the absolute values of the differences between each temperature's average spectra in relation to the average spectrum baseline of 25 °C for each frequency value were computed. Subsequently, the root mean square deviation (RMSD) was obtained with the purpose of scrutinizing the amplitude variations between each temperature's average spectrum in relation to the 25 °C average spectrum baseline. The RMSD values were computed in frequency steps of 1 kHz, spanning a frequency range of 20 Hz to 20 kHz, in alignment with the methodology proposed in the study of [7].

3. Results and Discussion

The absolute value of spectral differences at each temperature compared to the 25 °C baseline is presented in Figure 2. A preliminary analysis reveals significant differences between the spectra of all evaluated temperatures compared to the 25 °C spectrum across the entire frequency range analyzed. However, greater differences in absolute values are noted for certain temperatures within specific frequency bands, notably 2 kHz to 3 kHz and 4 kHz to 5 kHz. These observations corroborate the findings of [11]. Additionally, our analysis further demonstrates that the effect of bed temperature on the electret microphone's response is not merely prevalent across most of the device's response spectra, but it can rather show an increasing trend in signal amplitude within a specific frequency band, such as that observed from 2 kHz to 3 kHz.

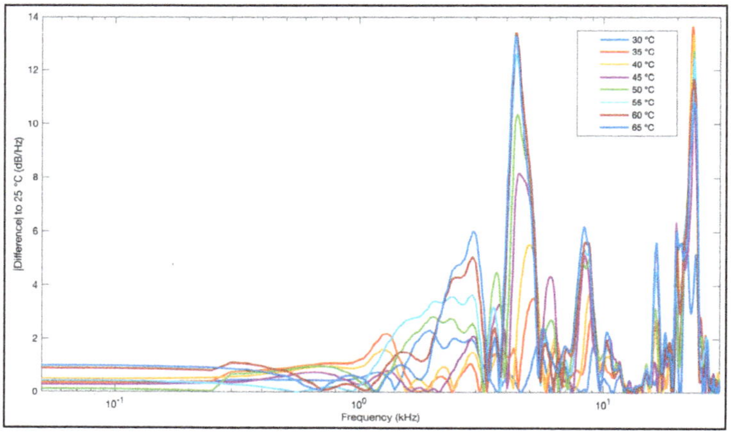

Figure 2. Absolute value of spectral differences at each temperature compared to the 25 °C baseline for each frequency.

The RMSD results of the spectral differences at each temperature, compared to the 25 °C baseline, are shown in Figure 3. The different colors in Figure 3 are representations of the temperature variation between 25 °C (dark blue) and 65 °C (yellow). An initial analysis, focused on assessing the RMSD values for each temperature separately, reveals noticeable differences within each 1 kHz frequency band. For instance, RMSD values at 35 °C generally increase compared to those at 30 °C, particularly in the frequency bands of 4.02 kHz to 5.02 kHz and 17.02 kHz to 18.02 kHz. However, this upward trend is not consistent across all frequency bands at 35 °C, as seen in the bands from 2.02 kHz to 3.02 kHz and 19.02 kHz to 20.02 kHz.

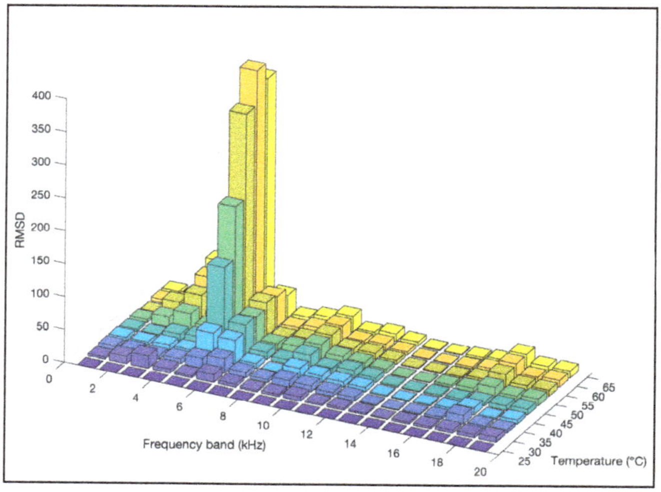

Figure 3. RMSD values for each temperature relative to the 25 °C baseline, evaluated across incremental 1 kHz frequency bands from 20 Hz to 20 kHz.

A follow-up analysis, looking at the RMSD values for each frequency band individually, also shows significant variations in how temperature affects the electret microphone's response. Specifically, the frequency bands from 2.02 kHz to 3.02 kHz and 4.02 kHz to 5.02 kHz display an increasing, yet non-linear, effect of temperature on RMSD values. This behavior is not uniform and diverges in some frequency bands, such as from 9.02 kHz to 10.02 kHz.

These findings suggest that temperature variations influence the electret microphone's response in a complex way, differing across various frequency bands.

4. Conclusions

This study was designed with the primary objective of investigating how the print bed temperature influences the response of a low-cost electret microphone that is affixed to the print bed for monitoring the FFF process. Utilizing the PLB method, we conducted tests at 5 °C intervals in the temperature range suitable for PLA filament printing, from 25 °C to 65 °C. The goal was to pinpoint the temperature at which the sensor's response begins to manifest significant changes and to evaluate whether these changes are confined to specific frequency bands.

The data generated from our experiments provide mixed results. The absolute differences between each spectrum and baseline and RMSD values, presented in Figures 2 and 3, respectively, do not yield a singular frequency range at which the sensor's response undergoes a general change due to temperature variation. Instead, our analysis revealed that the impact of temperature on the microphone's response varies across different frequency bands. Notably, distinct patterns of change are observed in the bands from 2 kHz to 3 kHz and 4 kHz to 5 kHz.

In addition, it was noted that there is not a consistent change in the transducer's response across all frequency bands, indicating that the influence of temperature is complex and depends on the specific frequency band under consideration. This complexity is further evidenced by the non-linear behavior of RMSD values for the evaluated temperatures, as it can be seen, for instance, in the bands from 2.02 kHz to 3.02 kHz and 8.02 kHz to 9.02 kHz.

Finally, the study clearly showed the change in the transducer's response according to the temperature variation and frequency, and the results can be useful when utilizing this type of transducer attached to a printing bed for monitoring purposes. Future research may explore alternative signal processing techniques, delve deeper into the intricacies of how specific frequencies are affected by temperature, and broaden the scope of temperature ranges.

Author Contributions: Conceptualization: T.G.L. and P.R.A.; methodology: T.G.L. and G.A.D.; software: C.S.J.; validation: P.R.A., R.G.d.O.J. and P.M.C.M.; formal analysis: C.S.J.; investigation: G.A.D.; resources: R.G.d.O.J.; data curation: P.M.C.M.; writing—original draft preparation: T.G.L. and P.R.A. writing—review and editing: T.G.L.; visualization: G.A.D. and C.S.J.; supervision: P.R.A.; project administration: P.R.A.; funding acquisition: P.R.A. All authors have read and agreed to the published version of the manuscript.

Funding: The authors would like to thank the São Paulo Research Foundation (FAPESP) for the grant number #2016/22038-8 and the National Council for Scientific and Technological Development (CNPq) for the grant number #306774/2021-6.

Institutional Review Board Statement: Not applicable.

Informed Consent Statement: Not applicable.

Data Availability Statement: Data are contained within the article.

Conflicts of Interest: The authors declare no conflict of interest.

References

1. Lopes, T.G.; Carmo, M.; Queiroz, L.M.; Aguiar, P.; França, T. Feasibility of Optical Profilometry for Quality Characterization of Monolayer Parts Obtained by Fused Filament Fabrication. In Proceedings of the 26th International Congress of Mechanical Engineering, Virtual Conference, 22–26 November 2021. [CrossRef]
2. Pinheiro, N.D.; Freire, R.T.; Conrado, J.A.M.; Batista, A.D.; da Silveira Petruci, J.F. Paper-based optoelectronic nose for identification of indoor air pollution caused by 3D printing thermoplastic filaments. *Anal. Chim. Acta* **2021**, *1143*, 1–8. [CrossRef] [PubMed]
3. Beniak, J.; Holdy, M.; Križan, P.; Matúš, M. Research on parameters optimization for the Additive Manufacturing process. *Transp. Res. Procedia* **2019**, *40*, 144–149. [CrossRef]
4. Wu, H.; Yu, Z.; Wang, Y. A New Approach for Online Monitoring of Additive Manufacturing Based on Acoustic Emission. In *Volume 3: Joint MSEC-NAMRC Symposia*; American Society of Mechanical Engineers: Blacksburg, Virginia, 2016; p. 8.
5. Wu, H.; Yu, Z.; Wang, Y. Experimental study of the process failure diagnosis in additive manufacturing based on acoustic emission. *Measurement* **2019**, *136*, 445–453. [CrossRef]
6. Lopes, T.G.; Aguiar, P.R.; Monson, P.M.d.C.; D'Addona, D.M.; Conceição Júnior, P.d.O.; de Oliveira Junior, R.G. Machine condition monitoring in FDM based on electret microphone, SVM, and neural networks. *Int. J. Adv. Manuf. Technol.* **2023**, *129*, 1769–1786. [CrossRef]
7. Lopes, B.G.; Alexandre, F.A.; Lopes, W.N.; Aguiar PR de Bianchi, E.C.; Viera, M.A.A. Study on the effect of the temperature in Acoustic Emission Sensor by the Pencil Lead Break Test. In Proceedings of the 2018 13th IEEE International Conference on Industry Applications (INDUSCON), Sao Paulo, Brazil, 12–14 November 2018; IEEE: Piscataway, NJ, USA, 2018; pp. 1226–1229.
8. Guianvarc'h, C.; Lavergne, T.; Risegari, L.; Honzik, P.; Rodrigues, D.; Gavioso, R.M. Temperature effects on the mechanical-acoustic properties of condenser microphones: Experimental characterization. In Proceedings of the International Congress on Acoustics, Aachen, Germany, 9–13 September 2019. [CrossRef]
9. Madarshahian, R.; Soltangharaei, V.; Anay, R.; Caicedo, J.M.; Ziehl, P. Hsu-Nielsen source acoustic emission data on a concrete block. *Data Brief* **2019**, *23*, 103813. [CrossRef] [PubMed]
10. *ASTM Standard E976-10*; Standard Guide for Determining the Reproducibility of Acoustic Emission Sensor Response. ASTM International: West Conshohocken, PA, USA, 2015. [CrossRef]
11. Barbosa, L.; Lopes, T.G.; Aguiar, P.R.; de Oliveira Junior, R.G.; França, T.V. Evaluating Temperature Influence on Low-Cost Microphone Response for 3D Printing Process Monitoring. In Proceedings of the 8th International Electronic Conference on Sensors and Applications, Online, 1–15 November 2021; MDPI: Basel, Switzerland, 2021; p. 67.

engineering proceedings

MDPI

Proceeding Paper

Structural Identification Using Digital Image Correlation Technology †

Samira Azizi [1,2,*], Kaveh Karami [1] and Stefano Mariani [2]

1 Department of Civil Engineering, University of Kurdistan, Sanandaj P.C. Box 416, Iran; ka.karami@uok.ac.ir
2 Department of Civil and Environmental Engineering, Politecnico di Milano, 20133 Milano, Italy; stefano.mariani@polimi.it
* Correspondence: samira.azizi@polimi.it
† Presented at the 10th International Electronic Conference on Sensors and Applications (ECSA-10), 15–30 November 2023; Available online: https://ecsa-10.sciforum.net/.

Abstract: Structural health monitoring has gained increasing research interest, particularly due to the societal concerns tied to the aging of current civil structures and infrastructures. By managing datasets collected through a network of sensors deployed over monitored structures, (big) data analytics can be executed. Traditional inertial sensors, such as accelerometers or strain gauges, necessitate intricate cable arrangements and lead to high maintenance costs. Lately, there has been a growing interest in non-contact, vision-based approaches to tackle these aforementioned issues. Among these methods, digital image correlation (DIC) can furnish a representation of tracked displacements at various points of a structure, particularly if physically attached targets are employed. In this study, a video capturing the vibrations of a structure was analyzed, with a focus on specific points, such as structural nodes where damage could be initiated or whose responses could be impacted by the mentioned damage. Displacement time histories were acquired, and a blind source identification technique was adopted to delve into the data and assess structural health. The proposed methodology demonstrates its capacity to accurately extract the vibration frequencies and mode shapes of the structure, even when they change in time due to damage.

Keywords: structural health monitoring; damage detection; vision-based methods; digital image correlation

Citation: Azizi, S.; Karami, K.; Mariani, S. Structural Identification Using Digital Image Correlation Technology. *Eng. Proc.* **2023**, *58*, 65. https://doi.org/10.3390/ecsa-10-16034

Academic Editor: Francisco Falcone

Published: 15 November 2023

1. Introduction

Monitoring the structural condition of civil structures is of critical importance. Modal analysis methods play an important role in identifying the dynamic characteristics of structures, e.g., by exploiting the vibrational response of the structures themselves. Blind source separation methods are a category of output-only measurement techniques [1] that require the installation of sensors at different locations over the structure to obtain the information to process. Throughout the structural lifespan, these sensors might deteriorate, leading to a decline in measurement accuracy. Among the latest non-contact measurement approaches, the utilization of video cameras for measurement collection has increased. Initially, the vibration of the structure is captured using a high-speed camera. By converting the video into a number of frames and through the application of various image processing algorithms, distinctive features are then extracted. Subsequently, modal analysis is executed to obtain the aforementioned dynamic characteristics. The merits of this approach are listed as follows: high-resolution measurements at all points, cost-effectiveness, and straightforward installation.

The digital image correlation (DIC) method, renowned for its efficacy in image processing, has progressively gained applications to quantify surface deformations. This is attributed to its capability to accurately quantify displacements and assess strains [2–10]. In early studies, stress analysis was carried out in a two-dimensional frame, but neglecting out-of-plane displacements can lead to errors in the calculation of the in-plane components. For this reason, studies have been conducted in a full three-dimensional context ever since [10–13].

Due to the potential damage of the target, methods have been proposed to extract image features that are invariant to intensity variations and rotation [14,15]. These features are tracked across different frames to obtain the displacements. These methods are feature-based. Youn et al. [16] initially obtained the frames of a vibrating six-story model and subsequently those of selected regions from the images. In the field of experimental solid mechanics, it is of interest to obtain mechanical properties and assess the deformation experienced by a specimen when subjected to an external load [17–20]. In this context, DIC can capture full-field data to enable the application of constitutive equations, facilitating the simultaneous tuning of multiple material properties [21–23].

Civil structures exposed to significant lateral forces, such as strong winds and earthquakes, are susceptible to damage due to a reduction in stiffness. By comparing the structural response with a baseline model, the identification of damage can be obtained [24–27]. In this paper, a video of a vibrating beam was adopted to investigate the sensitivity of displacements at various sampling points to structural damage. We employed the DIC technique to allow for a precise assessment of the displacement of the said locations. Additionally, we adopted the blind source separation method to extract relevant frequency information. The results demonstrate the feasibility of DIC data obtained at sparse locations as the input for an identification process. Furthermore, changes in the frequency of vibrations of this structure can be correlated to the damage pattern.

2. Digital Image Correlation Overview

To generate a deformation map using the DIC method, the initial step consists of the identification of the positions of sampling points distributed across the reference undeformed image. Then, an area surrounding each sampling point is employed; this is achieved by means of a square correlation window consisting of $p \times p$ pixels. By moving the second window to find the maximum correlation between the reference and deformed subsets, the surface displacement at that point can be determined. Generally, identifying the correspondence of a single pixel between two images is challenging. The grayscale value of a single pixel can match with numerous other pixels in the second image, resulting in a lack of unique correspondence. To assure the accurate application of DIC, a measurement surface with random texture or artificial targets is adopted.

By handling the frames at the initial and current time, a comparative analysis can be conducted between the deformed image(s) and the reference one. The regions of interest, which contain the sampling points, must be defined in the reference image. To achieve efficient, accurate, and robust subset matching, an objective function that combines a similarity or dissimilarity metric between the reference subset and its deformed counterpart has to be formulated. There are different correlation criteria in the literature, which can be categorized into the following four groups [28]: a cross-correlation, sum of absolute difference, sum of squared difference, and a parametric sum of the squared difference. In practice, the changes in intensity between images may be induced by various reasons. Thus, a robust correlation criterion should be used to assess the intensity variations in the deformed images; otherwise, significant displacement errors may occur due to the mismatch of the intensity change model. Based on this discussion, a zero mean normalized cross-correlation (ZNCC) criterion is used in this paper, according to the following:

$$ZNCC(u,v) = \frac{\sum_{x,y}\left(f_n(x,y) - \overline{f}_n\right)\left(f_d(x-u,y-v) - \overline{f}_d\right)}{\sqrt{\sum_{x,y}\left(f_n(x,y) - \overline{f}_n\right)^2 \sum_{x,y}\left(f_d(x-u,y-v) - \overline{f}_d\right)^2}} \quad (1)$$

where $f_n(x,y)$ and $f_d(x-u,y-v)$ are the intensity values; f is the pixel in the locations (x,y) and $(x-u,y-v)$ of the $p \times p$ square window in the first undeformed frame and the current deformed frame; $\overline{f}_n$ and $\overline{f}_d$ are the average of the intensity value of the $p \times p$ square window in the first undeformed frame and in the current deformed frame. The location with a maximum ZNCC provides the displacement.

3. Identification Procedure

To show how the vibration frequencies and mode shapes of the structure can be determined, a video of a vibrating beam was adopted. At a later stage, the elastic properties of the structure were decreased to simulate damage, and the DIC was applied to obtain the displacement field. By comparing the identified frequencies corresponding to the different levels of structural stiffness, we show that the displacement obtained with the DIC can be used for damage detection.

The number of excited modes used in this model is assumed to be smaller than the number of sampling points. We, thus, need a dimension reduction technique to reduce the dimension of the displacement matrix. By applying a singular value decomposition (SVD) to the displacement matrix $\boldsymbol{D}$, and by considering the corresponding non-zero singular values, the dimension of the problem is reduced. The complexity pursuit (CP) can be next adopted to obtain the modal frequencies and excited modes.

4. Numerical Investigation: Cantilever Beam

A model of a vibrating cantilever beam with a fixed end is considered as excited by an initial velocity here. The following beam properties were considered: the mass per unit length at 0.051 Kg/m, the length at 0.18 m, Young's modulus at 1.91×10^{11} Pa, and a moment of inertia of 0.124×10^{-12} m^4. A video of a vibrating beam is generated using three modes. The number of points at which the vibrations are measured and used to create a video is set to 37. By using the generated video, displacements are measured at 12 sampling points using the DIC. By adopting the identification procedure based on SVD-CP, the vibration frequencies and mode shapes are obtained.

Moving to the results of the analysis, in Figure 1a, a comparison is presented between the actual displacement used to generate the video and the displacement obtained via applying the DIC to the video. The displacement results tend to be in perfect agreement with their time evolutions. Figure 1b shows the results in terms of the displacement measurements at three locations at the top, middle, and bottom of the beam.

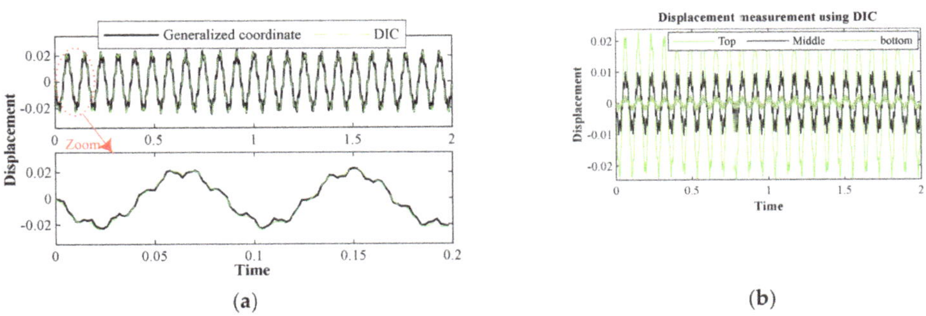

Figure 1. Displacement time histories for the vibrating beam: (**a**) comparison between the DIC results and the actual time evolution; (**b**) measurements at three locations.

As the purpose of this research was to identify mode shapes and vibration frequencies, Figure 2a shows the mode shapes in the actual state and those obtained via the SVD-CP, using the DIC at the said twelve locations along the longitudinal axis of the beam. It can be observed that the first and second modes have acceptable agreement, but this is not the case for the third mode. The reason for this could be the small amplitude of the vibrations in this case study. As can be seen in Figure 2b, the power spectrum associated with the vibration frequency of the third mode is, in fact, very small.

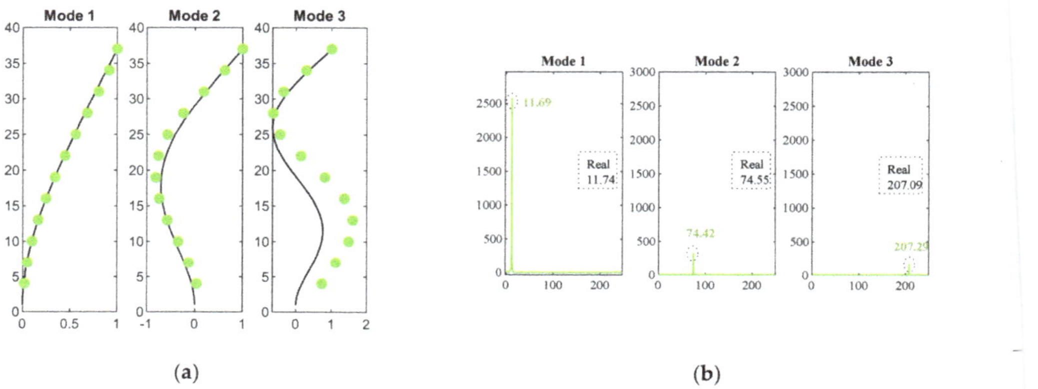

Figure 2. Comparison between the actual and estimated beam features using the DIC−SVD−CP-based method for the healthy state: (**a**) actual mode shapes and estimated ones using the 12 points (green circles); (**b**) frequencies of vibration.

To model the damage in the structure, the structural stiffness was reduced in two different scenarios by 15% and 25%, respectively. In Figure 3a, the identified frequencies were gathered for the undamaged and damaged scenarios, as obtained with the procedure based on the SVD-CP on DIC measurements. The singular values are reported in Figure 3b for these three scenarios. It can be seen that only the first three values are non-zero; through this outcome, it is possible to understand how to reduce the displacement matrix dimensionality to the number of excited modes.

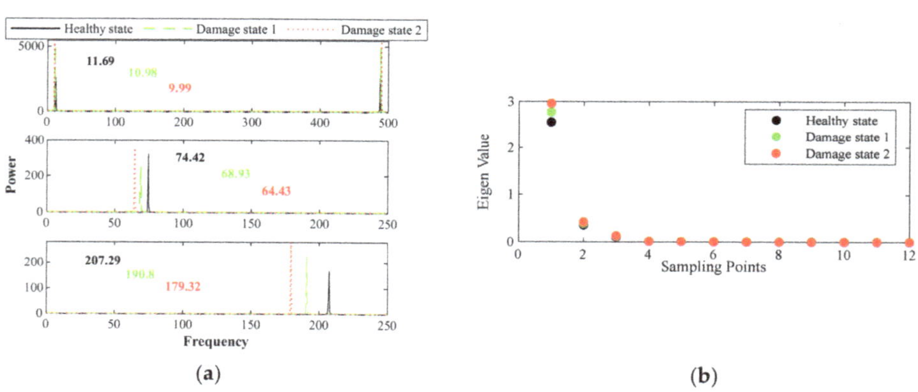

Figure 3. Comparison of the results relevant to the three considered states: (**a**) identified frequencies; (**b**) eigenvalues.

To establish a criterion to quantify the structural damage, the results are collected in Table 1. Here, the vibration frequencies corresponding to the first, second, and third modes are separately reported and divided by those corresponding to the healthy state and those subtracted from unity. Values in the last column show that, for the first damaged state, the criterion applied to the three modes led to an error amounting to 21%, 6.4%, and 1.2%, respectively. Hence, the results corresponding to the second and third modes prove to be more accurate. In the second scenario, the measured damage levels using the vibration frequencies lead to an error amounting to 5.2%, 0%, and 0.7%, respectively.

Table 1. Results provided by the damage criterion.

Ratio of Frequency	Mode 1	Mode 2	Mode 3	1-$\sqrt{\alpha}$
1—(DamageState1/Healthy)	0.061	0.073	0.079	0.078
1—(DamageState2/Healthy)	0.141	0.134	0.135	0.134

5. Conclusions

This research is devoted to data collection from a video of the vibrations of a structure and achieved through the digital image correlation technique. The acquired displacements were employed to discern the beam modal parameters and the mode shapes using a blind source separation approach. The findings showcase the effectiveness of this method in accurately measuring displacements. Furthermore, damage has been ad hoc introduced in the model of the structure by altering the material elasticity; with two different scenarios, an exploration of frequency and amplitude variations in the resulting vibrations becomes possible.

The estimation of displacement and frequency through video-based measurements underscores the potential of the DIC framework in the realm of structural health monitoring. In future studies, models capable of inducing damage at specific locations will be incorporated. Additionally, as the DIC method needs the judicious selection of the analysis window, a further in-depth investigation in this regard is necessary.

Author Contributions: Conceptualization, S.A., S.M. and K.K.; methodology, S.A., S.M. and K.K.; validation, S.A., S.M. and K.K.; formal analysis, S.A.; resources, K.K.; data curation, S.A.; writing—original draft preparation, S.A.; writing—review and editing, S.M.; visualization, S.M.; supervision, S.M. All authors have read and agreed to the published version of the manuscript.

Funding: This research received no external funding.

Institutional Review Board Statement: Not applicable.

Informed Consent Statement: Not applicable.

Data Availability Statement: The data presented in this study are available on request from the corresponding author.

Conflicts of Interest: The authors declare no conflict of interest.

References

1. Sadhu, A.; Narasimhan, S.; Antoni, J. A review of output-only structural mode identification literature employing blind source separation methods. *Mech. Syst. Signal Process.* **2017**, *94*, 415–431. [CrossRef]
2. Hild, F.; Roux, S. Digital Image Correlation: From Displacement Measurement to Identification of Elastic Properties—A Review. *Strain* **2006**, *42*, 69–80. [CrossRef]
3. Huang, J.; Pan, X.; Peng, X.; Zhu, T.; Qin, L.; Xiong, C.; Fang, J. High-efficiency cell–substrate displacement acquisition via digital image correlation method using basis functions. *Opt. Lasers Eng.* **2010**, *48*, 1058–1066. [CrossRef]
4. Cofaru, C.; Philips, W.; Van Paepegem, W. Improved Newton–Raphson digital image correlation method for full-field displacement and strain calculation. *Appl. Opt.* **2010**, *49*, 6472–6484. [CrossRef]
5. Luu, L.; Wang, Z.; Vo, M.; Hoang, T.; Ma, J. Accuracy enhancement of digital image correlation with B-spline interpolation. *Opt. Lett.* **2011**, *36*, 3070–3072. [CrossRef] [PubMed]
6. Tong, W. Subpixel image registration with reduced bias. *Opt. Lett.* **2011**, *36*, 763–765. [CrossRef] [PubMed]
7. Chu, T.C.; Ranson, W.F.; Sutton, M.A. Applications of digital-image-correlation techniques to experimental mechanics. *Exp. Mech.* **1985**, *25*, 232–244. [CrossRef]
8. Peters, W.H.; Ranson, W.F. Digital Imaging Techniques In Experimental Stress Analysis. *Opt. Eng.* **1982**, *21*, 213427. [CrossRef]
9. Sutton, M.; Mingqi, C.; Peters, W.; Chao, Y.; McNeill, S. Application of an optimized digital correlation method to planar deformation analysis. *Image Vis. Comput.* **1986**, *4*, 143–150. [CrossRef]
10. Peters, W.H.; Ranson, W.F.; Sutton, M.A.; Chu, T.C.; Anderson, J. Application Of Digital Correlation Methods To Rigid Body Mechanics. *Opt. Eng.* **1983**, *22*, 226738. [CrossRef]
11. Luo, P.F.; Chao, Y.J.; Sutton, M.A.; Peters, W.H. Accurate measurement of three-dimensional deformations in deformable and rigid bodies using computer vision. *Exp. Mech.* **1993**, *33*, 123–132. [CrossRef]

12. Helm, J.D.; McNeill, S.R.; Sutton, M.A. Improved three-dimensional image correlation for surface displacement measurement. *Opt. Eng.* **1996**, *35*, 1911. [CrossRef]
13. Park, S.; Park, H.; Kim, J.; Adeli, H. 3D displacement measurement model for health monitoring of structures using a motion capture system. *Measurement* **2014**, *59*, 352–362. [CrossRef]
14. Mair, E.; Hager, G.D.; Burschka, D.; Suppa, M.; Hirzinger, G. Adaptive and generic corner detection based on the accelerated segment test. In Proceedings of the Computer Vision–ECCV 2010: 11th European Conference on Computer Vision, Heraklion, Greece, 5–11 September 2010.
15. Bay, H.; Ess, A.; Tuytelaars, T.; VanGool, L. Speeded-up robust features (SURF). *Comput. Vis. Image Underst.* **2008**, *110*, 346–359. [CrossRef]
16. Yoon, H.; Elanwar, H.; Choi, H.; Golparvar-Fard, M.; Spencer, B.F., Jr. Target-free approach for vision-based structural system identification using consumer-grade cameras. *Struct. Control Health Monit.* **2016**, *23*, 1405–1416. [CrossRef]
17. Mirzazadeh, R.; Azam, S.E.; Mariani, S. Micromechanical Characterization of Polysilicon Films through On-Chip Tests. *Sensors* **2016**, *16*, 1191. [CrossRef] [PubMed]
18. Mirzazadeh, R.; Azam, S.E.; Mariani, S. Mechanical Characterization of Polysilicon MEMS: A Hybrid TMCMC/POD-Kriging Approach. *Sensors* **2018**, *18*, 1243. [CrossRef]
19. Torzoni, M.; Manzoni, A.; Mariani, S. A multi-fidelity surrogate model for structural health monitoring exploiting model order reduction and artificial neural networks. *Mech. Syst. Signal Process.* **2023**, *197*, 110376. [CrossRef]
20. Entezami, A.; Mariani, S.; Shariatmadar, H. Damage Detection in Largely Unobserved Structures under Varying Environmental Conditions: An AutoRegressive Spectrum and Multi-Level Machine Learning Methodology. *Sensors* **2022**, *22*, 1400. [CrossRef]
21. Wang, P.; Pierron, F.; Thomsen, O.T. Identification of Material Parameters of PVC Foams using Digital Image Correlation and the Virtual Fields Method. *Exp. Mech.* **2012**, *53*, 1001–1015. [CrossRef]
22. Grédiac, M.; Pierron, F.; Surrel, Y. Novel procedure for complete in-plane composite characterization using a single T-shaped specimen. *Exp. Mech.* **1999**, *39*, 142–149. [CrossRef]
23. Leclerc, H.; Périé, J.; Roux, S.; Hild, F. Integrated digital image correlation for the identification of mechanical properties. In Proceedings of the Computer Vision/Computer Graphics Collaboration Techniques: 4th International Conference, Rocquencourt, France, 4–6 May 2009.
24. Azizi, S.; Karami, K.; Nagarajaiah, S. Developing a semi-active adjustable stiffness device using integrated damage tracking and adaptive stiffness mechanism. *Eng. Struct.* **2021**, *238*, 112036. [CrossRef]
25. Karami, K.; Fatehi, P.; Yazdani, A. On-line system identification of structures using wavelet-Hilbert transform and sparse component analysis. *Comput. Civ. Infrastruct. Eng.* **2020**, *35*, 870–886. [CrossRef]
26. Capellari, G.; Chatzi, E.; Mariani, S.; Azam, S.E. Optimal design of sensor networks for damage detection. *Procedia Eng.* **2017**, *199*, 1864–1869. [CrossRef]
27. Capellari, G.; Chatzi, E.; Mariani, S. Cost-benefit optimization of sensor networks for SHM applications. *Proceedings* **2018**, *2*, 132.
28. Pan, B.; Xie, H.; Wang, Z. Equivalence of digital image correlation criteria for pattern matching. *Appl. Opt.* **2010**, *49*, 5501–5509. [CrossRef]

engineering proceedings

MDPI

Proceeding Paper

A High-Precision Robotic System Design for Microsurgical Applications †

Xiaoyu Huang, Elizabeth Rendon-Morales and Rodrigo Aviles-Espinosa *

Robotics and Mechatronics Systems Research Group, School of Engineering and Informatics, University of Sussex, Brighton BN1 9RH, UK; e.rendon-morales@sussex.ac.uk (E.R.-M.)

* Correspondence: ra408@sussex.ac.uk

† Presented at the 10th International Electronic Conference on Sensors and Applications (ECSA-10), 15–30 November 2023; Available online: https://ecsa-10.sciforum.net/.

Abstract: The introduction of robotic systems in medical surgery has achieved the goal of decreasing procedures' invasiveness, positively impacting the patient's prognosis by reducing the incision size, surgical infections, and hospitalization time. Nowadays, robotic surgery is used as an integral part of urology, gynecology, abdominal, and cardiac interventions. Despite its adoption in several surgical specialties, robotic technology remains limited in the area of microsurgery. In this paper, we present the development of a robotic system providing sub-millimeter motion resolution for the potential manipulation of fine structures. The design is based on linear delta robotic geometry. The motion, resolution, and repeatability of the developed system were simulated, followed by proof-of-concept experimental testing. The developed system achieved a motion resolution of 3.37 ± 0.17 μm in both the X- and Y-axes and 1.32 ± 0.2 μm in the Z-axis. We evaluated the system navigation, setting a zigzag trajectory with dimensions below those found in blood vessels (300 to 800 μm), and found that the system is capable of achieving a maximum resolution of 3.06 ± 0.03 μm. These results demonstrate the potential application of the here-presented robotic system for its use in microsurgical applications such as neurosurgery, plastic, and breast cancer surgeries.

Keywords: robot design; parallel robot; microsurgery; high precision

Citation: Huang, X.; Rendon-Morales, E.; Aviles-Espinosa, R. A High-Precision Robotic System Design for Microsurgical Applications. *Eng. Proc.* **2023**, *58*, 66. https://doi.org/10.3390/ecsa-10-16221

Academic Editor: Francisco Falcone

Published: 15 November 2023

1. Introduction

Within the last two decades, multiple efforts have been focused on introducing robotic surgeries in routine procedures within multiple surgical disciplines [1] spanning from ophthalmology, urology, gynecology, cardiology, etc. The benefits of such technology compared to open surgery include the reduction of incisions from centimeters to tens of millimeters, resulting in minimized patient trauma, reduction in surgical site infections, and shorter hospitalization time, providing benefits to the patient's recovery. Current commercial robotic systems allow the surgeon to manipulate the laparoscopic tools using a set of controls that translate the surgeon's natural hand and wrist movements into corresponding surgical tool movements, providing the millimetric precision required for general surgical procedures.

However, specialized procedures requiring sub-millimeter precision manipulation such as reconstructive microsurgeries, surgical anastomosis, vitreoretinal eye surgery, and neurosurgery have not yet benefited from robotic surgery [2,3], posing the need for the development of robotic systems specially designed for such applications. Surgical anastomosis is a surgical technique used to make a connection between tubular body structures such as blood vessels, with diameters ranging between 300 and 800 μm. The reconnection of such structures is key to the re-establishment of lymphatic flow required in vascularized tissue transplantation [4]. Commercial platforms such as the da Vinci robot have been used to attempt procedures requiring high-precision manipulation such as microsurgical anastomosis. However, the design specifications of such systems are best suited to perform key-hole surgeries that deal with structures having dimensions in the centimeters range [5], making challenging the provision of sufficient motion resolution to accomplish tasks at a micro-scale level. State-of-the-art examples of robotic systems capable of achieving high-precision actuation are limited. These have been mainly evaluated in research laboratories trialing microsurgical applications including blood vessels nerve tracts and other soft tissue structure reconnection. Initial first-in-human surgeries reported in the literature include lymph venous anastomosis required for breast cancer-related lymphedema treatment [6]. Here, the surgeon's movements can be downscaled down to ~70 μm motion precisions well beyond key-hole surgical robots. Other robotic systems developed for microsurgical applications include the MUSA robot [7] and MMI's Symani System with NanoWrist (Italy) [8]. These systems provide magnification of the surgical area allowing simplified implementations in the operating room to perform micro-surgical procedures. Such robots are based on serial link geometries designed to cover larger workspaces. This, however, possesses the challenge of cumulative backlash errors that are often compensated using pre-loaded differential gears impacting the robot complexity, often associated with price [9].

In this paper, we present the design and simulation of a surgical robotic system based on a parallel geometry aiming to reduce the design complexity while providing sub-millimeter motion resolution. Using this type of geometry, the robot link dimensions were tailored for a defined workspace and resolution suited for microsurgical tasks. The motion, resolution, and repeatability of the developed system were assessed using simulations followed by an experimental proof-of-concept high precision, mimicking a stitching task to validate its potential to be used for microsurgical procedures.

2. Materials and Methods

In a parallel robot geometry, the end-effector is connected to the base using multiple link sequences, forming a closed-loop system. This has been widely used in industrial pick-and-place applications [8]. The implementation of parallel structures can be used to simplify the robot design while achieving high precisions [9].

The design of the proposed surgical robotic system is based on a linear delta robotic geometry consisting of three linear actuators, three pairs of parallel legs, and twelve spherical joints, enabling high stability, low inertia, and high motion precision all required for microsurgical tasks. Thus, in this paper, we describe the robot design methodology consisting of three stages: geometrical structure design, robot simulation and validation, and robot proof-of-concept testing.

2.1. Geometrical Structure Design

The simplified geometrical model of the linear delta robot design is shown in Figure 1.

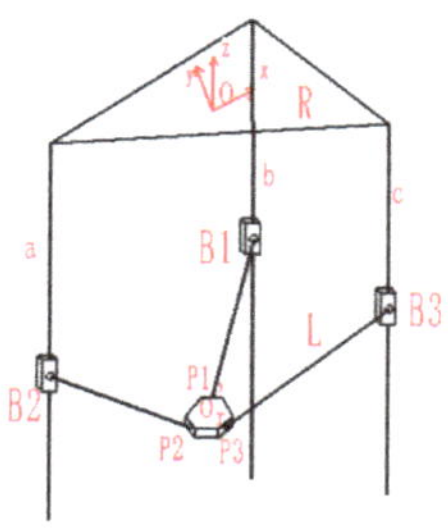

Figure 1. Geometrical construction of linear delta robot.

Here, the ball joints are fixed to linear sliders (B1, B2, B3) and the end-effector (P1, P2, P3). The coordinate of the base plane O is (0, 0, 0), the coordinate of the end-effector O′ is (x, y, z), and the displacement of the linear sliders are z1, z2, z3. This implies that the coordinates of the B1, B2, B3 are (0, $\sqrt{3}/2$, z1), ($-\sqrt{3}/2$, z2), and ($\sqrt{3}/2$, 0, z3), respectively. The geometrical relationship between arm length L, the radius of the based platform R, and the radius of the moving platform are obtained using Equations (1)–(4).

Equation (1) shows the spatial relationship between robotic arms and the center of the mobile platform using vector arithmetic, to express the arm length L.

Equations (2)–(4) are related to the robotic arm length L, which is obtained using end-effector coordinates (x, y, z), r, R, and the vertical motion of the three linear sliders z1, z2, and z3 respectively. Using mathematical simplifications, the equations of x, y, z or z1, z2, z3 can be obtained.

$$\overrightarrow{OO'} = \overrightarrow{OBi} + \overrightarrow{BiPi} + \overrightarrow{PiO'} \tag{1}$$

$$x^2 + (y + r - R)^2 + (z - z1)^2 = L^2 \tag{2}$$

$$\left(x - \frac{\sqrt{3}}{2}(r - R)\right)^2 + \left(y - \frac{1}{2}(r - R)\right)^2 + (z - z2)^2 = L^2 \tag{3}$$

$$\left(x - \frac{\sqrt{3}}{2}(r - R)\right)^2 + \left(y - \frac{1}{2}(R - r)\right)^2 + (z - z3)^2 = L^2 \tag{4}$$

2.2. Kinematic Simulation

Using Equations (1)–(4) while defining the motion limits of B1, B2, and B3, the motion state of the end-effector O′ was represented dynamically using simulations. The robot design and the kinematics modeling were simulated using MATLAB 2020 (Math Works, Natick, MA, USA) and CATIA V5-6R2018 (Dassault Systems, Vélizy-Villacoublay, France).

Following the geometric solution of forward kinematics, simulation results showed that the motion resolution considering the final arm length and the base platform radius (L = 250 mm, R = 235 mm, r = 20 mm) ranges between 0.24–0.625 μm. Thus, the minimum single-step motion resolution of a single motor is 0.625 μm considering micro-stepping of 1/8. Figure 2 presents an example trace resulting from the simulations outlining the temporal evolution of the XYZ-axis coordinates of the end-effector. Figure 2a shows the displacement of the *X*-axis and *Y*-axis coordinates over time, and Figure 2b shows the *Z*-axis coordinate displacement over time. From these results, it can be observed that the maximum achieved motion resolution under the final dimension parameters is ~0.24 μm.

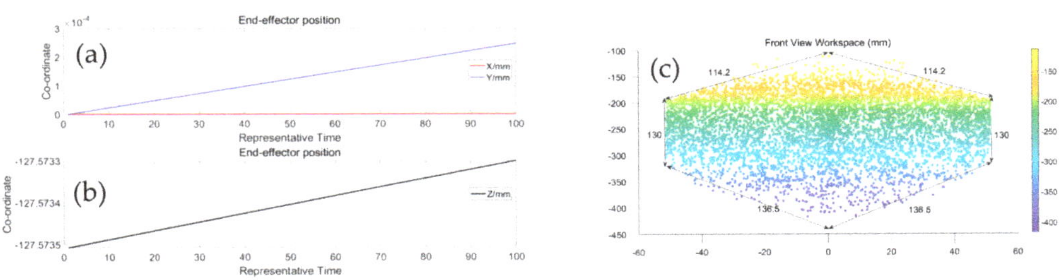

Figure 2. Temporal evolution of the XYZ-axis coordinates of the end-effector. (**a**) Coordinates' changes in X- and Y-axes. (**b**) Temporal resolution of Z-axis. (**c**) Front view of the workspace obtained using simulations.

To assess the robotic workspace considering the obtained dimensions (L, R, and r), all simulated trajectory points of the end-effector O′ were traced. Figure 2c shows the front view of the simulated full workspace having a pyramid-like structure with dimensions of ~114.2 mm × 114.2 mm × 110 mm and 130 mm depth.

The forward kinematic simulation shows that the end-effector resolution of the linear delta robot can reach a maximum of 0.36 µm/pulse considering 1/8 micro-stepping and 1.75 µm/pulse considering 1/2 micro-stepping. By integrating both the workspace and motion resolution, it can be concluded that using the design parameters obtained (R = 250 mm, L = 235 mm, r = 20 mm), it is possible to reach the required precision for microsurgical applications dealing with structures having dimensions between 300 to 800 µm [10].

The inverse kinematic (IK) simulation was used to determine the accuracy of the stepper actuator setting and to guarantee, using simulations, that the system configuration will match the sub-millimeter level performance required for microsurgical applications. The IK expressions are represented in a similar fashion to the forward kinematics. Here, the positions B1, B2, and B3 were expressed as the positions of end-effector (x, y, z). Following the geometrical construction of the linear delta robot shown in Figure 1, the coordinates of the linear sliders on the XY plane were considered to be fixed, then the change in the coordinates is reflected in the displacement of the *z*-axis coordinates relative to the initial position (z1, z2, z3). The real-time *z*-axis coordinates of B1, B2, and B3 can be obtained using Equations (5)–(7):

$$z1 = \pm\sqrt{L^2 - x^2 - (y + r - R)^2} + z \tag{5}$$

$$z2 = \pm\sqrt{L^2 - \left(x - \frac{\sqrt{3}}{2}(r - R)\right)^2 - \left(y - \frac{r - R}{2}\right)^2} + z \tag{6}$$

$$z3 = \pm\sqrt{L^2 - \left(x - \frac{\sqrt{3}}{2}(R - r)\right)^2 - \left(y - \frac{r - R}{2}\right)^2} + z \tag{7}$$

Note that Equations (5)–(7) show the vertical motion of the linear sliders z1, z2, and z3, expressed by the parameters L, R, and r, which are affected by the end-effector's position (x, y, z). The notation ± corresponds to the robotic motion direction, where a positive sign indicates a downward movement of the end-effector's position.

Figure 3 shows the IK simulation results of a zigzag trajectory. Figure 3a shows the zigzag trajectory in the XY-plane, consisting of horizontal and vertical displacements set to describe 150 µm and 15 µm segments, respectively. Figure 3b shows the coordinate changes in the end-effector in the XY-plane, while Figure 3c shows the changes in three robot sliders (Z1, Z2, Z3). In this case, a 15–150 µm displacement of the end-effector produces a displacement of 25.4–218.7 µm of the three linear sliders of the robot.

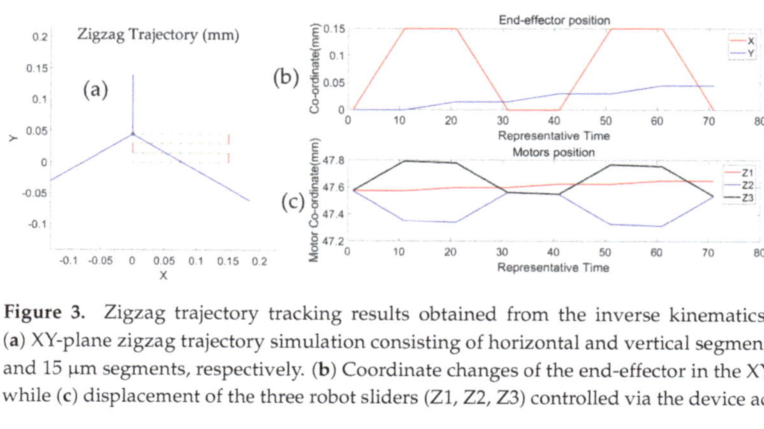

Figure 3. Zigzag trajectory tracking results obtained from the inverse kinematics simulation. (**a**) XY-plane zigzag trajectory simulation consisting of horizontal and vertical segments of 150 µm and 15 µm segments, respectively. (**b**) Coordinate changes of the end-effector in the XY-plane, and while (**c**) displacement of the three robot sliders (Z1, Z2, Z3) controlled via the device actuators.

3. Results and Discussion

Robot Proof-of-Concept Trajectory Testing

Considering the geometric design and simulation results described in Section 2, the proposed robotic system was built employing metallic beams (MakerBeam B.V., Utrecht, The Netherlands) and custom-made 3D-printed components. Figure 4a shows the physical prototype (top view) of the robotic platform. The drive system uses three 35 mm SH3533-12U40 stepper motors (Sanyo Denki, Tokyo, Japan) and a BSD 02.V motor driver (RTA Pavia, Marcignago, Italy). The robotic system is controlled with a custom-made GUI controlled though a MyRIO 1900 FPGA (National Instruments, Austin, TX, USA).

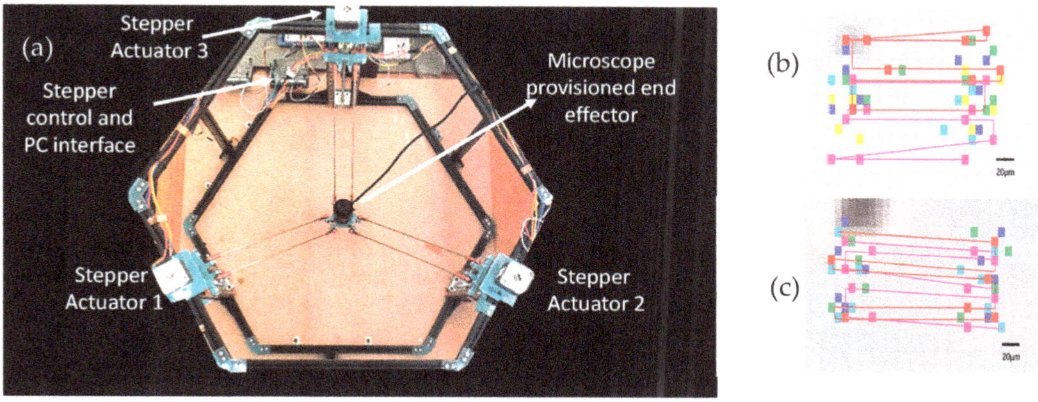

Figure 4. (**a**) Physical prototype of the developed robotic platform (top view), (**b**) zigzag trajectory without compensation, and (**c**) with compensation.

For the evaluation of the precision and kinematic performance of the linear delta robot manipulator, a noncontact metrology approach based on a bright field microscopy system was used. To reach the required micron-level resolution, the linear slider displacement was set to 20 microns/pulse, corresponding to an angular movement of 0.225°. This resulted in a motion resolution of 1.32 ± 0.02 μm for the Z-axis and 3.37 ± 0.17μm for the X- and Y-axes, indicating the system's potential to be used for microsurgical tasks [11]. The motion performance of the proposed robotic system was set up in a laboratory environment and evaluated using a proof-of-concept experimental evaluation shown in Figure 4. We define a zigzag/raster trajectory, which consists of five 180 μm horizontal lines and four 15 μm vertical lines. For each test, three forward trajectories and three backward trajectories were carried out.

The representative zigzag trajectory test was set up in LabVIEW software (v2021 National Instruments, Austin, TX, USA) considering the dimensions of fine structures such as skin cells with an average size of 50 μm [11]. For data analysis, the videos of the executed trajectory were recorded. By tracing a zigzag 2D trajectory, we aim to simulate a simplified stitching procedure, where the end-effector is to be moved from side to side to join the adjacent portions of tissue. The zigzag 2D trajectory is shown in Figure 4. Figure 4b is the executed zigzag trajectory without compensation, as can be observed in the resulting differences between the trajectory segments and figure. This is to be expected as the robot operates in an open-loop configuration.

However, further analysis of the traces indicated repeatable offset errors within the trajectory which can be further compensated. The results after trajectory compensation are presented in Figure 4c. Table 1 shows the detailed results of the compensated zigzag trajectory. It can be observed that the average trajectory error is improved from 40 ± 2.13 μm (uncompensated trajectory) to 5.64 ± 0.63 μm (compensated trajectory) with an average angular error reduction of $2.7 \pm 0.54°$. Assessing the resolution obtained within this proof-of-concept test resulted in an average motion resolution of 3.06 ± 0.03 μm.

Table 1. Test results of the compensated zigzag trajectory.

No.	Resolution [μm]	Angle Error [°]	Trajectory Error [μm]
Zigzag 1	3.17 ± 0.03	-2.52 ± 0.85	9.59 ± 0.91
Zigzag 2	3.08 ± 0.04	-2.30 ± 0.25	3.47 ± 0.65
Zigzag 3	2.94 ± 0.02	-3.28 ± 0.53	3.85 ± 0.34
Average	3.06 ± 0.03	2.7 ± 0.54	5.64 ± 0.63

Comparing the results obtained from the simulations and experiments, it can be concluded that the built system dynamics and the open-loop control affect the obtained motion accuracy. This is expected as the built system elements' specifications and tolerances differ for the ideal conditions set in the simulations. Nonetheless, the motion resolution was found to be less than 5 microns, indicating that the system offers a performance that goes well beyond that required for performing microsurgical operations.

4. Conclusions

In this paper, the design, simulation, and proof-of-concept evaluation of a linear delta robot device to be used for microsurgical applications has been presented. The robot design was focused on achieving high motion precision tasks considering the dimensions of structures such as blood vessels ranging between 300 to 800 μm.

This developed system had dimensions of R = 250 mm, L = 235 mm, r = 20 mm, and a pyramid-like workspace with dimensions of ~114.2 mm × 114.2 mm × 110 mm and 130 mm depth. This resulted in a motion resolution of 3.37 ± 0.17 μm in both the X- and Y-axes and 1.32 ± 0.2 μm in the Z-axis considering individual steps. The system was evaluated considering a zigzag trajectory with dimensions similar to those found in blood vessels. We found that performing the zigzag trajectory in a series of three experiments resulted in a displacement between these, resulting in an error of 40 ± 2.13 μm. With the addition of displacement and angle compensation, these errors were reduced by 85.9%. These results demonstrate the potential application of the here-presented robotic system for its use in microsurgical procedures.

Author Contributions: Conceptualization, R.A.-E.; investigation, X.H.; methodology, R.A.-E. and E.R.-M.; project administration, R.A.-E.; software and data collection, X.H.; supervision, R.A.-E. and E.R.-M.; writing, R.A.-E., E.R.-M. and X.H. All authors have read and agreed to the published version of the manuscript.

Funding: The authors would like to acknowledge the EPSRC-Engineering and Physical Sciences Research Council E-futures 2.0 Sandpit (award number EP/S032045/1).

Institutional Review Board Statement: Not applicable.

Informed Consent Statement: Not applicable.

Data Availability Statement: All the data presented in this study are contained in the article's main text.

Acknowledgments: X.H. acknowledges the University of Sussex scholarship for his Ph.D. studies. The authors acknowledge the in-kind contributions from EAO LTD.

Conflicts of Interest: The authors declare no conflict of interest.

References

1. Morrell, A.L.G.; Morrell-Junior, A.C.; Morrell, A.G.; Mendes, J.; Freitas, M.; Tustumi, F.; Morrell, A. The History of Robotic Surgery and Its Evolution: When Illusion Becomes Reality. *Rev. Colégio Bras. Cir.* **2021**, *48*, e20202798. [CrossRef] [PubMed]
2. Diana, M.; Marescaux, J. Robotic Surgery. *Br. J. Surg.* **2015**, *102*, e15–e28. [CrossRef] [PubMed]

3. Bhandari, M.; Zeffiro, T.; Reddiboina, M. Artificial Intelligence and Robotic Surgery: Current Perspective and Future Directions. *Curr. Opin. Urol.* **2020**, *30*, 48–54. [CrossRef] [PubMed]
4. Brodie, A.; Vasdev, N. The Future of Robotic Surgery. *Ann. R. Coll. Surg. Engl.* **2018**, *100* (Suppl. 7), 4–13. [CrossRef] [PubMed]
5. Lai, C.S.; Lu, C.T.; Liu, S.A.; Tsai, Y.C.; Chen, Y.W.; Chen, I.C. Robot-Assisted Microvascular Anastomosis in Head and Neck Free Flap Reconstruction: Preliminary Experiences and Results. *Microsurgery* **2019**, *39*, 715–720. [CrossRef] [PubMed]
6. van Mulken, T.J.; Schols, R.M.; Scharmga, A.M.; Winkens, B.; Cau, R.; Schoenmakers, F.B.; Qiu, S.S.; van der Hulst, R.R.; MicroSurgical Robot Research Group. First-in-Human Robotic Supermicrosurgery Using a Dedicated Microsurgical Robot for Treating Breast Cancer-Related Lymphedema: A Randomized Pilot Trial. *Nat. Commun.* **2020**, *11*, 757. [CrossRef] [PubMed]
7. Yang, U.J.; Kim, D.; Hwang, M.; Kong, D.; Kim, J.; Nho, Y.H.; Lee, W.; Kwon, D.S. A Novel Microsurgery Robot Mechanism with Mechanical Motion Scalability for Intraocular and Reconstructive Surgery. *Int. J. Med. Robot.* **2021**, *17*, e2240. [CrossRef] [PubMed]
8. Savastano, A.; Rizzo, S. A Novel Microsurgical Robot: Preliminary Feasibility Test in Ophthalmic Field. *Transl. Vis. Sci. Technol.* **2022**, *11*, 13. [CrossRef] [PubMed]
9. Aitzetmüller, M.M.; Klietz, M.-L.; Dermietzel, A.F.; Hirsch, T.; Kückelhaus, M. Robotic-Assisted Microsurgery and Its Future in Plastic Surgery. *J. Clin. Med.* **2022**, *11*, 3378. [CrossRef] [PubMed]
10. Usca, A.; Gabriela, R.U.S.; Birlescu, I.; Vaida, C.; Pisla, A.; Schonstein, C.; Gherman, B.; Tucan, P.; Pisla, D. Workspace Analysis of Two Innovative Parallel Robots for Single Incision Laparoscopic Surgery. *Acta Tech. Napoc.-Ser. Appl. Math. Mech. Eng.* **2022**, *65*, 25.
11. Huang, X.Y.; Rendon-Morales, E.; Aviles-Espinosa, R. Towards Cellular Level Microsurgery: Design and Testing of a High Precision Delta Robot for Medical Applications. In Proceedings of the Hamlyn Symposium on Medical Robotics 2023, Royal Geographical Society, London, UK, 26–29 June 2023.

engineering proceedings

Proceeding Paper

Machine Learning for Accurate Office Room Occupancy Detection Using Multi-Sensor Data †

Yusuf Ibrahim [1,*], Umar Yusuf Bagaye [2] and Abubakar Ibrahim Muhammad [2]

1 Department of Computer Engineering, Ahmadu Bello University, Zaria 810211, Nigeria
2 Department of Electrical and Electronics Engineering, Kaduna Polytechnic, Kaduna 800282, Nigeria; uybagaye@gmail.com (U.Y.B.); ibngarko@yahoo.com (A.I.M.)
* Correspondence: yibrahim@abu.edu.ng

† Presented at the 10th International Electronic Conference on Sensors and Applications (ECSA-10), 15–30 November 2023; Available online: https://ecsa-10.sciforum.net/.

Abstract: In this paper, we present a comparative study of several machine learning (ML) approaches for accurate office room occupancy detection through the analysis of multi-sensor data. Our study utilizes the occupancy detection dataset, which incorporates data from temperature, humidity, light, and CO_2 sensors, with ground-truth labels obtained from time-stamped images captured at minute intervals. Traditional ML techniques, including Decision Trees (DT), Gaussian Naïve Bayes (NB), K-Nearest Neighbors (KNN), Logistic Regression (LR), Support Vector Machines (SVM), Multilayer Perceptron (MLP), and Quadratic Discriminant Analysis (QDA) are compared alongside advanced ensemble methods like RandomForest (RF), Bagging, AdaBoost, GradientBoosting, ExtraTrees as well as our custom voting and multiple stacking classifiers. Also, hyperparameter optimization was performed for selected models with a view to improving classification accuracy. The performances of the models were evaluated through rigorous cross-validation experiments. The results obtained highlight the efficacy and suitability of varying candidate and ensemble methods, demonstrating the potential of ML techniques in enhancing detection accuracy. Notably, LR and SVM exhibited superior performance, achieving average accuracies of 98.88 ± 0.70% and 98.65 ± 0.96%, respectively. Additionally, our custom voting and stacking ensembles demonstrated improvements in classification outcomes compared to base ensemble schemes, as indicated by various evaluation metrics.

Keywords: machine learning; ensemble learning; room occupancy detection; multi-sensor data

Citation: Ibrahim, Y.; Bagaye, U.Y.; Muhammad, A.I. Machine Learning for Accurate Office Room Occupancy Detection Using Multi-Sensor Data. *Eng. Proc.* **2023**, *58*, 67. https://doi.org/10.3390/ecsa-10-16019

Academic Editor: Stefano Mariani

Published: 15 November 2023

1. Introduction

Occupancy detection refers to the process of determining whether a space or area is currently occupied by people or objects. This can be accomplished through various means and technologies and serves several purposes in different domains, including building management, safety, security, energy conservation, and automation. For instance, efficient energy management in office spaces is today a concern, where environmental sustainability and cost-effectiveness go hand in hand. Estimates indicate that precise office room occupancy detection can lead to energy savings ranging from 30% to 42% [1,2]. These savings can be further optimized, reaching up to 80%, when occupancy data are integrated into HVAC (Heating, Ventilation, and Air Conditioning) control algorithms [3]. Therefore, there is a growing need for accurate occupancy detection methods to harness the full potential of these energy-saving opportunities. This quest for precision in occupancy detection has led to substantial research efforts, especially in the application of ML models. Previous studies have shown that, with sufficient relevant data, the accuracy of occupancy detection can yield remarkable performance levels [4–6]. In this paper, we utilize multi-sensor data which are becoming increasingly popular in ML applications as they can provide more accurate and reliable results compared to using a single sensor. The significant contributions of this paper include:

1. Systematic comparison of a wide range of ML models, from traditional to advanced ensemble methods;
2. Optimizing hyperparameters of selected models in order to enhance performance;
3. Evaluating custom voting and multiple stacking classifiers and demonstrating their role in improving classification performance.

2. Related Work

Several ML-based data-driven techniques have been utilized for occupancy detection in buildings. Candanedo and Feldheim [7] assessed the accuracy of predicting office room occupancy based on data from light, temperature, humidity, and CO_2 sensors, using various statistical classification models in R programming language. They used three datasets for training and testing, considering whether the office door was open or closed during occupancy. The best accuracies (ranging from 95% to 99%) were achieved with Linear Discriminant Analysis (LDA), Classification and Regression Trees (CART), and RF models. The inclusion of time stamp information generally improved accuracy, and the LDA model achieved occupancy estimates of 85% and 83% using only the temperature predictor in two different testing sets. In their study, Yang et al. [8] employed KNN alongside various environmental and specialized sensors for identifying and counting the number of occupants. Their findings show the potential to attain accuracy levels ranging from 95.4% to 97.5% for binary occupancy detection. Additionally, when estimating the count of occupants, the root mean square error (RMSE) falls within the range of 0.121 to 0.79. Dong et al. [9] and Lam et al. [10] paired SVM with a sensor network to gauge the occupancy levels within an office building. Their investigations yielded a consistent accuracy rate of approximately 75% for detecting the number of people present. Zuraimi et al. [11] utilized a combination of CO_2 data and feed-forward neural networks (FFNNs) to estimate the number of occupants in a theater, resulting in an average accuracy of 70%. Similarly, Dong et al. [9] and Lam et al. [10] introduced an environmental sensor network testbed and demonstrated its utilization for detecting occupancy numbers within an office building. Their works employed a neural network to identify the number of occupants, achieving an accuracy rate of 75%. In their study, Kraipeerapun et al. [12] introduced two approaches for determining occupancy. The initial approach employed a combination of stacking and a multiclass neural network, while the second method fused stacking with a dual-output neural network specifically designed for occupancy detection. The validation outcomes demonstrated accuracy levels ranging from 68.87% to 91.18%. Kim et al. [2] introduced a label noise filtering method, which improves occupancy detection accuracy by eliminating noisy data collected from sensors. The results yielded an average accuracy increase of 1.5%, with the CART model showing a significant improvement from 94.3% to 97.6%. Dutta and Roy [13] developed the OccupancySense model, which addresses occupancy detection and prediction by fusing Internet of Things (IoT) indoor air quality data with static and dynamic context data, achieving higher forecasting accuracy using the CatBoost algorithm. The model outperforms other ML algorithms, and with a non-intrusive approach, accurately detecting occupancy, predicting headcount, and estimating room occupancy density at 99.85%, 93.2%, and 95.6% accuracy, respectively. Elkhoukhi et al. [14] highlight the limitations of batch learning techniques and introduces three non-stationary ML algorithms for stream data processing. The experimental results demonstrate that these algorithms, integrated into an IoT-based platform, can accurately predict the number of occupants in smart buildings with an accuracy exceeding 83% while efficiently utilizing computational resources.

3. Materials and Methods

3.1. Data Collection and Preprocessing

We utilized the publicly available occupancy detection dataset, which includes sensor data from temperature, humidity, light, and CO_2 sensors, as well as ground-truth labels obtained from time-stamped images captured at minute intervals [7].

3.2. Feature Engineering

We performed correlation analysis on the dataset to identify relevant features for the occupancy detection task. Most features have strong positive correlations with the target variable (occupancy) except for humidity and humidity ratio, with a relatively weak correlation with the target variable. However, we retained all features without thresholding any sensor data in our analysis. Truncating below a specific threshold and trying other feature combinations is left for future research.

3.3. Model Selection

For our analysis, we selected a set of traditional ML as well as advanced (ensemble) models for the comparative study. The traditional ML models include Decision Trees, Gaussian Naïve Bayes, KNN, LR, SVM, MLP, and QDA. The Ensemble methods include RF, Bagging, AdaBoost, GradientBoosting, and ExtraTrees. Furthermore, we tried several Custom ensemble methods as follows:

1. Voting Classifier, consisting of LR, RF, and SVM;
2. StackingClassifier1, consisting of LR, RF, and SVM as base estimators with LR as the final estimator;
3. StackingClassifier2, consisting of Decision Tree, KNN, and MLP Classifiers as base estimators with LR as the final estimator;
4. Stacking Classifier3, consisting of GaussianNB, SVM, and QDA as base estimators with LR as the final estimator;
5. StackingClassifier4, consisting of RF, MLP Classifier, and SVM as base estimators with LR as the final estimator.

3.4. Hyperparameter Optimization

In order to obtain better performance, we further performed parameter tuning via a grid search for RF, SVM, and KNN classifiers. Each grid search was performed with 5-fold cross-validation. For RF, the search was conducted over the number of estimators (10, 20, 30), maximum depth (15, 20, 30, 50), and criterion (gini, entropy). Also, the SVM was tuned over C (1, 10, 100) and kernel types (linear, poly, rbf, sigmoid). Finally, KNN was optimized by searching for the optimal number of neighbors (2, 3, 5, 10, 15, 20).

3.5. Model Training and Evaluation

Rigorous cross-validation experiments (using 5-fold cross-validation) were performed in order to assess the performance of the models. We then split the dataset into 70% training and 30% testing, retrained each model on the training set, and evaluated the models' performance on the test set using accuracy, precision, recall, and F1-score as performance metrics.

4. Results and Discussions

Tables 1 and 2, respectively, show the 5-fold cross-validation as well as the testing results for the traditional ML models, while Tables 3 and 4, respectively, show the 5-fold cross-validation as well as the testing results for the ensemble models.

From Table 1, LR and SVM achieved the highest validation accuracies of 98.88% and 98.65%, respectively. These models also demonstrated strong precision, recall, and F1-Score values, indicating their suitability for accurate occupancy detection. Also, for the test data (Table 2), SVM, LR, KNN, and DT models exhibit high accuracy levels above 99%. The ensemble methods (Table 3), particularly our voting and stacking models, show high performance, with stackingclassifer1 achieving the highest validation accuracy of approximately 98.89 $\pm$ 0.72%, outperforming others. Classification results on the test data (Table 4) indicate that most ensemble methods achieve high accuracy levels, with RF, Extra-Trees, and StackingClassifier1 being particularly notable achieving above 99.30% accuracy. These models also exhibit strong precision, recall, and F1-Score values, reflecting their effectiveness in making accurate predictions. Also, the voting ensemble, which recorded a slightly lower accuracy, still demonstrates a good balance between precision and recall. For the optimized models, we finally arrived at the following as the best hyperparameters for the respective algorithms: SVM (C = 10 and kernel = 'linear'), KNN (n_neighbors = 20), RF (n_estimators = 50, max_depth = 44, and criterion = 'entropy'). Utilizing these parameters, the test results presented in Table 5 were obtained. The performance improvements recorded for KNN and RF show that hyperparameter optimization can improve the predictive accuracy of ML classifiers.

Table 1. Cross-validation results for the traditional ML methods.

Model	Average Accuracy	Average Precision	Average Recall	Average F1-Score
SVM	0.9865 ± 0.0096	0.9500 ± 0.0368	0.9958 ± 0.0016	0.9720 ± 0.0193
LR	0.9888 ± 0.0070	0.9582 ± 0.0280	0.9958 ± 0.0029	0.9764 ± 0.0144
KNN	0.9639 ± 0.0102	0.9283 ± 0.0145	0.9151 ± 0.0582	0.9204 ± 0.0245
DT	0.8363 ± 0.1437	0.7511 ± 0.2595	0.8419 ± 0.1246	0.7477 ± 0.1506
NB	0.9368 ± 0.0249	0.7915 ± 0.0654	0.9983 ± 0.0011	0.8814 ± 0.0410
MLP	0.9699 ± 0.0162	0.9375 ± 0.0504	0.9377 ± 0.0806	0.9340 ± 0.0375
QDA	0.9482 ± 0.0393	0.8359 ± 0.1150	0.9954 ± 0.0024	0.9042 ± 0.0680

Table 2. Results of the traditional ML methods on test set.

Model	Accuracy	Precision	Recall	F1-Score
SVM	0.9906	0.9644	0.9958	0.9798
LR	0.9904	0.9650	0.9943	0.9794
KNN	0.9908	0.9735	0.9866	0.9800
DT	0.9911	0.9809	0.9802	0.9805
NB	0.9668	0.8748	0.9979	0.9323
MLP	0.9531	0.8311	0.9936	0.9072
QDA	0.9825	0.9342	0.9936	0.9630

Table 3. Cross-validation results for the ensemble methods.

Model	Accuracy	Precision	Recall	F1-Score
RF	0.8589 ± 0.1249	0.7783 ± 0.2526	0.8703 ± 0.1073	0.7797 ± 0.1361
Bagging	0.9337 ± 0.0167	0.8925 ± 0.0908	0.8360 ± 0.1174	0.8516 ± 0.0424
AdaBoost	0.9352 ± 0.0283	0.8180 ± 0.1023	0.9530 ± 0.0291	0.8754 ± 0.0479
GBoosting	0.9552 ± 0.0200	0.8962 ± 0.0949	0.9322 ± 0.0435	0.9084 ± 0.0326
ExtraTrees	0.9140 ± 0.0266	0.8075 ± 0.0496	0.8311 ± 0.1465	0.8115 ± 0.0740
Voting	0.9861 ± 0.0096	0.9488 ± 0.0368	0.9954 ± 0.0018	0.9711 ± 0.0194
Stacking1	0.9889 ± 0.0072	0.9593 ± 0.0285	0.9952 ± 0.0026	0.9767 ± 0.0149
Stacking2	0.9765 ± 0.0119	0.9359 ± 0.0473	0.9682 ± 0.0366	0.9505 ± 0.0244
Stacking3	0.9880 ± 0.0082	0.9579 ± 0.0335	0.9933 ± 0.0051	0.9749 ± 0.0168
Stacking4	0.9874 ± 0.0098	0.9541 ± 0.0385	0.9952 ± 0.0028	0.9738 ± 0.0199

Table 4. Experimental results of the ensemble methods on test set.

Model	Accuracy	Precision	Recall	F1-Score
RF	0.9935	0.9838	0.9880	0.9859
Bagging	0.9914	0.9823	0.9802	0.9812
AdaBoost	0.9903	0.9675	0.9908	0.9790
GBoosting	0.9908	0.9709	0.9894	0.9800
ExtraTrees	0.9932	0.9858	0.9844	0.9851
Voting	0.9906	0.9657	0.9943	0.9798
Stacking1	0.9932	0.9824	0.9880	0.9852
Stacking2	0.9921	0.9789	0.9866	0.9827
Stacking3	0.9885	0.9577	0.9936	0.9754
Stacking4	0.9930	0.9811	0.9887	0.9849

Table 5. Experimental results of the optimized methods on test set.

Model	Accuracy	Precision	Recall	F1-Score
Grid-SVM	0.9887	0.9683	0.9957	0.9818
Grid-KNN	0.9920	0.9703	0.9946	0.9823
Grid-RF	0.9939	0.9807	0.9924	0.9865

5. Conclusions

In conclusion, this paper has presented a comparative study of ML approaches for office room occupancy detection using multi-sensor data. Our findings indicate that LR and SVM achieved impressive performance. Furthermore, our custom stacking ensembles demonstrated significant improvements over most base ensemble schemes. The study provides a comprehensive insight into the potential of several ML techniques in the domain of room occupancy detection.

Author Contributions: Conceptualization, Y.I. and A.I.M.; methodology, Y.I. and U.Y.B.; software, Y.I. and U.Y.B.; validation, U.Y.B., A.I.M. and Y.I.; investigation, Y.I.; resources, Y.I. and U.Y.B.; data curation, Y.I.; writing—original draft preparation, Y.I.; writing—review and editing, U.Y.B. and A.I.M.; visualization, A.I.M.; supervision, Y.I.; project administration, A.I.M.; funding acquisition, Y.I., U.Y.B. and A.I.M. All authors have read and agreed to the published version of the manuscript.

Funding: This research received no external funding.

Institutional Review Board Statement: Not applicable.

Informed Consent Statement: Not applicable.

Data Availability Statement: The data used in this study was obtained from the UCI Machine Learning Repository (https://doi.org/10.24432/C5X01N). The dataset is publicly available and can be accessed at https://archive.ics.uci.edu/dataset/357/occupancy+detection (accessed on [illegible] June 2023).

Conflicts of Interest: The authors declare no conflicts of interest.

References

1. Erickson, V.L.; Carreira-Perpiñán, M.Á.; Cerpa, A.E. Occupancy modeling and prediction for building energy management. *ACM Trans. Sens. Netw. (TOSN)* **2014**, *10*, 1–28. [CrossRef]
2. Kim, Y.-M.; Lee, Y.-H.; Pyo, C.-S. Accurate Occupancy Detection via Label Noise Filtering Technique. In Proceedings of the 2020 International Conference on Information and Communication Technology Convergence (ICTC), Jeju Island, Republic of Korea, 21–23 October 2020; pp. 1381–1383.
3. Brooks, J.; Kumar, S.; Goyal, S.; Subramany, R.; Barooah, P. Energy-efficient control of under-actuated HVAC zones in commercial buildings. *Energy Build.* **2015**, *93*, 160–168. [CrossRef]
4. Zemouri, S.; Gkoufas, Y.; Murphy, J. A machine learning approach to indoor occupancy detection using non-intrusive environmental sensor data. In Proceedings of the Proceedings of the 3rd International Conference on Big Data and Internet of Things, Melbourn, Australia, 22–24 August 2019; pp. 70–74.
5. Zhao, H.; Hua, Q.; Chen, H.-B.; Ye, Y.; Wang, H.; Tan, S.X.-D.; Tlelo-Cuautle, E. Thermal-sensor-based occupancy detection for smart buildings using machine-learning methods. *ACM Trans. Des. Autom. Electron. Syst. (TODAES)* **2018**, *23*, 1–21. [CrossRef]
6. Toutiaee, M. Occupancy detection in room using sensor data. *arXiv* **2021**, arXiv:2101.03616. [CrossRef]
7. Candanedo, L.M.; Feldheim, V. Accurate occupancy detection of an office room from light, temperature, humidity and CO_2 measurements using statistical learning models. *Energy Build.* **2016**, *112*, 28–39. [CrossRef]
8. Yang, Z.; Li, N.; Becerik-Gerber, B.; Orosz, M. A systematic approach to occupancy modeling in ambient sensor-rich buildings. *Simulation* **2014**, *90*, 960–977. [CrossRef]
9. Dong, B.; Andrews, B.; Lam, K.P.; Höynck, M.; Zhang, R.; Chiou, Y.-S.; Benitez, D. An information technology enabled sustainability test-bed (ITEST) for occupancy detection through an environmental sensing network. *Energy Build.* **2010**, *42*, 1038–1046. [CrossRef]
10. Lam, K.P.; Höynck, M.; Dong, B.; Andrews, B.; Chiou, Y.-S.; Zhang, R.; Benitez, D.; Choi, J. Occupancy detection through an extensive environmental sensor network in an open-plan office building. *IBPSA Build. Simul.* **2009**, *145*, 1452–1459.
11. Zuraimi, M.; Pantazaras, A.; Chaturvedi, K.; Yang, J.; Tham, K.; Lee, S. Predicting occupancy counts using physical and statistical Co_2-based modeling methodologies. *Build. Environ.* **2017**, *123*, 517–528. [CrossRef]

12. Kraipeerapun, P.; Amornsamankul, S. Room occupancy detection using modified stacking. In Proceedings of the 9th International Conference on Machine Learning and Computing, Singapore, 24–26 February 2017; pp. 162–166.
13. Dutta, J.; Roy, S. OccupancySense: Context-based indoor occupancy detection & prediction using CatBoost model. *Appl. Soft Comput.* **2022**, *119*, 108536. [CrossRef]
14. Elkhoukhi, H.; Bakhouya, M.; El Ouadghiri, D.; Hanifi, M. Using stream data processing for real-time occupancy detection in smart buildings. *Sensors* **2022**, *22*, 2371. [CrossRef] [PubMed]

MDPI AG
Grosspeteranlage 5
4052 Basel
Switzerland
Tel.: +41 61 683 77 34

Engineering Proceedings Editorial Office
E-mail: engproc@mdpi.com
www.mdpi.com/journal/engproc

MDPI